Principles and Practices for the
Safe Processing of Foods

Comments on *Principles and Practices for the Safe Processing of Foods*

'This volume presents state-of-the-art information on the design, construction, and sanitary maintenance of food processing plants; it provides guidelines for establishing and implementing the Hazard Analysis Critical Control Points System and for training personnel in hygienic practices. The references are timely, reflecting the latest scientific and technological knowledge in the principles and practices of food-processing safety. An increase in our awareness of the problems of microbial safety and hygiene and a renewed approach to quality control, as presented in this book, can greatly enhance our ability to achieve a safe and wholesome food supply.'

John Kvenberg – United States Food and Drug Administration, Washington DC, USA

'This book is quite unique in that, perhaps for the first time, "traditional" food safety procedures are considered against a backcloth of modern quality management techniques. The multi-discipline nature of this book provides invaluable guidance on the wide range of information to be considered to ensure safe food processing. I believe that this book will become firmly established as a definitive text and essential reading for all those involved in food processing.'

Mike Stringer – Campden Food and Drink Research Association, Gloucestershire, UK

'This is one of the most valuable reference books ever written for the food industry – an invaluable source for everyone involved in production and purveying of food products.'

Cleve Denny – formerly of National Food Processors Association, Washington DC, USA

Principles and Practices for the
Safe Processing of Foods

edited by
David A Shapton
Norah F Shapton

WOODHEAD PUBLISHING LIMITED
Cambridge England

Published by Woodhead Publishing Limited
Abington Hall, Abington
Cambridge CB1 6AH, England

First published 1991 Butterworth-Heinemann Ltd
Paperback edition 1993
Reprinted 1994
Reprinted 1998, 2001 Woodhead Publishing Limited

British Library Cataloguing in Publication Data
A catalogue record for this book is available from the British Library.

ISBN 1 85573 362 5

Printed in Great Britain by Martins the Printers, Berwick upon Tweed, UK

Foreword

Pure, safe, wholesome food is the minimum expectation of today's consumer. For a food processing company, it must be the minimum standard for consumer acceptance. H. J. Heinz Company has made the purity and safety of its products a hallmark of the brand from the very beginning. Our Founder understood the desire of consumers for products they could trust. Today, we strive to continue the legacy of Henry J. Heinz by making the brand that bears his name a surrogate for safety in an often uncertain world.

As important as food safety is, consumers usually take it for granted. They often overlook the fact that the process of safety is complex and difficult. It requires constant vigilance on the part of every person involved in production. The modern consumer is willing to entrust his or her health and safety to food processors. We are obliged to remain ever faithful to that trust.

To do this, the food processor must do two things. First, he must acquire the requisite knowledge of the safest processing methods. Second, he must apply a range of management skills to insure that this knowledge is continuously applied.

Often, safe food processing is a collaborative effort. Even market competitors exchange knowledge and methods individually or through trade organizations. Governments and academia lend their support to this great collective effort.

This book is composed in that communal spirit. Originally, it was compiled to guide Heinz managers who are personally accountable for current worldwide production and responsible for the continuing improvements required by our Corporate commitment to Total Quality Management.

Heinz is now presenting this reference to a wider public. We do so in the hope that it will add to the sum of professional knowledge on the subject and offer management some useful guidelines to help turn that knowledge into action.

We offer this volume as a contribution to the continuing effort of the food processing industry to insure consumers everywhere a safe, pure and wholesome food supply.

Anthony J. F. O'Reilly
Chairman, President and Chief Executive Officer
H. J. Heinz Company

Editorial note and acknowledgements

This book originated from an initiative of Dr Lee S. Harrow and was developed for publication under the leadership of Dr Thomas A. MacMurray. The editors, David A. and Norah F. Shapton place on record the willing help received from many colleagues within Heinz together with associates in Research Associations and in other companies.

To progress the book an internal editorial and drafting group was established to assist the editors. Members of this group were:

Mr. M. Belleville – Star-Kist Foods, USA
Professor A. Casolari – H. J. Heinz Company, Plasmon, Italy
Ms D. Crosby – Ore-Ida, USA
Dr L. S. Harrow (retired) – Heinz World Headquarters, USA
Mr J. Hennekam – H. J. Heinz Company, Elst, Netherlands
Dr R. Laurita – H. J. Heinz Company, Plasmon, Italy
Mr M. P. R. Jones – H. J. Heinz Company, UK
Mr R. Otto – Heinz USA
Dr C. Pedretti – H. J. Heinz Company, Plasmon, Italy
Mr J. L. Segmiller – Heinz USA
Mr A. Zegota – formerly with H. J. Heinz Company – Canada

Drawings for this book were prepared by Mr Walt Bedel (retired – Heinz USA Engineering Department) of Pittsburgh, PA.

The libraries of Heinz UK, Heinz USA and of the Campden and Leatherhead Research Associations have provided invaluable help during the period of writing this book. Much useful advice and help have been received from those outside the organization and particular thanks go to:

Mr Cleve B. Denny – the National Food Processors Association, USA
Dr M. Stringer, Mr R. H. Thorpe, Mr D. A. Timperley and Dr S. J. Walker of the Campden Food and Drink Research Association, UK.

Special thanks go from the editors to Marcy McKenney at Heinz World Headquarters for preparing, altering and amending the manuscript during the period of development of this book.

It is recognized that longer-term development plans may begin with immediate, if partial, improvements. Few plants are totally state-of-the-art and most inevitably have examples of older and newer technologies. However, Total Quality Management applies, whatever the circumstance and a range of design ideas and practices is therefore offered to the reader.

In the spirit of Habitual Incremental Improvement, readers with suggestions or proposals for additions are invited to write to:

Dr Thomas A. MacMurray
Vice President, Technical Development
H. J. Heinz Company
600 Grant Street – 60th Floor
Pittsburgh, PA 15219
USA

Copyright acknowledgments

Campden Food & Drink Research Association (Technical Manual (TM) 1, 1968; TM 7, 1983; TM 8, 1985; TM 12, 1986; TM 17, 1987; TM 18, 1987; TM 19, 1987); Churchill Livingstone (from *The Examination of Waters and Water Supplies – Thresh, Beale and Suckling* (1949), 6th edn (ed E. Windle Taylor); *Hygiene and Food Production* (ed A. Fox, 1971); Ellis Horwood Limited, Chichester (*Hygienic Design and Operation of Food Plant* by R. Jowitt, 1980); Elsevier Applied Science Publishers Ltd, Barking, Essex, (from *Developments in Soft Drinks Technology* – 2 (ed H. W. Houghton), 1981; *Food Microbiology and Hygiene* by P. R. Hayes, 1985); Food and Drink Federation, 6 Catherine Street, London, UK; *Food Engineering Magazine* – A Chilton Company publication; *Food Processing* – a Putnam Publishing Co. journal; Food Processors Institute, Washington, DC (from *A Guide to Waste Management in the Food Processing Industry*, 1979), © FPI, for Section 2.1, page 53, Table 2.06; General Mills Inc.; International Association of Milk, Food & Environmental Sanitarians, Ames, Iowa (from *Journal of Milk and Food Technology*, 1975; **38**, (6) 370); Institute of Food Technologists, Chicago, Illinois (from *Food Technology*, April, 1984, 109); Institution of Chemical Engineers, Rugby, UK (from *Profitability of Food Processing*, 1984, 1. Chem. E. Symposium series no. 84, p.36); International Dairy Federation (from Duke, M. 'Good Manufacturing practices – an essential ingredient of Quality and Safety', in *Bulletin of the IDF* no. 229/1988); Lavrids Knudsen Maskinfabrik A/S (for Alfa-Laval data sheet shown here as Table 5.10); Ministry of Agriculture, Fisheries and Food (from *Food Hygiene Codes of Practice 10*, 1981; and data from MAFF project published in Campden Technical Memorandum No. 523, 1989; © Crown Copyright); McGraw-Hill, New York (from *Sanitation for the Food-Preservation Industries* prepared by the Association of Food Industry Sanitarians Inc. 1952); National Academy Press, Washington, DC (from *An Evaluation of the Role of Microbiological Criteria for Food Ingredients*, by the National Academy of Sciences, © 1985); The Pillsbury Co.; Society of Dairy Technology, Crossley House, 72 Ermine Street Huntingdon, Cambs PE18 6EZ England (for data from 'Table 4 Growth parameters of the natural bacteria flora of pasteurized milks of varying keeping quality' taken from a paper 'Modelling the Relation Between Bacterial Growth and Storage Temperature in Pasteurized Milks of varying Hygienic Quality' by M. W. Griffiths and J. D. Phillips published in the Society's *Journal* **41**, no. 4, November, 1988; and for Table 4 Comparison of Laboratory results for nitrate in skimmed milk powder from a paper The 'Application of HACCP System for Milk Powder Manufacture' by Martin Woodhall published in the Society's *Journal* **42**, no. 4, November, 1989); Technical Institute of Food Safety, 4135 Hampshire Avenue North, Crystal, Minnesota 55427, USA (from *Engineering for Food Safety and Sanitation* by Thomas J. Imholte, 1984); © University of Toronto Press (from ICMSF *Microorganisms in Food; 2 Sampling for Microbiological Analysis: Principles and Specific Applications*, 2nd edn, 1986); Van Nostrand Reinhold (from *Quality Cost for the Food Industry* by A. Kramer and B. Twigg © 1980 An AVI book published by AVI acquired by Van Nostrand Reinhold All Rights Reserved; from *Principles of Food Sanitation*, 2nd edn, by Norman G. Marriott © 1989 An AVI book published by Van Nostrand Reinhold All Rights Reserved).

Disclaimer

Contents

List of figures

List of tables

Chapter 1
Introduction

1.A Purpose

Each Heinz company, regardless of its location, must adhere to Corporate Policies that direct a wide range of activities. This book is designed to provide a microbiological interpretation of Heinz Corporate Policy on Quality Foods and Manufacturing Standards (Corporate Policy 3.12). The policy states that the Chief Executive Officer/Managing Director of each affiliate is responsible for establishing and maintaining proper facilities for the controlled production of wholesome foods in a sanitary manner. This applies equally to foods made 'in-house' as well as 'co-packed' – that is, made by another company to a Heinz specification.

This book establishes principles and practices that experience has shown useful, together with some suggestions for further reading. This will help users develop procedures and criteria, appropriate to their local situation, which will comply with Policy 3.12. These procedures and criteria should be used unless any local legal or contractual obligations require something more stringent, in either the country of production or the country of sale.

While there can be no guarantee that illness never will result from consumption of foods, food processors can minimize the risk of illness from food consumption by taking steps that conform to scientific and operational knowledge. Every product should be known to be safe and wholesome under specified conditions of manufacture, distribution and use. These specifications must account for potential for abuse during distribution and misuse by the consumer.

Properly controlled production and distribution practices minimize the risk of product spoilage. Therefore, in order to provide the required assurance of product safety and stability, we must show that we consistently apply comprehensive knowledge of the preservation system in the manufacture, distribution and storage of food in our care.

No product population or lot can be 'totally purified' by non-destructive examination and segregation. Therefore, the risk of food poisoning is minimized only by designing and implementing appropriate preventive measures. The food processors are, correctly, held accountable for doing this.

Accordingly, the organizational structures and methods required for minimizing the risk of food poisoning must incorporate a Total Quality Management (TQM) approach. This requires an emphasis on error-free operation rather than 'purifications' of the product after manufacture.

The purpose of this book is to assemble microbiological design and operational materials that experience has shown to be useful. Collecting from many sources not readily accessible, this book will facilitate the controlled production of clean, wholesome and safe foods in a microbiologically sanitary manner.

This book is not intended to address the other food safety issues, which are not less important, such as pesticide or residual toxicants on raw ingredients and packaging.

1.B Background

In many industrialized countries, there is an important paradox in the public's perception of food. There is the desire for natural or organic food, which is perceived as being produced without 'artificial' aid or interference with nature, and swiftly delivered from producer to the store or user. At the same time, there also is the desire for food

1

that is safe, convenient, readily available, storable for an indefinite period – all at the most affordable price.

When trying to resolve this paradox, the processor must give priority to the safety of the food. Because there is potential for human suffering or even death, there is also the possibility of harm to the reputation of the manufacturer and the food industry. Loss of public confidence in the safety of food is a serious matter that may have considerable financial consequences. For example, in 1989 egg producers in the UK lost millions of pounds sterling when the microbiological safety of eggs was questioned. Processors are aware that the length and complexity of the food chain – from farmer to consumer – causes unsafe food to have severe public health, financial and legislative consequences.

Food may become microbiologically unsafe if a disease-causing organism (pathogen) is present and causes an infection. Alternatively, a poison (toxin) may be produced, causing the illness usually referred to as an intoxication. Food with infections and intoxications must, therefore, be prevented.

Some business consequences of food-borne infections or intoxications

Usually, when the public obtains information on unsafe or potentially unsafe food, the manufacturer's knowledge is incomplete and the crisis situation is extremely unstable. Complete public health facts and financial consequences cannot be determined until the outbreak ends. A full evaluation takes time and all consequences may not be known for years. Pending legal proceedings may be lengthy and sales may take time to recover to previous volume levels.

Also, the true cost of a product recall, whether due to safety or spoilage, is not always recognized. Apart from the obvious direct loss of the recalled product, there may be a loss of current and future sales. There will be the cost of time spent dealing with the problem and other logistic costs incurred. Another hidden cost is the diversion of effort from the main purpose of the business, which is to deliver a safe product that can be sold at a profit.

An early example of financial impact occurred in March 1963. Following a widely publicized incident of botulism in Detroit (USA), sales of canned tuna that year were $50 million less than in the previous year.

This and other incidents led to legislative or regulatory actions. For example, due to a small number of cases of botulism during approximately a decade, the US Food and Drug Administration (FDA) made very detailed regulations for the canning of low-acid and acidified low-acid foods. Following the outbreak of Listeriosis in California (USA) in 1985, the FDA instituted a product recall to a range of dairy products, which is estimated to have cost the industry $66 million in 1986 and 1987.

Examples of Public Health and financial consequences over a five – year period

Incident	Date	Country	Number of illness cases	Cost in $US
1	1982	UK	245	268 000
2	1984	Canada	2700	10 million
3	1985	USA	16 000: 2 deaths	Dairy bankrupted. Over 30 million paid in settlements.
4	1985	USA	142: 47 deaths	Lawsuit of 800 million*
5	1985	UK	76 (48 infants)	37 million
6	1986	UK	54	285 000
7	1987	Switzerland	Reported 30–60 deaths	1 500 000

*Manufacturer bankrupt. Officers of company found criminally negligent and imprisoned.

Causative organism and level found in contaminated food

Incident	Food	Organism	Probable numbers eaten per person
1	Chocolate bar	*Salmonella napoli*	50–100
2	Cheddar cheese	*Salm. typhimurium*	Less than 10
3	Pasteurized milk	*Salm. typhimurium*	3 000
4	'Mexican Style' cheese	*L. monocytogenes*	Found in 25 g samples
5	Infant dried milk	*Salm. ealing*	1.6 organisms per 450 g of dried baby food were found
6	Pasteurized milk	*Salm. branderup*	Not known
7	Vacherin Mont d'Or cheese	*L. monocytogenes*	Found in 25 g samples

Outbreaks may, therefore, lead to legislation which helps the industry produce safe food, e.g. low-acid canned foods in the USA and the pasteurization of all liquid milk sold in Scotland. Outbreaks also may lead to official intervention, like the dairy product recalls in the USA, which is costly to the industry.

Legislative concern is usually intended to protect the consumer from food poisoning, injury from contaminants or dishonest practices. Spoilage, like waste, is regarded as the problem of the producer and/or distributor. This is an important problem, the magnitude of which only becomes apparent when the 'cost of quality' is accurately quantified.

Positive action for microbiological safety

The most effective way of achieving food safety is to adopt the philosophy that, in principle, food poisoning should be preventable. This means understanding its causes, determining the required preventive and remedial measures, and managing food handling so that these measures are always used. This applies throughout the food chain, from primary producer to consumer, not just to the food processor.

Traditional quality control methods, such as periodic monitoring of storage conditions and manufacturing processes and the testing of a small number of finished products, are simply not adequate for achieving food safety, controlling spoilage and meeting the needs of the marketplace. Corporately, Heinz believes it is necessary to adopt the TQM approach for product safety and quality.

This work ethic applies to everyone in the company. The TQM approach addresses planning, control and improvements on a project-by-project basis to achieve major improvements quickly. Using W. Edwards Demings' definition, TQM means integrating the efforts of a company to achieve a predictable degree of uniformity and dependability, at low cost, with quality suited to the market. It is a continual activity, led by management, in which everyone recognizes personal responsibility for safety and quality. This means aiming to 'get it right the first time', thus achieving safety and reducing the cost of quality. However, nothing in business remains the same, so it is essential to strive for continuing improvement. This idea is expressed as the principle of 'Habitual Incremental Improvement' (HII) and is very much a part of TQM.

For the purposes of this book, TQM requires:

- Soundly based technology;
- Carefully considered, clearly expressed and properly integrated systems and procedures dedicated to achieving product safety and quality goals.

Two key elements for implementing TQM are:

1. The recognition of the importance and power of the 'Hazard Analysis Critical Control Point' (HACCP) system; and
2. The use of audit techniques for safety and quality purposes.

Audits are used to achieve 'Root-Cause Corrective (Remedial) Action' for any deficiencies which are found; to assure management that gains which have been made are held; and to close the feedback loop which assures that each quality system is properly implemented.

1.C Uses of this book

This book is intended to be a resource for a variety of uses. It may:
- Present information which is of direct use, e.g. data in figures or tables.
- Help develop unified, comprehensive systems and procedures for the safe production and distribution of products. These include the requirement that full and proper records need to be kept not only for statutory or contractual reasons but also to provide data for analysis in improvement projects required by Habitual Incremental Improvement (HII).
- Help develop quality audits (see Chapter 2) which, when 'Root-Cause Corrective (Remedial) Action' is taken, make an essential contribution to HII.
- Aid in making Hazard Analysis Critical Control Point (HACCP) analyses (see Chapter 3) which are the basis for a positive assurance of product safety and stability.
- Help develop quality assurance procedures.
- Promote awareness of broadly based Good Manufacturing Practices (GMP).
- Be useful in developing training programs.
- Start a new line of thinking or a new approach to a problem.

Be aware that because of the principle of Habitual Incremental Improvement (HII), it should be used as a starting point and not as a finishing point. Remember that its purpose is to facilitate the controlled production of clean, wholesome foods in a sanitary manner. Use this book in any way that does this because it is not a textbook to be learned. However, individual circumstances and needs are unique, so this book cannot provide detailed answers for every need nor is it a collection of inflexible requirements.

To Heinz companies

This book challenges management and technologists who are responsible for current production and for making the on-going improvements required by our commitment to Total Quality Management. It also is a resource to be used in aiding our general progress.

To co-packers

Co-packers must have a clear understanding of what is expected and required when packing food under the Heinz label. A brief summary of pertinent microbiological and foreign material criteria must be part of the purchase specification. As a background to these criteria, this book gives an understanding of our approach to safe food production, Good Manufacturing Practices (GMP) and microbiological quality. This should provide a better understanding of the minimum standards which are acceptable to the local Heinz company. These standards should be expected to form part of any contract, either explicitly or implicitly. To avoid misunderstanding, discussion between the technical representative(s) of the co-packer and the appropriate Heinz technical department(s) is essential before any agreement is reached and any assurance of compliance is given or received.

To others

Companies, research institutions, academia and regulatory authorities collaborate in the promotion of food safety to a greater extent than may be realized. Processors must, however, take responsibility for their operations. This book offers insight into ways by which an individual processor may achieve microbiologically safe foods.

1.D A note on the layout of this book

The layout and proportions of material in this book are unusual because it is primarily intended to help the manager or technologist/microbiologist working in a food processing plant. Information must be easily accessible for ease of use. Those involved know that time is critical when close deadlines are to be met. This is the reason for the choice, arrangement and layout of material.

The length of the book may seem formidable, especially if it is not in the reader's native language. However, after considerable discussion, it was decided that this was the minimum amount of information needed.

The material is a compilation of principles, examples of pertinent detail, data, suggestions for layouts or systems as well as some references and additional reading. It is organized into chapters, each dealing with a particular topic. For the convenience of the user, cross-referencing has been kept to a minimum.

Also for user convenience, the summary of contents at the front of the book is supplemented by a detailed list of contents at the start of each chapter. This will give quicker access to a particular piece of information.

References may be given at the end of a section or at the end of several sections. The basis for this is to make access convenient.

Chapter 2
Assessment of an operation

Introduction

Informed and independent assessment made by auditing is an essential part of the application of Total Quality Management (TQM) philosophy, and whether applied to an 'in-house' operation or vendor or co-packer uses the same approach and applies similar criteria. The aim is to determine how and how well Corporate Policy 3.12 objectives are fulfilled. This Policy requires 'the controlled production of clean, wholesome foods in a sanitary manner.'

In making the assessment, the basic approach is to determine whether the processes and products are inherently safe. This is done by assessing how adequately the HACCP strategy is being applied for safety (and quality) to all aspects of the production and distribution of foods. This is because HACCP is a logical system and should, if consistently applied, be the most effective way to insure the safety as well as the quality of production.

It is also recognized that assessment using 'audit' ('quality audit') techniques gives an objective measure of how well TQM is being implemented as well as a positive measure of deficiencies which need 'root-cause corrective (remedial) action' together with independent confirmation that 'gains made have been held'.

Such assessments are an integral part in the application of ISO 9000 and BS 5750 quality systems which involve all employees in the business from top management to the most basic grade of operator. Such quality systems are clear, unambiguous and sharpen the individual's sense of accountability.

To be effective, auditors need expertise, experience and training. The last should be expected to include 'in-house' as well as appropriate external training. Local conditions vary, but increasingly there is a desire by processor and major customers for some form of accreditation of quality systems and hence of auditors.

In making the assessment it is necessary to consider:

- Management philosophy, attitudes and organization
- Plant, processes and procedures
- Quality auditing

2.A Management philosophy, attitudes and organization

Philosophy and attitudes

In reality, although there is an enormous variety of business in the food industries there are some common 'product safety and quality' principles which need consideration here.

One of the most important of these is management philosophy and attitudes. Only senior management can say with authority 'We hold certain things to be important; we will provide resources to enable specific ways of working; and we will insist that these things are done'. This principle applies equally to a large complex organization or to a small plant managed by its owner with a few helpers.

Some key questions

Some key questions about senior management philosophy and attitudes include:

- Is there an awareness of basic technical principles relating to the product/processing/packaging/distribution system under consideration? This includes, but is not limited to, generally recognized Good Manufacturing Practices (GMPs) and, of course, regulatory requirements.
- Is there an awareness of the need for and the methodology of technical control which can fairly and realistically be described as a positive attitude?
- In what ways are management philosophy and attitudes reflected in the organization? How does the organizational structure promote or hinder the achievement of Product Safety/Quality goals? Remember that it has been known for the effective decision-making structure to be different from that shown in the organization chart.
- Who *'owns'* product/process/packaging/distribution 'Safety and Quality'? Specifically, is it the *operational* or the *quality* departments?
- What is the evidence for adequate resources of plant, people and organization in order to achieve declared or implicit Safety/Quality goals?
- In what ways is the technical function independent of production?
- Is there an openness to new ideas and change?

Other useful and important questions are suggested in Campden Technical Manuals 12 and 16 [1, 2], which have a wider application than the cannery and frozen food factories in their titles.

The assessment of the inevitably partial answers to such questions is a matter for experienced judgment, and those making such assessments are accountable for and must stand by their judgments. These judgments, however, must be made, as it is totally unacceptable to use the price of a product as the sole effective criterion of management philosophy, attitudes and organization.

Organization

A discussion of the organization of Quality Control or Quality Assurance and Audit is outside the scope of this chapter. Heinz is committed to Total Quality Management, which includes Quality Assurance and Audit, because we believe it is the most effective and the most cost-effective organizational pattern. However, it is important to recognize that the expression of organizational principles differs in small and large organizations. In a small business, the personality of individuals becomes relatively more important and may well override organizational niceties. It is always important to be clear as to who makes decisions and, particularly with a small organization, what are the contingency plans for the absence of key personnel, irrespective of hierarchical status.

Some questions

It is necessary to address organizational issues in relation to such questions as:

- Who can make the decision to quarantine a suspect lot? Who can make the decision to release different types of quarantine?
- Who has the authority to stop and to re-start production after an incident with safety or quality implications?
- Who supervises a lot designated for destruction? Does the procedure deliver the required assurance of safe and proper destruction?
- Who would be involved in a product recall? Could the recall procedure deliver what is required within likely time constraints?

Other organizational questions include:

- What is the organization of technical personnel? What are their duties and responsibilities, their qualifications and experience, their hours of work, e.g. in relation to technical presence on night shift or when production overtime is being worked?
- What is the organizational relationship between technical and production personnel at management and other levels? Is there a training role for technical personnel?
- How is the availability of both information and procedures organized in order to make certain those with a 'need to know' have the required data when they need them?
- If services are contracted out, e.g. cleaning, pest control, analysis, what is the organization to manage and control the contracts?
- How are safety and quality data checked, presented, organized and used?

Other and more detailed questions are given in Campden Technical Manuals 12 and 16[1, 2] and in the auditing checklist given as an appendix to this chapter.

Warning For assessment of an operation, a checklist is valuable as a guide or framework. It should include 'open-ended' questions because it is not a substitute for critical thought or expert judgment when making an assessment. A useful strategy, on occasions, is to ask 'What is happening here? How do we know that is right (i.e. in control)? What happens if and when it goes wrong?'

To summarize, the key questions about organization are:

- If applied – can it deliver what is required?
- What evidence is there that it works well and is applied consistently?

2.B Plant, processes and procedures

Plant and processes

Scope

When the word 'plant' is used, it does not just mean the food processing equipment. Assessment of plant starts by considering:

- The environment (or grounds) – i.e. the site

then moves inwards to:

- Building – design construction and sanitation
- Food handling equipment – design, construction, installation, maintenance and sanitation
- Ingredients, including the supply chain to the processor
- Manufacturing and filling (together with sterilizing if food is canned). Don't forget that rated output becomes a quality factor if design limits are exceeded either in throughput or length of running before cleaning and sanitizing.
- Packaging
- Storage
- Distribution

Clearly, within the usual time constraints, only a relatively small sample of observations can be made. The assessor needs to decide in advance what are likely to be the key factors affecting the safety and quality of the specific product/processing/packaging/distribution system under consideration. Additionally, the assessor should look for those signs which indicate that GMPs are in place. This means that weight is given to the general impression of orderliness, good housekeeping, sanitation together with evidence of managerial and technical control. It must be stressed that general impressions are *not sufficient* and pertinent; specific items must be examined.

Some questions

As an example of questions that may be in the assessor's mind, consider the environment (or grounds), i.e. the site. The assessment is of the risk of contamination by harmful foreign material, e.g. dust or tainting substances, and the hazards due to pests, e.g. rodents, insects, birds. The questions may include:

- Are there any nearby sources of contamination?
- Condition of the perimeter fence. Does it trap litter? Evidence of effective pest control? If unsatisfactory, does management know or care?
- Condition of the site. Is it tidy? Harborage for weeds and pests? Roads and parking lots clear? Evidence of flooding or water standing in area? Is traffic routing likely to cause contamination? Evidence of trash (litter) from trucks waiting to load or unload? Does lighting attract unwanted insects?
- Are doors and windows left open to allow entry of dust or pests?
- Evidence of effective pest control at exterior of building?

Other questions are suggested by the Campden Technical Manuals 12 and 16[1, 2], by Shapton[3], and in the checklist in the appendix to this chapter.

Other and more detailed questions should be considered under the other headings as appropriate for the specific assessment.

Procedures

In essence, procedures are the 'rules of the game'. They are important for the user because they combine the function of an operating instruction with that of a criterion. To be effective they must tell the user clearly and unambiguously what to do, e.g. to operate a process, to make a judgment or to produce a product. Equally, the auditor should be able to tell from a procedure what ought to happen. Experience has shown that the following comments may be helpful.

Writing good procedures is more difficult than it may seem. It is not easy, for example, to use clear, unambiguous language or to organize text in a way that is 'user-friendly'. Sometimes, sheer length can be a barrier to both understanding and ease of use. If helpful to the user, reference can be made to material which could be used but is not necessarily maintained by the procedure system. An example of this is a good operating manual produced by an equipment supplier. However, it is essential that procedures which relate either to other procedures or any other documents form an integrated and fully referenced system.

Remember also that once procedures have been developed and accepted, they must be updated or maintained. One great advantage of electronic data handling to the point of use is that one entry can update *all* users. It is all too easy to overlook the practical difficulties of updating a procedure held in a number of locations *all* of which must be properly maintained if they are to be effective.

All procedures will need to be altered at some time. It is therefore essential to have a clearly understood 'contrary to' mechanism. The rule should be that compliance with procedures is mandatory and changes to procedure should only be made by the issue of a duly authorized 'Contrary to Procedure Instruction' or by withdrawal and re-issue of all or part of a procedure. It should also be the rule that 'Contrary to Procedure Instructions' should only be issued for a limited, defined time.

A 'Policy' may be defined as a statement of intent, aim or goal. A 'Procedure' is therefore a device or way of implementing a 'Policy' and if followed should be capable of delivering the required result. To avoid misunderstanding, it is therefore important to have both policies and procedures in writing and available to all those with a 'need to know'. A 'Policy' statement should include:

- A statement of purpose or intent – what should it achieve?
- A statement of scope – where does it apply? For example, it may only apply to one particular location or department.
- The policy statement – as briefly as possible.
- Accountabilities for implementation – by office or title (not by name).

A 'Procedure' should include:

- A statement of purpose or intent deriving from an identical policy
- A statement of scope – where does it apply?
- The structure of the procedure. This should consist of numbered paragraphs and may be as long as necessary. It is important to make reference ('cross-reference') to any 'interface document' such as other procedure(s), instructional or other types of manual, personnel training records, e.g. certification for operating sterilization equipment. Where applicable it may contain data, e.g. target and limits for temperature, ingredient weights. Where alternatives have been approved as part of the process, they should be given in the procedure. This part of the procedure may also indicate corrective actions.
- Accountabilities for issue and maintenance of procedures – by office or title.
- Accountabilities for implementation – by office or title.

Auditing experience shows that procedures fail to deliver the required results because:

- They are not correctly followed.
- The procedures are incorrect, e.g. temperatures or weights are wrongly stated.
- The procedures are inadequate or incomplete.

When 'system audits' are first introduced, which inevitably challenge the ability of a procedure system to delivery what is required, it may be found that the existing system is seriously inadequate. It is a better use of resources to put good procedures in place rather than to continue showing or attempt to justify the inadequacy of the current system.

2.C. Quality audits

Background

It is sometimes a real problem to know what is meant when the term 'quality audit' – often shortened to 'audit' – is used. This is because the term is used to mean different things particularly when quality audit systems are first introduced to a company. It is not uncommon in the early days to hear the word 'audit' used interchangeably with the term 'inspection' – because it seems more impressive. While it is true that any quality audit will involve 'inspection' (or 'examination') it is, in fact, quite different from 'inspection'.

For convenience the term 'quality audit' is used in this chapter. It covers both 'safety' and 'quality' aspects since similar principles and methodologies are used.

It may be helpful to recall that in ancient Rome there were those who were employed to hear (financial) accounts. They would compare what they were told by the different officials who were dealing with the matter. From these early beginnings, the practices and profession of 'Internal (financial) Auditing' developed.

The principles and ways of working of quality audits have a surprising amount in common with 'Internal (financial) Audits' but there are important differences. One of these is that internal auditing is a recognized profession with well-established schemes of training. Quality auditing has a way to go in this respect. Another very important difference is that when a quality auditor assesses a deficiency as critical, major or minor, this does not automatically translate into amounts of currency at risk. In quality auditing terms it is an assessment of the significance to the quality system of a particular finding. For example, a deficiency would be regarded as more serious if it were clearly indicative of a number of other deficiencies not yet explored than if it was a 'one-off' finding. Note that in some audits this assessment may not be made, but auditors must always think through the implications and significance of their findings. Another example is that procedures in a specific operation may be inadequate to deliver what they are supposed to do. If the personnel concerned are both very experienced and careful there may not be any immediate financial losses because of the lack of good procedures. However, in the long term, when personnel, methods, and levels of manning or dedication change, the cost of not having proper procedures to 'get it right the first time' can be expected to be significant.

The introduction of a quality audit system

The transition from the traditional quality control to a quality assurance and audit system is not easy. Apart from the scale of organizational changes, do not underestimate the educational and training needs or the difficulties individuals may have in changing long-established ways of working. As examples of this, management may confuse 'audit', 'inspection' and 'investigation' and this must be expected to cause confusion for those attempting to carry out management wishes. Another example is that quality control inspectors may have, and often enjoy, an unofficial role as 'Technical Foremen' and may well have difficulty in adapting to a role as 'Factory Auditor'.

This is partly because the 'quality control department inspector' often finds familiar ways of working hard to change. It is also because production personnel find it equally hard to change familiar ways and to accept fully the implications of the fact that they are now primarily responsible for the quality of their output. There can also be a reluctance to accept that the factory auditors (i.e. the ex-inspectors) are there as a 'second pair of eyes' whose primary purpose is to assess the performance of procedures and systems instead of providing immediate answers to problems. It also takes time to appreciate the obvious and true proposition that audit is not adversarial nor is it concerned with the trivial or unimportant.

When introducing an 'in-house' quality audit system, it is important to achieve a demonstrable success. This requires careful planning, cooperation from an auditee who understands and supports the concepts of auditing together with concentration of resources to achieve results in a reasonable time. Quality audit must be understood as a management tool. In-house, it has a facilitating role and its value therefore depends on timely 'root cause corrective (remedial) action' being taken. If this is not done, then audit is being misused and is a waste of resources. Audit can also provide an assurance that improvements or gains which have been made are being maintained. With vendor or co-packer audits the primary purpose is to assess the ability of another company to produce consistently to Heinz criteria.

Quality audits – definitions and outline of procedure

Definitions

For the purpose of this chapter, audit definitions and procedures have been generalized, and what follows could be regarded as a broad consensus, rather than applicable to a particular company or department. This is because audits can be done at various levels and are appropriate if they deliver what is necessary for the particular purpose(s). In essence, all audits are an independent appraisal and assessment of an operation, activity or control.

Note – Definitions are not listed alphabetically, but in the order in which they may well be considered when thinking about audits.

Audits

Two useful definitions of audit are given in ISO 8402[4] and by Sawyer[5]. Note that Sawyer's definition is for financial internal audits.

ISO – 'A systematic and independent examination to determine whether quality activities and results comply with planned arrangements and whether these arrangements are implemented effectively and are suitable to achieve objectives.'

Sawyer – 'An independent appraisal of the diverse operations and controls within an organization to determine whether acceptable policies and procedures are followed, established standards are met, resources are used efficiently and economically and the organization's objectives are being achieved.'

It is important to remember that audit observations, like a census, are a sample of on-going events from which conclusions are drawn which form the basis for action. Although audits are an integral part of Total Quality Management (TQM) and are invaluable as a basis for Habitual Incremental Improvement (HII), audits, because they are a sample, can *never* guarantee that *all* production is satisfactory. Be aware that whatever the type of 'quality audit', its aim and purpose is to facilitate 'root-cause corrective (remedial) action' to correct deficiencies.

Types of audit

There are many descriptions which could be given, but for practical purposes, the following four categories may be used: systems audit, compliance audit, investigative audit, vendor (supplier), or co-packer audit.

Systems audit

This is a thorough, comprehensive, systematic independent review, appraisal and assessment of all or part of a system, control or procedure, e.g. of a manufacturing operation; ingredient or product handling; documentation; records or administration system. The purpose is to determine whether there are deficiencies and inadequacies which require remedial action. Since not everything can be done at once, priorities must be assigned. See also the definitions in ISO 9000 and BS 5750.

Compliance audit

This is an independent inspection, appraisal and assessment of compliance with a company procedure or control. Much 'in-house' factory auditing is in this category.

Investigative audit

This is an independent investigation, appraisal and assessment of quality hazards and risks associated with current or proposed practices. It could be argued that this category does not fit easily into either the ISO or Sawyer definitions. However, senior management does ask for this type of investigation and wants them called 'audits'. They have therefore been included.

Vendor (supplier) or co-packer audit

This is an independent examination, appraisal and assessment of the ability of another company to consistently deliver to Heinz safety and quality requirements (criteria).

Deficiencies (critical, major, minor)

These are conditions or actions which are not as they should be. They should prompt the auditor to determine the full significance and implications of the deficiency and to question the ability of the system to deliver the required effective control. The primary value of a deficiency rating is to the auditor and for 'in-house' management information. Auditors should be alert to the possibility of other deficiencies not necessarily within the strict boundaries of the audit. This should be prompt the auditor to ask 'What else is wrong?'

The three categories of deficiency recognized as standard usage in a quality audit report are:

- Critical. This is judged likely to cause serious damage to the business because it is likely to be unlawful; to be of significant public health concern; to be a substantial infringement of a Code of Practice or Good Manufacturing Practice (GMP); to cause a major consumer complaint or to be seriously unsafe.
- Major. This is likely to prevent an activity, function or unit from meeting a substantial part of its objectives or goals. This may be because, for example, of bad manufacturing practice and/or because of errors or omissions in authorized procedures.
- Minor. This is judged to require reporting by the auditors because it may hinder the achievement of an objective or goal. It needs to be corrected within a specified time and is therefore *not* trivial or unimportant.

Hazard

This is a potential to cause harm to the consumer (which may also be thought of as safety) or to the product (which may also be thought of as spoilage).

Hazard analysis

This is any system which analyses the significance of a hazard to consumer safety and/or product acceptability. It is an integral part of any audit and can be of considerable value in the discussions between auditors and auditees. However, it is not usual for the analysis to be formally and specifically reported as part of an audit report.

Risk

This is the chance, which may be expressed as a probability, that a hazard will be realized (or occur).

Risk assessment

This is any system which assesses the probability of a hazard occurring. Usually in an audit, because of the absence of good factual data, this becomes largely a matter of an individual's judgment or opinion, e.g. the risk is 'low', 'moderate' or 'high'. It is therefore usually prudent to concentrate on hazard analysis in order to determine the appropriate priority for action on a deficiency.

Categories for priority of action

Consideration of hazards, together with some notion (estimate) of the risk(s), determines the priority of response to a deficiency. It is simplest to use two categories:

- Immediate response – priority 1. The operation or activity must be stopped unless the deficiency can be immediately corrected or sufficiently improved. This priority will rarely be used. Although the action will have been completed by the time the report is issued, it *must* be reported.
- Planned response – priority 2. This requires a realistic but prompt completion date to be given by the addressee or assignee(s). The time scale will vary according to the seriousness of the deficiency, the nature of the hazard(s), the perceived risk and the feasibility of the corrective action in relation to e.g. equipment or a specialist labor availability.

Addressee

This is the person to whom the audit report is addressed, and who is responsible for coordinating the follow-up in his or her area, e.g. within a plant or home office department or area.

Assignee

These are person(s) other than the addressee who is/are responsible for some designated follow up. For example, in an in-house report addressed to the plant manager, some home office action may be required from a home office assignee, e.g. product development manager or process engineer.

Independence (of auditors)

The requirement for independence of the audit means both that the auditors are free from pressures or inducements which could reasonably be held to bias their report and also that they are not so familiar with the subject of the audit that they are biased by this very familiarity. Auditors should also be independent in that they do not have to depend on the auditee for analysis of data. 'In-house', this means that auditors will always be independent of the line, function or area management being audited. Independence also means that auditors should not be responsible for detailing corrective or remedial action.

Outline of procedure

Consideration of the very wide variety of audits means that it is not possible, or indeed desirable, to give a 'recipe' or 'checklist' for audit procedures. It is always sound strategy to go back to 'first principles' and approach the subject with an open, logical mind.

However, audits consist of several stages which may usefully be described as:

- Planning
- Observation
- Appraisal and assessment
- Reporting
- Follow-up

Planning

The first and in many ways the most important step is to be clear about the purpose of the audit, i.e. what must it deliver or accomplish. Although this sets the initial boundaries it should be recognized that these may need to be altered either as a result of preliminary work, e.g. on documentation, or during the course of the audit. This is particularly true of systems audits.

After this it is usual to review and assess past reports (if any) and to collect and examine critically the appropriate procedures, records, reports and any other documentation. It is sometimes helpful to have blank as well as completed records or forms. With completed records it is helpful to have them relating to a similar period of time. Where work extends over more than one shift it is useful and sometimes important to have records of the same type over the other shifts. It is not unknown for 'days' and 'nights' to do things differently. Time spent on this evaluation is seldom wasted and usually suggests useful lines of enquiry. For example, if recorded times of observations or values seem to be 'rounded' rather than actual, this suggests an obvious lead although it may or may not be important in a specific situation.

Particularly when system audits are being introduced, it is quite likely that system inadequacies or inconsistencies will be found at this preliminary stage. Remembering that the aim of audit is to facilitate 'root-cause corrective (remedial) action' it may not be necessary or advisable to continue further with the audit at this time.

From the preliminary review, it is usual to develop lines of enquiry or themes and to formulate key questions. No later than this stage, the debate and discussions between auditors is invaluable in identifying probing questions and areas needing further study.

The next stage is to prepare working documents and obtain any necessary calibration or test equipment, making sure that it is of a suitable standard of accuracy. At this stage it is prudent for auditors to insure that they have facilities for analysis and presentation of data both to the auditees and for the audit report as the needs may well be different. Experience teaches that it is better not to rely on these being available locally.

The next stage is to decide on who are the key personnel contacts; if there is anyone who can adequately substitute for them; and what documents, records etc. are needed 'on-site'. At this stage the agenda and dates of the 'opening' and 'close-out' meetings may be set, although it is wise to build in flexibility since the 'close-out' meeting can be significantly altered by findings 'on-site'. After this stage, travel arrangements may be made!

Observation

Before getting to the building(s) a preliminary assessment may usefully be made of the location and condition of the external areas. It is good practice to do this whether the audit is 'in-house' or not.

Before starting observations it is important to hold an 'introduction' or 'opening' meeting which may be formal or informal as appropriate. This is where the auditors meet the auditees at the appropriate management level(s), introduce themselves and explain the reason(s) for the audit. It is sometimes an opportunity for the auditors to stress that audit is *not* a fault-finding adversarial process measured by scoring points but an attempt to identify, as a preliminary to removing, obstacles to greater efficiency and so improving 'quality' as part of our TQM philosophy. Experience shows that an important contribution made by an auditor is that of 'bringing a fresh pair of eyes' to a situation. The auditor will, of course, bring much else in the way of experience and expertise and should have the confidence of senior management to report fairly and honestly. Providing that the auditor takes a wise and sensitive approach, the auditees soon appreciate the value of 'the fresh pair of eyes' in making improvements. Rightly and properly, the auditees should be given full recognition (or credit) for the ideas, observations and other contributions they make. Senior management and the experienced quality auditor should recognize that 'in-house' there is a significant 'facilitator' role for the audit process.

The 'opening' meeting is always an opportunity for the auditees to make the auditors aware of any pertinent information, e.g. any sensitive issues in the area or any special or unusual problems.

Because auditors need to be shown what happens, they will need both to observe and talk with operators, supervisors and management. Their presence should not be obtrusive as auditors have the duty to make 'professional' and effective audits and show due care, expertise and diligence. They should therefore be able to take an impartial or objective view of a situation or problem, produce timely work, make unobtrusive but accurate and shrewd observations, produce sufficient detail(s) to support recommendations and prepare effective reports. Experienced auditors understand the importance of numbers. For example, '14 broken tiles' on a wall carries more weight than 'a lot' or 'many'; or there were 'two open (unscreened) windows in the processing area' is more telling than 'some of your windows are open!

Significant and sometimes vital information comes from the answers to auditors' questions. Auditors therefore need to be able to ask 'open-ended' questions, e.g. 'What is the usual range of temperatures?' rather than 'What temperature is it at now?' The answers to 'what if' or 'what happens when' questions are sometimes revealing but remember, all answers need to be considered and assessed. In this assessment a comparison of what is seen and what is heard is important.

Auditors will want to check documentation and/or recording currently in progress. Sometimes, as in an 'in-house' compliance audit, this may be the primary purpose. Associated questions may be, for example: Are records made at the specified time? Are any records being made, in part, ahead of time? Are records made on the correct paperwork? (It has been known for records to be made on scrap paper before being entered on 'official' paperwork.)

Auditors may need to check on calibration by 'spot checks' and/or examination of calibration records.

Auditors may also wish to check on training records and on cross-referenced procedures for adequacy and completeness.

Appraisal and assessment

Experience shows that it is helpful for auditors to 'process' observations as they go along by making a rough draft of the audit report. This is particularly helpful in a complex systems audit in concentrating attention on important and significant points and in making a critical and dispassionate appraisal or assessment.

The quality of this appraisal and assessment is the foundation upon which the audit report is built. Auditors know that the principle of Habitual Incremental Improvement (HII) applies to them and that this provides an on-going challenge and stimulus.

The appraisal and assessment need to be organized for the 'close-out' meeting, whether this is formal or informal.

Reporting – 'close-out' meeting

The first reporting is made at the 'close-out' meeting between auditors and auditees before the auditors leave the site. It ought to be purely a discussion meeting on the auditors' findings which enables the auditors to confirm that the audit has been sufficiently comprehensive and acquaints the auditees with what will be reported.

Quite specifically, it is not a meeting to edit the report – that remains the sole responsibility of the auditors. Neither is it a meeting to agree accountabilities for remedial action – that is the responsibility of the addressee and assignee(s). It is also the last chance for the auditors to be sure that they have fully understood the implications of the audit subject and that the correct addressee and assignee(s) have been identified.

It is right that auditors give praise where it is due and this can be done as appropriate, both at the close-out meeting, e.g. repainting has made a great improvement to the entrance to the department, and/or in the final report.

Audit report

The purpose of an audit report is to place on record the findings, i.e. what has been observed or discovered, what deficiencies have been found together with recommendations and the appropriate priority for action to achieve 'root-cause corrective (remedial) action'. In addition to the subject being audited, the report should include any other observations and deficiencies which have been noted. These include Good Manufacturing Practice (GMP) issues and Sanitation or Good House-keeping items which are conveniently placed in an Annex or Appendix to the main report.

As it is important that audit reports are recognized as being factual and objective, this should be reflected in the style of writing. It is worth remembering that it is the quality not the quantity of words that is important.

In order to achieve an effective follow-up it is necessary to have clear accountability for remedial action. Thus the audit report must have a clearly identified addressee and, where applicable, assignee(s). Copies should also be sent to those who have a 'need to know' and as additionally decided by the auditors' line management or as requested by the addressee.

Attached to each recommendation is the priority for action (Priority 1 – immediate response, Priority 2 – planned response where timing is specified by the addressee or assignee). In some companies, it may be the practice for the auditor to have both the appropriate seniority to designate a time limit for an improvement and the authority to take prohibitive action if this limit is exceeded. However, it should always be possible for an auditor to challenge what appears to be an unduly slow response.

Because of the range of audit subjects, e.g. from simple compliance audits to complex system audits, there is a wide range of appropriate report styles and forms. These can range from a simple pre-printed form to a complex and detailed report. They will all include the following pertinent items:

- Addressee and circulation list
- Findings (including other observations, e.g. GMP, sanitation issues)
- Action plan with who was informed of findings, action to be taken and priority or deadlines as appropriate.
- Whether action has been taken and, if so, whether the deficiency has been remedied. If the corrective action is judged not to be adequate, what further action was taken.

With a complex systems audit, the report is likely to be formal and structured and may include the following:

- Title page with addressee and circulation list
- Contents
- Outline – introduction, aims, summary of findings, conclusions (in effect, a management summary).
- Action Plan. See Figure 2.1 for layout. This lists the recommendations by number and their categories; gives a brief description of the recommendations; indicates the kind of remedial action which is needed in general terms, e.g. administrative, engineering, R&D.
- Background – this gives the reasons why the audit was undertaken.
- Findings – in as many sections as are appropriate and may include individual 'conclusions' at appropriate places.
- Appendix or appendices as needed
- Annex for other observations, e.g. GMP, sanitation/Good Housekeeping issues.

ACTION PLAN

AUDIT REF. NO.

Remedial Action On Audit Recommendations **Addressee To Complete And Return To Auditor**

By / / Addressee Or Assignee To Inform Auditor When Action is Complete.

RECOMMENDATION					DATE		
NO. AND CAT.	DESCRIPTION	PRIORITY	HOW	BY WHOM	ACTION PLANNED BY	COMPLETED	AUDITORS INFORMED

PRIORITY
1. IMMEDIATE – AS THE REPORT IS WRITTEN AFTER THE EVENT, THE TIME OF CORRECTION SHOULD BE GIVEN
2. PLANNED RESPONSE – THIS IS TO BE COMPLETED BY THE ADDRESSEE OR ASSIGNEE(S) UNLESS THE AUDITOR HAS ISSUED AN IMPROVEMENT ORDER SHOWN, SHOWN AS "COMPLIANCE WITH IMPROVEMENT ORDER" IN THE HOW COLUMN. THE ADDRESSEE OR ASSIGNEE IS ACCOUNTABLE FOR REPORTING COMPLIANCE.

Figure 2.1 Example of an 'action plan' layout

Follow-up

This is an essential part of the audit and can be done in a variety of ways adapted to individual circumstances. However, it is a sound principle that those taking corrective (remedial) action are responsible for informing the auditors either that the work has been completed by or before the forecast date or (and this should always be done as soon as possible) that the forecast date cannot be met and provide a new forecast date. The onus should be on those taking action to provide good and sufficient reasons for the delay.

Auditors should, of course, make enquiries if a response has not been received by the due date. On occasions, the job may have been done but the auditors not informed because of administrative omission.

If the auditors are dissatisfied with the response which has been received, then the matter should be taken up with the most senior management involved. This is not necessarily limited to those shown on the circulation list of the audit report.

There is another aspect of 'follow-up' which should be remembered. One purpose of audit is to determine whether 'gains have been held'. This is sometimes known as a 're-audit' or 'follow-up' audit. Judgment is needed to decide how best to use limited audit resources but the claims of 'follow-up' audits should not be overlooked.

Notes on 'in-house' audits

1. It is worth stressing again that audit is of significant value in promoting timely 'root-cause corrective (remedial) action' and in providing assurance that 'gains have been held'. If senior management are not committed to this, audits are a waste of resources.
2. Audits can be done at several levels, but an important component of success is to have auditors of sufficient caliber. Those with experience of auditing know that it is a difficult and demanding job. The danger of overfamiliarity with a plant is a very real one which prevents the auditor bringing 'a fresh pair of eyes' to the subject. It is therefore prudent to plan for a succession of auditors. Persons of the right caliber may well be found among those approaching retirement age, or the young 'high flyers' who need an insight into how the business actually works. The leader of the audit team (or team manager) in these circumstances consciously needs to maintain continuity and standards and to supervise audit training and experience.
3. For straightforward compliance audits and for some vendor (supplier) or co-packer audits a single technical (quality) auditor is considered sufficient. For more complex audits, e.g. system audits, or when developing audit skills more than one auditor is needed. This is partly because too much may be happening at any one time for a single person to make the necessary close observations. It is also because dialogue, discussion and mutual prompting between auditors is often the source of discovering the real, underlying deficiencies. Remember that audits must be done in sufficient detail and with sufficient rigor if they are to deliver what is required of them.
4. When introducing system audits, it is important to achieve an early demonstrable success, and this requires concentration of resources. There can be considerable debate on what is best to choose. It is worth remembering at such a time that it is usually profitable to concentrate on a narrow topic with clear boundaries, e.g. can seam control in a single filling department; glass handling on a single line; or the operation of a single procedure such as manufacture of a single soup. This is because the number and experience of auditors may well be the principal constraint.
5. The question is sometimes raised as to whether audits should be unannounced. Experience shows that this depends on the type of audit and on its purpose. 'In-house' compliance audits are often made unannounced. System or investigative audits where it is essential to have discussions with particular individuals or to observe a particular part of a process will need to be pre-planned in detail and may well be done after giving prior notice.

Note on laboratory audits

It is not always appreciated how wide ranging and flexible quality audits can be made. Experience has shown the usefulness of audit in revealing weaknesses as well as providing an assurance of strengths.

Obviously, the audit needs to be designed for individual needs but should at least include questions on such topics as:

- Management – policies, procedures, staffing structure, extent of manning (single, double or triple shift, week-end and public holiday working).
- Job description accuracy and adequacy.
- Facilities – building, equipment, special facilities, e.g. incubation rooms.
- Safety facilities, equipment, procedures and features including first-aid and fire drills. Are individual responsibilities clearly understood and stated?
- Methods manual – basis of choice of method, adequacy of presentation, training program, status as reference laboratory, internal validation of methods.

- Adequacy of records and procedures for decision making on basis of records, preservation of records, action taken following unsatisfactory results, analysis of trends and review of what these mean for the business.
- Checks to insure that all samples have been taken and tested according to established procedures.
- Records for servicing and calibration of instruments, including scheduled date of next service.
- Clarity and compliance with laboratory cleaning and disinfection procedures.
- Written instructions for the operation and checking of each item of equipment. Do these include what to do if things go wrong? Are they in sufficient detail, e.g. for incubation – is there a thermometer, how often is it check calibrated, is there a record of checks on indicated temperature, is there a detailed procedure for changing the temperature which specifies the minimum time before the incubator is used after a temperature change? Is the incubator cooled if operating near ambient temperatures? Is there a cleaning procedure and how often is cleaning done?
- How does the laboratory relate to (interface with) its 'customers' whether internal or external?

Notes on vendor (supplier) or co-packer audits

1. The usual experience is that there is a wide spectrum of operations going from the very good and thoroughly professional, with management who know their job and are able to demonstrate that they operate consistently to high standards to the few of whom the reverse is true. Remembering that the purpose of the audit is to assess the ability to produce consistently, i.e. through each and every contract run, to Heinz safety and quality requirements, there is usually no difficulty in placing operations which are at either end of this spectrum. The difficult area is where there are deficiencies but where it is reasonable to believe that some, if not all, could be corrected in a predictable time.

   Because what is wanted is a sound technical basis for a commercial relationship, the 'positive approval' approach is used by many companies. This means that a quality audit is made before the vendor or co-packer is put on an 'approved list'. Commercial dealings are then made on the basis of this listing. When dealing with an uncertain situation it is as well to remember that the best collective judgment is required. The honesty, integrity, knowledge, experience and good judgment of the auditor(s) is critically important and must be given full weight. Remember, the quality auditor is also a businessman.

   A key issue is an assessment of auditee management philosophy, attitudes and their real willingness to make changes. To make what can be a very difficult decision simpler, some companies use a 'points rating system'. It is as well to remember that allocation of points itself requires judgments to be made. It should therefore be used as an aid to judgment and not as an 'easy way out' or substitute for the collective judgment which has to be made. It is the quality of this collective judgment which is important and the principle of Habitual Incremental Improvement (HII) applies just as much here as to other parts of a business.

   An approach which may be taken is to make a HACCP analysis jointly with the vendor or co-packer. This can then be made the basis for audit questions. If this approach is taken, the HACCP analysis needs to be re-run if changes are made in the ingredient/process/product/packaging or distribution system.

2. Invariably, there is a shortage of time in making an 'on-site' assessment of a vendor (supplier) or co-packer so careful planning for a visit is needed and attention must be concentrated in *all* key areas. While looking for direct evidence of technical control, the experienced auditor places weight on the general impressions of orderliness, good housekeeping, sanitary status as well as attitudes of management and personnel.

   Because time 'on-site' is a major constraint, use may sometimes be made of a questionnaire sent in advance of the visit to help in forward planning. As stated in Campden Technical Manual 12[1], 'The response to questions, particularly of a subjective nature need to be interpreted with some caution and claimed practices should always be validated. Questionnaires are not a substitute for on-site inspection.'

References

1. Campden R. A. (1986) Technical Manual No. 12. *Guidelines for the Establishment of Procedures for the Inspection of Canneries*, Campden Food Preservation Research Association, Chipping Campden, Gloucestershire GL55 6LD
2. Hall, L. P. (1986) Technical Manual No. 16. *Guidelines for the Establishment of Procedures for the Inspection of Frozen food Factories*, Campden Food Preservation Research Association, Chipping Campden, Gloucestershire GL55 6LD
3. Shapton, D. A. (1986) Canned and bottled food products (soups, mayonnaise and sauces). In *Quality Control in the Food Industry*, Vol. 3, 2nd edn, edited by S. M. Herschdoerfer, pp. 261–322. London/Orlando, FLA: Academic Press
4. ISO 8402 (1986) *Quality – vocabulary*. Available as British Standard *BS4778 Quality Vocabulary Part 1: 1987 International terms*. London: British Standards Institution. (Note. Also available in BSI Handbook 22)

5. Sawyer, L. B. (1983) *The Practice of Modern Internal Auditing*, 2nd edn. Altamonte Springs, FLA: Institute of Internal Auditors

Further reading

There are useful chapters in the *Bulletin of the International Dairy Federation* No. 229/1988, titled *Contemporary Quality Assurance*, which include:

M. Promisel – Chapter 11, *Quality Standards and Quality Auditing – an Industry Program*, which gives a series of checklists.
J. T. Funkhouser – Chapter 13, *Costs and Benefits of Quality Assurance*, which asks some very pertinent questions of management.
R. Wooden – Chapter 8, *HACCP Approach to Product Safety*, which shows checklists centered on the product.
BSI (1987) *BSI Handbook 22, Quality Assurance*, Third revision, London: British Standards Institution

Note. This contains
BS 4778 Quality Vocabulary: ISO 8402–1986
BS 4891 A Guide to Quality Assurance
BS 5233 Glossary of Terms Used in Metrology
BS 5750 Quality Systems: ISO 9000, 9001, 9002, 9003, 9004 all 1987
BS 5760 Reliability of Constructed or Manufactured Products, Systems, Equipments and Components
BS 5781 Measurement and Calibration Systems
BS 6143 Guide to the Determination and Use of Quality Related Costs

This handbook is a convenient collection of British Quality Standards similar to, or identical with, International (ISO) Standards. In the USA, the American National Standards Institute, in association with the American Society for Quality Control, has published the ISO 9000-9004 standards as Q90, Q91, Q92, Q93 and Q94, ANSI, 1430 Broadway, New York, 10018 NY.

Appendix

Heinz USA vendor and co-packer auditing checklist

Vendor/Co-packer Name   Production Location(s)

Ingredients/Products   Ingredient
purchased from this   specification number
location   Factory

Date of audit   Date of report

Heinz auditor(s)   Plant representatives

Heinz purchasing agent   Certification status

Note: Answer the following questions, if applicable, by placing a checkmark in the proper column. All items which are checked with a 'no' response should be detailed on the Form L-395 Audit Report.

I. Audit preparation prior to visit   Yes No

1. Has the vendor/co-packer been informed of the audit date? (Vendors handled via the Purchasing Department.)
2. Do you have a copy of all pertinent Heinz specifications, i.e. ingredient specifications, co-pack specifications?
3. Do you have copies of any necessary specific Heinz USA checklists (e.g. factory sanitation, warehouse sanitation, sterilization, hazard analysis, etc.)?
4. Do you know what Government Agency this establishment operates under? For example, USDA, FDA regulations.

If pertinent:

- Do you have a copy of the FDA inspectional methods for ingredients from chapter 6 of the FDA inspectors operations manual (Heinz QCM volume 23)?
- Do you know what USDA inspection requirements are applicable (USDA inspectors manual)?

Heinz USA vendor and co-packer quality checklist

II. Quality assurance program   Yes No

1. (a) Does the facility have a QA Manager or other designated individual responsible for quality?
   (b) What is the reporting relationship of the QA function?
   _____

2. (a) Are the quality tests accomplished on raw materials, in-process materials, and on the finished product?
   (b) Are the test results recorded?
- What is the frequency and type of tests and the significance of the results?
   _____

- List the chemical, physical, and micro-biological tests performed.
Chemical _____
Physical _____
Microbiological _____
3. Are these quality tests performed on an adequate frequency basis to assure uniform good quality and adherence to the Heinz ingredient specifications and/or co-packer specifications?
4. Are quality tests performed by:
   (a) Plant personnel
   (b) Contract laboratories
   (c)
   Other _____
5. (a) Does this facility have a copy of the Heinz ingredient/co-pack specifications?
   (b) Does the specification reflect current standards and practices at this facility?
6. Does this location have a self-auditing program (accomplished by QA)?
7. (a) Is there a net weight checking program in effect and enforced?
   (b) Is it consistent with current State and Federal requirements?
8. Is each finished product bag, drum, etc. labeled as to the identity of the material, name and address of the manufacturer, quantity and batch/lot number?
9. (a) Is there a finished product coding system?
   (b) Does a documented plan exist for product recall?
   (c) What is this locations' definition of a lot/batch? _____
10. (a) Do the plant employees follow good manufacturing practices?
   (b) Does a documented program exist and is it enforced?

11. (a) Does the quality assurance group Yes No
have copies of recent regulatory inspections?
(b) Are these kept on file?
(c) Are they available for review?
12. Have appropriate actions been taken to correct all deficiencies noted on these regulatory reports?
13. (a) Is the plant water safe and of suitable quality?
(b) Is it tested microbiologically?
(c) Is it checked for residual chlorine?
(d) Are records of potability kept?
(e) What is the frequency of testing?
14. Are responsible plant personnel knowledgeable in pertinent regulations and standards as outlined in Heinz specifications (e.g. FDA, USDA, Codex Standards, grade standards, pesticide regulations, trade standards)?
15. Is there some manner of finished goods inspection program to confirm integrity of packaging prior to shipment?
16. Is there a defined documented system of 'critical control points' for all operations?
17. Are the following quality systems clearly documented?
(a) Ingredient and packaging material specifications.
(b) Quality test methods.
(c) Processing procedures.
(d) Finished product test standards and frequencies.

III. Manufacturing building and grounds

(Refer to Heinz USA factory sanitation checklist)

1. Is the manufacturing building properly designed to house the process and so that it can be cleaned and maintained in a sanitary manner?
2. Are the interior floors, walls and ceiling constructed of materials which are compatible with the process and are they properly cleanable?
3. Is the building properly designed: (a) To prevent entry by dirt, dust, insects, birds, rodents, and other foreign substances?
(b) So as to separate the processing area from support services and personnel areas (e.g. maintenance shops, employee cafeteria)?
4. (a) Is the processing area properly designed to provide for prompt removal of odors, vapors, and/or dust accumulation?

(b) Is area condensation controlled Yes No adequately?
5. (a) Are walls and equipment surfaces properly maintained?
(b) Is there any evidence of flaking paint, rust, or other potential contaminants?
6. Do the plant grounds have proper slope and adequate drainage?
7. (a) Is the plant waste disposal system adequate?
(b) Is there any potential for inadvertent product contamination?
8. Are electric lights and other glass equipment properly located and shielded to guard against glass breakage and resultant product contamination?
9. (a) Is there a product reclaim area?
(b) How is rework and scrap product handled?
(c) Is rework tested prior to re-entering processing?
(d) Who is responsible for disposition of rework and scrap? _____
(e) Is the reclaim area properly located?
10. (a) Are there suitable personnel facilities available?
(b) Are there separate areas available for changing clothes, breaks, and lunch rooms?
11. (a) Are lavatories properly designed and adequate for the number of employees?
(b) Are the lavatories properly maintained?
(c) Do they meet all health regulations?
(d) Are toilet paper, wash hands signs, self-closing doors, adequate soap and towels, and running water available?
12. (a) Are waste disposal units located throughout the plant?
(b) Are these units properly covered and maintained?
(c) How frequently are these waste receptacles emptied and cleaned? (Every 24 hours recommended)

IV. Processing equipment

1. Is the processing equipment adequately designed for the specific process or operation?
2. Does the equipment layout permit for adequate cleaning and maintenance?

3. Is the equipment constructed of proper Yes No material for:
   (a) Product contact surfaces?
   (b) Structural members, equipment supports, and access structures?
   (c) Process support equipment – non-product contact surfaces?
4. Is the equipment designed to accommodate adequate clean-up procedures (e.g. are there any dead spaces in equipment or product piping or conveyors)?
5. Are processing tanks, vats or other holding equipment properly enclosed or covered to prevent product contamination?
6. (a) Are all equipment motors, gear-boxes, agitators, and shafts properly mounted to keep bearings and grease seals outside the product zone?
   (b) If bearings or seals are in the product zone, are they sealed or self-lubricating?
7. Are all food contact lubricants on the USDA approved chemicals list (USDA MID inspected plants only)?
8. Are storage racks available for storing pipe equipment parts, utensils and other product contact equipment?
9. (a) Are sifters, screens, filters, magnets and other similar units installed to detect and remove foreign materials as appropriate?
   (b) What is the frequency of cleaning?

   _____

10. Are adequate measures taken to prevent *Geotrichum* (slime mold) buildup or other microbiological contamination of processing equipment?

V. Processing operation

1. Is the processing operation separated from raw materials and finished goods storage?
2. Is the equipment cleaned and sanitized as often as necessary to prevent contamination due to residual buildup during operation?
   Note. This includes fumigation procedures in grain products and related handling equipment. Also USDA meat and poultry plant mid-shift cleanup.
3. (a) What happens when a processing breakdown occurs?

   _____
   _____
   _____
   _____

(b) Is the product susceptible to mi- Yes No crobial growth and spoilage?
(c) Does physical/chemical deterioration occur?
(d) Are there prescribed startup procedures?
4. (a) Who has authority to alter processing parameters?

   _____

   (b) Are deviations recorded?
5. Are there opportunities for unintentional additives to enter through equipment, cleaning supplies, etc.? (e.g. Is there a separate storage area for cleaning chemicals and chemical compounds?)
6. (a) As applicable, are all in-process temperature control mechanisms (for both heat and cold) being complied with?
   (b) Is there a routine monitoring system?
7. *Co-packers only*: Are they in conformance with all Heinz specifications as detailed in the Heinz Co-packers Manual?

VI. Personnel work habits and practices

1. Are smoking, eating, chewing gum and tobacco prohibited except in designated areas away from production?
2. Are adequate measures taken to prevent contamination of the product (e.g. proper clothing, hairnets, gloves)?
3. Is there a standard jewelry policy?
4. Are production areas free from lunch boxes, sweaters, coats, purses and other personal gear?

VII. Pest control program

1. Is there an established pest control program for:
   (a) Rodents
   (b) Insects
   (c) Birds
   (d) Other pests?
2. Is the work accomplished by:
   (a) In-plant personnel or
   (b) Pest control operator (PCO)
   (c) Name of PCO _____
3. (a) Are the program requirements documented?
   (b) Are records kept?
4. Are all insecticides and rodenticides approved for use in this type of facility?

5. (a) Are all insecticides and rodenti- Yes No
cides applied properly?
(b) Is their coverage adequate and
timely?
6. Do all pesticide application practices
preclude the possibility of product
contamination?
7. Are all windows and open areas
screened or otherwise protected to
prevent insect and pest entry?
8. Are all bait boxes in good condition,
properly placed and serviced?
9. Is there a map depicting the location of
all numbered bait stations?
10. (a) Are insect electrocuter units, air
curtains and other protective devices
used where needed?
(b) Are they properly located?
(c) Are they serviced routinely?
11. Is there any evidence of pest activity in
or around the plant?

VIII. Clean-up program

1. Is there a cleaning program with a
sanitation supervisor or other designated responsible individual?
2. (a) Are there sound cleaning procedures?
(b) Is there a procedures manual?
3. Are chemicals and cleaning compounds on the USDA approved chemicals list (USDA MID inspected plants
only)?
4. What sanitizers are:

(a) used?	(b) when?	(c) where?	Are they used properly? Check *yes* or *no*
_____	_____	_____	_____
_____	_____	_____	_____
_____	_____	_____	_____
_____	_____	_____	_____
_____	_____	_____	_____

5. (a) Is there a separate storage area for Yes No
chemicals and chemical compounds?
(b) Is this area removed from raw
material and finished product storage?

IX. Warehousing and distribution
(*Refer to Heinz USA Warehouse Sanitation
Checklist*)

1. Is the warehouse construction and
maintenance adequate?
(a) • Do doors, windows, and other
closures fit properly?
• Are there any openings large
enough for rodent or other pest
entry?
(b) • Are floors kept clean?
• Is there a scheduled cleaning
program?
• Are damaged goods and product spillage cleaned up routinely?
(c) • Is all product stored at least
450 mm (18 in) from the warehouse walls?
• Is there adequate room for
proper cleanup and pest control?
2. (a) Is there good control of storage
temperature?
(b) Where cold storage is necessary
are adequate records of temperature
control kept?
3. (a) Is there a program to assure clean
pallets?
(b) If so, where applied?
4. (a) Is ventilation in the warehouse
adequate?
(b) Do condensation, dust, or other
sources of contamination exist?
5. (a) Are all carriers inspected and
cleaned before loading?
(b) Are they properly sanitized and/or
fumigated when required?
6. Are proper inventory procedures followed (i.e. first in, first out procedure)?

Chapter 3
Establishment and implementation of HACCP

3.A Establishment and use of HACCP

Introduction

Every company which intends to remain in business understands the importance of financial control and manufacturers also know the importance of controlling processes, both for cost and 'Quality' reasons. 'Quality' in this context means producing a product which will satisfy the customer. What determines 'Quality' will depend on the product from any particular industry or factory, but to achieve the objective of quality throughout the whole of production then Total Quality Management (TQM) must be applied to every aspect of operations within a company. Underpinning TQM is the philosophy that it is the management system which prevents the production of poor quality goods; it does not achieve its objective by selecting good quality products from among a mixture of poor and good quality ones. In other words, it is a 'Right First Time' management system, and implicit in TQM is the concept that the whole of the processing and distribution system needs analysis in order to achieve quality in the final product.

As far as the food industry is concerned, the quest for quality has been recognized for many years, and 'Quality Control' has been a traditional approach for the production of 'Quality' food. This typically consisted of inspection of ingredients and processes together with a limited amount of end ('finished', UK) product testing. 'Quality Control' differs from TQM in that it is not a comprehensive management system. However, there is one aspect of a TQM program which demands consideration above all other aspects, and that is food safety. What, then, is safe food? It is food which presents the minimum risk of illness to the consumer, whether illness results from:

- Pathogenic (disease-causing) microorganisms present in the food.
- Toxins (poisons) in the food, arising from microbial action.
- Chemical residues deriving from farm production methods or from factory processing.
- Injurious foreign material.

When the incidence of illness arising from unsafe food is considered, illness caused by microorganisms far exceeds that caused by chemical residues and toxicants (including pesticides), food additives, natural poisons or toxic substances and foreign materials. Assessment of risks associated with food safety presented at a conference in 1978[1] gave the ratio of 100 000 to 1 for microbial contamination to pesticide residues. Therefore to address the issue of food safety the main emphasis must be to reduce the hazards caused by the presence in food of unwanted pathogens, including those able to produce microbial toxins.

Material given in Chapter 1 illustrates the considerable cost implications which can arise from the loss of safety in manufactured foods. The size of some of these outbreaks is a reflection of the present structure of the food industry or, more accurately, food industries. Much of the food eaten in industrialized countries is not locally produced but comes from a factory, or a fast food outlet (see Chapter 1). This means that any one type of food bought from supermarket shelves is likely to have come from a small number of factories which may be domestic or located abroad (overseas). The manufacturer must, therefore, exercise all due care with his production (including bought-in materials) in order to overcome the hazards from pathogenic microorganisms and foreign materials; only by doing this will the food present a minimum risk of illness to consumers.

How is the manufacturer to apply 'due care'? This is a term which encompasses *what* is needed, but gives no guidance either to management or to technologists regarding what is the most effective action to take to achieve safety in manufacture. Traditionally, approaches have included:

● Education and training
● Inspection of premises and processes
● Microbiological testing of plant and product

Each manufacturer is likely to have applied, to a greater or lesser extent, all three of these approaches. A unified, comprehensive systematic approach to the whole question of minimizing hazards has not been applied by many food manufacturers until very recently, even though such a concept had been developed in the 1960s in the USA. This fundamental, powerful technique called the 'Hazard Analysis Critical Control Point (HACCP) System' was the result of a joint effort, within the US Space Program, of the Pillsbury Company, the National Aeronautics & Space Administration and the US Army Natick Laboratories to apply a zero defects philosophy to food production for astronauts. It is based on an engineering system, the Failure Modes Analysis Scheme, which consists of examining the product and all of the components and processes used to make that product and asking – what can go wrong within the total system?

It was realized that the system had wider applications than the Space Program, and following its presentation in the US in 1971 it was adopted in 1973 by the Food & Drug Administration in relation to the inspection of low acid canned food. In the mid-1970s the US Department of Agriculture used it in meat plants. Then in 1980, the World Health Organization (WHO) produced a report which was prepared in collaboration with the ICMSF[2]. This report concluded that 'the HACCP concept is a desirable alternative to traditional control options. It can be applied at a better cost/benefit ratio in comparison to other approaches as it is based upon a more systematic and logical approach to the avoidance of food hazards' WHO saw application in both developed and developing countries.

'Guidelines to the establishment of Hazard Analysis Critical Control Point (HACCP)'[3] were published in 1987 as Technical Manual 19 by Campden Food & Drink Research Association (then CFPRA). It gives an explanation and description of the HACCP system, its terminology, and includes seven examples of applying HACCP to different food products. Then in 1988, the ICMSF

published *The Application of the Hazard Analysis Critical Control Point (HACCP) system to ensure microbiological safety and quality*[4]. This book details what HACCP is, including definitions of components of this system, and the background information that is needed before it can be successfully applied. Outline analyses of 19 factory products are described together with the application of the HACCP system to marketing and retail stores, food service outlets and finally the home, recognizing that everyone has a contribution to make to food safety.

It is important to realize that the abbreviation HACCP, like the word 'audit', can be understood in a number of ways and used for a number of purposes. Clear, careful, logical consideration of all possible options are required if the analysis is to deliver what is expected and required of it. It is helpful to remember that essentially it is a systematic, multi-disciplined, task force way of working. The familiar principle applies that the results are only as good as the quality of the work put into the analysis.

In particular, it is important and not just pedantic to be clear and rigorous in the use of terminology. During the running of an analysis it is easy to lose sight of the primary objective if this is not done.

HACCP definitions

The terms used within the HACCP System need to be defined before consideration is given to the way in which the system is applied. This is not quite as simple as it may seem, and while it may be tedious, it is essential. As an example, both *Webster's* and the *Shorter Oxford English Dictionary* define hazard as 'risk' (*Webster's*) or 'risk of loss or harm' (*Shorter Oxford English Dictionary*). This is reflected in everyday use of these two words as they are most often used interchangeably. However, within the HACCP System 'hazard' and 'risk' have their own separate and distinct meaning, and they must be defined and used precisely if the analysis is to be of real use. Similarly, the other terms must be understood and used correctly by *everyone* who is working with the HACCP System.

Hazard

Hazard is the potential to cause harm to the consumer – the safety aspect; or to the product – the spoilage or quality aspect – and is present at *any* stage in the life of the product where unacceptable microbiological contamination or where growth or survival of unwanted microorganisms may occur. The term is used in this way in this book and can be applied to foreign material and chemical residues as appropriate.

Risk

Risk is the probability that a hazard will be realized or will happen. In principle, risk may be quantified mathematically, but most microbiological safety failures are anticipated to occur at such low probabilities that mathematical probabilities may not be helpful because of inadequate data. Therefore, risk may well be ranked as low, medium or high, based on judgment or experience.

Critical control points(CCPs)

These are points identified in the location, process, or product formulation which minimize or prevent safety hazards being realized. Pasteurization is an example of a step which theoretically minimizes, but in practice prevents, a hazard occurring.

Be aware that CCPs are not necessarily identical with Regulatory or Quality Control Points, but in certain cases they are the same. For instance, in pasteurized milk the pasteurization stage is both a safety CCP, as the time/temperature combination was that chosen to render milk free from Mycobacterium tuberculosis (and in fact insures sufficient destruction of other vegetative forms of pathogenic bacteria) and a regulatory CCP, since in many countries milk must be shown to have received this specified heat treatment to satisfy legislation. However, pasteurization is only one of several quality control points, *since although spoilage psychrotrophs are destroyed by pasteurization, contamination with these organisms can occur at the filling stage if the Cleaning in Place regime is not adequate. Although filling should not be overlooked as a* safety *CCP (Listeria monocytogenes is ubiquitous and grows at chill temperatures), inadequate control of cleaning is more likely to cause a spoilage problem than a loss of safety.*

Concern

Concern relates to the seriousness resulting from any failure to achieve control of the process, and is derived from the knowledge of the effect of a hazard not being controlled.

When safety is the focus of the HACCP system, then epidemiological information and knowledge of the anticipated pathogen or pathogens forms the basis of ranking the severity of the concern. Levels of concern may be 'High', 'Medium', 'Low' or 'Negligible'. These levels of concern *must* be based on a technically informed judgment which reflects the severity and/or extent of illness resulting from loss of control. Remember – a low risk associated with a severe hazard gives rise to high concern.

Predictability

It must be emphasized that even when a hazard has been identified and the risk assessed, the date or time when that hazard will be realized due to loss of control *cannot be predicted.*

Remember that a low, or negligible, risk can happen at any time. If this unlikely event is associated with a high concern, that is, a life-threatening effect on consumers, then such a low risk must be addressed with real care. It is obvious, of course, that care must be given to any high risk which is found within the analysis.

Do not fall into the trap of thinking that a probability of one in 5000 defective packs means that every 5000th pack is, in fact, defective.

Botulism illustrates well the interdependence of concern, risk and predictability. The illness is very serious, and is often accompanied by death among its victims, but in the UK is regarded as a low risk. Between 1957 and 1987 there were only two outbreaks with a total of six cases, although two deaths occurred in one of these outbreaks. The previous largest outbreak had been in 1922, when eight people died from eating wild duck paté. However, in 1989 an outbreak involving 27 cases occurred, and one person died. Hazelnut yogurt proved to be the source of botulinum toxin, and it was canned hazelnut puree which contaminated the yogurt.

By the addition of this puree, yogurt – a food which is widely regarded as both 'safe' and 'healthy' by the public – had become a vehicle for a large food poisoning outbreak because the controls necessary to maintain a canned low-acid product as a low risk were not applied. The consequence of this action was that suddenly, a product of presumed low risk and concern had become associated with severe illness.

HACCP application

Introduction

HACCP should be applied as a systematic approach to hazard identification, risk assessment and hazard control throughout any product/process/packaging line at a manufacturing site, and includes the distribution system. The likely abuse of the product should also be considered. Each stage of the process should be examined as an entity as well as in relation to other stages. The analysis should include ingredients as well as finished product, and also the environment of production since it is realized that this contributes to microbial as well as to foreign material contamination.

This focusing on environment can be illustrated by the baby milk *Salmonella* contamination which occurred in the UK in 1985, although the importance of the environment has been known in the USA from the mid-1960s in connection with dried milk plants. In the baby milk, a particular and rather unusual species of *Salmonella*, *Salm. ealing*, was found in the powder. This species was found in only one of the many samples taken at the production site, in the insulation of the dryer next to a crack in the dryer wall. It should be noted that this incident resulted in a loss of £49 million to the company concerned. As another example, the close association of production environment and pathogens is also found with *L. monocytogenes* in factories producing mold-ripened soft cheese. *L. monocytogenes* has been recovered from brining baths, from brushes in the processing areas, from shelves in the ripening rooms and from floors and floor drains.

Principles of stepwise analysis

How is the HACCP systematic approach applied in practice? The following steps should be taken:

1. Prepare a flow diagram of the process, from ingredients through to the customer. The full details of ingredient specifications, the packaging system, the product formulation and processing must be known.
   An example of a flow diagram using conventional process engineering symbols at each process stage is given in the New Zealand Institute of Food Science and Technology publication (see Further Reading).
2. Identify the hazards: then assess the severity of these hazards and the accompanying risk and level of concern for each stage of the process, including 'bought-in processed' as well as raw ingredients.
3. Determine/identify the Critical Control Points (CCPs) at which the hazards can be controlled. Then, select the control option which must be in place at each CCP.
4. Specify the criteria that indicate whether an operation is under control for each CCP. Thus, if heat is the control option, the exact temperature and time of heating must be specified, and the tolerances which can be allowed must be set. Or the control option may be chemical, such as salt to inhibit pathogen development; or acetic acid, which will result in death of pathogens as well as inhibition of spoilage organisms such as yeast. In these cases, the salt concentration must be specified, or the acetic acid strength, together with their tolerances. For a combination of factors (hurdles), always determine the effect of the 'worst case' combination.
5. Establish and implement procedures that monitor each CCP to check that it is under control. The procedures should measure accurately the chosen factors which control a CCP; should be simple; and give a quick result. Appropriate records are needed as part of a positive assurance of safety.
6. Specify and record what corrective action is necessary when the monitoring result shows that a CCP is not in control.
7. Verify that the HACCP system is working, by use of supplementary information. This is where microbiological examination of product during and/or after processing and packaging has its place in the HACCP system.

Pathogens

These above steps give the framework to HACCP application, but an input must then be made of the microbiological aspects of food safety. This information relates to the pathogenic microorganisms themselves and can be considered under the following headings:

- What are the most usual pathogens?
- How do they cause illness?
- What epidemiological information is available regarding their association with illness?
- Where are any of these pathogens most likely to occur, both outside and inside the food factory environment?
- What properties of these pathogens need to be known so that the correct, most effective, control option may be applied at each of the critical control points?

Chapter 10 considers these points and includes profiles of the more usual pathogens. Each profile is the starting point for hazard evaluation indicating the control options to be applied at the CCPs, whether associated with ingredients, product, processing or distribution. Be aware that research may alter the known limiting values of growth parameters.

Product formulation

The formulation of the product may well be a CCP in its own right, or more usually be a component of a CCP within a process system. For example, the a_w determines microbial inhibition in dried fruit but a pickle formulation depends upon the pH, the type of acid, the salt concentration and pasteurization conditions to prevent microbial growth at ambient temperatures; whereas in the case of cottage cheese it is the pH of the cheese in association with storage at chill temperatures which inhibits pathogen growth.

From this it is seen that a formulation which prevents pathogen development within the shelf life of the product must not be changed without challenge testing to confirm that the change in recipe has not adversely affected safety.

It must be remembered that it is not only a change in quantity of an ingredient that can affect safety, it may also be an alteration in type of ingredient. For instance, a change from acetic acid to a different acid may still retain the original pH value but will almost certainly affect the multiplication rate of bacteria as well as yeasts and molds in the product. Similarly, changing from a canned ingredient to a fresh, chilled or frozen ingredient, e.g. with vegetables or chicken, is likely, because of a different microbial population in the ingredient, to result in an alteration in the shelf life and possibly the safety of the product.

These examples show that any change in product formulation must not be introduced commercially until HACCP has been carried out. The definitive HACCP can only be carried out when all the changes have been decided upon, so a full specification for any proposed new ingredient must be available in order to apply HACCP correctly.

Product hazard characteristics and risk categories

Following recognition in the USA in the late 1950s and early 1960s that Salmonellosis was 'one of the most important communicable disease problems', the Committee on *Salmonella* of the National Academy of Sciences – National Research Council (NAS/NRC) considered firstly, how to assess the degree of hazard of foods, feeds and drugs and, secondly, what sampling plans would provide adequate assurance that the food offered a minimal hazard to the consumer. In their report published in 1969[5] they defined three hazard characteristics which foods could possess, and also five risk categories of foods – the category depending upon how many hazard characteristics a particular food contained. Note that a category was created for foods intended for the infants, aged and infirm based only on containing a sensitive ingredient. Table 3.1 from Olson[6] relates the three hazard characteristics to the five risk categories.

In this original use of hazard assessment, the category to which the food belonged decided which sampling plan would be applied to the lot or batch of food concerned. However, in 1974 at a Symposium on Microbiological Considerations of HACCP systems held during the annual meeting of the US Institute of Food Technologists, Peterson and Gunnerson[7] used the NAS/NRC scheme to classify ingredients used in frozen food products. The hazard category (or risk category as defined by NAS/NRC) given to an ingredient formed the basis of applying 'sampling and analysis . . . to provide statistical reliability' as part of a HACCP program. The HACCP Working Group of the National

Advisory Committee on Microbiological Criteria for Foods[8] has increased the number of hazard characteristics to six, and the risk categories now number six. This report makes the point that the categories help to recognize the risk associated with ingredients and also how the ingredients 'must be treated or processed to reduce the risk for the entire food production and distribution sequence'.

Risk categories may also be applied to food products as well as ingredients, as is made clear by Corlett in his chapter in *Food Protection Technology*[9].

It should also be realized, although this point is not specifically dealt with in the earlier papers[5,6], that classification of risk category may be used as part of Supplier Quality Assurance (SQA) within a

Table 3.1 Categories of food products based on product hazard characteristics

| Type of food | Hazard characteristic[a] | | | Category |
	A	B	C	
Intended for infants, the aged, and the infirm	+	+ or 0	+ or 0	I
Intended for general use	+	+	+	II
	+	+	0	III
	+	0	+	III
	0	+	+	III
	+	0	0	IV
	0	+	0	IV
	0	0	+	IV
	0	0	0	V

[a] A = Product contains sensitive ingredient
B = No destructive step during manufacture
C = Likelihood for growth if abused

HACCP program. Classification need not and should not be confined only to sampling plans and, through these, to meeting specific microbiological criteria. It may be used in a more cost-effective way to insure that the processing received in the supplier's factory is effective, through the identification of the CCPs in the process and the correct control at each CCP.

It needs to be appreciated, though, that two very different ingredients or products may both belong to the same risk category yet require very different control options applied at the CCPs in their processing. In the NAS/NRC categories HTST liquid milk and a chilled processed meal would both be classed as risk category III, i.e. open to recontamination after processing and before packing; have a potential for abusive handling; and have no terminal heat process after packaging or when cooked at home. However, in normal processing HTST milk would have minimal exposure to environmental contamination between processing and packing and in a well-designed, maintained and cleaned plant would not be exposed to contamination by pathogens from the process plant. A chilled meal, in contrast, would be exposed to the atmosphere, and possibly to handling by personnel in the assembling of the meal itself. It would therefore be subjected to a higher risk of contamination, including such pathogens as *L. monocytogenes*, *Salmonella* spp. and *Staphylococcus aureus*, unless the area of assembly was designed and managed specifically as a 'high care' area. For details on 'high care' areas see the section 'Prevention of cross-contamination' in Chapter 6.

Assigning risk categories to ingredients or foods can, therefore, be helpful when applying a HACCP program. However, it is only the beginning (or first stage) of risk assessment and unless the analysis is continued in detail, so that both the specific pathogen or pathogens and the correct controls applying to that food are identified, the safety of the food will not be insured.

Common process features

It must be clearly understood that there are common features in most processes, and these must be analyzed within the HACCP system as thoroughly as the particular product in the process under scrutiny. These features are as follows:

- The factory environment, to include the environment of the area within the factory used for a particular product or process.
- Design – including buildings (fabric) and equipment.
- Hygiene of buildings and equipment
- Maintenance of buildings and equipment
- Quality of water supplies
- Segregation of raw material and finished product
- Cleaning and sanitizing procedures
- Personnel traffic and hygiene

Each of the these features needs to be separately considered, and the following seven steps must all be applied.

- Identify the hazards
- Determine the critical control points (CCPs)
- Specify control criteria at each CCP
- Monitor the control at each CCP
- Determine the action necessary to correct any loss of control at a CCP
- Devise the recording system that documents the HACCP plan
- Detail the system-verification procedure.

Several of these features are, within a single manufacturing site, common to more than one

process or product but others, such as equipment design, maintenance, and cleaning procedures, may in fact only apply to one product. Each product requires its own analysis for those features which are not common. Additionally, quality control points and control options with respect to spoilage organisms should be made as part of the TQM program. Note: Remember that the effect of line stoppages may affect a HACCP analysis.

Examples of HACCP application to processes to achieve microbiological safety

To illustrate the application of the HACCP system, three processes have been chosen, each manufactured according to a different technology.

1. Powdered formula milk (see Figure 3.1)
2. Yogurt with added fruit or nut puree (see Figure 3.2)
3. Frozen soft-filled bakery products (see Figure 3.3)

The specification for each process is used to prepare a flow diagram, which includes all stages of processing. The flow diagram is the essential first step to take before beginning the detailed analysis.

On each flow diagram the CCPs have been marked as well as the most likely sites of microbiological contamination. Not every CCP will be examined in detail. However, one or two CCPs within each process will be discussed. Each has a particular function to perform with respect to the specific technologies chosen.

The CCPs selected are:

- The environment for powdered formula milk in example 1.
- The incubation (culture fermentation) for yogurt in example 2.
- Fruit/nut purees processed 'off-site' for yogurt in example 2.
- The pasteurization of ingredients for soft-filled bakery products in example 3.

However, it must be remembered that every processing stage is linked to the other stages, and therefore *all* the stages need their controls functioning correctly for the production of safe food to be achieved.

Powdered formula milk

The hazards associated with this process are that pathogens may be present either in the raw milk or other ingredients, or they could gain access to the powder from the environment, or from the addition of an ingredient at the powder stage. These pathogens present a high level of concern, since the powdered formula milk is intended for feeding small

27

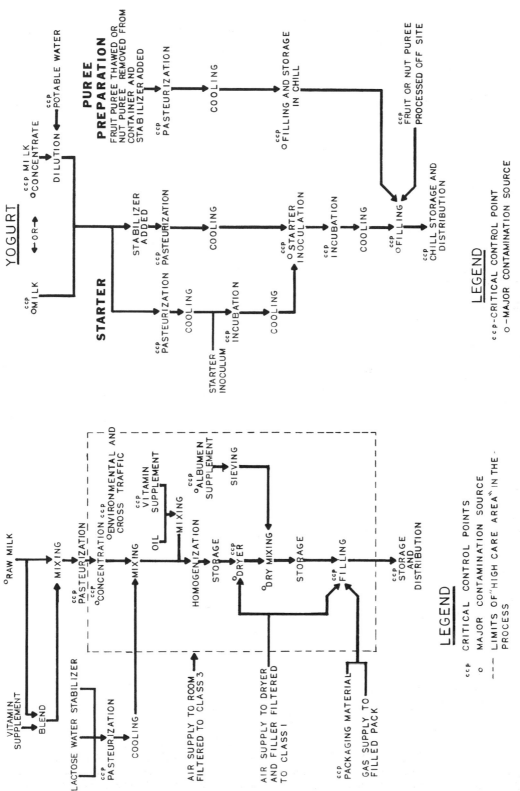

Figure 3.1 Flow diagram for powered formula milk

Figure 3.2 Flow diagram for yogurt with fruit or nut puree

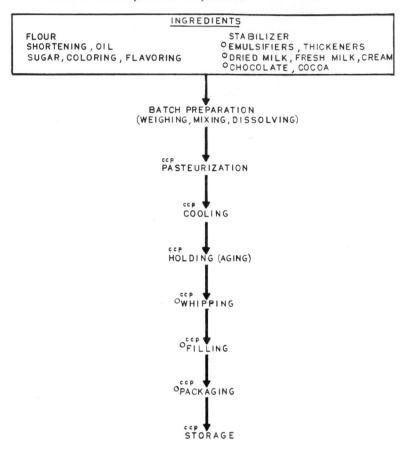

INGREDIENTS

FLOUR	STABILIZER
SHORTENING , OIL	○EMULSIFIERS , THICKENERS
SUGAR, COLORING , FLAVORING	○DRIED MILK, FRESH MILK, CREAM
	○CHOCOLATE , COCOA

BATCH PREPARATION
(WEIGHING, MIXING, DISSOLVING)

ccp
PASTEURIZATION

ccp
COOLING

ccp
HOLDING (AGING)

ccp
○WHIPPING

ccp
○FILLING

ccp
○PACKAGING

ccp
STORAGE

LEGEND

ccp-CRITICAL CONTROL POINT
○-MAJOR CONTAMINATION SOURCE

Figure 3.3 Flow diagram for frozen soft-filled bakery products

babies, and so pathogens must be either destroyed or prevented from contaminating the powder following the destruction stage. Destruction of pathogens is effected by pasteurization of the raw milk and as many of the other ingredients as is technically possible. Attention must also be given to the hazard of toxin-formation in the milk before it is pasteurized, and to the possibility of toxin production occurring in the pasteurized concentrate due to *Staph. aureus* being present as an equipment contaminant in poorly cleaned or maintained plant.

Prevention of powder contamination is achieved by control of the drying room environment, which includes transport air, and the following outline shows how the environment CCP needs to be approached. The headings indicate the main points to be considered – in every case the object is to

prevent pathogenic microorganisms being carried into the drying room, whether by means of air, people or materials.

CCP: Environment of drying/filling room

The following controls should be implemented to prevent pathogen contamination of the environment.

● Design of drying/filling room

Roof of room
1. It should prevent any accumulation of water or powder residues.
2. The air outlet from the drying room should be positioned away from the air inlet to the drying room.

3. The storm water drainage system should be constructed outside the drying room.
4. Birds should be prevented from settling on the roof, or on the air outlet or inlet pipes.

Room
1. There should be no openings directly to exterior. All doors should be double, and no windows should be openable.
2. The only entrance for operatives should be through a changing area.
3. The fabric of floor and walls to be smooth and impervious. This prevents the accumulation of water and moist powder residues on the floor and lodging of powder on walls.
4. Drains must be designed, installed and maintained to prevent the back-up of contaminated effluent.
5. Girders should ideally be absent. If present, they should present the minimum surface area on which powder could accumulate. (See Chapter 4 – girder design.)
- Working procedures in drying/filling room
  1. The air to the room should be filtered to Class 3 (BS 5295: Part 1)[10], and there must be sufficient air flow through the filter assembly to maintain a positive pressure in the room.
  2. The air to the powder transport system should be filtered to Class 1 (BS 5295: Part 1)[10], as should the air to the dryer and to the filling stations. Filtration should be considered for air into powder storage tanks if a vacuum is created while the powder is being emptied from the tanks.
  3. Cleaning must be done regularly and frequently to prevent powder residues accumulating. The main cleaning should be carried out with a vacuum, and local cleaning by dedicated and sanitized brushes or scrapers. Color marking of brushes and scrapers helps to insure that these items remain dedicated, as they can be easily identified if they are removed from the dryer room.
  4. Roof and room need to be kept pest free.
  5. Drains need to be regularly sanitized, and cleaned as required.
  6. The roof drainage system needs to be checked for leaks.
  7. Tools required for maintenance of equipment should be dedicated, and not leave the room. They should also be sanitized.
  8. Fork lift trucks should be cleaned and sanitized immediately before entering the drying room. They should only enter the drying room for the purpose of equipment maintenance or repair.
  9. Only dedicated pallets, preferably made from plastic, should be present in the production area. Incoming materials should be protected by plastic film, and before being transferred to the dedicated pallets the film should be cleaned and then removed. Finished, packed dried product should be palletized outside the drying/filling room; the production line should be designed to insure this.
  10. Packaging should be removed from outer cases immediately before entering the room. The first 4 meters of packaging should be discarded from the spool before it is fed into the filling chamber.
  11. All personnel entering the room must change into dedicated protective clothing, including footwear. Changing is to take place in the specified changing area, and the dedicated protective clothing should be discarded in this area and not be worn in the rest of the factory.
  12. Access of personnel to the drying/filling room should be restricted to operators, with other technical and maintenance personnel having access as required. It would be prudent not to allow visitors into the room.

Yogurt

The main hazard with yogurt is that it may act as a carrier of either pathogens or toxins. By its nature, a fermented milk at pH 3.9–4.2 held in chill, it does not permit growth of pathogens, and has been shown to cause death of *Campylobacter jejuni*. However, a poorly made yogurt with a higher pH value may allow pathogen survival, if not growth. Should a preformed bacterial toxin or a mycotoxin be introduced into the yogurt, then the toxin will not be destroyed during the shelf life of the yogurt. The control options are, firstly, pasteurization of milk to destroy any vegetative pathogens, then rapid fermentation to achieve a pH of 4.3 which will inhibit any contaminating pathogens; secondly, control of added fruit or nut purees so that no pre-formed toxin is added to the yogurt. The two CCPs used as examples are the yogurt fermentation stage of the process, and the fruit/nut puree bought in as a processed ingredient for yogurt manufacture.

CCP: Yogurt fermentation

The following are the control options which must be carried out at this CCP to achieve rapid acid production.

1. The milk for yogurt manufacture must be free from penicillin, and is checked for penicillin on receipt at the site.
2. The milk is heated to 90°C for 15 minutes before inoculation. The heating, as well as destroying

pathogens, is necessary for the starter culture to produce acid at its optimum rate, and also destroys the bacteriophages active against *Lactobacillus delbrueckii subsp. bulgaricus* and *Streptococcus salivarius subsp. thermophilus* – the two bacteria which comprise the yogurt culture.

3. The milk must be cooled to 43°C before starter inoculation.
4. The yogurt starter culture must have the *L. delbrueckii subsp. bulgaricus* and *S. salivarius subsp. thermophilus* components in the correct proportion – usually a ratio of 1:1. It must be free from contaminating microorganisms, and initiate acid production within 30 minutes of inoculation. The inoculation of the milk must be done aseptically, to prevent contamination of the fermentation vat.
5. The incubation must be at the correct temperature for rapid acid production. The rate of acid production is monitored during the 3.5–4 hours incubation.

CCP: Fruit/nut purees processed by supplier (off-site)

The hazard is that the purees could carry pathogen or toxin if the processing was insufficient. Since no processing is given by the yogurt manufacturer, the pathogen or toxin would be added to the yogurt.

The fruit and nut purees need to be considered separately, as the different pH value of each puree will affect pathogen behavior.

Fruit puree

Fruit puree would be expected to present a low risk for bacterial pathogens since the pH of fruit purees is too low to allow growth. There could be a mycotoxin risk if mold contamination was present. The heat applied to fruit purees by the supplier is primarily to destroy yeasts, the main spoilage organism of yogurt, and is sufficient to destroy vegetative pathogens, but not spores of *Clostridium botulinum*. However, the low pH would not permit growth of *Cl. botulinum* and no toxin production would occur. In addition, the puree is generally held in chill following the heat processing, so that growth of any spoilage yeasts which survived the heating stage would be inhibited.

Nut puree

Nut puree could be expected to carry both toxins and pathogens if the processed puree was held at ambient temperature following the heating stage. This is because the puree has a pH value which permits growth of pathogens, and if a botulinum process ($F_o = 3$) had not been given any *Cl. botulinum* spores in a can of puree could germinate,

grow and produce toxin. It should be remembered that a heating process to destroy yeasts is very much less than $F_o = 3$.

Mycotoxin (aflatoxin) might be present if the nuts had not been correctly harvested or stored after harvesting. Should the nuts become moist, then mold growth would be expected to occur, and if toxinogenic strains were present, mycotoxins in the nuts could result. These toxins would not necessarily be destroyed by the heat processing given to the puree.

Control options for puree

1. The specification of the processing applied to the purees by the supplier must detail the temperature of heating and the time of holding or the lethality of the process given to the purees. For the neutral nut purees this heating must at least be a botulinum process for an ambient stable product with a shelf life of some months. If outgrowth of spores is prevented by formulation, e.g. by a low a_w, then this should be explicitly stated in the specification. This will act as a warning when making a change to the formulation which alters the a_w of the puree. A different control option may then need to be used in the product or process to maintain safety.
2. The supplier must have been audited to satisfy the yogurt manufacturer that the process meets the specification and that the process controls are sufficient to insure that at all times the supplier can deliver a correctly processed puree. Controls should include preparation and storage of the puree before heat processing and extend to pertinent Quality Systems and their management.
3. A certificate of heat processing and container integrity could accompany each batch.
4. Mycotoxin (aflatoxin) presence should be prevented by insuring that the nuts did not become moist during harvesting and storage. Monitoring for aflatoxin should be applied to each batch of nuts before making into puree.

Frozen soft-filled bakery products

The principal hazard associated with these products is that pathogens in the filling will cause illness to the consumer when the product is thawed. Because customer abuse may occur, and the product may therefore be held at ambient temperature following thawing, an 'absence' of pathogens in the filling is the requirement for the manufacturer. This can be achieved by the pasteurization stage of the process, although it must be remembered that after pasteurization the product is very liable to contamination at the whipping and filling stages. The freezing stage, and frozen storage, cannot be relied upon to destroy pathogens which may have recontaminated the product.

CCP: Pasteurization

The control option here is to heat the prepared ingredients to destroy any pathogens which they may carry. The ingredients dried milk, fresh milk, cream, chocolate and cocoa all fall into Category II, that is, 'sensitive ingredients that can have a pronounced effect on food safety', so must be regarded as potential sources of pathogens. The temperature/time combination selected to heat the prepared batch must be sufficient to insure a 4 log cycle kill of *L. monocytogenes*. Other pathogens, e.g. *Salmonella* spp. and *Staph. aureus*, must be considered, and an appropriate reduction in numbers achieved. It must not be forgotten that microorganisms can have enhanced heat resistance in the presence of fat or oil. It must also be remembered that the whole of the batch must receive the specified heat treatment, and that if the heating is carried out as a batch process in a kettle covered only with a loose lid, then special attention needs to be given to insure that the surface of the batch has attained the correct temperature. This has been highlighted in the US FDA and Milk Industry Foundation International Ice Cream Association 'Recommended guidelines for controlling environmental contamination in dairy plants'[11] which cites 'improper equipment design, lack of proper outlet valves, lack of proper air space thermometers and improperly operated air space heaters' as all being observed in vat pasteurization systems. They recommend that the airspace temperature between the product and the top of the vat (that is, where the cover is placed) must be maintained at 2.8°C (5°F) above the minimum pasteurization temperature.

If the heating is carried out in a scraped-surface heat exchanger, which is the other most likely pasteurization equipment to be used for this product, then the flow rate of the product must be correctly controlled, thermometers correctly placed, and these thermometers must be regularly checked against a reference thermometer, and the temperature of processing recorded. Since flow diversion valves are not usually fitted to scraped-surface heat exchangers, alarm devices should indicate when the minimum pasteurization temperature has not been reached and also when the product or hot water flow rates have exceeded the tolerances allowed in the process specification. Ideally, such equipment should have a flow diversion valve fitted to insure that under-heated product is excluded from the finished product.

The other CCPs in the process must be controlled, with special reference to the environment at the whipping, filling and packaging stages.

HACCP application to processes to achieve microbiological stability/quality

The application of HACCP to the three specific processes considered above has been mainly concerned with insuring the safety of the food, and the controls have all been those which would insure the destruction, or inhibition, of pathogenic microorganisms. It is true to say that for most spoilage organisms the same controls would apply, but there are some spoilage organisms which are more resistant than pathogens. These are mainly among the spore-forming bacilli; *Bacillus stearothermophilus* is a good example. Its spores are among the most heat resistant known.

In canned products the heat process selected is that which insures destruction of mesophilic spoilage spore-forming organisms. The exact process will depend on the product, type of retort and likely spoilage organisms but the lethality will be more, perhaps considerably more, than a botulinum cook. Another very resistant spoilage microorganism is *Zygosaccharomyces bailii* which has enhanced resistance to heat, acid and preservatives as well as to chemical disinfection used in plant cleaning. This organism is also known as *Saccharomyces bailii*. Any fruit juice or acid food process must consider the properties of this microorganism in relation to the identified CCPs as well as the quality control points.

What must also be considered when applying the HACCP system to spoilage microorganisms is what numbers can be tolerated in the finished product. As far as food safety is concerned and the realization that *Salmonella* at low levels can cause illness (see Chapter 1) the processor is working towards pathogen numbers of at least absence from 1 g, and increasingly towards absence from 25 g. For certain products, similar low levels are being applied by the processor to specific spoilage organisms and Table 3.2 illustrates this point. It shows the relationship between psychrotroph numbers and shelf life in

Table 3.2 Relationship between initial psychrotrophic count in pasteurized milk and the shelf life at 6°C

Geometric means of initial psychrotroph count	Shelf life in days[a]
1 per 100 ml	>10
1.1 per 10 ml	8–10
0.84 per ml	6–8
13.8 per ml	5–6
80.2 per ml	<5

[a] Shelf life calculated as the time for the bacterial count in the milk to reach log $10^{7.5}$ per ml at 6°C.

Adapted from Griffiths and Phillips[12].

pasteurized milk, and is work carried out by Griffiths and Phillips[12].

Similarly, in cottage cheese the aim is to reduce the level of the spoilage yeasts and *Pseudomonas* spp. to less than 1 per 100 g of product, and yogurt manufacturers are aiming to achieve absence of yeast from a 125 g retail container.

These low levels are because of the shelf life expected of these products in the market place. A number of microorganisms which are contaminants of specific products, and cause spoilage when the numbers reach 10^7 to 10^8/ml or g, are capable of rapid multiplication at temperatures of 6° to 7°C (43° to 45°F) and these temperatures are not unusual in the UK during handling in the retail chain and in customers' refrigerators. Therefore, to prevent spoilage during the shelf life of the product the manufacturer should control these contaminants at each CCP in the process.

HACCP application to foreign material contamination

As well as applying HACCP to processes and products in order to minimize microbiological risks, both from safety and spoilage (quality) aspects, HACCP may be applied in order to minimize risk of foreign material contamination in products. The same logical, thorough, multi-disciplinary method of working is necessary for each identified foreign material hazard in each process. Here are some examples of foreign materials which may be found in a processing factory, either from ingredients, equipment, or as a result of the procedures applied on the process floor.

From ingredients:
Fruit pits
Twigs, stems or wood pieces
Field insects
Glass
Stones
Torn pieces of packaging
Nails
Staples
Rodents or their hair
Antibiotics
Pesticide residues
Fungicide residues
Animal hormone residues

From equipment:
Grease
Metal shavings (most likely from worn threads or swarf)
Glass – if glass sight glasses or mercury-in-glass thermometers are used
Rubber or plastic from worn gaskets
Nuts/bolts/rivet heads/screws/nails
Wood splinters from pallets

From procedures:
Sanitizing solution from filler bowl in the first packs of product
Bristles from brushes
Plastic from worn squeegees or scrapers
Pieces of packaging
Threads of cloth – from cleaning cloths which have been re-used
Pens/pencils from outside pockets of operatives' overalls
First-aid plasters which have become detached
Hair
Jewelry from operatives
Buttons

Each of these and other possible hazards must be approached from the point of view of assuring their absence by suitable control options at the critical control point(s) with which they are most likely to enter the product. In the case of certain metals, it is possible to use metal detectors to screen each pack of finished product, and to reject any packs containing such metals. For most foreign materials, however, this is not possible, and the way to eliminate them is through good design and appropriate controls forming an integral part of Good Manufacturing Practice.

3.B Implementing the HACCP system

It is important to appreciate that while guidance can be given on general principles, each application is unique in its circumstances and needs its own 'package' put together in order to deliver what is required of it. It is particularly important to appreciate the detail that is needed since overlooking consequences of loss of control brings about a false sense of safety. In extreme cases this could end a business.

For any process or product it is obvious that there is a considerable amount of information to be gathered and assigned to its correct position within the framework of the HACCP system, i.e. points 1 to 7 as already described. Expert judgments need to be made on which control options to apply in light of the information relating to each product or process. It is important that a group or team with expertise is formed to implement the HACCP system, and that the core members of the group comprise a microbiologist, an engineer, a production manager and a cleaning specialist; with representatives from purchasing, product development, packaging, distribution and marketing forming part of the group as necessary.

The group should formulate a HACCP plan for each specific process or product being analyzed. The HACCP plan includes the objective of the analysis, whether this is safety, spoilage or foreign material control; identification of the CCPs; the critical limits which should be in place at each CCP (including the tolerances on these limits); monitoring procedures; corrective action in the event of deviation from control procedures; verification; and the preparation of a report on the analysis.

The sections of the HACCP plan report should include:

- The purpose or aim of the HACCP, e.g. microbiological, foreign material, etc.
- The members of the group applying the HACCP system, including the job title of each member. People change jobs within an organization, and it is the job holder at the time of a review who would form part of the new group, not necessarily the person who was part of the original group
- A summary and conclusions
- Flow diagram or diagrams – highlighting the CCPs in the process
- Body of report which details the hazards at each CCP and the necessary control options, using the information gathered by this group

- Cross references to other analyses; for example, the equipment maintenance schedules, the cleaning schedules, the Company procedures or GMPs which apply to the process or part of the process
- Action taken as a result of the analysis

The report forms the record of the HACCP plan, so a summary of the information and recommendations it contains should be presented in a way that is readily available to anyone who needs to consult or use the report. It will be of special value when any changes are proposed for the process or specification concerned. A matrix is one way which allows this, and suggested column headings are:

CCP number
Process/storage stage of this CCP
Description of the stage
Hazard(s) associated with the stage
Hazards controlled by
Control limits
Deviations, and how they have been (or may be) corrected
Planned improvements

Table 3.3 gives an example of such a matrix applied to 4 CCPs selected from those applied to powdered formula milk (see Figure 3.1).

Table 3.3 Report of CCPs and control options for powdered formula milk (selected CCPs)

CCP No.	Process/ storage	Description	Hazard(s)	Controlled by	Control limits	Deviations and their corrections	Planned improvements
4	Concentration of vitamin-enriched milk.	Two-stage evaporator 1st stage: 60°–70°C 2nd stage: 50°–60°C	Halt in production leading to drop in temperature. Bacteria could grow in held product.	Correct feed to evaporator. Temperature controlled within specification for both stages.	If temperature of 2nd stage falls below 48°C, product is quarantined.		Equipment changed so that halts in production do not occur.
7	Mixing of oil and vitamin.	Mixed at 20°C in tank.	Pathogens in raw materials if incorrectly handled.	Supplier quality assurance of these materials.	Freedom from pathogens when agreed testing schedule applied.		
12	Transport of product.	Air used to carry powder through air ducting and into storage/ transit tanks.	Contamination carried in cooling air. Cracks in ducting can allow contamination.	Filtration of (transport) air. Inspection for cracks and imperfect joints.	Class I filters on cooling air inlets. Production halted if leaks discovered.		
14	Packaging of product.	Product placed in sachet made from impermeable laminate N₂ flushed before heat sealed sachet placed in box.	Impermeable laminate could be contaminated. Air supply to filler could be contaminated. Product residues in filler could contaminate fresh product.	Discard outermost layer of laminate on spool. Filtration of air to filler. Prevent build-up of residues by dry cleaning filler.	First 4 meters of spool discarded. Class 1 filters on air supply to filler.		Consider UV treatment of laminate to disinfect it.

Other deficiencies and cross references should be noted against the appropriate CCP No. and recorded on this matrix.

The group may well have highlighted deficiencies of equipment or procedures. Some of these are likely to have been corrected before the group has completed its work, and may not, in fact, form part of the report. It is however, prudent to note or crossreference them. What will also be highlighted are improvements which may be made to the process in the form of new equipment or better instrumentation, even though the existing process is not showing a deficiency. It is useful to list these improvements in the report, and they can then form part of the scheduled capital and revenue reviews carried out by management, and so upgrade the processing facility.

3.C Summary and conclusions

The Hazard Analysis Critical Control Points (HACCP) system is a systematic, rational, documented process of identifying the hazards, then estimating and reducing the risks associated with the processing and marketing of a given food product. An analysis should be done for all new as well as existing products and also if there are any modifications to the formulation, equipment, processing, or intended end use of a product.

HACCP was first introduced to the food industry at the 1971 Conference on Food Protection. In 1973, the use of the HACCP in the USA was mandated by law in connection with the production of low-acid canned foods. Despite the benefits of HACCP, its use by industry in the control of hazards other than in low-acid canned foods has been sporadic until recently.

In 1985, the US National Research Council published a report entitled 'An Evaluation of the Role of Microbiological Criteria for Foods and Food Ingredients'. This suggested that HACCP is the most rational approach to the control of microbiological hazards in foods and recommended that a HACCP system be required by regulation. Following this recommendation, HACCP has been suggested as a mandatory program by the National Advisory Committee on Microbiological Criteria in Foods (US)[8]. The foods included so far are ready-to-eat meat, chicken, shrimp and crabmeat. Although the main purpose of a HACCP system is to insure food safety, it may also be used to establish and monitor controls for contaminants, raw materials, the process, consumer use directions, and storage conditions that may contribute to unsatisfactory product quality. Critical control points are those key points in the production of food from the raw materials to finished product where the loss of control could result in an unacceptable food.

The HACCP system offers benefits to the regulator, the processor, and the consumer. The regulator and processor can concentrate on factors directly related to controlling hazards. By monitoring the critical control points, the regulator can also understand the effectiveness of the documented control methods. This system also provides the regulator with a complete history of the operation instead of observing conditions on a single inspection day. Similarly, the processor, as well as controlling his operation on a continuous basis, can also prevent hazards rather than reacting to them. The consumer benefits by receiving a product produced under conditions wherein hazards have been identified and controlled.

It is important to educate, train and retrain employees in the use of the HACCP system. With employee turnover, there may otherwise come a time when very few people in a plant will understand what the HACCP system is or understand the need for the various controls that have been established; therefore training should be done on a semiannual basis or as necessary. It is worth observing that had the NAS/NRC *Salmonella* control principles of 1969[5] been applied to other pathogens and processes it is at least arguable that many outbreaks would have been avoided and costly crisis management would have been replaced by cost-effective control.

References

1. Truswell, A. S., Asp, N. G., James, W. P. T. and McMahon, B. (1978) *Conclusions, Proceedings of Marabou Symposium on Food and Cancer*, p. 112. Stockholm: Caslon Press
2. World Health Organization (1982) *Report of the WHO/ICMSF Meeting on Hazard Analysis, Critical Control Point System in Food Hygiene*, Geneva (9–10 June 1980) VPH/82.37 Geneva
3. Campden R. A. (1987) Technical Manual No. 19. *Guidelines to the Establishment of Hazard Analysis Critical Control Point (HACCP)*, Campden Food Preservation Research Association, Chipping Campden, Gloucestershire GL55 6LD
4. ICMSF (1988) *Microorganisms in Food 4 Application of the Hazard Analysis Critical Control Point (HACCP) System to ensure Microbiological Safety and Quality*, International Commission on Microbiological Specifications for Foods. Oxford: Blackwell Scientific
5. NAS/NRC (1969) *An Evaluation of the Salmonella Problem*, National Academy of Sciences – National Research Council, Committee on *Salmonella*, National Academy of Sciences, Washington, DC
6. Olson, J. C., Jr (1975) Development and present status of FDA *Salmonella* sampling and testing plans. *Journal of Milk and Food Technology*, **38**, (6), 369–371
7. Peterson, A. C. and Gunnerson, R. E. (1974) Microbiological critical control points in frozen foods. *Food Technology*, **28**, (9), 37–44
8. NAC (1990) *Hazard Analysis and Critical Control Point System*, HACCP Working Group of the National Advisory Committee on Microbiological Criteria for Foods, USDA/FSIS, Room 3175, South Ag. Building, Washington, DC

9. Corlett, D. A. Jr. (1987) Selection of microbiological criteria based on hazard analysis of food. In *Food Protection Technology*, edited by C. W. Felix. Chelsea, MICH: Lewis Publishers

10. BSI (1976) BS5295: *Environmental Cleanliness in Enclosed Spaces Part 1*. Specification for controlled environment clean rooms, work stations and clean air devices, London: British Standards Institution

11. *US Food and Drug Administration/Milk Industry Foundation International Ice Cream Association* (1988) Recommended guidelines for controlling environmental contamination in dairy plants. Dairy & Food Sanitation, **8**, (2), 52–56

12. Griffiths, M. W. and Phillips, J. D. (1988) Modelling the relation between bacterial growth and storage temperature in pasteurized milks of varying hygienic quality. *Journal of the Society of Dairy Technology*, **41**, (4), 96–102

Further reading

Bauman, H. E. (1987) The Hazard Analysis Critical Control Point concept. In *Food Protection Technology*, edited by C. W. Felix. Chelsea MICH: Lewis Publishers. This chapter should be required reading for all senior management as it encapsulates the system.

NZIFST (1984) A *Guide to the Preparation of a Quality Assurance Plan*, New Zealand Institute of Food Science & Technology, New Zealand

Chapter 4
Buildings

4.A. Site and environment

4.A.1 Plant location

Introduction

A new development may mean either that a new location, sometimes referred to as a 'green-field site', is chosen or that an existing site is developed. Remember that an unsuitable location has product and cost implications throughout the working life of the plant.

Either decision involves factors which are not directly linked with the product(s), e.g. tax incentives, planning (zoning) restrictions, opportunities for lease or purchase of buildings, availability of specific finance, etc. However, a discussion of these factors is outside the scope of this book.

Factors which are linked with the products may be considered under the following headings:

- Physical
- Geographical
- Infrastructure
- Management

Physical factors

These can include but are not necessarily limited to the following.

Geology

The geology of the site; its stability, the load-bearing ability of the subsoil (which could increase construction costs), the slope of the land (absence of a steep slope is helpful in reducing construction costs), freedom from flooding, earthquakes, snow or mud slides.

Size of plot

The size of the plot; is it adequate for present needs and future possible or probable expansion? As Imholte[1] states 'all too often, the parcel of land purchased for the best of building plans becomes too small. Once a food manufacturing plant has outgrown itself, all kinds of compounding product safety problems can occur. Crowding limits the options for an efficient and sanitary plant layout.'

Access to the plot

- By road – are there any traffic problems, or obstacles such as low bridges or weight restrictions on routes likely to be used?
- By rail – does the site permit easy grades and curves? Are there any out-of-gauge problems in connecting to the trunk railroad?
- By water – is the depth of water adequate for present and future vessel sizes? Are there restrictions on berthing facilities, problems with tides, locks, etc.?

Water supply

Are there adequate supplies of potable water available throughout the year for present and future needs? Is the water quality always adequate for all processing needs? Remember that city supplies may be drawn from different sources at different times of the year. Also remember that potable water means simply that it is suitable for human consumption – it may *not* be suitable for your products. For instance, humic and fulvic acids from peaty waters used to produce culinary steam result in chlorophenol taints in products.

Waste disposal

Plant waste water usually contains significant amounts of organic matter, which increases the Biological Oxygen Demand (BOD) and Chemical Oxygen Demand (COD). This can cause problems at the off-site sewage plant and means increased costs for sewage disposal. Discharge into rivers is usually tightly controlled, therefore space may be required on the site for appropriate treatment before discharge. Solid waste handling needs careful planning if it is not to become a sanitation problem. Apart from conforming with local regulatory requirements, space should be allocated for storage far enough away from the plant to permit effective pest control (i.e. of rodents, insects and birds). Particular attention must be given to the disposal of plastic waste, both to avoid litter and for environmental reasons.

Energy supply

Energy is required for heating, lighting and power. When considering availability, thought should be given not only to the economics of supply but also to the consequences of disruption of supply, especially of electricity and gas. Consider the effects of disruption on both processes and storage facilities. Both current and possible future needs should be assessed in relation to the distribution of electricity and gas within the plant site.

Relationship to the immediate neighborhood

Detailed assessment needs to be made of pollution risks by and to adjacent areas. This can be very difficult if the plant is sited on an Industrial Park or Zone which is only part completed. Contaminants such as smoke, dust, ash, solvent vapors, e.g. from an automobile paint shop, are of obvious concern. Less obvious risks are illustrated by a food plant which found unwanted numbers of air-borne bacterial spores. These were traced to another plant upwind which used sawdust to clean animal furs. The waste sawdust was incompletely burned and carried the spores. Remember that control of exhaust stack emissions and effluents may be diligent but even the best systems are neither totally effective nor are they completely free from malfunction. Nearby 'sanitary' landfills or city rubbish dumps (tips) and neighboring plants where there is frequently surface water, storage of discarded machinery, boxes, crates or rubbish which harbor rodents or breed insects can be expected to provide sanitation hazards.

Warning notes
The proximity of a poultry rearing or packing plant upwind may be a source of *Salmonella* spp., etc. This is particularly important where food is made or packed for susceptible consumers, e.g. dried baby foods, geriatric packs.

Proximity to residential areas can cause problems apart from any complaints which may be made about plant noise or traffic. Zoning, planning, or building regulations may establish industrial area boundaries. However, manufacturing process changes or alterations to industrial zone boundaries may create problems for both plant operators and residents.

Effects of climate

Risks of storm damage, e.g. by wind, rainfall, hail, snow or ice, should be considered in relation not only to the plant, power supplies and communications but also to the supply of agricultural produce, water supply and waste disposal.

Effects on the site itself

The prevailing winds should not be allowed to blow dust or rubbish into the plant building, particularly into food preparation or handling areas. This may have implications for a proposed layout.

When considering physical factors affecting the location of the plant it is prudent to consider cost implications of likely or possible changes, i.e. to calculate an 'insurance factor'. For example, currently the water supply may be adequate but future residential or industrial building may restrict supply or change the sources of supply which may alter the properties of water used in the processes.

Geographical factors

These factors concern the proximity and convenience to potential markets as well as sources of ingredients, e.g. fruit and vegetables. Remember the importance of time lost in making and receiving deliveries.

Infrastructure

These factors include not only the availability and cost of labor of the required grades or types but also the availability of technical services, delivery of parts, the quality of contract services and the speed of response in getting assistance in dealing with technical problems. Cultural or religious factors which could affect the working of the plant should also be considered.

Additionally, consideration should be given to computer linkage requirements for Computer Integrated Manufacture (CIM), both present and future.

Management

These factors concern issues of organization and control in relation to the 'home' or 'head' office and other plants within the Company.

4.A.2 Plant environment

Introduction

The immediate environment of the plant may be considered under the following headings:

- Landscaping
- Traffic
- Service areas and parking lots
- Pest and weed control

Landscaping

This can be done as an aid to control rodents by depriving them of places to live (harborage) and reducing dust. Raising outside equipment 23 to 30 cm (9 to 12 in) clear of the surface prevents it from being a rodent harborage. The landscaping should not include small streams or lakes which may attract insects or birds. Troller[2] describes visiting a newly constructed plant located less than 100 m (100 yd) from a wildfowl sanctuary. Inevitably routine environmental samples were positive for *Salmonella* spp. Troller also advises keeping shrubbery at least 10 m (30 ft) away from the building and warns against on-site planting of trees or shrubs which provide food or harborage for birds. Remember that some flowers and shrubs can attract insects to the processing plant. Both Troller[2] and Katsuyama and Strachan[3] advise a grass-free strip 0.6 to 0.9 m (2 to 3 ft) covered with a layer of gravel or stones 2.5 to 3.8 cm (1 to 2½ in) deep around the building(s). This helps to control weeds and rodents and is very convenient for the sanitarian inspecting rodent bait stations or traps placed against the building. Lawns should be kept cut to discourage insects and other pests that find harborage in tall grass. Employees should not eat on the plant grounds because dropped food attracts birds, rodents and some insects. Putting food down for stray cats etc. on the site should be prohibited, although it is recognized that this may be difficult to enforce.

Aesthetically pleasing surroundings to a plant contribute to the Company image as a good neighbor, have a beneficial effect on employee morale and can be planned with low first cost and upkeep.

Traffic

The layout of roads around the site buildings should be planned for safe movement of all vehicles and pedestrians. Cross traffic with pedestrians should be minimized. For some plants, it may be necessary to restrict heavily soiled vehicles to particular routes in order to minimize cross contamination of the finished product(s). Roads should be maintained in good condition to reduce insects or microorganisms breeding in standing water, to reduce product damage in transit and minimize the chance of personal injury.

Service areas and parking lots

Service areas such as rail sidings, parking lots for trucks and cars, collection and disposal of trash and waste, boiler rooms and surplus equipment dumps can easily become sources of contamination or pests. To minimize the risks, such areas should not only be kept in good order but also be well drained either by providing a drainage system or by sloping (grading) the ground. Good drainage is needed to prevent contamination to food products by seepage, foot-borne dirt or providing breeding grounds for pests. It is insured by maintaining the correct slope on service areas and parking lots and providing adequate surface water drainage even during storm conditions.

Pest and weed control

A good perimeter fence is important and not only for security reasons. It discourages or prevents children from entry into the grounds, which is important if unlocked bait boxes are placed outside the plant. Troller[2] recommends, where local codes allow, a chain-link fence at least 2.1 m (7 ft) high with appropriate entry gates. These fences filter pieces of paper and other debris. Because they 'collect' litter, they must be part of the 'good housekeeping program'. Grass should be trimmed around fences. If land or buildings on the outside of the perimeter fence present a hygiene problem or hazard, extra precautions will be needed and the help of regulatory authorities may have to be invoked. Imholte[1] recommends a strategy of keeping rodents outside buildings. This requires two lines of defense, an inner one of bait stations at the foundation walls of buildings at 15 to 21 m (50 to 70 ft) intervals with a few mouse traps near the building entrances. The outer line is at the perimeter or about 50 m (54 yd) from buildings. Bait stations are again at 15 to 21 m (50 to 70 ft) intervals.

When reducing insect pests it is useful to remember that only about 1 in 20 are seen. The first strategy is to avoid creating breeding and feeding areas on the site. Insects are carried by winds, so they will inevitably be present at the plant. This is relevant to unauthorized opening of doors and windows and siting of protection against flying insects. The type of lighting used can affect the numbers of nightflying insects. The ratio of insects around mercury vapor lamps to high-pressure sodium lamps is about 100 to 1. This is because an ultraviolet component of the light attracts insects and more of this is produced by the mercury vapor lamp. Thus high-pressure sodium lamps should be used within 50 m (54 yd) of plant buildings. It has been suggested that mercury vapor lamps of 400 to 1000 watts might then be used further out to attract nocturnal flying insects away from the plant.

Bird infestation can become a serious problem. On the plant site the strategy is to make it unattractive by denying birds food and harborage. Care must be taken with waste disposal as open containers are, in effect, feeding trays. Spillages of food, e.g. seed, pulse (dried peas, beans or lentils), grain or pasta, should be cleared promptly from roadways or truck docks.

Weed control is important not only to reduce harborage for insects and rodents but also to prevent air-borne seeds getting into the plant. Normal gardening methods should be used to control weeds. Agricultural chemicals which sometimes might be recommended for weed control may be toxic and are a potential source of chemical contamination in the food produced at the factory. An *approved only* chemical use program should be implemented and very strictly enforced.

4.A.3 Plant layout

Introduction

The primary purpose of the building structure is to protect ingredients, the processing equipment and food products. Ideally, it should not impose constraints on the process or plant layout.

Single- or multistory buildings

There is no such thing as an 'ideal' design for food plants. Both single- and multistory designs have advantages and disadvantages.

Single-story design has some advantages:

- It simplifies material handling and the supervision and deployment of personnel.
- It allows a simple 'straight line' flow from ingredients through processing to finished packs, which reduces the risk of product contamination.
- It is easier – and cheaper – if heavy floor loadings are required.
- It is easier to arrange flexibility for equipment changes and for expansion of the building.

The major disadvantage is that where gravity flow systems are needed for efficiency, multistory structures are required.

An appropriate compromise may be a single-story building with varying head room and the opportunity of a mezzanine floor, to allow for gravity flow where this is necessary.

External features

In principle, simple designs of external details are best. Ledges and other architectural features which may attract birds are to be avoided.

Careful consideration needs to be given to the cleaning of roofs over food manufacturing areas. Exhaust stacks are usually vented through the roof and sometimes process equipment extends through or is mounted on the roof. Imholte[1], in his Chapter 3, gives examples of sanitary exhaust stack designs, shown as Figure 4.1. Leakage of product from roof-mounted process equipment may be reduced by enclosing it in a penthouse. Particles, especially of hygroscopic powders, will deposit on the roof if this is flat, and if left unattended may attract birds, rodents or insects which are known *Salmonella* spp. and *Listeria monocytogenes* carriers. The presence of pools of water will also encourage these pests, and Troller[2] recommends a minimum slope of 1 mm in 96 mm (1 in per 8 ft) to insure positive drainage. It should also be remembered that in rainy areas, flat roofs have a higher maintenance cost than pitched roofs if they are to be kept in good, serviceable condition.

Air intakes may present a problem if they are incorrectly sited so that contamination from exhaust

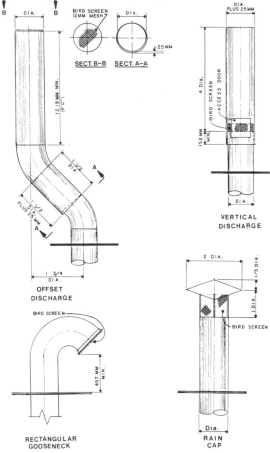

Figure 4.1 Sanitary stack designs (from Imholte[1])

treatment on roofs to reduce widely varying internal temperature.

Corrugated 'galvanized iron', ribbed sheet or metal pan roofing are also not advised for food plants. This is because the high rate of heat transfer and expansion of metal creates problems with condensation and the maintenance of properly sealed joints. Condensation and poor joints are obvious contamination hazards.

Katsuyama and Strachan[3] state that gypsum roofs have been used in warehouses and other dry storage areas, but stress that this material should not be considered for wet processing areas. This has obvious implications if a change of use is proposed for the area.

The selection of building materials is difficult because of the large number which are available from which one must be chosen for the job to be done. Higher-quality materials usually cost more, last longer, and have lower maintenance costs. By considering the food product risks, the proposed cleaning method and the design life of the structure, good decisions can be made. This is a good subject for use of Hazard Analysis Critical Control Point (HACCP) techniques. (See Chapter 3.)

Plant layout

Troller[2] summarizes the United States FDA Good Manufacturing Practices (GMPs) and the USDA Meat Inspection Program as requiring that the plant shall provide:

1. Adequate space for equipment installation and storage of materials.
2. Separation of operations that might contaminate food.
3. Adequate lighting.
4. Adequate ventilation.
5. Protection against pests.

The second point on separation of operations may be expanded to stress the importance of preventing the contamination of cooked or semi-cooked foods by raw (uncooked) foods. Other countries, e.g. within the EEC, have regulations such as the Meat Directive 77/79 which affect what may be done.

To put these principles into practice, flow sheets are the key documents. These can show the flow of materials and products from department to department as well as the sequence of the manufacturing process. Imholte[1] shows how they can be used to study department flows and to identify operations or departments which present a potential hazard to the food product. Appropriate action can then be planned to minimize the hazard, e.g. by separating ingredient storage from the processing area. He lists areas which need to be separated and the UK Food Hygiene Codes of Practice No. 10[4] gives an alternative and rather similar list when considering

stacks or roof-deposited debris, including fecal material from birds, can be taken into the process area. For high-risk products, all air intakes should be filtered with microbiological filters which are easily accessible for maintenance and cleaning.

As an illustration of the detail that needs to be considered, Imholte[1], in his Chapter 3 on Buildings, commends the use of smooth asphalt for roofs, together with various durable felt roofing papers and insulating materials in areas where spillage is anticipated. To prevent blockage of downspouts he recommends that they be fitted with bullet-nose grates (gratings) that project upwards. It is also unwise to route run-offs from roofs through production areas. He also warns against product contamination of stream or river waters if the downspouts are connected to storm sewers, since usually these discharge directly into natural water tributaries.

Another example of the detail needed in hot climates is the desirability to use external reflective individual departmental areas. It is obvious that HACCP has a direct application in these studies and has been successfully used by several companies. The key principle is to do whatever is necessary to protect the food and the processes. This means that the local environment within the building must not only comply with all legislative requirements but also:

- Not be responsible for contamination, deterioration or damage to packaging materials or ingredients, to products being manufactured, to machinery or finished packs.
- Achieve and contribute to the effective maintenance of appropriate sanitary (hygienic) standards, e.g. of cleanliness, of space for installation, operation, and maintenance of machinery, of lighting, ventilation and pest control.

In considering specifications and costs it is wise to consider the implication of possible future changes in use, e.g. the expansion of manufacturing areas into what is currently planned or used as a storage area.

Additionally, the layout should insure that:

- Ingredients should move from 'dirty' to 'clean' areas as they become incorporated into food products.
- Conditioned (e.g. chilled) air and drainage should flow from 'clean' to 'dirty' areas.
- The flow of discarded outer packing material should not cross the flow of either:
  - unwrapped ingredients
  - finished product
- There is sufficient space for plant operations including processing, cleaning and maintenance. Space is also required for movement of materials and pedestrians. When space is limited it is particularly important to provide, and mark clearly, traffic lanes for vehicles and employees. This is to reduce the risk of both accidents and damage to goods and equipment.

- Operations are separated as necessary. There are clear advantages in minimizing the number of interior walls since this simplifies the movement of materials and employees; makes supervision easier; and reduces the area of wall that needs cleaning and maintenance.

It should be noted that minimizing the number of interior walls is not in accordance with European regulatory thinking, which tends to want areas separated by walls. This EC requirement relates to the wish to separate what would otherwise be a common airspace for ingredients, storage, manufacture, processing rooms and filling halls. However, separation of different operations by time, particularly in small plants, may be acceptable. This separation is a critical control point. The question of what is essential therefore has a regulatory as well as technological aspect.

References

1. Imholte, T. J. (1984) *Engineering for Food Safety and Sanitation*. Crystal, MINN: The Technical Institute of Food Safety
2. Troller, J. A., (1983) *Sanitation in Food Processing*. New York: Academic Press
3. Katsuyama, A. M., and Strachan, J. P. (1980) *Principles of Food Processing Sanitation*. Washington, DC: The Food Processors Institute
4. DHSS (1981) *Food Hygiene Codes of Practice No. 10. The Canning of Low Acid Foods – A Guide to Good Manufacturing Practice*. London: HMSO. (Note: This is currently being revised and is out of print).

Further reading

Codex Alimentarius Commission (1988) *Volume A. Recommended International Code of Practice. General Principles of Food Hygiene*, CAC/Vol.A – Ed 2 Second Revision, Rome: FAO/WHO. (1985).

4.B Design and construction

4.B.1 Design

Introduction

When sanitary design features are incorporated into new or altered building construction they will improve the appearance of the structure, reduce cleanup time and costs and decrease product contamination risks.

Good sanitary design requires great attention to detail. For this reason, each building job needs a separate, individual, study of sanitary design aspects.

Sources of information

There are a number of sources of information on good sanitary design which may be used, depending on local conditions. These may be categorized as:

- Government regulatory and advisory agencies.
- Industry research and trade associations who may also know of architects and/or contractors who have experience of jobs similar to the one under consideration.
- Universities or technical institutions.
- Technical and trade publications.

Information on sanitary design is scattered and no single source is likely to have all the answers to your specific problems. For this reason, when meeting with architects and contractors it is important to assess their knowledge, experience and enthusiasm for sanitary design. This book cannot offer detailed engineering specifications. It does set forth some of the important principles with indications of the depth of detail that needs to be considered in discussions with architects and contractors.

These will be considered under the following headings:

- Frames
- Walls
- Windows (External and Internal)
- Doors and Docks
- Ceilings
- Floors
- Drains
- Stairs
- Elevators
- Lighting
- Ventilation
- Insulation

Note: For information on utilities see Section 4.E.

Frames

Frame construction means that the external walls are a skin to protect personnel, materials, and machinery, and thus are usually made of light materials, although brick etc. may be used for aesthetic reasons. Machinery should not be suspended from or connected to the building framework. Machines should have separate supports capable of withstanding the stresses and vibration which occur during their use.

Frames may be:

- Reinforced concrete with reinforcing rods in the concrete beams and columns
- Composite structures where the reinforcement is with steel H or I beams (rolled steel joists or RSJ, UK)
- Steel framing with H or I beams and columns

Reinforced concrete and composite methods of construction have the advantage that they eliminate ledges which collect dust and require cleaning, which is an on-going cost. However, when the forms (or shuttering, UK) are removed from the set concrete, the surface is unlikely to be smooth enough either to be left bare or for further treatment such as painting. At times quite extensive honey-combing may occur. Defects in the surface may harbor insects and should be remedied while the concrete is still uncured or green. This is done in the US with a rubbing stone and a mixture of plain cement and water mixed to the consistency of very thick paint filler (Imholte[1]) or 'bag rubbed' in the UK.

Steel framing is commonly used and has the advantage over reinforced concrete of being quicker to construct and is also easier to modify after completion.

Unless care is taken, a beam may be located too close to the wall. If this happens an inaccessible linear cavity is created between the web of the beam and the wall and which therefore cannot be cleaned. Imholte[1] recommends allowing 100 to 150 mm (4 to 6 in) between the beam and the wall, boxing in the beam on one side (or both) or filling in the void with either polyurethane foams sealed with two thick coats of epoxy paints to prevent entry of insects, or, alternatively, lightweight concrete. Remember, however, that shrinkage may cause cracks in the concrete which need to be sealed with a caulking material. Figures 4.2. through 4.4. illustrate ways of enclosing beams.

Ideally, *all* beams would be sealed, but providing that there is no contamination or pest hazard, it may not always be necessary. Positive reasons, however, are required showing that a hazard does not exist.

Troller[2] states that for non-product areas, exposed structural members are satisfactory, as is wood framing, providing that these structures can be kept reasonably neat and dust free. However, it should be remembered that ingredients are 'product' and wood is a potential hazard in the presence of insects or other pests which attack wood. Any

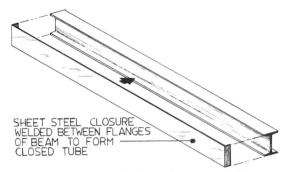

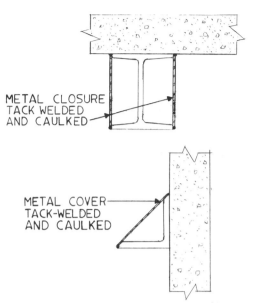

Figure 4.2 Some ways of enclosing beams – by welding sheet metal closure (from Imholte[1])

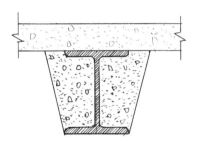

Figure 4.4. Some ways of enclosing beams – support member treatment in a product zone (from Troller[2]

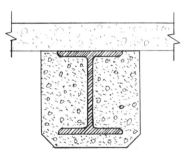

Figure 4.3. Some ways of enclosing beams – enclosed channel beams for sanitary roof construction (courtesy of The Pillsbury Company)

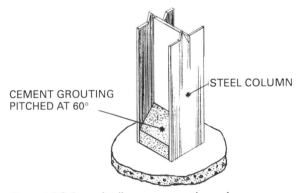

Figure 4.5 Column details – proper grouting makes cleaning much easier (from Imholte[1])

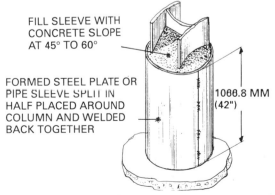

Figure 4.6 Column details – extra protection for warehouse columns to prevent lift truck damage (from Imholte[1])

preservatives used must not taint or adversely affect ingredients or stored products. Troller[2] recommends using steel pipe or tubular columns rather than concrete, which becomes chipped and is difficult to maintain.

Column floor junctions with H or I steel sections can create small, difficult to clean areas. These areas should be grouted with a slope of 60° towards the web of the column as shown in Figure 4.5.

Columns in warehouses are liable to damage from (fork) lift trucks and there are cleaning problems at the column/floor junction. Imholte[1] presents a solution to these problems (see Figure 4.6). The column is surrounded by a 1.07 m (42 in) protective pipe sleeve. The void between steel column and sleeve is filled with a non-shrinking concrete mixture, sloped upwards at 45° to 60° angle at the top of the web and beam flanges. Note that this angle does not prevent dust settling, but does make it more obvious and prevents the ledge being used as a dump for trash.

As an alternative to boxing in steel H or I section columns placed too close to the wall, a pilaster may be constructed around the column – see Figure 4.7.

Channel wall girders should be installed as an inverted 'U', i.e. with the flanges down to avoid a full length cup-like cavity which gathers debris and is difficult to clean.

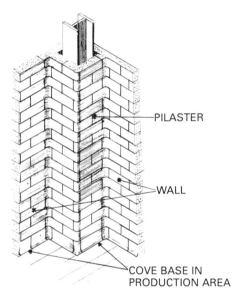

Figure 4.7 Use of pilaster to prevent column creating uncleanable areas (from Imholte[1])

Square or rectangular tubular framing offers sanitation advantages, e.g. on beams the dirt-collecting surfaces are eliminated and on columns the difficult to clean floor/column junctions are eliminated. Round tubing or pipe is also a sanitary design for structural columns. When any tubular shape is used it must be sealed, e.g. when drilling holes for any attachment. Imholte[1] advises that the holes are threaded then closed with a screw or that other positive measures are taken to seal the holes so that insects cannot enter. Unused holes should also be sealed.

In wet manufacturing areas where steam, aggressive soil from ingredients or products, and heavy-duty cleaning or sanitizing agents are likely to come into contact with structural steel, galvanizing or plating, e.g. with zinc, should be considered. For practical reasons this is usually used for new additions or major remodeling projects.

External walls

The prime requirement of external walls and their foundations is that they should be water, insect and rodent proof. When considering rodent proofing or rodent stopping the AFIS book[3] gives the reminder that rodents may:

- Gain entrance through a 12.7 mm diameter (½ in) hole (rats) or 6.4 mm (¼ in) hole (mice).
- Walk along or climb up vertical wires.
- Climb the outside of vertical pipes not more than 76 mm (3 in) in diameter.
- Climb the inside of vertical pipes not more than 102 mm (4 in) or less than 38 mm (1½ in) in diameter.
- Climb the outside of vertical pipes of any size if the pipe is within 76 mm (3 in) of a wall or other continuous support for a rat.
- Jump 660 to 915 mm (26 to 36 in) both horizontally and vertically from a flat surface.
- Drop 15 to 25 m (50 to 80 ft) without being killed.
- Burrow vertically in earth to a depth of 1.2 m (4 ft).

Because Norway rats can burrow to a depth of 1.2 m (4 ft) they can tunnel under the building foundations. They do this by digging vertically downward along the face of the foundation until they reach the bottom of the footing, when they make a horizontal and then upward tunnel to reach interior floors or sub-floor.

However, a rat, if it reaches an impenetrable object while digging vertically down, will not dig away from the wall to get around the object but will cease burrowing on that tunnel and go elsewhere. Entry of rodents can be prevented according to Katsuyama and Strachan's[4] recommendations either, in a new building, by pouring a concrete L-shaped foundation 610 mm (24 in) to 915 mm (36 in) below grade (or ground level UK), with a horizontal lip extending 300 mm (12 in) out from the base, or by providing an existing building with a curtain wall of concrete or 26 gauge galvanized sheet metal of the same dimensions. Note: 26 gauge is approximately 0.475 mm thick (0.019 in).

Considering the exterior surface further, concrete block walls which are not finished with brick need sealing to prevent leakage of wind-driven rain. There are a number of 'water-proofing' methods including cement paint, stucco and shotcrete. If brick is used, it should be of a type to which dust does not cling.

Exterior walls are likely to have some pipes or wires attached to them. The AFIS book[3] gives diagrams showing still recommended designs of sheet metal rat guards which are reproduced in Figure 4.8.

For the convenience of the reader, other anti-rodent counter-measures are given as an appendix

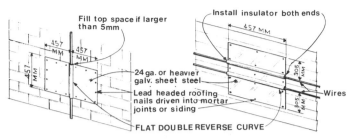

FLAT GUARD FOR SMALL PIPES AND WIRES

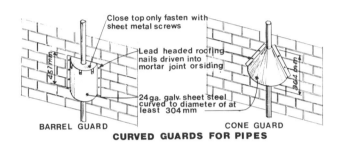

CURVED GUARDS FOR PIPES

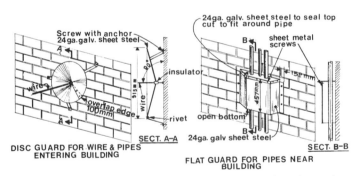

Figure 4.8 Several types of sheet metal rat guards installed on pipes and wires (designs from US Department of Health, Education & Welfare, Communicable Disease Center, in AFIS[3]

to this section. as they have a wider application than 'walls'.

Brick walls may sometimes be constructed so that there are vertical air spaces between the brick layers. In others, walls may be made with hollow tiles or masonry blocks. These spaces can become breeding places and harborage for insects, and rodents may enter and move around the building through broken hollow blocks. Katsuyama and Strachan[4] report that rodents have been known to gnaw through concrete cinder blocks, sun-dried adobe bricks, exposed aggregate in concrete and concrete which has not completely hardened.

Therefore, solid wall construction only is recommended for food plants.

Internal walls

The prime requirement of internal walls is that they should be smooth, flat, resistant to wear and corrosion, impervious, easily cleanable and white or light colored. Detailed sanitary performance specifications for materials are hard to find, thus work needs to be done to develop them for individual applications. See also 'Further Reading'.

Katsuyama and Strachan[4], as well as giving

some standards of sanitary construction, make the point that *internal* walls should be proof against rodent and other pests. Additionally, panelling or finishes should allow for easy replacement if mechanical or other damage occurs.

Imholte[1] has a section on walls which is thoroughly recommended to those concerned with their design, construction, or alteration. It begins:

> Selecting wall surfaces is difficult because of the variety of available materials. In addition, the finished surface requirements of these walls will differ according to their location in the plant, and the type of abuse they suffer from cleaning chemicals. Wet process manufacturing areas require wall surfaces that are impervious to moisture and will stand up against the harsh actions of cleaning chemicals. Dry, dusty manufacturing processes require a hard, smooth finish that is free of pits, checks, and cracks. Warehouse walls require a surface that will not harbor insects and other vermin. All wall surfaces must be built with materials that are easy to clean.
>
> There are several wall materials that can be used alone or in combination with a number of surface coatings to obtain a wall surface that is smooth, non-porous, non-absorbent, and impervious to moisture. A vast number of new wall-coating systems and facing-tile materials are available to upgrade the walls of many older and well-established food plants. Prepare the surface carefully to receive these new materials. In some cases, removing all of the old surface material, repointing the joints of blocks or brick walls, and patching the cracks and holes will be necessary.

In addition to stressing the importance of careful preparation of the wall to receive the chosen finish, Imholte[1] also underlines the need to anticipate the maintenance costs of the proposed finish.

Experience shows that the correct grade of ceramic glazed tile properly applied is a cost effective, hard-wearing, attractive and easily cleaned surface for wet processing or microbiologically 'clean' areas. However, the grade of tile and the method of application are critically important. Be sure that tiles are attached so that there is no cavity behind them which could harbor insects, e.g. Pharaoh's ants.

There are, of course, a number of surface finishes which can prove satisfactory. It is important to be clear on the duty of the surface and on any possible changes during its working life. It is also important to take expert advice when making the choice and to accept that it is most unwise to attempt to cut or shorten recommended 'curing' times.

Imholte[1] draws attention to the sanitation problems associated with metal walls, which are due to their expansion and contraction making sealing difficult, and their high heat transfer rate which causes potential or actual condensation problems. Unless effectively caulked, insect infestation may occur and rodent infestation can be a problem with prefabricated or 'field-assembled' sandwich panel walls. The solution he proposes is to make tightly fitting end caps for the top and bottom of each panel and then spot weld or rivet followed by sealing to prevent insect infestation. This is illustrated in Figure 4.9.

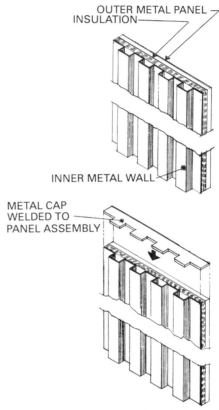

Figure 4.9 Rodent proofing of panel – by capping panels top and bottom to prevent rodent entry (from Imholte[1])

Wall angles, corners and junctions of walls and floor should be imperviously sealed and, for ease of cleaning, be rounded.

As well as coving of wall/floor junctures, it is also recommended where the floor meets structural support columns, curbing, concrete pads supporting equipment, or where equipment is grouted to the floor. The radius of curvature of concave coving is usually 25 mm (1 in) to 150 mm (6 in). Alternatively, 'angle' coves may be used at floor/wall and floor/column joints. The slope of the angle is usually 45° to 60°, as is shown in Figure 4.5.

Walls in areas where (fork) lift trucks with baskets or tote bins are used may need protection, particularly if tiled. Sealed tubular guards or curbing may be used (Figure 4.6.)

It is standard practice to cast curbs integrally with the floor. The exact size of a curb depends on the particular application, but usually a height of 100 to 150 mm (about 4 to 6 in) by 50 to 75 mm (about 2 to 3 in) wide is adequate. Curbs should, of course, be coved at the juncture with the floor.

Curbs may be used to protect walls from fork lift traffic. For this purpose, Katsuyama and Strachan[4] advise that a channel iron 'bumper' is incorporated into the curb. These curbs may be 200 mm (8 in) or more high and up to 450 mm (18 in) wide.

Note that in storage areas, rodents will treat a protective curb as an additional wall, and this influences the positioning of traps and baits.

Windows

These deserve more attention than they sometimes receive, as they are comparatively costly, requiring both cleaning and maintenance. Where appropriate, glass used in windows should be unbreakable, e.g. laminated, reinforced, or polycarbonate sheet. In production areas unscreened open windows invite insect contamination of food. It is obvious that window frames should be corrosion resistant and that external sills should slope to keep the rain clear of the walls. Inside sills or ledges should be minimized and sloped at 20° to 45° to prevent their use as a 'temporary' storage place for pencils, containers, tools, etc.

A good case can be made for high-hazard operations being in a windowless area. As Imholte[1] wisely observes: 'Regardless of the sophistication of the heating, ventilating and air-conditioning systems, there are always those who are uncomfortable with the surrounding temperature conditions and feel the need to open a window or a door.' External open windows need to be screened to prevent insects, dust or birds getting into the plant. In their determination to open windows, personnel have been known to break or remove screens, thus raising the risk of product contamination and increasing maintenance costs. If windows are required for natural light, the use of glass blocks means that they cannot be opened and maintenance costs are low.

It should be noted that many modern food factories do not have windows in any food processing area. This absence also reduces foreign body contamination.

Screening

Imholte[1] advises that if screens are located within 1 m (3 ft) of the ground, they should be reinforced with a heavy-gauge wire 6 mm (¼ in) mesh screen to keep out rodents. Suitable materials for rodent screens would be:

- Woven wire cloth; 6 mm × 6 mm (¼ in × ¼ in) #19 gauge wire or heavier
- Expanded metal; diamond mesh 6 mm (¼ in) maximum width #20 gauge sheet or heavier
- Perforated metal; perforations no more than 6 mm (¼ in) diameter width #24 gauge or heavier

Note: Screening materials are usually available in stainless steel, aluminum, bronze or galvanized steel. Use stainless steel for highly corrosive areas with high humidity and acid vapors from products or ingredients.

Gauge is a measurement given as weight or thickness, and the exact value varies, depending on the type of metal and whether or not it is galvanized. The gauge numbers quoted in this text are all US standard gauges; other countries may have different but equivalent standards.

For normal weathering and non-corrosive environments, aluminum is the material of choice alternatively, bronze or galvanized steel may be used.

Insect screens need to keep out small flying insects, and are usually made of woven wire cloth. Suitable materials and sizes are:

- Stainless Steel 18 × 18 mesh of 0.23 mm (0.009 in) diameter wire.
- Aluminum, bronze or galvanized steel – 18 × 14 mesh or 18 × 16 mesh of 0.28 mm (0.011 in) diameter wire. 14, 16 and 18 mesh mean that number of strands per inch (25.4 mm). Other materials can be used, e.g. 18 × 14 mesh 'glass fabric', but they must be cleanable and of adequate mechanical strength.

In warm climates, where natural ventilation is required on a south facing wall, concrete louvers external to screened windows are effective in reducing solar heat gain.

Doors and docks

The purpose of these is to let people and/or goods pass through a dividing wall while preventing, as far as possible, the entry of pests and dust, i.e. contamination. Remember, overhanging roofs for weather protection too often provide insanitary harborage for birds. Doors will therefore need to be close fitting with a maximum clearance of less than 6 mm (¼ in) and preferably less than 3 mm (⅛ in). Door jambs (frames) should be metal and rust resistant in wet areas. Jambs should not be hollow, but if they are, the voids may be filled with mortar which will prevent infestation. The frames should also be caulked at the junctures with walls. Wood is unsuitable for jambs as it is vulnerable to rodent attack. Thresholds should be caulked. Doors for personnel are preferably of hollow metal with good tight spot-welded seams since any loose fitting seams allow insect infestation. Solid-core doors are available and should be clad in stainless steel in wet areas, but remember, wood is vulnerable to rodents. Suitable plastic doors are becoming more widely used.

With doors which open to the outside and which will be used at night, it is good practice to locate lights 9 to 12 m (30 to 40 ft) away to attract insects away from the entry doors.

Doors should be self-closing and should be prudently designed to withstand expected use and misuse. Where heavy fork lift traffic is expected, automatic opening and closing doors are advantageous even with non-rigid, (e.g. plastic) doors. The same opening mechanism can be made to work alarm lights for the safety of passing personnel.

Strip-doors are usually made of overlapping heavy-gauge plastic. As a back-up for external doors they are effective in keeping out birds and help to exclude insects and dust unless they are deliberately tied back. Internally, they act as thermal barriers in chill rooms and freezers where the outer doors need to be left open for long periods of time.

Air curtains

Air curtains or air doors may be used, especially where truck doors must be left open, to prevent loss of heat or conditioned air where the plant is under positive pressure and also entry of insects and dust. The effectiveness of air curtains depends on the velocity of the air, the width, thickness, and angle of the curtain, internal air temperatures and pressures together with the absence of strong winds. An example of requirements for service entries and dock openings is summarized in Table 4.1 below.

Air discharge nozzles need to be as wide as the opening so that the air-stream covers the entire opening from corner to corner and top to bottom. A 127 mm (5 in) thickness nozzle is preferred for all installations and should be used for door widths of 2.4 to 3.7 m (8 to 12 ft). The air velocity should be at

least 488 m (1600 ft) per minute measured at a distance of 910 mm (3 ft) above the floor. For personnel entrances, the air stream should be continuous across the entire width of the opening, with a thickness of at least 254 mm (10 in) but preferably 419 mm (16½ in) The desired air flow pattern is obtained by discharge through adjustable louvers. The air velocity should be at least 503 m (1650 ft) per minute measured 910 mm (3 ft) above the floor.

External doors

External doors should not open directly into production departments. They should always be self-closing to keep out rodents, birds and insects. Further insect proofing is given by a strip door or, as Robson and Barnes[5] suggest, at busy entry points a double door arrangement incorporating a lobby with electrical trapping or insecticidal vaporizer. The sanitary design of truck doors and docks is discussed in an excellent section by Imholte[1] (pages 68 to 76).

Truck doors should be either of the metal roll-up or the hollow-panel lift-up types. Remember that mice need only a 6 mm (¼ in) hole to gain entry so there must be a good fit at the door frame or door track. Roll-up doors may have a sheet metal enclosure above the door frame to contain the door in the 'up' position. This traps dust and dirt; Imholte's[1] solution is either to order a door with no enclosure or to cut away the lower portion of sheet metal.

Docks and rodent proofing

Dock levelers may be permanently installed in a pit which is difficult to rodent proof. Imholte[1] recommends casting a steel box directly into the concrete for new installations (see Figure 4.10). On existing pits add a 203 mm (8 in) wide sheet of 18 gauge (this is 1.18 mm, 0.0478 in) thick sheet metal on the front face of the pit opening and on the pit walls, using an epoxy cement to bond the metal

Table 4.1 Summary of air curtain requirements

Item	Service entrances and docks	Personnel entrances
Nozzle thickness	Doors less than 2.4 m (8 ft) wide Optimum 127 mm (5 in) Minimum 76 mm (3 in)	Optimum 419 mm (16.5 in) Minimum 254 mm (10 in)
	Doors 3 to 3.7 m (10 to 12 ft) wide Minimum 127 mm (5 in)	
Air velocity per minute	488 m (1600 ft) minimum	503 m (1650 ft) minimum

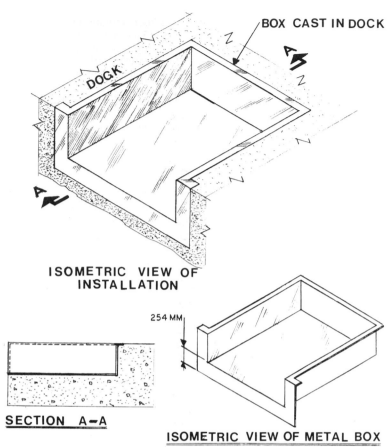

BOX CAST IN DOCK

DOCK

ISOMETRIC VIEW OF INSTALLATION

254 MM

SECTION A-A

ISOMETRIC VIEW OF METAL BOX

Figure 4.10 Rodent proofing of dock leveler pit using a steel box (from (from Imholte[1])

sheet vertically. These devices work because rodents have difficulty climbing slick (smooth UK) metal surfaces.

Dock doors on receiving and shipping docks are often of the roll-up (Track-well) type. These are usually corrugated and roll up and down in a vertical track and are difficult to seal in the closed position, particularly where railroad sidings are built into the dock. Figure 4.11 illustrates a solution to this problem. Note, however, that excess clearance, over 6 mm (¼ in), in the vertical door track may allow rodents to enter. Imholte[1] recommends using tubing to reduce this clearance. He also suggests an alternative method using a steel channel threshold at the same elevation as the lower portion of the railroad track bulb section with railroad track gaps directly in line with the fall line of the door.

All dock walls, whether interior or exterior, should be rodent proofed. The need to rodent proof interior docks is because doors are left open for quite long periods of time while maneuvering rail or road vehicles. Figure 4.12 from Imholte[1], describes the preferred construction of a new dock with a 300 mm (12 in) overhang or alternatively with a 305 mm wide strip of light gauge sheet metal.

RUBBER WELTS CEMENTED TO ASTRAGAL TO FIT SNUGLY INTO RAIL WHEEL FLANGE GROOVE

ROLLING DOOR

RUBBER ASTRAGAL

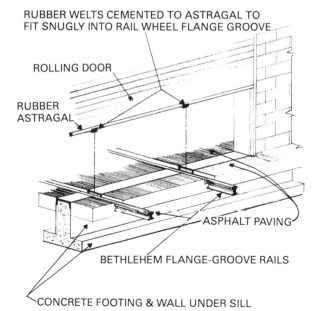

ASPHALT PAVING

BETHLEHEM FLANGE-GROOVE RAILS

CONCRETE FOOTING & WALL UNDER SILL

Figure 4.11 Rodent proofing of roll-up door over railroad track (courtesy of General Mills Inc.)

SECTION A–A

DOCK WITH PLATE

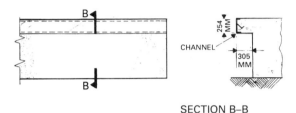

SECTION B–B

DOCK WITH OVERHANG

Figure 4.12 Rodent proofing of interior dock (from Imholte[1])

Docks may have steps to the higher level which rodents can easily climb. Imholte recommends a design of metal gates which prevents this (Figure 4.13). Exterior docks should be sheltered, but the structural steel frame which supports the canopy provides excellent roosting and nesting places for birds with inevitable potential for contamination of ingredients or finished product. Imholte[1] states that the way to eliminate this problem is to totally enclose the canopy with tightly fitting facing panels at the ceiling, on the ends and at the front. He warns that if the panels are not tightly fitting, rodents will gain access.

The design of exterior doors to refrigerated storage, whether finished goods warehouses or transit chilled stores, should enable sealing around the vehicle being loaded or unloaded. This is becoming a basic requirement for new premises designed to meet EC directive 77/99.

Ceilings

These should be non-porous, easy to clean and light colored with a finish that has desirable light-reflectance characteristics. However, ceilings tend to accumulate dust, condensate, droplets from manufacturing processes, etc. which can lead to mold growth and insect infestation. Moisture also encourages the growth of bacteria, e.g. *Listeria* spp.

Careful selection and application of materials and attention to construction details significantly influence the sanitary status of the ceiling. Sound-absorbent tiles, especially if they are porous, can present real sanitation problems. The use of sound-absorbent materials built into a polypropylene tile 'sandwich' is suggested as an alternative where noise levels must be reduced.

Suspended ceilings

Suspended ceilings are discussed by Troller[2], who believes that the advantages only outweigh the

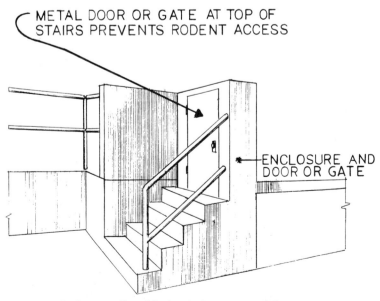

Figure 4.13 Rodent proofing of dock stair (from Imholte[1])

disadvantages when there is a serious problem with dust, flaking paint, etc. from above. They are likely to be required for high-hazard operations. Hayes[6] recommends the use of a service floor which houses ducts, pipework, cables, etc. with the floor forming the ceiling of a production area as a means of reducing to the minimum overhead equipment and thus the risk of contamination of product.

Whatever ceiling is used, the junctures between walls and ceiling, if rounded and sealed against dust and water, will be easily cleaned and not become insect harborages.

Corrugated metal

Imholte[1] discusses the problems of using corrugated metal ceiling panels in production areas. Cavities formed by the corrugation at supporting beams are difficult to clean. Additionally, the joints at centers can harbor insects and are hard to keep sealed. Metal panels have the further disadvantage that because of the high heat transfer rate they can cause considerable condensation problems. Imholte[1] advises against the use of metal ceilings in production areas, but if they are to be used, he recommends use of smooth flat panels. Remember that condensation drips may contaminate ingredients, equipment or products, and increased ventilation can be costly and may not be fully effective.

Concrete

Imholte[1] recommends poured concrete slabs or ceiling panels, as they allow for suspension of pipework and other fittings. Formed-and-poured-in-place concrete ceilings are preferred for production areas of the plant since precast concrete panels require caulking. The risk with caulking is that vibration and repeated cycles of heating and cooling loosen the caulking material, which can then be a contamination hazard to food products.

Precast panels are suitable for warehouse, equipment maintenance and similar areas. All concrete surfaces should be made smooth by grinding or filling before treating with a coating to provide a smooth finish. Paint should only be applied to concrete after it has been conditioned (cured, UK) but care should be taken to avoid peeling and subsequent contamination of food materials. Epoxy paint is recommended by Troller[2]. Other materials include a sprayed-on plastic coating. This, however, will 'bubble' if water seeps in from above or behind the coating. In selecting ceilings remember the consequences of a change in use to a manufacturing area.

Warning about the use of gypsum plasterboard

Gypsum board ceilings are satisfactory in some warehouse applications and Imholte[1] discusses a ledge-free framing arrangement. *Do not use this material in production areas* as it has neither the strength to support fittings nor is it impervious to moisture. Plaster ceilings are only suitable for limited domestic applications, e.g. administrative areas, locker rooms, etc., as they are not impervious to moisture and are not durable.

Floors

Floors play an important part in plant operation, sanitation and costs. Wise selection of materials, and specification of the way in which they are to be used, is based on knowledge and understanding of operations which take place (or may take place in the future) in a particular area, as well as on the properties of the materials.

Ideally, floors should be:

- Impervious to spillages of product, use or spillage of cleaning materials and solutions, hot and cold water, disinfectants including chlorine, pesticides, lubricants, etc.
- Durable, i.e. resistant to impact or scuffing by containers, equipment, pipework, tools; damage by traffic, e.g. (fork) lift trucks; resistant to attack physically or chemically by products, cleaning materials, water, disinfectants including chlorine, pesticides, lubricants, etc.
- Slip resistant to personnel and traffic
- Non-toxic and non-tainting in use
- Easy to clean and disinfect
- Of good appearance, i.e. neat, clean and look well maintained
- Repairable as a portion or a section if damage or excessive wear has occurred

Remember that floors have to withstand use, cleaning and abuse. Cleanliness results from the ability to remove surface contamination easily and quickly. If this cannot be done quickly some contamination will be left. Contamination of any kind which penetrates into flooring materials becomes more difficult or impossible to remove and results in sanitation and/or operational problems.

Slope of floors

Floors should be sloped to avoid residual pools which are both microbiological and safety hazards. The usual rule is that floors should be sloped to drains or gutters at 1 in 40 to 1 in 60. It should be remembered that the slope needed for good drainage depends on the surface roughness. A rough surface tends to 'hold' water. However, safety and practical considerations for traffic, e.g. trolleys and bins, limit the slope. Katsuyama and Strachan[4] give 6 mm per 305 mm (¼ in per ft) and state that gutters and drains should be located at sufficiently close intervals so that the distance from high point to gutter or drain averages 3 m (10 ft) and never exceeds 4.6 m (15 ft). This allows expansion joints to be placed in the driest areas mid-way between gutters or drains.

In wet areas, and particularly in upper production floors, an impermeable membrane is required. Care should be taken to insure that this membrane remains intact when the surface flooring is repaired or replaced. Be aware that, in practice, this can present considerable difficulties.

It is advisable that process floors should be located above ground level or grade to reduce both risk of flooding and drainage problems. To make cleaning easier and quicker, areas where spillage is likely may be contained by curbing, and coved to 102 mm (4 in) radius corners at the floor/wall and other permanent junctures, e.g. column or equipment supports. Remember that the floor of the curbed area needs to be sloped to drain to prevent puddles.

Note, however, that in warehouse and storage areas the practice of using a concrete curb about 440 mm wide × 150 mm high (18 in × 6 in) to insure that pallets are not stacked close against the wall complicates the rodent control program. This is because rodents have poor sight and feel their way along walls by their whiskers. To the rodent, therefore, there are two walls – the curb and the real wall. Both will be used as runways which must therefore be trapped.

It is worth remembering that disruption caused by defective flooring may be very costly as well as inconvenient, and that it is difficult to make really satisfactory partial repairs. It is also true that defects relate more often to poor planning and preparatory work or lack of 'curing' time than to the finishing material.

Choice of materials

While the choice and application of flooring materials requires expert local advice, the following notes on some commonly used materials may be useful.

Plain concrete

This is commonly used and is very suitable for warehouse, storage, maintenance and some manufacturing floors. It is chemically attacked by acids, either from foods or produced by microbial action on sugars etc. in foods. Heavy traffic or running or dripping water will also cause deterioration. The use of dense mixes (low water content) and proper curing increases strength and makes the concrete more impervious. Surface-hardening compounds may tend to glaze and become slippery. Imholte[1] has an excellent section on concrete floors which is useful for the technologist.

Metal plate

These floors are used for very heavy traffic or where severe impact, mechanical abuse or thermal shock occurs (for example, in shipping areas and loading docks, in front of cannery sterilizing retorts, or places where heavy wheeled vehicles turn). Such floors are made by using square plates about 300 × 300 mm (12 × 12 in) which have a raised pattern for traction. These plates are usually bonded with an epoxy adhesive to a concrete sub-floor. The epoxy material must be formulated to absorb stresses resulting from different coefficients of expansion of metal and concrete. Chemical corrosion is the usual cause of deterioration.

Wood

Wood floors have the advantage that they are comfortable where operators stand for most of the day. They are used in warehouse or storage areas where there is only light traffic but should not be used in wet areas. In dry production areas great care must be taken to avoid wood splinters contaminating the product. Urethane sealers are often used to give a hard surface. Remember that under EC Directive 77/99, wood is not allowed within the manufacturing plant.

Vinyl

Vinyl or vinyl asbestos or asphalt tile or linoleum are not suitable in food processing or storage areas, as water or cleaning solutions seep between the tiles and loosen them. Food particles in the eroded spaces may create microbial or insect pest problems. Vinyl sheet with welded seams may be suitable for some applications.

Bituminous/asphalt

Bituminous/asphalt floors are used for interior railroad car loading and unloading docks as the pliable surface flexes with the movement of the train. However, steam, lubricants and other oils weaken the surface so it must be sealed to produce a smoother and easier-to-clean finish. Repairs are easily made using cold asphalt mixtures.

Chemical resistant

Chemical resistant is the best description for the types of floors used in processing areas. They require a sound sub-floor which has been properly prepared for the surface finish (or 'topping'). Knowledge, experience and care are needed in the:

- Analysis of the chemical, physical and thermal conditions which the floor must withstand
- Selection of the surface finish
- Preparation of the sub-floor
- Application of the surface finish

Chemical resistant floors are of two main types; monolithic surfacings, and bricks, pavers or tiles set in a bonding bed or laid over an impermeable membrane.

Monolithic surfacings

Monolithic surfacings consist of a layer or 'topping' of uniform composition bonded directly to the sub-floor which is often concrete. There are a number of materials of different chemical types although epoxy resins, polystyrene and polyurethanes are commonly used. Some epoxy materials have a perceptible odor and these should obviously be avoided in ingredient or food contact areas.

Diagrams showing the layers and other details are given in British Standard BS 8204 Parts 1 and 2 for *in-situ* floorings. (See Further Reading.)

There are two main types of epoxy resins:

- *Amide* hardened – these are easier to lay, tolerate wet conditions during application and are decorative. Because of their comparatively low chemical resistance they are only suitable for use in non-food handling areas, e.g. offices.

- *Amine* hardened – these need expert care in selection and application to achieve a good floor. They shrink while curing, which causes stresses that need stress relief joints at a maximum of 6.1 m (20 ft). These joints are illustrated in Figure 4.14.

Monolithic surfacings may slowly absorb liquids, which, if acid, will eventually attack the concrete sub-floor. They are therefore not recommended for use under continuously wet conditions where the pH is less than 6.5. Epoxy resins, like other materials, have thermal limits which should not be exceeded. It is prudent to seek expert local advice on the monolithic surfacings which are currently available. Where the proposed use of a material approaches or is near the limits for that material, it is wise to ask the supplier if there is a user who has experience of these conditions. If so, ask to speak with the user and see whether their experience is valid for your application.

Remember that monolithic floors are unacceptably damaged by weld-metal splashes, so must be laid following any welding work on structure or equipment.

In the choice of monolithic surfacings, the chemical resistance to likely food products, cleaning and disinfection materials together with their temperatures over the expected operating range should be considered. Table 4.2 from Beauchner and Reinert[7] gives an idea of the properties of commonly used types of materials. For properties of newer materials see Cattell in the Further Reading.

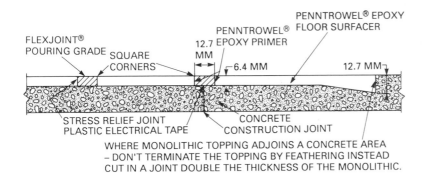

Figure 4.14 Typical flexjoint installation in monolithic topping (courtesy of Atochem North America Inc., formerly Pennwalt Corp.)

56

Table 4.2. Chemical resistance of resinous flooring materials (Beauchner and Reinert[7]

Chemical	Epoxy	Furan	Phenolic	Polyester
Acetic acid up to 10%	yes	yes	yes	yes
Acetic acid glacial	no	yes	yes	yes
Ammonium hydroxide	yes	yes	no	no
Animal oils	yes	yes	yes	yes
Bakery products	yes	yes	yes	yes
Beer	yes	yes	yes	yes
Boric acid	yes	yes	yes	yes
Butter	yes	yes	yes	yes
Butyric acid	no	yes	no	no
Casein	yes	yes	yes	yes
Cheese, all	yes	yes	yes	yes
Chlorine water	no	no	no	yes
Chromic acid up to 5%	no	no	no	yes
Cider	yes	yes	yes	yes
Citric acid	yes	yes	yes	yes
Citrus fruits	yes	yes	yes	yes
Coffee	yes	yes	yes	yes
Corn syrup	yes	yes	yes	yes
Egg yolk	yes	yes	yes	yes
Ethyl alcohol	yes	yes	yes	yes
Formic acid	no	yes	yes	yes
Fruit extracts	yes	yes	yes	yes
Glucose	yes	yes	yes	yes
Horse radish	no	yes	yes	yes
Hydrochloric acid	yes	yes	yes	yes
Ice cream	yes	yes	yes	yes
Jams and jellies	yes	yes	yes	yes
Lactic acid	yes	yes	yes	yes
Ketchup	yes	yes	yes	yes
Margarine	yes	yes	yes	yes
Malic acid	no	yes	yes	yes
Malt	yes	yes	yes	yes
Malt liquors	yes	yes	yes	yes
Milk, fresh	yes	yes	yes	yes
Milk, sour	yes	yes	yes	yes
Mineral oil	yes	yes	yes	yes
Muriatic acid	yes	yes	yes	yes
Molasses	yes	yes	yes	yes
Mustard	yes	yes	yes	yes
Nitric acid up to 5%	no	no	no	yes
Oleic acid	no	yes	yes	yes
Olive oil	yes	yes	yes	yes
Pectin	yes	yes	yes	yes
Phosphoric acid	yes	yes	yes	yes
Pickles	yes	yes	yes	yes
Potassium hydroxide up to 30%	yes	yes	no	no
Salad oils	yes	yes	yes	yes
Shortening	yes	yes	yes	yes
Sodium bicarbonate	yes	yes	yes	yes
Sodium chloride	yes	yes	yes	yes
Sodium hydroxide up to 30%	yes	yes	no	no
Sodium hydroxide 30% and over	no	yes	no	no
Sodium hypochlorite up to 3%	yes	yes	no	yes
Soft drinks	yes	yes	yes	yes
Soft drink concentrates	yes	yes	yes	yes
Soups	yes	yes	yes	yes
Soya oil	yes	yes	yes	yes
Stearic acid	yes	yes	yes	yes
Sugar	yes	yes	yes	yes
Sulfuric acid up to 50%	yes	yes	yes	yes

Table 4.2. Continued

Chemical	Epoxy	Furan	Phenolic	Polyester
Syrup	yes	yes	yes	yes
Tannic acid	yes	yes	yes	yes
Tartaric acid	yes	yes	yes	yes
Tea	yes	yes	yes	yes
Trichloroethylene	no	yes	yes	no
Trisodium phosphate	yes	yes	yes	yes
Vinegar	yes	yes	yes	yes
Yeast	yes	yes	yes	yes

Typical physical properties of resinous materials

Property	Epoxy	Furan	Phenolic	Polyester
Tensile strength, psi	2 000	1 000	1 000	2 000
Compressive strength, psi	10 000	8 000	7 000	9 000
Bond strength, psi	250	150[a]	250	250
Modulus of rupture, psi	6 000	3 500	2 500	4 000

Note: All information is based on 'Intermittent splash and fume' service. For specific information contact the mortar manufacturer.

[a]Higher bond strength furans are available.

The conclusion is, therefore, that monolithic surfaces are best:

- In dry or only intermittently wetted areas which have a good slope to prevent puddle formation.
- Where there is adequate resistance to the foods and the cleaning and disinfecting materials used.
- Where knowledge, skill and care have been taken in preparation of the sub-floor and in the application of the surfacing. However do not underestimate the considerable degree of skill needed to lay a monolithic surface to prevent 'ponding', i.e. to avoid puddles forming when liquids are used or spilled.

Brick/paver/tile

Brick/paver/tile floors, states Imholte[1], 'remain popular for the wet processing areas of a food plant. The key to successful installation lies with selecting the proper bed-setting cements and grouting mixtures'. By usage in the USA bricks or pavers are thicker than 38 mm (1½ in) and the brickmason (bricklayer, UK) 'butters' the brick and lays it in position. Tiles are thinner than bricks and the tilesetter spreads a leveling and bonding bed and sets the tile in it. There are two types of tile used in the UK, quarry tiles and ceramic tiles. Quarry tiles are cheaper than ceramic tiles. They are easy to clean and have:

- A slip resistant surface
- A good resistance to impact
- Usually excellent resistance to chemical attack

They are also noisy to walk over. Ceramic tiles have similar properties to quarry tiles and are slip resistant because of materials added during manufacture. They offer an attractive and durable finish.

The ways in which bricks/pavers/tiles are set into the bed are described by Beauchner and Reinert[8] as resin bed, sand–cement bed and membrane construction and shown in diagrammatic form in Figure 4.15. Sheppard, in Katsuyama and Strachan[4], gives a detailed but slightly different account of the alternative methods of installation.

Only the membrane construction gives a liquid-tight barrier under chemical-resistant flooring and the plastic nature of asphalt allows for slight movement, although normal building expansion and contraction requires the installation of 'control' or expansion joints. Note that because of slight movement or flexing, bricks/pavers/tiles should be at least 24.5 mm (1 in) or 30 mm (1³⁄₁₆ in) thick to have sufficient strength. This is the older US practice. In Europe, somewhat less is considered sufficient, and 20 mm (¹³⁄₁₆ in) thick tiles are used for heavy duty areas.

Membrane construction is normally used for very wet areas where high chemical resistance is needed and heavy traffic occurs.

Floor openings

Holes will need to be cut in floors for pipes, ducts or other equipment to pass through. These should be sleeved or curbed as appropriate.

Better designs for wet manufacturing areas where the floor is likely to be frequently hosed are shown in Figure 4.16 from Campden Technical Manual 17[9], Section 5, and in Figure 4.17 from Imholte[1]. Imholte[1] also gives a sketch for larger openings in Figure 4.18.

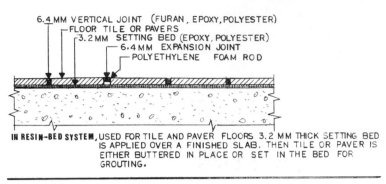

IN RESIN-BED SYSTEM, USED FOR TILE AND PAVER FLOORS 3.2 MM THICK SETTING BED IS APPLIED OVER A FINISHED SLAB. THEN TILE OR PAVER IS EITHER BUTTERED IN PLACE OR SET IN THE BED FOR GROUTING.

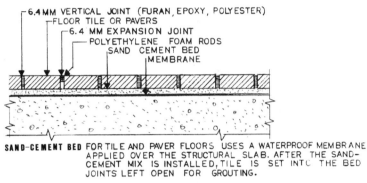

SAND-CEMENT BED FOR TILE AND PAVER FLOORS USES A WATERPROOF MEMBRANE APPLIED OVER THE STRUCTURAL SLAB. AFTER THE SAND-CEMENT MIX IS INSTALLED, TILE IS SET INTO THE BED JOINTS LEFT OPEN FOR GROUTING.

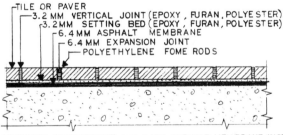

MEMBRANE CONSTRUCTION A 6.4 MM ASPHALT MEMBRANE, INSTALLED OVER THE FINISHED SLAB, CAN BE REINFORCED WITH ASPHALT-IMPREGNATED GLASS CLOTH A RESIN-CEMENT BED IS APPLIED TO THE MEMBRANE, AND BRICK OR PAVER IS BUTTERED AND PAVED INTO THE BED.

Figure 4.15 Design of beds and membrane (from Beauchner and Reinert[8]

Drains

Drains may be run to remove either process waste *or* sanitary (domestic) waste. Process waste is from the plant operations and includes e.g. spilt liquids or small solids such as vegetable pieces. Remember, solid waste adds to BOD and COD and is better removed before it reaches the drain.

Capacity

The drains and drainage system should be capable of effective operation at the maximum load generated at the plant which may well be during cleaning and simultaneously under storm conditions. This capacity avoids 'back-up' and consequent contamination of working areas. Note that local legislative requirements or codes may specify particular design, construction or installation, although these are unlikely to have been drafted with reference to food plant needs.

Sanitary sewers from toilets etc. which contain fecal material should *not* be routed through food process areas or storage areas. Plant drainage and sanitary sewers should not be cross-connected. The outfall arrangement of plant drains, sanitary sewers

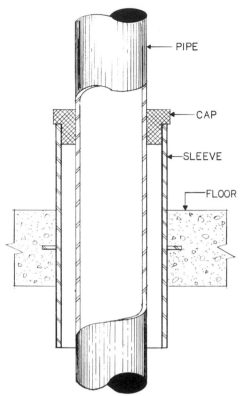

Figure 4.16 Design for pipe sleeve (from Campden Technical Manual 17[9]

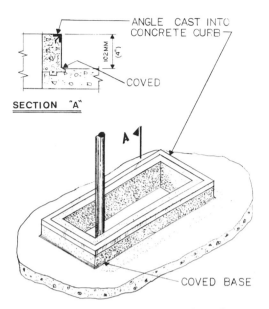

Figure 4.18 Floor openings for pipework, equipment, ducting, etc. should be curbed – particularly in wet manufacturing areas (from Imholte[1])

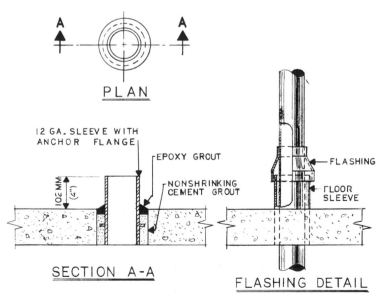

Figure 4.17 Alternative sleeving for floor pass-through (from Imholte[1])

and storm sewers should always be such as to prevent any hazardous material backing-up and contaminating the plant areas.

Overhead drainage lines should be carefully routed to minimize contamination if and when they leak in food process or storage areas. In planning always make the *worst case* assumptions.

Unblocking drains and grease interception

Be sure to fit sufficient clean-out (unblocking) plugs on drainage lines. Drains, sooner or later, will block and need rodding, high-pressure hosing or other remedial treatment. The clean-out plugs must be sited so that they are not a contamination risk to food ingredients or products. Drain lines must be vented as required to prevent the back-up of sewer gases into plant areas. Vents must *not* be near fresh air intakes.

Where required, drains can be fitted with large, removable buckets to trap small quantities of sediment or solid waste. These traps should be positioned well away from processing equipment.

Large quantities of solids will need to be removed by a screen. Interceptors can be fitted to remove grease or oil. They should be installed where they may be easily maintained and cleaned. Flow control valves are fitted upstream of an interceptor so that the effluent enters without turbulence. By passage over a series of separator baffles, oil, grease and fats rise to the top of the interceptor and accumulate until removed. The waste water flows on into the drainage system. Imholte[1] shows a typical arrangement on his page 94.

Adequate drainage is essential throughout the food handling area except for freezer rooms and floor drains in most dry storage areas. Remember, if water is used in processing or cleaning, drainage is required.

Drains may be considered to be of two types; closed floor drains (pipe drains, UK) and open-trench or gutter drains. Trench drains are preferred where large amounts of water are used, where solid product, e.g. vegetable pieces, or packaging material such as can ends are expected or where frequent washdowns or cleaning is required. They are also, when properly constructed, easy to keep clean.

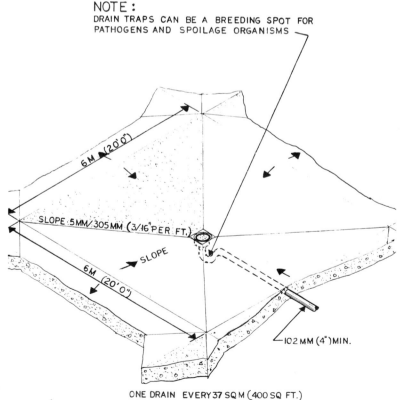

NOTE :
DRAIN TRAPS CAN BE A BREEDING SPOT FOR PATHOGENS AND SPOILAGE ORGANISMS

6 M (20'0")

SLOPE:5MM/305MM (3/16"PER FT.)

SLOPE

6 M (20'0")

102 MM (4")MIN.

ONE DRAIN EVERY 37 SQ M (400 SQ FT.)

3MM (1/8") DROP PER 305MM (1'-0")
TO GIVE 610 MM PER SEC. (2 FT. PER SEC.)
RUN-OFF (TAKE AWAY)

Figure 4.19 Design of floor drains – adequate installation can assure proper drainage (from Imholte[1])

Floor drains

Floor drains should be set in the floor so that the top of the grating is 3 mm (⅛ in) below the level of the floor. Note that, in contrast, gratings for channel drains are set slightly above floor level. Drain holes or weep holes can be installed above the drain seepage pan to allow any liquid that seeps alongside of the drain to enter the drain. This is shown in Beauchner and Reinert[8].

It is important to insure that floor drains have adequate capacity for both existing and any future proposed use of the area and that adequate plans have been made for unblocking the drains when this is necessary.

Floor drains should have a minimum water seal of 46 mm (3 in) but full 'S' traps, bell traps and crown-vented traps should *not* be used. A removable secondary strainer prevents the entry of rodents and cockroaches. A large removable bucket may be fitted at a convenient point where small quantities of sediment or solid waste are expected. Usually, unless extremely wet conditions are likely, there should be at least one floor drain every 37 m² (400 ft²) of floor. Floors should be sloped to drain as shown in Figure 4.19. For other information, see Cattell in the Further Reading.

Gutter or trench (channel) drains

Gutter or trench (channel, UK) drains should have smooth vertical walls, rounded or coved bottoms (see Figure 4.20) and a minimum slope of 1 in 100. Usual sizes are 152 to 304 mm (6 to 12 in) wide and 152 to 203 mm (6 to 8 in) deep. Deeper gutters than this may be difficult to clean. The pipe drain take-away must be adequate to clear quickly the maximum volume of water used, so will need to be more than 102 mm (4 in).

Gutters should be covered with a strong open metal grating rugged enough to withstand any unavoidable heavy traffic including (fork) lift trucks which may be used to transport equipment as well as foods in the area. These gratings should be in short

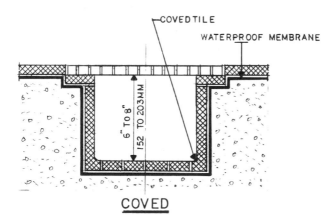

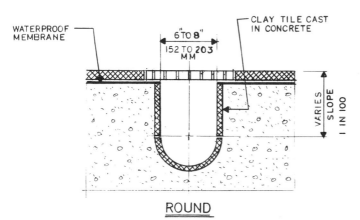

Figure 4.20 Trench drains – good design reduces sanitation problems and facilitates cleaning (from Imholte[1])

sections of less than 1 m (39 in), so that they can be easily lifted for cleaning. To avoid solids being jammed in the openings, the holes should be wider at the bottom than the top. The gratings should be set 3 mm (⅛ in) above the surrounding floor level to minimize damage to the edges of the gutter. The shoulders of the gutter may be protected by Z angle irons as described by Katsuyama and Strachan[4].

Gratings should rest on ridges or buttons to allow water and small particles to drain away. They should *not* be supported on rods or studs set into the gutter wall.

There is an interesting design idea by AFIS(3), who suggest a steeper angle of discharge for lateral gutters into a main gutter. This gives an increased speed to water flowing at the end of the lateral gutter which therefore carries debris into the main channel more readily.

Stairs

It is sometimes forgotten that contamination, e.g. mud, cigarette butts (ends, UK), can be brought into the food plant on footwear. Stairs should not only be safe to use but be easy to clean, and designed with thought given to such contamination risks. Troller[2] states that 'closed-tread stairs with single support post either attached to a base plate or preferably embedded in concrete, are acceptable for product areas [Figure 4.21]. Stringers should be of steel plate or tubular construction. Channels are not acceptable.'

Where stairs are used to give access to a platform (walkway) above a product stream the Campden RA Technical Manual 7[10] gives a sound design. This is shown as Figure 4.22. here, which is their 'Figure 72 and text'.

Stairwells (staircases, UK) should be well lit and kept clean. Protected light fittings are not necessary in these areas.

Elevators

The most significant sanitary aspect of elevators is the well or 'foot' which is at the base of the shaft. It is often poorly lit, rather inaccessible and harbors debris which can feed pests. Inspection at least twice a year and regular cleaning are advised by Troller[2]. Although unlikely, accumulation of paper and other trash could be a serious fire hazard which might result in injury or death.

Lighting

It is axiomatic that adequate lighting is essential for effective work. Good lighting will:

- Be sufficient for the work or function.
- Be of suitable color where this is important, e.g. in tasting rooms, for color matching, or inspection areas.

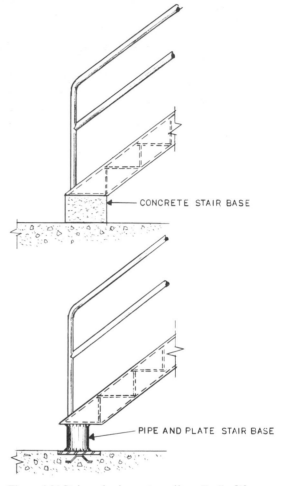

Figure 4.21 Stair anchoring systems (from Troller[2]

- Provide suitable contrast between the work and the background – especially important for fine work.
- Give minimal glare either from the light units or by reflection from other surfaces.
- Give soft shadows, if required, but not deep shadowing.
- Have shaded light sources if they have high surface brightness.
- Come from fixtures (fittings, UK) designed and installed to:
  - avoid contamination of food or equipment in the event of breakage.
  - be insect- and dust-tight.
  - be easy to clean and maintain.

Some recommended lighting levels

Published tables give a range of values and do not always take into account factors such as color, glare

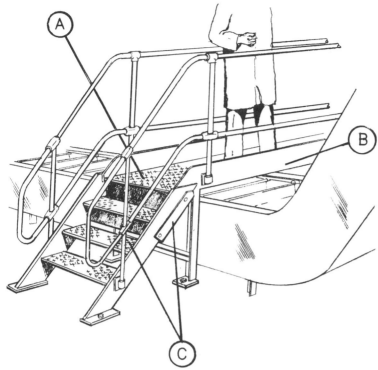

Figure 4.22 Design for stair and platform (walkway) over open food (from Campden Technical Manual 7[10]. A. Decking should be made from solid plate, i.e. chequer plate which has a raised anti-slip surface. B. Kick-plate minimises the chances of dirt being transferred from footwear. Where possible, the kick-plates and decking should be made as a one-piece construction. C. Staircase risers above the level of the product flow should be encased to prevent dirt being transferred into the product. Also steps should be made of the same anti-slip material as the decking

and shadowing. Expert advice should be sought and taken for specific applications. Some examples of recommendations are given in Table 4.3.

As can be seen from Table 4.3, lighting levels have generally increased over the last three decades. This is due to technical developments and increased efficiencies, especially with fluorescent lamps, which are used where there is economic justification and no technical objection. They are not recommended where they will be frequently switched on and off, as this greatly reduces lamp life, nor for use in cold or freezing conditions unless suitable precautions are taken. It should also be noted that differences in color composition of light make it inadvisable to use mixtures of certain lamps, e.g. discharge and metal filament.

Light loss

Lighting levels will be at the maximum when an area has been equipped with new lighting and been recently decorated or cleaned. Imholte[1] states that

within six months as much as 50% of the original output may be lost. This is due to:

- Normal depreciation   20%
- Dirt on walls and ceilings   10%
- Dirt on fixture reflection   20%
- *Total*:   50%

Because of this, lighting systems should be oversized by 30% to 40% and effective cleaning (using detergents) and maintenance schedules established and used.

Design of fixtures

Good design features are shown in Figure 4.23. which help to minimize dust settling on the fixture and make it easy to clean. It is an advantage to have hanging fixtures which can easily be unhooked and to have the electrical supply through a plug so that the unit is easily removable for cleaning or maintenance. This is important where lights are above open foodstuffs. Alternatively, installing

Table 4.3 Recommended lighting levels in lux (foot-candle values shown in parentheses)

Ingredient area	AFIS[3] 1952 Minimum		Imholte[1] 1984 Maintained level		Troller[2] 1983 Minimum		HEINZ (UK) 1988 Data Illumination		Glare index
Dock	–		–		220	(20)	200	(18)	–
Receiving	110	(10)	220–320	(20–30)	–		200–300	(18–28)	25
Warehouse	–		220–320	(20–30)	220	(20)	200–300	(18–28)	25
Bulk Storage	–		320–440	(30–40)	–		–		
Product Preparation	220	(20)	–		–		350–500	(33–47)	25
Product Inspection	220	(20)	–		–				
Inspection Tables	220	(20)	1184–1340	(110–130)	540–750	(50–70)	–		
Process Department	220	(20)	590–700	(55–65)	540–645	(50–60)	400–500	(37–47)	25
Packaging	–		750–860	(70–80)	220–320	(20–30)	–		
Food Displays	–		–		320	(30)	–		
Warehouse Storage	55	(5)	–		220	(20)	200–300	(18–28)	25
Warehouse Shipping	110	(10)	220–320	(20–30)	220	(20)	–		
Maintenance Areas	320	(30)	750–860	(70–80)	540–645	(50–60)	–		
Building Entrance	–		–		110–320	(10–30)			
Offices:									
Intermittent reading and writing	320	(30)	–		–		500	(47)	19
Prolonged close work	540	(50)	645–970	(60–90)	540	(50)	500	(47)	19
Laboratory:									
Ordinary Tasks	320	(30)	–		540–645	(50–60)	–		
Product Exam.	540	(50)	–		540–645	(50–60)	–		
Locker and Restroom	110	(10)	320–540	(30–50)	220–320	(20–30)	150	(14)	–
Cafeteria Kitchen	220	(20)	–		320	(30)	300–500	(28–47)	22
Cafeteria Dining	110	(10)	485–540	(45–50)	110–320	(10–30)	300–500	(28–47)	22
Stairways, elevator	110	(10)	–		–		–		
Corridors	55	(5)	–		–		150	(14)	22
First Aid Room	320	(30)	–		–		–		

Notes
1. Units given as lux first, foot-candles in parentheses. The conversion factor of 10.764 lux = 1 foot-candle has been approximated in this table.
2. Different sources used different description and categories – for this table the nearest category has been chosen.
3. The UK DHSS Code of Practice No. 10 (1981) gives minimum recommended levels of 540 lux (50 foot-candles) at inspection points and 220 lux (20.5 foot-candles) in other areas and is an example of an insufficiently detailed recommendation.
4. Glare Index. This is calculated from fixture mounting height, room dimensions, view of the fixtures endways/sideways, view of the room etc. A value of 19 is a lesser glare than one of 25.

SECTION

Figure 4.23 Good design features of light fitting

lighting units flush with the ceiling minimizes dust settling in production departments.

Glare and contrast

Glare can cause discomfort, increase risk of accidents, interfere with vision, and lead to eye fatigue. Direct glare is caused by excessive brightness or brightness contrast in the line of vision, e.g. a bright unshielded lamp against a dark ceiling. Reflected glare is caused by light reflected from walls, ceilings, machines, wet floors or belts, table tops etc. within the line of vision. It is often worse than direct glare because it is so close to the line of vision that the eye cannot avoid it. Glare is thus an important design consideration.

Heavy contrast in lighting levels is undesirable because the eye has difficulty in adjusting to the different levels and becomes tired, with consequent eyestrain. Strong shadows are an example of such a contrast.

Emergency lighting

Emergency lighting is required in an area if the normal lighting system should fail. This unlikely event may be caused, for example, by:

- Main supply failure (power cut, UK), e.g. storm damage to power line
- Local failure, e.g. power cable cut by drill during building work
- Sub-station malfunction
- Lighting circuit fuses blown

Emergency lighting is of two types:

Escape lighting – this both shows and lights escape routes and enables orderly evacuation as well as assisting emergency services by locating firefighting and first-aid equipment.

Standby (auxiliary) lighting – this enables normal activities to continue. It may be planned so that it is switched on if the failure is likely to be prolonged in specific areas.

For further information consult specialists and publications such as those of the Illuminating Engineering Society, York House, Westminster Bridge Road, London SW1, UK.

Ventilation

Imholte[1] describes ventilation as the exchange of unwanted air for a controlled and conditioned supply of healthy fresh air. There are two aspects of ventilation:

- General air supply within the plant
- Specific systems for local, specialized application

General air supply

General ventilation performs three functions. It:

- Removes obnoxious odors, heat and moisture from plant areas. It will also, for example, eliminate the products of combustion from LPG (fork) lift truck operation.
- Introduces fresh air, free of any foreign odors and product contaminants. Rates of air change to achieve this vary according to circumstances. The USDA Handbook No. 191[11] recommends a design of six changes per hour. Hayes[6] gives 1.5 for stores and offices and 20 changes per hour in cooking areas. Imholte[1] suggests a 'few' to 60 changes per hour. Clearly this is an area where professional advice should be sought from those with both knowledge and experience. Imholte[1] adds that a strong case can be made for the design where food manufacturing areas are maintained under a slightly positive air pressure.
- Provides a comfortable environment for employees. Hayes[6] quotes temperatures of 21° to 22°C (69.8° to 71.7°F) for sedentary work (UK) although 24°C (75°F) might be regarded as comfortable in some other places. He quotes 13° to 14°C (55°C to 57°F) for manual work. Usually, optimal humidity is given as being within the range of 30 to 70%.

Specific air supply

Specific ventilation or air-conditioning provides a safe local environment for a high-risk product such as dried formula milk for babies or portioning and packaging of meat delicatessen or other chilled neutral products which receive no further processing before being consumed. Such requirements vary according to need, e.g. Class III air for the environment but Class I air for transport of dried formula milk. *Warning* Instances are known where a high extraction rate and an insufficient input rate of conditioned air have resulted in a partial vacuum being formed. This was sufficient to draw unwanted air from preparation and storage areas through double doors intended to act as an effective barrier.

Note on Legionnaires' disease

This is a type of pneumonia caused by a bacterium – *Legionella pneumophila* – that can almost universally be found in cold or hot water systems. The organism grows best in tepid water. It is only when the organism is carried by fine droplets as an aerosol that it will enter the lungs. Aerosols may be generated by splashes from cooling systems, taps, flushing toilets, shower baths, etc. Fortunately, however, only a very small proportion of the organisms appear to be pathogenic. Typically, outbreaks have occurred when the aerosol produced by an evaporative cooling tower entered the intake of an air-conditioning system and thus was spread to parts of a building. This can happen intermittently when specific and unusual wind speeds and directions occur. Note that chlorinated cooling water systems – for example in a cannery – should not be a risk as the organism is sensitive to chlorine provided that it is not protected by a film of growth, or 'slime', on the tower surfaces. This is why chlorination before cooling is, in principle, better practice than chlorination after cooling.

The DHSS Code of Practice[12] is an up-to-date, comprehensive document which applies generally and not just to Health Care premises. The key to control is to:

• Avoid dead-ends or places where water may be stagnant in hot and cold water systems (in principle, this includes fully charged firefighting lines, but to date, no cases are known from this cause).
• Clean and disinfect heat exchangers and cooling towers.

Additionally:

• Hot and cold water lines should be insulated, as needed, to maintain cold water below 20°C (68°F) and hot water above 62°C (143°F).
• Shower outlets and personnel washing outlets should be capable of achieving 50°C (122°F) within one minute of first running the water, in order to minimize the risk of generating aerosols of contaminated water.

In a system in which process water is heated to 50° to 55°C (122° to 131°F) and stored in a large tank there may be the possibility of some of the water cooling and remaining static (or stagnant) for some days. These conditions could allow growth of *L. pneumophila*. If this water is used for cleaning with hoses the aerosols formed may be contaminated. Heating the entire tank to at least 70°C (158°F) at least once during the week would kill *L. pneumophila*.

A less serious but occasionally troublesome problem is the 'sick building syndrome', where personnel suffer minor ailments, e.g. epidemics of common cold-like illness. This may be related to humidifiers. Those that use steam rather than water appear to be trouble-free in this respect. Remember that aerosol generation is more common than is usually thought. The CIBSE Technical Memoranda TM13 [13] states 'Legionnaire's disease has been linked with the accidental spraying of contaminated water during the commissioning period of such [firefighting] sprinklers.' It is obvious that any system designed to distribute and recirculate conditioned air will also distribute aerosols.

Insulation

Imholte's[1] comments on duct insulation – 'Insects can live in almost all types of insulating materials and must be sealed out by the insulation jacket' – apply widely. Foam insulation may also trap and hold water which, if it has contact with plywood, will cause rotting. Instances are known where rockwool insulation has become wet which resulted in high microbial counts that caused spoilage of product. For insulation of pipework see Section 4.E on Utilities. Remember that there may be local regulatory requirements for insulation of hot or very cold lines. An example of this is on steam lines of more than a specified diameter.

References

1. Imholte, T. J. (1984) *Engineering for Food Safety and Sanitation*. Crystal, MINN: The Technical Institute of Food Safety
2. Troller, J. A. (1983) *Sanitation in Food Processing*, New York: Academic Press
3. AFIS (1952) *Sanitation for the Food-preservation Industries*. New York: McGraw-Hill
4. Katsuyama, A. M. and Strachan, J. P. (eds) (1980) *Principles of Food Processing Sanitation*. Washington, DC: The Food Processors Institute
5. Robson, J. N. and Barnes, G. (1980) Plant design to minimize non-microbial contamination. In *Hygienic Design and Operation of Food Plant*. Edited by R. Jowitt. pp. 121–135. Chichester: Ellis Horwood
6. Hayes, P. R. (1985) *Food Microbiology and Hygiene*. London/New York: Elsevier Applied Science
7. Beauchner, F. R. and Reinert, D. G. (1972) How to select sanitary flooring Pt.1. *Food Engineering*, **44**, (9), 129–134. A Chilton publication
8. Beauchner, F. R. and Reinert, D. G. (1972) How to select sanitary flooring, Pt. 2. *Food Engineering*, **44**, (10), 120–122 and 126. A Chilton publication.
9. Thorpe, R. H. and Barker, P. M. (1987) Technical Manual 17. *Hygienic Design of Liquid Handling Equipment for the Food Industry*, The Campden Food Preservation Research Association, Chipping Campden, Gloucestershire GL55 6LD
10. Campden R. A. (1983) Technical Manual No. 7. *Hygienic design of food processing equipment*, The Campden Food Preservation Research Association, Chipping Campden, Gloucestershire GL55 6LD.
11. USDA (1970) *US Inspected Meatpacking Plants, a Guide to Construction, Equipment, Layout*, Agriculture Handbook No. 191, Washington, DC: US Government Printing Office
12. DHSS and the Welsh Office (1988) *The Control of Legionellae in Health Care Premises – A Code of Practice*. London: HMSO
13. Anon (1987) *Minimizing the Risk of Legionnaires' Disease*, Technical Memorandum TM13, The Chartered Institution of Building Services Engineers, Delta House, 222 Balham High Road, London SW12 9BS

Further Reading

Legionella Control Measures, Health and Safety Executive (1987), *Legionnaires Disease*, Guidance Note EH48, London: HMSO

DHSS (1989) *Report of the Expert Advisory Committee on Biocides*, London: HMSO

Construction information

Miller, H. (ed) (1976) *Guide to the Choice of Wall and Floor Surfacing Materials – A Cost in Use Approach*, London: Hutchinson Benham

Cattell, D. (1988) *Specialist floor finishes: design and installation.* Glasgow and London: Blackie (Note this book includes a chemical resistance table)

BSI (1987) *BS8204: Part 1. In-situ floorings Part 1. Code of Practice for Concrete Bases and Screeds to Receive in-Situ Floorings*. London: British Standards Institution

BSI (1987) *BS 8204: Part 2. In-situ floorings Part 2. Code of Practice for Concrete Wearing Surfaces*. London: British Standards Institution

The Federation of Resin Formulators and Applicators Limited publishes application guides of which No. 4 is *Flooring Guide – Synthetic Resin Floor Screeds* and no. 6 is *Flooring Guide – Polymer Flooring*. The address is: The Secretary, The Federation of Resin Formulators and Applicators Limited, 16 Courtmore Avenue, Fleet, Aldershot, Hampshire GU13 9UF, UK

Appendix – Rodent proofing

Introduction

Every processing plant must have an effective and comprehensive pest control program. Rodents are pests of considerable Public Health and economic significance and it is therefore important to discourage their entry. Because they are attracted by food and harborage (shelter), good housekeeping practices go hand-in-hand with rodent proofing measures designed to deter entry into the plant.

Rodent behavior, especially that of rats, can seem to be almost intelligent in the human sense but is rationally explainable on the basis that they are very shy or secretive, wary animals which adapt rapidly to changes in their environment and have considerable physical capabilities. Remember that rodents must gnaw, so they will attack structural materials such as wood, plastic, cinder blocks and aluminum as well as food.

This appendix provides some starting points for further detailed consideration of the subject of your particular proofing needs.

Behavior

Rodents will:

- Climb vertical and horizontal wires, ropes, cables, pipes and also trees and rough exterior walls.

- Enter through small holes – 6 mm (¼ in) mice; 12.5 mm (½ in) rats.
- Jump 900 mm (3 ft) vertically and 1200 mm (4 ft) horizontally (rats). Mice are not so agile.
- Drop without being injured – about 2.4 m (8 ft) for mice and 15.2 m (50 ft) for rats.
- Swim well (rats, especially the Norway or sewer rat).
- Burrow in earth. (Norway rats will burrow 1200 mm (4 ft) vertically. However, if they meet with an impenetrable obstruction, they will cease burrowing that tunnel.)

Materials

A variety of materials may be used to prevent the free movement of rodents. They must be resistant to gnawing and be maintained in good order as necessary. Examples of 'traditional' materials are as follows:

Cement mortar; a 1:3 mixture or richer. May be mixed with broken glass.
Concrete; a 1:2:4 mixture or richer
Sheet metal; galvanized – 26 gauge or heavier
Perforated metal; rust resistant – 24 gauge or heavier. The diameter of perforations not to exceed 6 mm (¼ in)
Expanded metal; rust resistant – 28 gauge or heavier – openings not to exceed 6 mm (¼ in)
Iron (steel) grill; equivalent in weight to above – slots not to exceed 6 mm (¼ in)
Hardware cloth; galvanized or rust resistant – openings not to exceed 6 mm (¼ in)

Explanation of metal gauges:

19 gauge is 1.06 mm (or 0.0418 in)
24 gauge is 0.61 mm (or 0.0239 in)
26 gauge is 0.45 mm (or 0.0179 in)
28 gauge is 0.38 mm (or 0.0149 in)

Applications

These must be designed to suit the individual circumstances. The following examples give an idea of ways in which rodent proofing has been applied:

Foundations – Make use of the fact that burrowing rodents meeting with an impenetrable obstruction abandon the tunnel. A curtain wall 600 mm (2 ft) deep with an outward lip of 300 mm (1 ft) at the base will provide a suitable obstruction. This lip should be made of 24 gauge galvanized sheet metal or concrete, as shown in Figure 4.24. Corrugated siding (corrugated iron, UK) – This is not a desirable material but may be found in 'agricultural' storage sheds. Because the corrugations provide easy entry for rodents, this entry needs

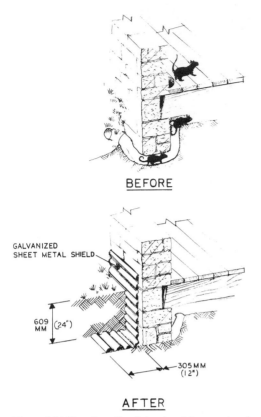

BEFORE

GALVANIZED
SHEET METAL SHIELD

609 MM (24")

305 MM (12")

AFTER

Figure 4.24 Curtain walls of sheet metal prevent rodent burrows under building foundations (from AFIS[3])

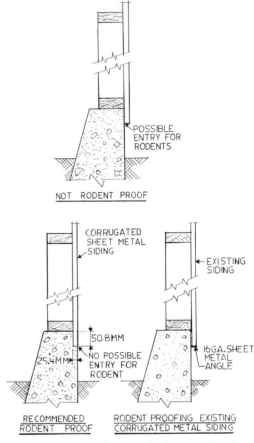

POSSIBLE
ENTRY FOR
RODENTS

NOT RODENT PROOF

CORRUGATED
SHEET METAL
SIDING

EXISTING
SIDING

50.8MM

25.4MM

NO POSSIBLE
ENTRY FOR
RODENT

16GA.SHEET
METAL
ANGLE

RECOMMENDED
RODENT PROOF

RODENT PROOFING EXISTING
CORRUGATED METAL SIDING

Figure 4.25 Suggested means of rodent-proofing the openings between foundation and corrugated metal siding (Kabak Co., San José, California, in AFIS[3])

to be stopped. A suggested means of doing this at a juncture between foundation and siding is shown in Figure 4.25.

Wall construction – New walls should be solid and resistant to rodent attack. Some rodent proofing measures can be applied to older double-wall construction as shown in Figure 4.26.

Wall openings – These may be needed for utility pipes and should be rodent proofed. One way of doing this is shown in Figure 4.27 using 6 mm (¼ in) hardware cloth around steam pipes.

Pipes and wires – Inevitably some pipes or wires will need to be attached to the walls. Several types of sheet-metal rat guards are shown in Figure 4.28.

Doors – There may seem little value in rodent proofing doors which may need to be opened for periods during a shift. Rodents, however, are unlikely to enter while there is activity around the doors especially during daytime, or even at night if the lighting is bright. Doors need to be closed when not in use and in addition should be rodent proofed. A method of rodent proofing an older type of sliding door is shown in Figure 4.29.

Roofs – Roof openings are usually considered to need *bird* proofing which means blocking access holes with wire or cement. Openings which are accessible to rats from pipes or wires also need treatment. Metal collars can be fitted to the pipes or wires and the openings covered with 6 mm (¼ in) hardware cloth.

Abandoned pipes or sewers – These should be capped or blocked with concrete.

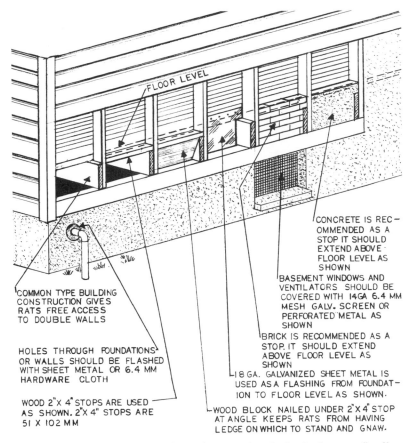

FLOOR LEVEL

CONCRETE IS REC-
OMMENDED AS A
STOP IT SHOULD
EXTEND ABOVE
FLOOR LEVEL AS
SHOWN

BASEMENT WINDOWS AND
VENTILATORS SHOULD BE
COVERED WITH 14GA 6.4 MM
MESH GALV. SCREEN OR
PERFORATED METAL AS
SHOWN

BRICK IS RECOMMENDED AS A
STOP. IT SHOULD EXTEND
ABOVE FLOOR LEVEL AS
SHOWN

18 GA. GALVANIZED SHEET METAL IS
USED AS A FLASHING FROM FOUNDAT-
ION TO FLOOR LEVEL AS SHOWN.

WOOD BLOCK NAILED UNDER 2" X 4" STOP
AT ANGLE KEEPS RATS FROM HAVING
LEDGE ON WHICH TO STAND AND GNAW.

COMMON TYPE BUILDING
CONSTRUCTION GIVES
RATS FREE ACCESS
TO DOUBLE WALLS

HOLES THROUGH FOUNDATIONS
OR WALLS SHOULD BE FLASHED
WITH SHEET METAL OR 6.4 MM
HARDWARE CLOTH

WOOD 2" X 4" STOPS ARE USED
AS SHOWN. 2" X 4" STOPS ARE
51 X 102 MM

Figure 4.26 Double wall construction and suggested methods of rodent proofing (from AFIS[3])

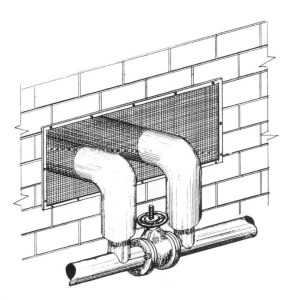

Figure 4.27 Wall opening for steam pipes rodent-proofed with hardware cloth – 6 mm (¼ in) mesh (from AFIS[3])

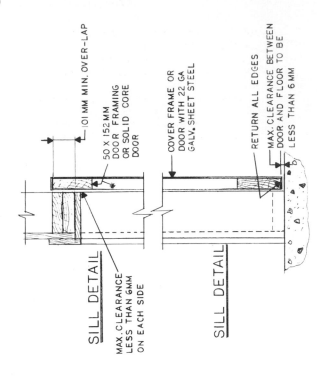

SILL DETAIL

101 MM MIN. OVER-LAP

50 x 152 MM DOOR FRAMING OR SOLID CORE DOOR

MAX. CLEARANCE LESS THAN 6MM ON EACH SIDE

COVER FRAME OR DOOR WITH 22 GA GALV. SHEET STEEL

RETURN ALL EDGES

MAX. CLEARANCE BETWEEN DOOR AND FLOOR TO BE LESS THAN 6MM

SILL DETAIL

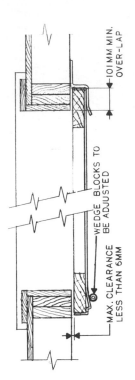

JAMB SECTION

101MM MIN. OVER-LAP

WEDGE BLOCKS TO BE ADJUSTED

MAX. CLEARANCE LESS THAN 6MM

Figure 4.29 Rodent proofing sliding door with wedge-type door stops (Kabak Co., San José, California, in AFIS[3])

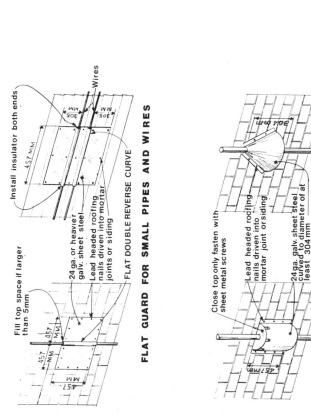

Install insulator both ends

Wires

305 MM

305 MM

457MM

24 ga. or heavier galv. sheet steel

Lead headed roofing nails driven into mortar joints or siding

FLAT DOUBLE REVERSE CURVE

Fill top space if larger than 5mm

457 MM

457 MM

457 MM

FLAT GUARD FOR SMALL PIPES AND WIRES

Close top only fasten with sheet metal screws

304MM

Lead headed roofing nails driven into mortar joint or siding

24 ga. galv. sheet steel curved to diameter of at least 304mm

CONE GUARD

457mm

BARREL GUARD

CURVED GUARDS FOR PIPES

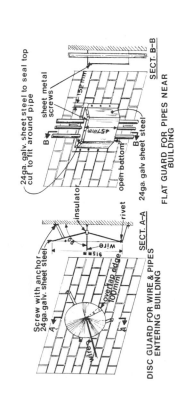

24ga. galv. sheet steel to seal top cut to fit around pipe

sheet metal screws

152mm

457mm

open bottom

SECT. B-B

24ga. galv sheet steel

FLAT GUARD FOR PIPES NEAR BUILDING

Screw with anchor 24ga. galv. sheet steel

insulator

wire

90°

915mm

rivet

SECT. A-A

overlap edge 100mm

wire

DISC GUARD FOR WIRE & PIPES ENTERING BUILDING

Figure 4.28 Sheet metal rat guards – installed on metal pipes and wires (designs from US Department of Health, Education and Welfare, Communicable Disease Center in AFIS[3])

Further reading

AFIS (1952) *Sanitation for the Food Preservation Industries*, pp. 38–45. New York: McGraw Hill

Katsuyama, A. M. and Strachan, J. P. (1980) *Principles of Food Processing Sanitation*, pp. 177–181. Washington, DC: The Food Processors Institute

Troller, J. A. (1983) *Sanitation in Food Processing*, pp. 268–273. New York: Academic Press

Scott, G. H., and Borom, M. R. (1968) *Rodent-Borne Disease Control Through Rodent Stoppage*, US Department of Health, Education and Welfare, Public Health Service, Communicable Disease Center, Atlanta, GA. Note, this book is now out of print but may be available

Glossary

The following are terms which those whose first language is not English have found unfamiliar and are not included or are difficult to find in easily available dictionaries. Note that what is given is a description rather than a definition of the term.

Angle discharge (specifically applied to lateral gutter (trench) drains into main gutter drains) – The downward slope or pitch of the lateral or side drain is increased near the discharge or junction with the main drain. That is, the angle is altered near the discharge to increase the speed of the water so that it carries debris more easily into the main drain.

Beams – These are horizontal structural members or supports of a floor or ceiling. They are distinct from columns, which are vertical. Beams may be of steel, reinforced concrete or other suitable material.

Bonding bed – Floor tiles or bricks are set onto a layer of material over a concrete base (called a bed). The bedding material must be suitable for its intended use and compatible with concrete. It must stick to both concrete and tile or brick. A resin-bed system which bonds or 'sticks together' the concrete and tile or brick may be referred to as a bonding bed.

Brick mason (USA) or bricklayer (UK) – The name of the person who lays flooring bricks. Also applied to the method used to lay flooring bricks as distinct from the thinner tiles. With tiles, a different method, called the tile setter's method, is used.

'Butters' – A brick mason 'butters' a brick before laying it, i.e. applies the adhesive rather like spreading butter on bread, hence the term.

Cast with the floor – A concrete mix is a liquid material which becomes solid when it 'sets'. Using the analogy of molten metal, the concrete is cast or poured into a mold. To make a curb or raised edge (see 'Curb' below), it should be integrally cast with the floor whenever possible, i.e. it should be made at the same time as the floor. This gives the strongest, most satisfactory result.

Caulked – Originally a ship-building term, when a material was driven between planks to stop leakage of water. It is used to describe the filling in of the spaces between e.g. two sheets of metal with an adhesive material which may or may not be flexible. Also used to describe the stopping up of crevices, e.g. of windows. The term 'grouting' is used for tiles.

Chute or shute – A steeply sloping channel or tube which conveys ingredients or food to a lower level, e.g. from a conveyor belt or hopper. Details of construction are given in Campden Technical Manual 7[10].

Clad – A dictionary definition is the past tense of the verb to clothe. A common usage is to state that a surface, e.g. a brick wall, has been clad (or covered) with another material, e.g. stainless steel. The covering material is termed 'cladding'.

Columns – These are vertical structural members, supports or girders which make the load-bearing frame, e.g. for walls. They may be of steel, reinforced concrete or other suitable material. Horizontal members are called 'beams'.

Concrete pad – A pad is a small pillar or box on which equipment stands. When made of concrete it is called a concrete pad.

Corrugated – Means a sheet, e.g. of metal or other structural material, shaped into equally curved ridges and grooves which are regular, parallel and straight. Corrugated iron is galvanized steel which can be used for roofs and walls. Even when care is taken it is not a recommended material for sanitary designs.

Cove, coving – A cove is a concave surface or molding. In building work the radius is often from 25.4 mm (1 in) to 152 mm (6 in). The term is sometimes used to describe an angled joint of about 45° to 60° between floor and wall or floor and steel column. Joints between the coving and the wall and floor should be filled with waterproof caulking to maintain a sanitary finish.

Curb, curbing – A raised edge or low wall used to protect building structures and equipment from damage, or to keep water, product or other materials confined to a designated area. Typical sizes around equipment and floor openings are 102 to 152 mm (4 to 6 in) high and 51 to 76 mm (2 to 3 in) wide. Curbs to protect building structures are usually larger than this. 'Curbing' is the material of a curb and is usually concrete.

Dock – This is the part of the building where ingredients or finished products are unloaded or loaded onto trucks or rail cars. The term originally refers to the place where ships or barges were unloaded or loaded.

Duct – A large pipe of more than 200 to 300 mm (8 to 12 in) diameter. It is also a sheet metal fabrication used to conduct ventilation or conditioned air around a building. The term is also applied to an enclosed electrical wireway.

Expanded metal – Perforated sheet metal usually pressed into a pattern which gives additional strength and rigidity. It is used as an alternative to a wire mesh or hardware cloth for excluding rodents or birds.

Flexjoint – Also known as an expansion joint. These are needed in brick or tile floors to allow for swelling which occurs with time, as well as thermal changes. (See Figure 4.14.)

Formed and poured in place – This is where a concrete construction is made *in-situ* (in-place). For ceilings in manufacturing areas this method of construction is preferred to the use of precast concrete ceiling panels which require the use of a caulking or sealing material. The presence of caulking material over production equipment is a contamination hazard.

Gasket – This is a suitably shaped piece of non-rigid material, e.g. neoprene or teflon, placed between the two faces of a joint to connect parts of a pipeline and prevent leakage. It is important that the gasket material retains its design properties at use temperatures of product and in the presence of cleaning and/or sanitizing agents; does not obstruct the flow and does not form crevices in the joint where food can lodge and microorganisms multiply.

Girders – Strong, structural support members of a building or heavy equipment, e.g. beams. Usually of steel either made in a single piece, e.g. in H or I section, or built up from plates, bars, etc.

Grouting – This is the material (also termed 'grout') which is used to bond the spaces between bricks or tiles and to make a level surface. The choice of the correct material is very important for achieving a level, smooth and durable surface.

Gutter – A narrow channel or ditch, e.g. as in gutter drain or, alternatively, trench drain. Such open drains should be covered by a grating. Also used to describe the trough used to collect rainwater from a roof.

Insulation – A material which prevents the conduction or transfer of electricity, heat or sound. Unless sealed, insulation must be expected to harbor pests or microorganisms. Mechanical damage to insulation is a contamination hazard. Heat insulation material is called 'lagging' in the UK

Jamb (or frame in the USA) – An upright or post which forms the side of a doorway. The word is often used as the plural; jambs. The door will be hinged to (or alternatively hung from) one jamb and will fit into the other.

Membrane (construction of) – In wet food processing areas a waterproof membrane is necessary because in time, water which may contain acids produced by microbial action together with cleaning and sanitizing solutions will seep through the brick, tile or grouting. The aqueous materials corrode the concrete substrate causing mechanical damage with the risk of infestation. Waterproof membranes are usually made from an asphalt material, which allows for a small movement while remaining waterproof,

together with suitable reinforcing, e.g. woven glass cloth.

Paver – A brick or tile used to lay a floor.

Plain concrete – A mixture of cement, sand and water used as a mortar. Broken stone or other aggregate is added to the mixture to make the building material which may then be poured or cast. Variations in proportions (including the water content) have a significant effect on properties of the mixture when it has 'set' or hardened. 'Plain concrete' can mean a concrete surface which has not been painted or treated, e.g. with a surface hardener or with a coating such as a monolithic urethane or epoxy finish.

Poured concrete slabs – Formed or cast panels made from poured concrete.

Quarry tiles – Flooring tiles with a slip-resistant surface. This slip resistance is obtained in a different way from 'ceramic' tile in that 'ceramic' tiles have the slip resistance derived from material added during manufacturing.

Rodding – May be used to describe the strengthening or reinforcing rods used in reinforced concrete. Also used to describe the action of clearing a blocked drain by the use of rods.

Saucer slope – By analogy with the saucer of a domestic cup and saucer, this is a feature of a floor which is sloped or dished at regular intervals. A floor drain is installed at the lowest point in each dish. See Fig. 4.19.

Scuffing – The action of abrading, wearing or somewhat damaging a surface by repeated, fairly light, glancing blows. Thus paintwork or a tile surface may be scuffed by the repeated impact of traffic.

Sleeved – By analogy with an arm within a shirt or jacket sleeve, a pipe is said to be 'sleeved' when it is positioned inside a tube embedded in a wall or floor. This enables material carried in the pipe to be transported between production areas. The pipe is not cemented in the tube in order to allow for expansion and contraction. Sleeving through a floor is intended to prevent water, dirt, etc. from falling to the area underneath.

Solid core door – A door made from solid material throughout, e.g. wood or plastic. It is in contrast to one made by fastening sheet material over a frame. A 'core' means the central part of any object (including a door).

Strand – Material in the form of a fiber or thin wire. Usually used as the plural, e.g. strands of wire. Strands from an electric cable may be a contamination hazard.

Suspended ceiling – A ceiling which is significantly below the level of the structural or load bearing beams is said to be suspended. It enables light fittings to be fitted flush with the ceiling surface, and ventilation and other utilities to be installed above a smooth, easily cleaned ceiling surface.

Tack welding (spot welding, UK) – Is a type of welding in which the weld is not continuous. The crevices between the surfaces thus formed may harbor food residues which support microbial growth or pests.

Trench drain – A drain in the form of a narrow channel or ditch. It is an open drain as distinct from a pipe drain and should be covered with an open metal grating.

Unblocking plug – A plug set into a pipe drain usually at a right angled bend or at the base of a U-bend to enable the drain to be unblocked by the use of rods or hoses.

Web and flanges (of a beam) – In the H beam the central, (horizontal) part is the web, in an I beam the central (vertical) part is the web. In a U or channel section the base of the U or channel is the web. Other parts, which are not the web, are called flanges.

4.B.2 Note on construction

Assuming that the building design is sound and incorporates sanitary principles, the quality of the construction work can make or mar the plans. Obviously, the ideal is that the building should be done to an agreed standard within time and cost limits. In order to get somewhere near this ideal, much depends on the management and technical skills of the contractors and their experience of the needs of *food* plants. It is important to remember that, in reality, the cheapest job is the one that meets the required standards at lowest cost. This may not be the lowest bid. Because so many day-by-day decisions and choices have to be made, technical expertise of the contractors as well as that of the Company project engineers is an important factor in getting a job done which will prove satisfactory in the long term.

At some stage, even with a 'green-field' site, commissioning and/or production will be going on at the same time as building work. Problems associated with this are discussed in Section 4.C.2.

4.C Maintenance and alterations

4.C.1 Maintenance

'Prevention is better than cure'

Maintenance is carried out to insure that buildings function as designed. It must be done in a way which avoids contamination of ingredients, products or packaging materials and minimizes the risk of pest infestation.

Building maintenance is sometimes regarded as an expense which can conveniently be deferred. Before accepting this argument, answer the following 'By how much' questions:

- Is the risk of product contamination increased?
- Does it increase cleaning costs?
- Does it reduce hygiene standards?
- Does deferring the work add to the eventual cost?

Remember, the arguments for preventive maintenance of equipment apply also to buildings. Imholte[1] assesses the situation in this way:

Building maintenance involves timely repair of building materials as the need occurs. Deferring building maintenance can compound sanitation problems. Vermin gain access through and harbor in holes or cracks in the structure. Structure damage readily supports dust and dirt, and creates areas which are difficult to clean. Areas in need of repair frequently increase the risk of direct product contamination from such things as peeling paint, torn insulation, and broken light fixtures. Building repairs should be given a high priority, and they should be made as close to the date of damage as practical.

Floors

As an example of a practical approach to preventive maintenance, Sheppard's paper[2] prompts the general questions:

- What can go wrong?
- Why?
- What systematic and regular inspection and tests should be made to trigger action? (An example from this paper is given below in Table 4.4.)
- What remedial action(s) can be pre-planned?

Miscellaneous

Imholte[1] deals briefly with floors, walls, ceilings and roofs; making the point that leaking roofs can produce contaminated water well away from the original leak. This is also true of floor leaks in multistory buildings and of some pipeline leaks.

There is more to maintenance of electric lamps than changing them when they fail. This is dealt with in considerable detail in Technical Report No. 9 of the UK Illuminating Engineering Society[3].

Table 4.4 (from Sheppard[2]) Check these trouble points at recommended frequencies

	Inspection interval (months)
Acid-proof brick or tile floors	
Expansion joints	3
Drains and intercepts of pipes into pits and trenches	3
Trenches, walls and bottoms, especially at changes in direction	3
Pump bases	6
Equipment foundations, especially those passing through floor surfaces and structural members	6
Juncture points between floor and trench walls, especially where gratings or other covers rest on trench tops	6
Changes in direction – corners, curbs, etc.	12
Low points in floor areas, such as shallow sumps, especially if there is no drain at that point	12
High points	12
Wheeled-traffic lanes	12
Balance of the floor	12
Monolithic floors	
Expansion joints	2
Drains	2
Pump bases	4
Vertical walls, especially adjacent to pump bases, etc.	4
Equipment foundations	4
Areas adjacent to heat sources, or in which there are differences in thermal conditions	6
Very wet areas	6
Juncture points of floors and trenches, and other changes in direction	6
Low points	12
Balance of the floor	12

Note The maintenance engineer should mark up a plan drawing of the various areas and assign key numbers to those points that he feels are the most sensitive. A checklist with these key numbers can then be supplied to the inspectors at regularly stated intervals. When the inspections are made, the report is dated and signed.

Painting

Painting has other contamination risks apart from the obvious one of paint flaking and falling into open ingredients, products or packaging materials. Surfaces must be properly prepared and this may include sandblasting. Materials used in a painting job, e.g. thinners, may cause toxicity or taint problems if they splash onto ingredients, products or packaging materials or if their odor is absorbed. Consideration therefore needs to be given to all aspects of the painting operation.

Remember Imholte's[1] caution on page 129: 'Paint does not stand up well to adversities particularly when it has not been properly selected or when the surface has been improperly prepared.' When on-site (*in-situ*) surface preparation and painting of buildings or equipment is necessary, remember to protect:

● Food contact surfaces
● Processing equipment
● Controls, gauges, instruments, etc. in the vicinity of the painting work

Drop cloths or plastic sheeting are used and sealed with tape as necessary to prevent dust, rust, scale, paint spatters or overspray contaminating food contact surfaces or equipment. Ingredients and packaging materials should, of course, be moved from any area to be painted.

Before applying paint, materials which prevent adhesion, e.g. oil, grease, chemicals, scale, *must* be removed. This is often done with solvents and brushes but may require other hand tools or power tools. Wet blasting and vacuum blasting are preferable to sandblasting, which should not be used in areas where there is a large amount of equipment or if there is any doubt *whatsoever* that complete protection from sand infiltration can be achieved.

Should sandblasting be necessary, the area in which it is done should be completely isolated from other areas and equipment by dustproof partitions. Equipment which cannot be removed must be covered and sealed with drop sheets, plastic sheeting and tape to prevent sand infiltration. Ventilation systems may also need attention. Sandblasting should only be done during non-production 'shutdown' periods.

Prior to painting, all residual dust, rust, scale, sand, etc. resulting from preparation of the area should be removed by vacuum and clean brushes. Do not use compressed air or steam which simply redistributes the unwanted material.

Remember, only paint suitable for incidental food contact or classed as food-grade should be used on product zone surfaces. Epoxy-based paints or paints that do not contain toxic or tainting materials may be used for non-food zones if they are acceptable to the regulatory agencies. Examples of these paints are alkyl-, epoxy-, acrylic- or vinyl-based materials. Paints containing lead, mercury or other heavy metals should *never* be used. Great care should be taken in the selection and use of anti-mold paint.

Adequate air circulation and ventilation are needed during painting and until after the paint is dry and the smell of it dispersed. Be sure that exhausted air goes to atmosphere and not near air intakes or to areas where ingredients, products or packaging materials may be affected.

Obviously, paint spatters or droppings should be cleaned up and brushes, tools, equipment, paint and painting materials returned to the storage area when no longer needed for the job.

Waste materials, which includes paint and thinners which cannot be further used, should be removed from the area and disposed of in containers and/or receptacles intended for this purpose. For safety reasons waste paint rags should be discarded into containers with tight covers to prevent spontaneous or other combustion.

Prohibit absolutely the use of floor drains, sewers, ingredient, or product containers or packaging materials for disposal of painting wastes or materials.

4.C.2 Alterations

Introduction

Similar building technology is used for construction and alteration and, therefore, hygienic design, construction and installation follow the same principles. However, it is obviously more difficult to achieve hygienic standards and minimize contamination during alterations than when building on a 'green-field' site.

Imholte[1] asserts that good management makes the difference between a job well done and one which is not. In particular, the construction engineer in charge constantly judges the quality of the work and decides on many engineering details. Careful attention to detail is the key to cost-effective operation and cleaning as well as minimizing contamination risks. But the construction engineer cannot work on his own and needs the support of the entire management team to achieve the desired goals.

Further information and comments based on practical experience are given by Imholte[1] in his Chapter 6 ('Construction').

Contractors

It is very easy for sanitary standards to slip either during construction or in the usual situation of major building and installation works in a plant that must continue food production. Very positive management action must therefore be taken to maintain sanitary status.

Significant numbers of contractors and sub-contractors will be working in and around the plant. Pressure to finish a job on time may mean sub-contractors being brought on-site, perhaps from other 'non-food' jobs, for only a day or two. Unless contractors and their workers appreciate the additional constraints of working in a food factory, incidents which are serious and potentially or actually damaging to your business can be expected. For example, contractors were working in a department which was shut down. Unpleasant smells came from a drain near a filling machine. To overcome this, they brought in a 'pine'-smelling, concentrated, phenolic disinfectant which they dispensed using product containers which later were replaced on the line. Fortunately, they were recovered and removed prior to being filled. However, even the presence of a phenolic disinfectant in a food factory which uses chlorine as a disinfectant in the CIP system has been shown to cause sporadic chlorophenol taint in product manufactured on equipment where there was no direct contact.

Imholte[1] has wise words about contractors. He writes of contractors' supervision: 'These supervisors will be responsible for directing construction activities. Their familiarity with sanitary needs will result in the job being done as it was intended.' Of contractors' personnel he writes:

> Most construction personnel have little or no exposure to the sanitation requirements of the food industry. Indoctrination of key construction personnel will be very beneficial to the job at hand. Familiarize them with the problems, the consequences, and the solutions. In essence, put them on your team, make use of their many abilities, and take advantage of their knowledge. Above all, get them to work with you.

He also highlights the need for management skills for both parties to the contract in the section titled 'Job Organization'. This is important and is well worth reading in the original text. As an experienced engineer, Imholte sets forth the importance of both a good working environment and good organization and management. He wisely states: 'Remember, a clean, neat and well-organized job is a well-run job.'

It is clear, therefore, that both the work and the way in which it will be done need to be carefully planned. Part of this planning requires a clear expression of hygiene and good housekeeping requirements to the contractor. Table 4.5 is an example of rules which form the basis for indoctrination and supervision of contractors and their employees.

Table 4.5 Specimen rules for sanitation (hygiene) and housekeeping for contractors and their employees

General

You are now working in a food factory making food which people like you and your family will be eating. Help keep this food pure and uncontaminated by obeying these rules:

- Go to your work area by the most direct route (from the perimeter of the building) *but not* through a production area. If in doubt, check with the Company representative.[a]
- Cover any equipment in and around your work area with dustproof sheeting (thick plastic) sealed with tape as necessary to make a dustproof seal thus preventing contamination.
- Enclose the work area from floor to ceiling with dustproof sheeting (thick plastic) and seal all edges if building work is liable to create dust or loose debris. Check that the ventilation system is isolated or sealed to prevent dust spreading.
- In multistory areas, work on floors or ceilings must include appropriate protection of the adjacent floor levels.
- Any surface or substance likely to come into contact with ingredients, foodstuffs or packaging materials must be of food contact grade, non-tainting and inherently unlikely to cause contamination in use, e.g. by splintering.
- All materials used, whether temporary or permanent, including cleaning materials, glues, paints, etc. must be Company approved. This is to prevent inadvertent use of toxic or tainting materials.
- Replace all trunking covers after use. The use of electrical cable, tape or string to secure or support trunking is forbidden.
- *Use only* the waste containers intended for this purpose. *Do not* dispose of building, construction, painting or similar waste materials through product drains or sewers.
- Remove waste and rubble from the area by the most direct route through *non-production* areas.
- Notify the Company representative[a] immediately of any incident which you think might possibly contaminate or adversely affect a product. This will enable appropriate action to be taken as soon as possible, e.g. isolation, re-work of product, destruction.
- Contact the Company representative[a] immediately if you are unable to clean up after a job, e.g. between pipe runs. Where specialized cleaning is required you must inform the Company representative[a] of this before leaving the site.
- *Do not* use ingredient material containers, food product containers or packaging materials for *any other* purpose.
- Always take any special safety or hygiene precautions that are necessary for the job. If in doubt, check with the Company representative.[a] This may include getting a 'Permit to Work'.

Note. A 'Permit to Work' system for critical tasks may or may not be locally mandatory but is always good practice. The categories include:

- High voltage permit
- Digging permit
- Company permit – this is required where the contractors' work may directly affect or interface with production.

The simplest 'Permit to Work' administration that gives adequate protection to personnel and products is the best.

- The use of food or packaging materials for commissioning purposes must be authorized by the Company representative.[a]
- When the work is completed, leave the area clean and tidy with protective screening and sheeting removed.

[a] It is essential that contractors and their employees know who the Company representatives are and how they should contact them, e.g. by telephone at any time that work is in progress.

The indoctrination for Contractors should include:

- Orientation – locker rooms, cafeteria (canteen) storage areas, offices, etc.
- Security – including entry and leaving site, lockable tool rooms and compounds, responsibility for personal belongings, etc.
- Site rules – Sanitation (hygiene) and good housekeeping
- Site rules – Personal Safety – including fire procedures together with other pertinent items and the booklet of rules for contractors.

If the job is a major alteration/extension, consider the use of visual aids such as a film or video, e.g. the video produced by Voss, *Hygiene and the Engineer*, is a good example for this purpose and is obtainable from Voss Training Services, 2 Price Street, Birkenhead, L41 6JN, UK.

Personal standards

It is important that everyone working in a food factory, and this includes contractors and their employees, understand that this imposes obligations precisely because food is handled. Marriott[4] has an excellent chapter entitled 'Personal Hygiene and Food Handling' and there is also useful material in ICMSF[5]. Clearly, with contractors' employees who may not believe that they are 'food handlers' because they may not handle food directly, there is a need to stress that they can be, unless they are careful, the cause of contamination of food. Experience has shown that where sensible rules are applied and are firmly enforced to everyone, there is a respect for the company concerned and the rules are not resented. Discrimination either way in the enforcement of rules will most certainly cause resentment.

Table 4.6 gives some specimen rules which may need to be supplemented or adapted to local regulatory requirements or individual needs.

Delivering satisfactory standards of sanitation (hygiene) and good housekeeping

Insuring that a satisfactory standard of hygiene and good housekeeping is maintained by contractors and their employees is difficult, especially when major works are being done. This is particularly so if a substantial proportion of people on-site are contractors, some of whom, especially as deadlines approach, may only be on-site for one or two days. It is, however, essential to keep control of the situation. Consider, for example, the consequences if a consumer swallows a piece of glass. Even if it can be *proved* to have been broken by a contractor – which is likely to be difficult – who needs this kind of publicity?

Table 4.6 Specimen personal hygiene and safety rules

Important

We are all working in a food factory so we should all obey the site rules to comply with the law, to protect ourselves, and the products:

- Be clean, tidy and work safely.
- Maintain a high standard of personal cleanliness. Always wash your hands before starting work and after using the lavatory. This minimizes the spreading of bacteria (or germs).
- Cover hand wounds with a Company and Regulatory approved color plaster or bandage when working in or near production areas, open ingredients, or packaging materials. A domestic, flesh-colored sticking plaster is *not* adequate as it may be lost and become a serious product contamination hazard since it may be small and not easily seen.
- Report all injuries to the medical department (or First Aid room as applicable). If food handling equipment is involved, report full details to the Company representative who will insure prompt, thorough, clean-up.
- You should not remain at work on this site if you have a stomach upset, diarrhea, skin irritation or boils. (In certain countries, this is mandatory.) A medical certificate will be required to confirm that the illness has been cured before recommencement of work.
- Wear full, clean, protective clothing. Boiler suits or overalls should be regularly laundered, close-fitting with zips or metal studs. Buttons or loose fasteners are not acceptable. Hair must be kept covered either by a hard hat (where appropriate) or a soft cap. Beards must be covered by a net. Ear protection must be worn in designated, high-noise areas.
- No jewelry or watches are to be worn in production areas except for a plain wedding ring and 'sleeper' or 'keeper' ear rings.
- Spitting is forbidden anywhere on-site.
- Smoking, chewing tobacco, eating or drinking is allowed only in tea-bars, eating areas and other specially designated areas.
- Consumption of alcohol on-site is forbidden (unless, as in some countries, there are specifically designated areas and times).
- 'Horse play' or misuse of equipment, hoses, etc. is forbidden.
- Contractors must not handle ingredients, packaging materials or finished goods. If they need to be moved, contact the Company representative.
- Obey any specific hygiene requirements in your working area.

Although each job is different, the following should apply to whatever system of management is devised:

- Contractors' supervision should be expected to control their personnel in a way appropriate to work in a food factory.

78

CONTRACTORS NON-COMPLIANCE REPORT						PAGE ___ OF ___
DATE	TIME	AREA	NON-COMPLIANCE ISSUE	CONTRACTORS COMPANY	INDIVIDUAL(S) CONCERNED	ACTION TAKEN

Figure 4.30 Specimen log sheet – Contractor's Non-compliance Report

HYGIENE / GOOD HOUSEKEEPING CHECKLIST   PAGE ___ OF ___

CHECKED OUT BY _____

JOB (DESCRIPTION) _____   REF _____   PROJECT _____ PRODUCTION
FOR _____ AREA   IMPLEMENTATION _____ CONTRACTOR
LOCATION _____   ENGINEER _____ SANITARIAN

ITEM(S) INSPECTED	DATE	CONDITION	OUTSTANDING ISSUES	ISSUES RESOLVED/NOTES

FURTHER ACTION (IF NEEDED)

N.B. ALWAYS INSPECT ENCLOSED ELEVATORS, HOPPERS ETC. INTERNALLY AND EXTERNALLY.

Figure 4.31 Specimen checklist – Contract Work –
sanitation (hygiene)/good housekeeping

- Any company employee should be able to challenge a contractor's employee anywhere on-site who is breaking the rules. Any challenge should be courteous but firm.
- Company management and supervision have primary responsibility for their own areas. Contractors must either comply or stop work – refusal to stop or comply should be reported *immediately* to the Company Manager responsible for the work.
- A record of all non-compliances should be maintained by a designated person. Remember, *no news* is not necessarily *good news* and the purpose of the record is to help pressure for improvement to be directed where it is most needed. Records should therefore be kept of action taken and its effect.

A log sheet for use by area supervision could be based on the example given as Figure 4.30.

Additionally, confirmation is needed that hygiene and good housekeeping standards are being achieved during the work at each stage:

- Preparatory – are ingredients, packaging materials, finished products, utensils and equipment either removed or adequately covered and sealed as necessary?
- During the work – is there a good liaison with the contractor and the company to insure that engineering details enhance sanitation standards?

Are leaks and spillages cleaned-up promptly, and are cleaning needs discussed with the designated Company representative?
- When the job is complete, i.e. ready for hand-over, has thorough cleaning taken place and are there any outstanding hygiene or good housekeeping issues?

Adequate records of the hygiene state of the area are needed. If the job is complex or there are a number of jobs, a checklist may be considered. An example is shown in Figure 4.31.

References

1. Imholte, T. J. (1984) *Engineering for Food Safety and Sanitation*. Crystal, MINN: Technical Institute of Food Safety
2. Sheppard, W. L. (1974) Plant floors thrive on preventative maintenance. *Food Engineering*, **46**, (9), 81–88
3. IES (1967) *IES Technical Report No. 9. Depreciation and Maintenance of Interior Lighting*. London: The Illuminating Engineering Society
4. Marriott, N. G. (1989) *Principles of Food Sanitation*, 2nd edn. An AVI book published by Van Nostrand Reinhold, New York
5. ICMSF (1988) *Microorganisms in Foods. 4. Application of the Hazard Analysis Critical Control Point (HACCP) System to Ensure Microbiological Safety and Quality*. International Commission on Microbiological Specifications for Foods, Oxford: Blackwell Scientific

4.D Pest control and disinfestation

4.D.1 The pests

Permanent and occasional pests

When considering the problem of infestation in the food processing industry there are essentially two primary types of infestants (biological contaminants), i.e those which are:

- Permanent
- Occasional

The first type – *permanent* infestants – is definitely the most dangerous of the two since these live within the foodstuffs, reproduce extremely fast, and are capable of causing substantial damage.

In contrast, the second type – *occasional* infestants – are casual visitors, do not colonize the foodstuffs themselves but (in the case of some species) establish themselves in production or storage premises, thus creating a risk of ongoing contamination.

There is not necessarily a definite demarcation between the two types of infestants. It should always be remembered that pests are living organisms and:

- Infestants may well adapt to what are normally hostile environments
- There is a constant transmigration from one part of the world to another made possible by the trading of all types of ingredients and finished products.

Given this, any list and classification of infestants must be constantly reviewed and updated, since there is always the possibility of a species not currently included becoming of significant importance later on.

With this in mind, the list presented below includes the main types and species that must be known by the food processor *today*. There are four main classes of infestants:

- Insects
- Arachnids
- Mammals – Rodentia (rats and mice)
  – Chiroptera (bats)
- Birds

Remember, in a food plant any domestic animals (cats, dogs, etc.) are to be considered infestants.

Insects – permanent infestants

Insects are the most common infestants and undoubtedly the ones that give cause for most concern. They include two orders which are the most frequently found in infestations, i.e. Coleoptera (beetles) and Lepidoptera (butterflies and moths). These include a number of *permanent* food pests. This section starts with a detailed description of the various families in these two primary orders, and is then followed by notes on other classes and families of *occasional* infestants.

Coleoptera (beetles)

In this insect order 11 families are listed and their main physical and behavioral characteristics are described below:

a. Tenebrionidae

Tribolium castaneum and *Tribolium confusum*. The *Tribolium* genus – which is cosmopolitan (i.e. found throughout the world) – includes species which attack cereal derivates, particularly flour. *Tribolium castaneum* is commonly found in Italy in many imported food products.

The Tribolium attack cereals and derivatives, groundnuts, dried fruit, spices, oily seeds, milk chocolate and many other foods. They are unable to perforate the outer casing of whole seeds and thus are found either together with other insects which can do this, such as the *Sitophilus granarius*, or in seeds broken by machines, or in cereal products.

Tenebrio molitor and *Tenebrio obscurus*. These beetles can be as much as 18 mm (¾ in) long. Their larvae are omnivorous, while the adults feed mainly on cereals and cereal derivates. They live mainly in humid, dark places and their presence indicates poor hygiene. They are very frequently found in mills, especially inside the bases and hollow sections of machinery.

b. Silvanidae

Oryzaephilus surinamensis and *Oryzaephilus mercator*. These cosmopolitan insects attack cereals and their derivatives, pasta, dried fruit, dehydrated meat, sugar, tobacco, spices, etc. They are the cause of damage to cereals in storage and are also found in packed food products since their larvae easily manage to perforate cardboard, polyethylene, etc.

c. Curculionidae

Sitophilus granarius – Sitophilus oryzae – Sitophilus zea-mays. The first, commonly called the corn weevil, has been known for thousands of years. It attacks any type of whole cereal and pasta as well. Its development depends on the egg being deposited in the kernel.

The rice and maize weevils are found throughout the tropical zone and are commonly found in imported grain. Since, in contrast with the corn weevil which is wingless, both these weevils fly well, they can easily cause 'cross' or 'cross-over' infestations.

d. Bostrichidae

Rhysopertha dominica. This family includes wood-worms but *Rhysopertha* is also a cosmopolitan infestant of stored grain. The new-born larvae penetrate the kernel and larval development takes place inside the individual grains.

e. Trogositidae

Tenebroides mauritanicus. Well known as a grain infestant, this insect attacks cereals before and after milling, potatoes, nuts, spices, and dried fruit. The larvae bore tunnels in any wooden parts of machinery and these are later used as hiding places by other infestants.

f. Bruchidae

This family includes the so-called 'caterpillars', e.g. the *Acanthoscelides obtectus* or bean caterpillar, which attack the seeds of leguminous vegetables such as peas, beans, lentils, etc. The eggs stick to the seeds which are then penetrated by the larvae that subsequently develop.

g. Dermestidae

Trogoderma granarium. This insect originated in India and was spread to Europe by means of imported food during the second decade of this century. It is a serious and damaging pest for the cereal and beer industry. It is a terrible grain parasite that is difficult to remove even when using the most severe types of fumigation. This is due to the diapause phenomenon (suspended growth) which enables the larvae to withstand temperature and otherwise prohibitive conditions for a significant length of time. Diapause can in fact last for several years.

Dermestes maculatus. This infestant is typically found in tanneries and in factories producing bone or fish flour. The larvae bore tunnels in wood, cases, floors, and walls, causing serious damage.

h. Anobidae

Stegobium paniceum. This cosmopolitan pest infests cereal-based products, spices and powdered foods. It is frequently found in wholesale and retail stores. It moves from the originally infected substances to others close by. As it can easily be confused with *Anobium punctatum* (a woodworm) it often leads to the wrong type of disinfestation action being taken.

Lasioderma serricorne. A typical tobacco infestant, this pest also attacks a wide variety of seeds (oily and dry) and is the worst pest of cocoa seeds stored in West Africa. Although it is cosmopolitan, this infestant does not survive a European winter in an unheated environment.

i. Ptinidae

These infestants, commonly known as 'spider beetles' due to their appearance, include two species commonly found in warehouses and shops – *Ptinus tectus* and *Ptinus fur*. Their larvae feed on dry vegetables or organic material, and rubbish of all types, but are also capable of attacking ingredient materials and packed food products. They are highly resistant to insecticides and difficult to eliminate. Due to the fact that they are polyphagous, they can also contaminate other stored products and are also very difficult to remove.

j. Nitidulidae

Carpophilus spp. A typical example is *Carpophilus hemipterus*, an infestant of dried fruit – and of dried figs, raisins and currants in particular – but also of nuts, grain, and spices. This type of infestation is generally associated with fruit going moldy as a result of incorrect storage.

k. Cleridae

Necrobia rufipes. This infestant of tropical origin attacks copra, bone, cheese, meat, palm seeds and other oily seeds. Metallic bluish-green in color, it nips the skin if provoked. When infested substances are unloaded from a ship, swarms of the insects fly out from the hold, usually creating a certain amount of panic among those present.

Lepidoptera (butterflies and moths)

As far as Lepidoptera are concerned, the key point to note is that major food infestants in this class mainly belong to two 'superfamilies' – Pyralidoidea and Tineidea.

a. Pyralidoidea

This superfamily – consisting of rather colorless insects, whose adults are mostly yellowish or greyish in color while the larvae are yellowish with brown markings – comprises two families of particular interest as regards infestation:

Phycitidae which includes four species that are especially harmful to foodstuffs:

Ephestia eleutella. This is *Enemy No. 1* of food products traded internationally. It infests grain, oily seeds, cocoa, dried fruit, nuts and tobacco. This species not only attacks products in storage but is also present in factories, particularly in confectionery establishments, where it causes problems throughout processing, sometimes even reaching the shops inside the packs of finished products.

Ephestia cautella. This is a very serious pest of cocoa immediately after harvesting, e.g. in Nigeria. It is transferred with this raw material to temperate zones, i.e. in areas unsuitable for its development, and is then replaced by *E. eleutella*.

Ephestia kuhniella. This moth is typically found in the milling industry but also attacks dried fruit, seeds and spices. In the mills it spreads inside the machinery where the larvae secrete silky filaments forming a web that absorbs flour, excrement and exuviae (skins shed by insects). In some cases these webs become so thick as to block the flow of cereal. Major infestations of this type can cause grain or flour to have an acidic taint.

Plodia interpunctella. This infestant mainly attacks dried fruit (California (USA), Australia, and Mediterranean areas) and stored nuts, and also maize, chocolate, cereal derivatives and various spices in hot climates and subtropical regions.

Galleridae

Corcyra cephalonica. This pest, tropical in origin, has spread to Europe but is unable to survive the winter unless it happens to be in heated surroundings. It infests rice, cereal derivates, oily seeds, nuts, dried fruit and various types of spices.

Paralispa gularis. Although most usually known as the tinea infesting stored walnuts, this infestant also attacks almonds, hazel nuts, and ground nuts. In addition it can withstand cold surroundings.

b. Tineidea

This is the second superfamily of the Lepidoptera class that infests food products. The Tineidea are small tinea that reproduce on a wide range of dry materials of animal and vegetable origin and include, among others, the following families:

Gelechidae

Sitotroga cerealella. This is the true corn tinea, that attacks right from the time of harvest. The larvae develop inside the individual grains and thus it is very difficult to identify the presence of this infestation. *Sitotroga cerealella* is very common in southern Europe, Africa, South America and in the southern regions of the USA.

Tineidae

Nemapogon (*Tinea granella*). This is less important than the species mentioned above and attacks grain and its derivatives. It is widespread in Europe, North America and Japan.

Insects – occasional infestants

These include the four families which are described below:

Blattidae (cockroaches)

Large numbers of cockroaches are present in human surroundings when hygiene standards are low. They have been found in homes, offices, hospitals, warehouses, shops, cellars, restaurants, ships, planes, and rubbish dumps.

The development of international trade has fostered the worldwide spread of the main types of cockroaches, even as far north as Lapland, where they attack salted fish stocks. Of the 3500 species known to date, about twelve are harmful and two are truly cosmopolitan. The three primary species are the following:

Blattella germanica. This is a rather small cockroach, about 10 to 15 mm (½ in) long, of a yellowish-brown color with black lengthwise streaky markings. It is the most common cockroach worldwide. It prefers hot, moist premises, typically those which are centrally heated, such as hospitals, restaurants, food factories, hotels and ships. It has suckered feet and is able to walk on perfectly smooth vertical surfaces and on ceilings. Given these facts, although it is not a flying insect, any disinfestation work must take in the whole room and not just floors.

Blattella orientalis. This is the largest *Blattella* species of cockroach – about 20 to 24 mm (1 in) long – and is colored dark reddish-brown verging on black. Once considered to be the 'common cockroach' in Europe, it is now less common than the *Blattella germanica*. It is found in kitchens, drains, underneath sinks and, generally speaking, near a source of water. Its habits are similar to those of mice. It is a vehicle for microorganisms harmful to man.

Periplaneta americana. This is the largest cockroach of all, reaching a length of some 28 to 44 mm (1 to 1¾ in), and has a brownish-red colored back. Having been spread by trade, it is very common in tropical, subtropical and temperate zones. In the USA it is present in all the warmest Southern states. It is well established in the city of New York and is continuing to spread northward.

As with *Blattella* species, it prefers a hot, moist habitat. It migrates from one place to another, driven to do this by hunger. In some countries, e.g. Colombia, it is possible to see the mass migration of hundreds of these insects. During disinfestation one can see it move from one building or from one room to another. It is thus difficult to eliminate.

Orthoptera (including crickets)

These insects are close relations of the cockroach, and include the common domestic cricket (*Acheta domestica*). During summer months the latter lives in the open. In towns and cities its primary habitats are rubbish dumps and from these it can migrate to industrial premises, causing serious infestations.

Hymenoptera (including ants and wasps)

This order includes ants, wasps, bees, sawflies and related forms. In food premises, the most important are:

Monomorium pharaonis (Pharaoh's ant) and *Iridomyrmex humilis* (Argentine ant). They are attracted by proteins, fats, and sugars, and build nests that are difficult to reach, using hidden tracks. They are often difficult to identify and require tropical conditions – such as the hot and humid environment often present in food factories – for survival.

Bees and wasps. Attracted by sugar-based products, bees and wasps not only contaminate food but may also sting personnel. The nests and hives must be identified and premises must be protected by means of netting over windows, etc.

Diptera (flies)

This is the order comprising flies, which are carriers of pathogenic microorganisms and thus of disease, and therefore a *serious problem* for food producers.

Flies are attracted by an extremely wide variety of foods, but in particular by decomposing animal and vegetable waste and by excrement. They are thus primary carriers of human disease (typhoid, cholera, dysentery, etc.). The most important species in this respect are:

Musca domestica (house fly). This is the most common species and is found throughout the world.

Calliphora erythocephala (blue fly or blue bottle (UK)). This is often found in meat-processing facilities where it does considerable damage. It is the most common cause of bad meat both at home and in shops.

Lucilia spp. These also are found in meat-processing facilities.

Piophila casei. The larvae of this fly infest cheese, meat, ham and smoked fish by boring deep tunnels.

Drosophila spp. These are pests typically found in the fruit, jam, vinegar, and beer industries. They feed mainly on acidic or fermenting substances and are sometimes known as the 'vinegar fly' or 'fruit fly'.

Arachnids

Acarids (mites)

The members of this class that are most important in terms of food infestation are some species of acarids (mites). During the nymphal stage they are strongly resistant to chemicals, can be easily carried by the wind, insects or other animals, and thus are subject to passive as well as active spread.

They infest cereals and other products in granaries, mills, pasta factories, bakeries and the home, and also sugar, cheese, and dried fruit. Moreover, they can be found in carpets and wallpaper, and in blankets, etc. They can irritate the human skin and trigger allergic reactions.

The specific species are:

Acarus siro (flour acarid) – the most common cereal infestant
Tyrophagus casei (cheese acarid)
Carpolyphus spp. (dried-fruit acarid)
Glyciphagus destructor (warehouse acarid)

They all normally prefer humid surroundings.

Mammals

Mammals – Rodentia (rats and mice)

Rats and mice are capable of colonizing a wide variety of environments and especially – due to the presence of abundant food – warehouses and premises used for processing food for human and for animal consumption. If rodents' nesting places are not eliminated the concentration of rodents can reach a density of 1 per square meter (about 1 per square yard). They are a health risk and cause economic damage.

- *Health risk*: rats and mice can transmit to humans – either directly or indirectly – many diseases, among which the most important are:
  - Plague (*Pasteurella pestis*)
  - Murine typhus (*Rickettsia prowazcki*)
  - Rickettsial pox (*Rickettsia akari*)
  - Leptospirosis (*Leptospira icterhaemorrhagica*)
  - Trichinosis (*Trichinella spiralis*)
  - Salmonellosis (*Salmonella* spp.)
  - Virus lymphocytic coriomeningitis
  - Weil's disease (*Leptospira* spp.)

The causative microorganism is in parentheses.

- *Economic damage*: Rats and mice damage much more food than they actually eat (a daily intake equivalent to 10% of their body weight), both because they attack the food in several different places and because they contaminate it with urine and faeces (on average, each rodent produces some 20 faeces and several milliliters of urine daily),
  The species most commonly involved in infestation in the food processing industry are:
- Brown rat (*Rattus norvegicus*)
- Black rat (*Rattus rattus*)
- Domestic mouse (*Mus musculus*)

Of these, the *domestic mouse* is the worst of all since it proliferates and spreads extremely fast. A special watch must be kept on poor design features such as hollow dividing walls, walls filled with polystyrene, pillar casings, and false ceilings. These are all places where the domestic mouse can live, moving from one location to another without being particularly noticeable, and reproduce at a frightening rate.

The mouse differs from the rat in several respects, e.g.

- Eating habits – the mouse is an erratic eater, typically consuming very small quantities of nourishment in several different places.
- Territory – the mouse tends to move in a smaller territory than the rat, and this territory becomes increasingly smaller as the habitat becomes more favorable. For example, if there is plenty of food and also hiding places, then the mouse can easily survive in an area of just a very few square meters (yards) and thus nest inside production departments.
- Size – the domestic mouse is much smaller than the rat and therefore is quite capable of entering very small cracks and crevices, e.g. 5 mm (³⁄₁₆ in). It therefore finds it much easier than the rat to enter a factory or any other type of building.
- Patterns of movement – the mouse does not follow fixed paths and tends to move around erratically. This makes it very difficult to eliminate and it is essential to discover each nesting location which in turn implies careful inspection and monitoring.

Mammals – Chiroptera (bats)

Bats are mammals potentially constituting a serious problem in food factories and warehouses, where they lodge in the upper parts of rooms and buildings. They heavily contaminate the environment and foodstuffs due to the vast quantities of excrement left in the locations below where they are lodged, and to the drops of urine excreted during flight.

The danger is both health-related and economic. Bats disseminate many parasites capable of transmitting pathogenic germs to humans. They spread hematophagous and non-hematophagous Acari, hematophagous bugs, fleas and fly larvae. In addition, bats easily catch rabies and are considered to be major and dangerous carriers of this disease. Moreover, they are habitual carriers of Spirochetes (including that of leptospirosis) and of many fungi that are pathogenic to man. Remember, however, that in some countries, some species may be protected by law.

In Italy the types of bats most frequently found in cases of infestation in the food industry belong usually to three genera:

- *Miniopterus*
- *Myotis*
- *Pipistrellus*

and, less often, to:

- *Rhinolophus*
- *Myctalus*
- *Plecotus*

Their presence is noticed not so much by direct observation of the bats themselves as by finding their excrement on the floor or on the materials stored. Bat excrement must not be confused with that of the domestic mouse or of small birds such as sparrows:

- Bat excrement – Typically contains numerous fragments of the shell casings of the nocturnal insects on which they feed. It is easily recognizable since it disintegrates into dust as soon as it is touched.
- Rat and mice excrement – Contains hairs consumed by these animals when they clean their coats. It is sticky to the touch and does not disintegrate if subjected to pressure.
- Bird excrement – Contains a white material which is simply urine (mainly the crystalline salts of uric acid) combined with the darker-colored faeces.

Birds

Domestic sparrows and other birds

It is nearly always *domestic sparrows* (*Passer domesticus*) that cause problems of infestation in the food industry and in warehouses, while pigeons (*Columbia livia*), starlings (*Sturnus vulgaris*) or swallows (*Hirundo rustica*) are only found in occasional, specific circumstances.

Besides causing economic damage due to the deterioration and contamination of foods caused by their excrement, birds accumulate a large quantity of straws and twigs inside buildings for use in building nests – thus causing a hazard not only in health but also in terms of fire risk, e.g. in electric control panels, switch boxes, etc. Note – Insect infestations may also start from insects brought into the plant by birds. However, the biggest risk is the human health hazard, since birds can transmit a number of diseases to humans, either directly or indirectly, i.e.

- Mycosis (*Aspergillus* spp., *Candida* spp.)
- Ornithosis (Virus)
- Encephalitis (Virus)
- Pseudotuberculosis (*Pasteurella* spp.)
- Salmonellosis (*Salmonella* spp.)
- Toxoplasmosis (*Toxoplasma gondii*)

Remedial action for pest infestation

The discovery of an infested food or food product should always be followed by identification of the specific infestant and of how it has managed to penetrate or contaminate the foodstuff. Once this information has been collected, it is possible to use a *specific and focused* approach to insure successful elimination of the cause of infestation.

Thus, recognition and identification of specific infestants is vitally important, and can be done following the basic screening outline described above. However, another essential part of the identification process is comparison of the infestant or evidence of infestation found with the specific characteristics of each species in reference books providing exhaustively detailed descriptions and photographs. Such books are listed below.

Further reading

Adequate reference books should be regarded as key items of equipment for pest control laboratories. Good material should ideally be available in the local language, but the following English texts have been found to be useful in achieving good control of pests.

Kurtz, O'D. L. and Harris K. L. (1962) *Microanalytical Entomology for Food Sanitation Control*. Washington, DC: Association of Official Agricultural Chemists

Meehan, A.P. (1984) *Rats and Mice: Their Biology and Control*, Rentokil Library, East Grinstead, UK

Ebeling, W. (1978) *Urban Entomology*, Division of Agricultural Sciences, University of California, Berkeley, CA

Truman, L. C., Bennet, G. W. and Butts, W. L. (1982) *Scientific Guide to Pest Control Operations*, 3rd edn Duluth, MINN: Harcourt Brace Jovanovich; Cleveland, OH: Harvest Publishing Company

Katsuyama, A.M. and Strachan J. P. (eds) (1980) Food plant pests. In *Principles of Food Processing Sanitation*, pp. 167–224. Washington, DC: The Food Processors Institute

Imholte, T.J. (1984) Pest control systems. In *Engineering for Food Safety and Sanitation*, pp. 209–232. Crystal, MINN: Technical Institute of Food Safety

Troller, J.A. (1983) *Sanitation in Food Processing*. New York: Academic Press

Busvine, J.R. (1966) *Insects and Hygiene*, 2nd edn. London: Methuen & Co

Bond, E.J. (1984) *Manual for Fumigation for Insect Control*, FAO Plant Production and Protection Paper No. 54, Rome: FAO

Lawson, T.J. (ed) (1987) *Stored Products Pest Control*, British Crop Protection Council, Thornton Heath, UK

4.D.2 Guidelines for the exclusion and elimination of infestants

Background

It is relatively easy for insects, rodents, and birds to invade production premises. These undesired and undesirable guests normally adapt very quickly to their surroundings and thus it is not unusual for them to install themselves inside factories, even close to processing lines or in areas used to store food products. Generally speaking, these premises provide an abundance of the foods liked by these infestants, and thus infestation is likely to be protracted, accompanied by rapid proliferation of the species concerned.

Elimination of this infestation is achieved using a variety of approaches as described later on, but special attention *must* be given to insuring perfect application of *preventive measures*, i.e. measures designed to *exclude* infestants from premises since these are undoubtedly the key to winning the battle against food infestation.

In the following sections the basic guidelines relating to each group of infestants, i.e. insects, rodents, and birds, are discussed.

When developing a pest control program it is essential:

- To have expert advice – whether 'in-house' or external
- To put the program in writing
- To record results in detail so as to permit analysis
- To review the program at appropriate frequencies and record action taken or changes made.

Note. Local legislative requirements and Codes of Practice must be known and followed closely.

Insects

Introduction

Insects can invade production premises and from here install themselves in the products being made or in finished goods. They are typically able to perforate a large variety of packaging materials other than cans or glass jars, lodge inside a pack and from there start to colonize other packs.

Clearly, this means that contamination spreads very rapidly, causing serious damage. To avoid this ever-present risk, various types of action must be taken – besides actual disinfestation, described later. Consideration is first given to the preventive measures – in terms of *exclusion beforehand* of insects – and standards of hygiene required to minimize the risk of infestation by insects.

Exclusion measures

Points of insect access must be eliminated or – if elimination is impossible – reduced to a minimum in food production and storage premises.

To do this, the following action must be taken in the following areas.

External measures

- Outer boundaries of factory areas. The walls around the building should be surrounded by a tarmac or a well-paved path preferably 7 to 8 m (7½ to 8½ yd) wide – which must be kept as clean and empty as possible. The parking of pallets, goods, and equipment in this area must be kept short. In addition, this area must be well drained with appropriate drainage channels and grooves. Puddles of stagnant water provide an ideal breeding ground for many types of insects.

- Windows. These must be protected by anti-insect netting preferably using galvanized iron of 18 mesh. If this netting is not available, the windows must be kept permanently and tightly closed.
- Air (intake) vents. These too must be equipped with anti-insect netting. If the vents are located in critical production areas, they must be equipped with special absolute (HEPA) filters.
- Ventilation of premises. Condensation must be avoided. Condensation drips cause puddles of water to form on floors, or, at best, undesirable moist patches. *Water* means *life* and consequently the creation of conditions suitable for the proliferation of infestants.
- Lighting. It is important to remember that sources of light are a strong attraction for several species of insects although they act as a repellent for others. However, the lighting of operational areas should be such as to permit very accurate work, and also efficient inspection. The positioning of lights directly over areas where the products are processed should, however, be avoided, in order to prevent the product from being exposed to flying insects which sooner or later will gravitate towards the source of light. The positioning of lights directly in line with doors or windows should also be avoided. Remember, lights attract the insects to and within the food operations area.
- Doorways leading into production departments. The doorways that directly link production departments with the outdoor environment are the preferred means of access of flies and mosquitoes to invade the factory. When this problem exists, the solution is a double door where, between the two doors, a timed insecticide nebulizer is fitted – usually using pyrethrum, which when properly applied is non-toxic for humans. The doors must be equipped with a self-closing system. *Warning!* Pyrethrum sprays or other types of insecticides *must on no account* be used inside production departments during manufacturing, since the insects hit by the spray can fall anywhere, and that includes into the product.

Internal hygiene and cleanliness of premises

Of all the possible weapons against insects, hygiene and cleanliness are the most powerful and effective.

Indeed, often they are the *only* ones able to overcome certain problems. To achieve a good standard of hygiene the following list includes requirements that must be met:

- Production premises. These must be kept extremely tidy. Various types of material must not be allowed to accumulate and product leftovers or rejects must not be left standing in the area, but removed at frequent and regular intervals.
- Unused space. For example, corners and spaces behind and below machinery, between machines and between machines and walls must be kept scrupulously clean. Production leftovers must be removed at the end of the day. All of these *hidden points* should be frequently inspected to confirm that they are really clean. Remember, they are a favorite nesting place for insects.
- Refuse and waste. This must be collected in properly designed lidded containers. These should ideally be located outside production departments in separate rooms which should have, for example, tiled walls and floors for easy maintenance and perfect cleanliness.
- Walls and floors. In all premises these should always be kept clean. To make this easier, care should be taken to insure that there are no cracks, crevices or holes since these tend to retain dirt.
- Junctions between walls and pavements. These should be curved (coved) as this makes total removal of dust possible. The points where dust typically tends to deposit are skirting boards, window sills, and lamp fixtures (fittings). Frequent cleaning should be arranged for all these items, which must always be included in routine inspections.
- Electric control panels and wireways. These should be regularly cleaned by removing dust, etc.
- Drainage channels. These should be kept free from rubbish and regularly washed down.
- The space immediately below the roof. This may be overlooked but must be kept clean and disinfested. These spaces, which may not be easy to move about in, are a place where dust and food residues can remain undisturbed for considerable lengths of time, thus providing an ideal breeding group for insects. Given this, they must be regularly cleaned and frequently inspected.
- Damaged materials or products. These, particularly if already infested by insects, must be removed immediately from the production area. Clearly, this means regular and sufficiently frequent inspection of raw materials. (See the discussion pages in Section 4.D.3 on 'Monitoring and Inspection'.)
- Production machinery and equipment. This should ideally be set above the ground at least 200 mm (7½ in) using metal legs. Where this is not

possible, machinery or equipment should be sealed to the plinth or floor. Machinery should not contain any incompletely 'closed' parts where the powdery dust normally produced by food products might settle. When this happens, the parts tend to become a central infestation location that is also difficult to see. It is therefore much better that the powder falls directly on to the floor thus permitting easy cleaning after each production run.

The slots housing the levers of monoblocks should be protected by slightly overlapping rubber edges which, while permitting the levers to move freely, reduce the penetration of dust into the monoblock itself (which is difficult to reach during routine cleaning operations).

Machinery casings (or guards) often create channels where dusty residue accumulates. Unless these casings are strictly necessary – for reasons of accident prevention – they should be removed. Ideally, production machinery should be open and easy to clean and inspect.

Electric control panels may be installed on walls provided that the junction is sealed. Otherwise, they should be placed at least 500 mm (20 in) away from the wall. Push buttons should extend at least 50 mm (2 in) from equipment to allow for cleaning.

Wireways should be avoided when possible, using simple clamps instead. Remember, electric cables and wires tend to collect dusty food residues thus becoming a breeding ground for insects.

With overhead cables, the use of wireways is inevitable. The wireways should be of the open type which is easy to inspect and clean. If wires or cables are boxed, this provides passageways for insects and mice and makes disinfestation difficult.

The edges of conveyor beds should not be too high or made in such a way as to retain dust.

Rodents

Introduction

Rodents – rats and mice, and in particular the domestic mouse (*Mus musculus*) – are especially dangerous infestants due to their ability to colonize a wide range of different locations.

Here we will outline the basic guidelines concerning the prevention of access, hygiene requirements and in addition describe the use of chemical substances and traps in preventing proliferation of rodents should they manage to invade premises.

Exclusion measures

In principle, as with insects, access must be eliminated or reduced to a minimum in food production and storage premises.

External measures

The domestic mouse, as mentioned earlier, is capable of entering through cracks as little as 5 mm (3/16 in) wide. Thus, in principle, if production and storage premises have no cracks gaps or holes of 5 mm (3/16 in) or over in any part of their perimeter, they can be considered to be impenetrable.

In contrast, the rat – a much larger animal – can enter premises by tunnelling in the ground and bypassing building foundations.

Both rats and mice are excellent climbers and are also capable of gnawing and penetrating an enormous range of hard materials.

Given all this, the primary exclusion measures to be taken are as follows:

- Foundations. These should be in concrete and go down to a minimum of 300 mm (1 ft) below solid ground level; 600 mm (2 ft) is recommended.
- Potentially gnawable walls. In this case smooth sheet metal at least 300 mm (1 ft) wide should be used to cover the base of the walls. Rodents are unable to climb or gnaw smooth, flat surfaces since they need a rough surface to start their attack. In fact the most vulnerable points are doorframes, window casings, etc.
- Entry doors. These should be equipped with automatic self-closing devices.
- Openings around the pipes leading into and away from the premises. These must be protected, e.g. sealed with wire netting which is then filled and covered with cement. Alternatively, cement mixed with fragments of glass can be used.
- Surrounding perimeter area. As for insects, this space should be kept totally empty for as wide an area as possible. Any heaps of materials against walls are excellent hiding places and entry points for rodents.

The tarmac path around the building described for insect prevention is also important in the case of rats and mice. This path should ideally be 7 to 8 meters (7½ to 8½ yd) wide. No grass or weeds or bushes should be allowed to grow under the walls as these allow rodents to move about unnoticed.

When there is no tarmac path, the growth of weeds must be prevented by carefully scheduled use of weedkiller and the ground should be kept as flat and level as possible. The growth of vegetation can also be quite effectively prevented by using gravel to cover the ground.

- Drains. These are a favorite way of access for mice. Drainage channels and gratings should also be kept empty and clean. Drainage gratings and covers should be easy to remove for routine maintenance but at the same time must be firmly set in place with drainage holes designed to prevent the access of rodents.
- Inside the factory and warehouses. Remember,

untidiness must be avoided. Often lack of space leads personnel to store materials and/or equipment close to walls on a more or less permanent basis. The end result is that all these areas become difficult to clean and inspect, and an ideal habitat for mice.

- Prefabricated buildings. These often use plastic joints that can be gnawed by rodents. In these cases, careful ongoing inspection and control is imperative. When damaged plastic joints are found with signs of gnawing, these should be immediately replaced and sealed.
- Use of barriers. This is sometimes required to prevent rats and mice from reaching certain positions from where they could easily invade the factory or warehouse.

For further information see Section 4.B.1, Appendix on rodent proofing.

Internal hygiene and cleanliness of premises

Nearly all the guidelines indicated in the case of insect prevention also apply in the case of rats and mice with special emphasis on:

- Systematic removal of production leftovers and rejects.
- Use of tightly lidded containers made in such a way as to prevent gnawing.
- Constant tidiness, cleanliness and regular inspections.

Chemical baits

Should rats or mice gain entrance to the factory/warehouse grounds or premises, proliferation can be prevented by the use of rodenticides which are poisonous chemical substances and available in two main types.

- Acute, fast-acting type used in single doses.
- Delayed-action type used in multiple doses.

All rodenticides have to be incorporated into edible substances in order to make food baits capable of attracting the rodents. As an edible material, cereal in pellet or flake form is preferred, but grains are also useful. Mice also like chocolate but not cheese. These baits must not be spread around on a random basis, but positioned in special containers used exclusively for this purpose. This is because of the need to avoid (a) dangerous environmental pollution by loss or dispersal of poisoned bait and (b) rapid deterioration of the baits causing them to become less *appetizing* for rodents and therefore useless.

Powdery material should *never* be used as bait. The baits should always be in the form of flakes, pellets or grains.

Baits should be placed near the nests or the tracks used by the rodents and, whenever possible, between the nest and the usual or likely source of food. They should be renewed fairly frequently and when actually taken (bitten) by a rodent should be replaced daily. Any baits which have clearly deteriorated should be removed and destroyed. It is good practice to date the bait containers (trays) to monitor usage. The number of bait containers at each location should be recorded (logged) together with the dates of changing or replenishing.

Chemical baits can also be used as a *warning signal* (see Section 4.D.3 – Monitoring of Infestants) i.e. even when no infestation has taken place. In this case chemical baits are positioned on a periodical basis and systematically replaced.

By using this technique it is possible to:

- Monitor premises and highlight any processes of infestation that happen to occur.
- Vary the number of baits according to the intensity of the rodents present.

An important point to remember is that the poison used in baits should not be used more than twice during any given year. This is to avoid the risk of rodents developing resistance to a poison with the subsequent growth of permanently resistant strains.

When used outside, these baits must be protected from the weather since moisture causes them to deteriorate rapidly and become unappetizing. It is a good idea to place them in plastic bags since rats and mice will, in any case, gnaw these when seeking to reach the food. Good basic materials for these baits are corn flakes or whole or fragmented cereal grains in good condition.

Chemical baits should only use delayed-action, multi-dose rodenticides, e.g.

- Bromadiolone – a coumarinic derivative that is usually accepted by rodents. Consumption of this type of bait just once is capable of causing 100% mortality in the case of the brown and black rats, while it is slightly less effective with the domestic mouse. Bromadiolone in oil solution (0.25%) must be mixed to cereal grain in the ratio of one part to 49, giving a final concentration of 0.005%.
- Chlorophacinone – an indanedione derivative, used in 0.005% concentration. It is moderately toxic for dogs and cats but highly effective in the case of both the brown and black rats and the domestic mouse, for which it is usually lethal even after a single dose.

These poisons are anti-coagulants; they block availability of Vitamin K in the liver thus impeding coagulation.

The baits must be used liberally in an area until the rodents stop eating it. A campaign of this type can last from a week up to 3 to 4 weeks.

The degree of appeal of baits for rodents and of

likely deterioration depends on how they are preserved and the conditions of use.

In summary, poisonous chemical baits must be:

- Clearly marked as being *poisonous*.
- Collected and destroyed once they have deteriorated or have been taken.
- Frequently replaced during long-term use.
- Carefully stored in special places out of reach of people not authorized to use them.

The corpses of dead rats and mice must be removed and destroyed immediately, ideally by electric incineration.

Traps and repellents

In the paragraphs below the use of mousetraps, ultrasonic repellents, glues and gas is discussed:

- Mousetraps. These are useful and also *preferable* in cases where there is any risk of poisonous chemical baits contaminating foods. However, they can only be used in small areas and also have the disadvantage of requiring the use of labor and a certain amount of skill. Nevertheless, they do have the advantage of preventing rodents' corpses from ending up in inaccessible places with the risk of decomposition and infection. The effectiveness of traps depends largely on how well they are positioned. Mice are attracted by chocolate, and pieces of 'Mars™ Bar' make excellent bait.
- Ultrasonic repellents. These devices transmit sounds at a frequency of over 15 kHz and, if appropriately positioned, act as rodent repellents. In theory this system is undoubtedly interesting because it solves the rodent problem in a clean manner. In practice, however, the ultrasonic devices currently available to us provided only a weak initial repellent effect that never lasted long. Rats and mice soon got used to the signals and then ignored them.
- Vibrations. Some good results are obtained applying vibration generators to floors or walls of the area to be defended. Generators need to have frequent changes of vibration frequencies and they must be applied to walls directly. Vibrations don't permit nesting of the rodents.
- Glues. These are useful in the same circumstances as traps but remain active only as long as environmental conditions permit them to do so. The adhesive strips are rendered useless by dust, excessive heat or cold, and processing residues.
- Gas. As described in the section on disinfestation substances and products, gas – when its use is possible and desirable – is equally effective for both insects and mice. Methyl bromide has been the most commonly used type.

Keeping storage areas well lit has been successful in discouraging rodent infestation, as they prefer dark, quiet places.

Birds

Introduction

Birds are very difficult infestants to eliminate. Once they invade a factory or warehouse they are capable of producing extensive damage and, moreover, spread bacteria and viruses and sometimes insects and/or mites. The primary aim must be exclusion but standards of cleanliness are important as highlighted below. There are also some notes on specific techniques to be used to prevent proliferation, should birds enter premises.

Exclusion measures

Generally speaking, birds tend to gain access to factories via the upper parts of buildings. More or less the same basic guidelines indicated for rats and mice apply in the case of birds, with special emphasis on the following points:

- Main doorways and entrances. Self-closing doors are preferable. If proper doors cannot be fitted for reasons of cost, strips of heavy-duty rubber should be hung across the doorway or entrance. When doors are shut, there should be no gaps wider than 5 mm (³⁄₁₆ in) between the frame and the door itself.
- Roof. It is often easy to find gaps and cracks in a roof, and these should be sealed. If there are too many of them to do this, anti-bird netting can be fixed along the edge of the roof and anchored to the walls below. A tiled roof should ideally be built with flat tiles so that there are no gaps for birds to enter or nest in.
- Building as a whole. Ideally, the building should be as free of window sills, ledges and protuberances as possible, thus not giving birds a chance to lodge or nest.

There are, however, points where protuberances or ledges inevitably exist, especially with older buildings. In most cases it is sufficient to fill in the ledges, preferably using a smooth, slippery material, as shown in Figure 4.32.

If continued resolutely over time (birds are particularly persistent), repeated washing of nesting places with strong jets of water helps dissuade birds from lodging. Alternatively, nests may be removed by hand.

Nesting of birds can, however, be avoided. For example, in the case of swallows' nests two methods have been tested and shown to be effective. Swallows build their nests beneath the eaves of buildings, adhering to the vertical wall face and a portion of the eave. By stringing a wire along the junction of the wall and the roof overhang, and draping a 300 mm (12 in) curtain of polyethylene or aluminum foil over it, the swallows are prevented from attaching their nests to the wall because the

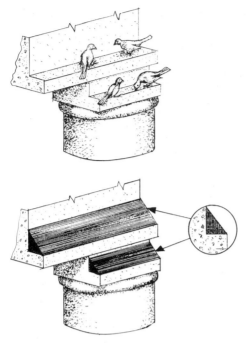

Figure 4.32 Example of how ledges can be eliminated

new surface is too smooth. The only disadvantage of this method is that it is unsightly.

The second method is slightly more complex but does not require the installation of unsightly material under the eaves and will keep the swallows away for several years. Two fine nichrome wires are installed under the eaves parallel to the wall and 25 mm and 75 mm (1 in and 3 in) away from it. The wires are then connected to a cattle-fence charger. When swallows attempt to build their nests, they receive strong – but non-lethal – shocks from the wires and will rapidly move to some other nesting site.

In covered loading and unloading docks, 12.5 mm (½ in) wire mesh positioned under the roof can prevent birds flying up to the beams, rafters, etc. They cannot perch or build nests.

Cleanliness

In the area surrounding the factory and warehouse, no material or shallow pools of water which are attractive to birds should be left in the open. If numerous birds surround the buildings, sooner or later they will enter, although bird proofing of the building greatly reduces the frequency and therefore the problem. Inside the building measures can be taken to reduce the number of places where birds can perch or nest, e.g. in storage sheds for ingredients.

Inside the premises it is important to remember that:

- Systematic removal of production leftovers is as important in the case of bird prevention as in insect prevention. Uncovered waste bins are the equivalent of large feeding trays.
- Birds need water, and thus all premises should be kept scrupulously dry, avoiding puddles of condensation, leaking taps, etc.

Specific techniques

Generally speaking, there are very few specific products for the control of birds. Birds in fact are generally protected by very severe laws. Given this, *they must only be caught for subsequent release in an area a long way away from the plant.*

Examples of the few products available on the Italian market are as follows:

- Chemical contraception. This is a method of chemical birth control designed for pigeons. The technique is based on the use of a bait coated with Ornitrol (azachosterol dihydrochloride).
- Avitrol 200. This is a compound (the main ingredient is 4-aminopyridine) that is also toxic for other animals and humans if consumed or inhaled. This particular treatment should be authorized by the health authorities and carried out by expert personnel. Once again baits are used; these are dosed with 0.5%–1% of product and placed close to the places where the birds tend to perch. After treatment, any unutilized baits must be removed.

A conventional approach to exclusion of birds from factory and warehouse premises should be based on use of the following techniques:

- Bird nets or mist nets (UK). These nets have a mesh with each side of about 20 mm (¾ in) to catch sparrows and small birds and 50 mm (2 in) for pigeons, starlings and larger birds. The nets are attached to ceilings and to the side walls and are at least 2 m (approx. 2 yd) wide. Ideally they should be set up in poorly lighted areas frequented by birds. Once the birds remain trapped in the nets, they should then be disentangled and freed a long way away from the factory/warehouse buildings.
- Traps. These must be built in such a way as to permit the birds to enter and remain for quite some length of time, since it would not be feasible to check the traps more than once a day. An example of one type of trap is shown Figure 4.33A. A cone-shaped configuration of wire guides the birds through an opening into a large cage that they can enter but from which they cannot escape. Figure 4.33B. shows an alternative; a balanced and divided cage roof allows

birds to enter, but prevents their escape. Grain may be placed in the cage as bait, but sometimes a container of water is even more effective.

Another type of trap uses a principle employed for pigeon lofts. The entrance to a screen cage consists of vertical rods that are hinged at the top. A bird trying to get into the cage to reach the food or water can easily swing the rods inwards, but once inside the cage, the rods return to their original

positions. Figure 4.33C. shows the door mechanism from *inside* the cage.

- Poisonous baits. Even although this technique, when its use is permitted by law, is generally considered to be the most effective, some difficulties and risks are involved. It is normally advised to place the same type of *unpoisoned* bait (cereals in the form of grain or pellets, corn flakes, etc.) for 4 to 5 consecutive days in places where the birds usually lodge or perch. After this period, when consumption of the baits is satisfactory, *poisonous* bait should then be put in the same places. Materials known to be effective but which may not be lawful in your country are:
  - Strychnine sulphate
  - Thallium sulphate
  - Tribromoethanol which is also used as a narcotic.

Other methods which may be legal are:

- Narcotic baits. Where poisonous baits are not allowed, narcotic baits may be permitted. Katsuyama and Strachan[1] state that the most widely used is tribromoethanol. After pre-baiting for several days, preferably just after dawn, the narcotic bait is used. This takes effect 10 to 15 minutes after it has been eaten and lasts for 1 to 3 hours. Anesthetized birds are collected and may then be released well away from the plant. α-chloralose may also be used at 0.5% concentration on a bait of rice or lentils.
- Shooting. Where this is legal, it has been described as moderately effective against roosting birds using a fine shot in a small-caliber gun. It is only successful when the dead bird has been recovered and removed. If used, full safety precautions must be taken to protect personnel, materials and equipment in the area.
- Glues and pastes. Various types of adhesive pastes and glues can be used to exclude birds. Although their effectiveness is limited, they are useful when dealing with limited infestation by birds. These substances are glues that do not dry or harden, and are applied to perfectly smooth surfaces. The layers of glue or paste should be at least 12 mm (½ in) thick and should not be more than 70 to 100 mm (2¾ in to 4 in) apart so that the birds cannot land between the various strips.
- Gas. The methyl bromide treatment used for insect disinfestation is also very effective for birds.

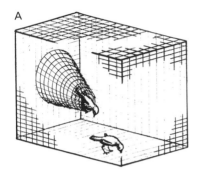

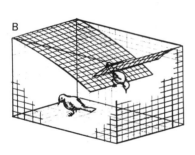

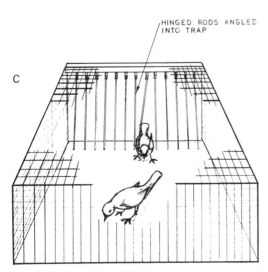

HINGED RODS ANGLED INTO TRAP

Figure 4.33 Examples of bird traps

Reference

1. Katsuyama, A. M. and Strachan, J. P. (eds) (1980) Food plant pests. In *Principles of Food Processing Sanitation*, pp. 167–224. Washington, DC: The Food Processors Institute,

4.D.3 Monitoring of infestants

Background

'Monitoring' means systematic inspection or surveillance and the recording of the species of infestants present, their numbers and where they were found. The filing and review of data concerning the identification and count of species of infestants found to be present, makes it possible to:

- Intervene immediately when abnormal situations occur.
- Follow the effectiveness over time of the various corrective and preventive measures taken.

The easiest way of collecting these data is to use a record form for each monitoring point. A record is made of each species of infestant found, specifying the number of pests captured or identified and the date/time of capture or identification. See Figure 4.34.

If done promptly, the recording and identification of infestants is one of the best means of preventing large-scale harmful infestations that are difficult and costly to eliminate. In a food plant, infestation can occur:

- Directly due to the infestant's entry of premises from outside through unprotected openings (which, in some cases, are also unnecessary).

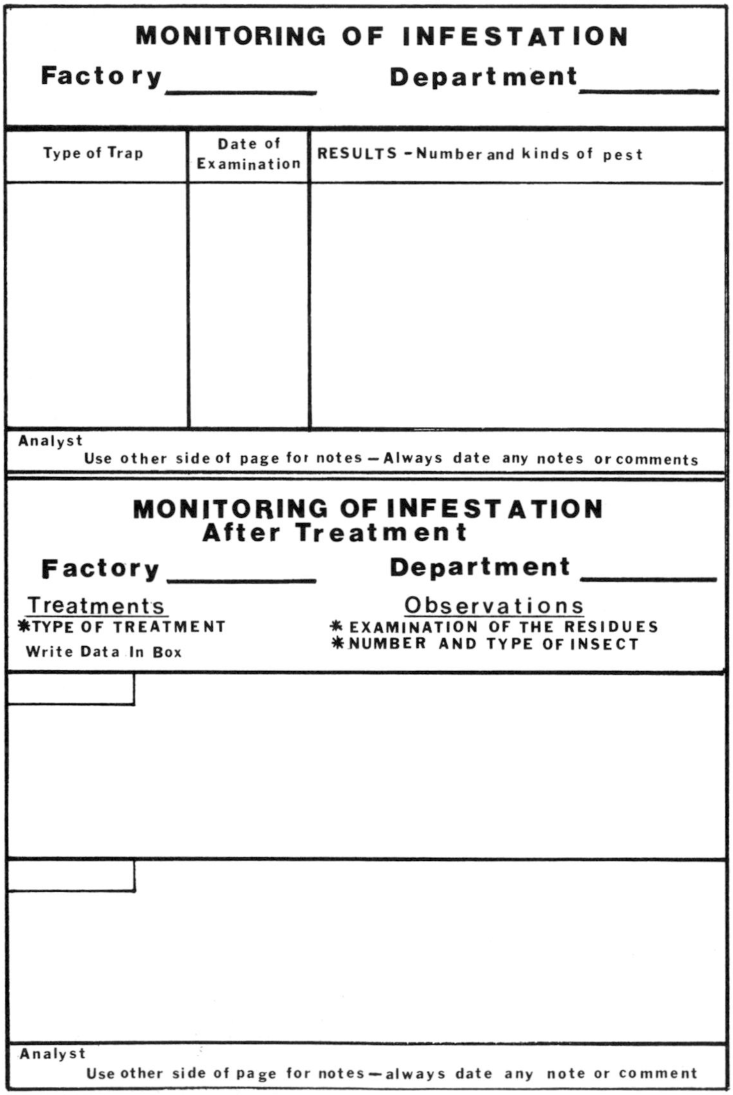

MONITORING OF INFESTATION

Factory_____   Department_____

Type of Trap	Date of Examination	RESULTS – Number and kinds of pest

Analyst
Use other side of page for notes – Always date any notes or comments

MONITORING OF INFESTATION
After Treatment

Factory_____   Department_____

Treatments
*TYPE OF TREATMENT
Write Data In Box

Observations
* EXAMINATION OF THE RESIDUES
* NUMBER AND TYPE OF INSECT

Analyst
Use other side of page for notes – always date any note or comment

Figure 4.34 Example of monitoring records

• Indirectly with infestants carried by raw or packaging materials, or various types of equipment (pallets, machinery, etc.).

The implication of this is that monitoring of infestants must include not just ingredients and other materials but also production and storage premises.

But even this is not enough. As already indicated, the areas surrounding factory and warehouse buildings must be controlled, together with all those areas of the factory that are in any way adjacent to the food storage and processing departments – cellars and attics in particular. It would clearly be a big mistake to neglect those very areas where infestants are most likely to enter and lodge. Indeed, it is in these areas that particularly effective preventive measures must be concentrated.

However, in spite of preventive measures, there is always the risk of infestation occurring due to the introduction of infestants by what is called the 'Trojan Horse' – i.e. ingredients, packaging materials, and other items entering the factory and warehouse.

Thus comprehensive monitoring is of ongoing importance and is discussed under the following headings:

• Materials entering the factory
• Outside areas
• Inside areas

Materials entering the factory

There are two main categories of materials – raw (ingredient) materials and packaging/other materials.

Raw materials

When possible, monitoring must start from the suppliers' factories, using the same criteria described below for the processor's establishment. In the processor's establishment, when raw materials are received but before they enter the warehouse, inspections must be performed and samples taken for laboratory tests. Inspections are basically divided into three parts, i.e. inspection of (a) vehicle used to transport materials, (b) packaging, and (c) the raw materials themselves.

Going into greater detail; ingredients are usually delivered in containers which can be manually handled (sacks or bags) or in bulk, e.g. by tanker.

• Sacks or bags – It is important to look for signs of insects or rodents (excrement, hairs, urine, evidence of gnawing) in the vehicle and on the sacks. Pallets, too, should *always* be inspected. Examine the folds of the sacks very carefully. Leave a certain number of sacks open for about

15 minutes, thus allowing any insects present to move into the open in search of air. Long-wavelength (365 nm) ultra-violet light from a hand-held lamp may be useful for inspection as it causes rodent urine to fluoresce.

Using the appropriate sieve, sift an adequate sample of product and examine the residue both visually and, if necessary, under the microscope for the presence of insects and insect fragments, mouse hairs, etc.

Use a 20-mesh sieve (45 denier/cm^2) for fine powder products and a 10-mesh sieve (13.4 denier/cm^2) for granular products. Examine residue on a white cloth.

A simple test for the presence of mites in starches or flours is to pour about 50 g (2 oz) into a small pile. If mites are present the sides of the pile will be seen to move within 2 minutes as the mites travel outwards to the light.

• Tanker – When the tanker arrives, inspect the raw materials through the upper loading hatches to see whether there are any live insects moving up to the surface. Take an appropriate sample for sieving and visual/microscopic examination.

Filth tests

The laboratory testing of samples must be performed using a suitable filth-test method. The exact type of method depends on whether the substratum being examined is:

• Completely soluble in water – and thus a liquid.
• Insoluble in water – powder or flakes.

The methods can be found in *Official Methods of Analysis* (14th edition) from paragraph 44,001 to 44,225 published by the AOAC (Association of Official Analytical Chemists, Inc. 111 North 19th Street, Suite 210, Arlington, Virginia 22209, USA) or in other specialized microanalytical method publications.

Regardless of the method used, the insects found, or insect fragments, eggs, larvae, chrysalises, etc. must be identified, counted and recorded and the same must be done for rodent hairs and excrement (the problem of identification is discussed later). The trend over time of these data will eventually permit immediate pinpointing of dangerous infestations or of the appearance of abnormal variations.

Another analytical technique that should not be ignored is the use of X-rays although it may be a 'research' rather than 'routine' tool. This in fact avoids the disintegration and chemical treatment normally required for filth-test methods and which tend to make microscopic examination long and difficult.

The X-ray apparatus concerned is based on the generation of soft radiation by an X-ray tube equipped with a beryllium 'window'. This technique is well-suited to entomological research on raw materials (grain, dried fruit and vegetables, freeze-dried vegetables, etc.) and finished products (biscuits, crackers, pasta, chocolate, etc.).

Radiography makes it possible to discover simultaneously:

(a) The extent of the infestation
(b) The development stages of the insects present
(c) Their degree of vitality

It is also possible to detect the presence of fragments of acarids and insects in processed foods as long as the skin has not deteriorated as a result of processing and cooking.

Another detection technique useful for monitoring – although not on a routine basis – is scanning electronic microscopy (SEM) which makes it possible to obtain full-view images (virtually three-dimensional) that are very clear, together with significant enlargement. For SEM examination, the samples are stuck to Duralumin slides, treated with OsO_4 gas and then vacuum covered with a thin layer of gold platinum that makes them opaque to the electrons.

Use of the SEM technique means that the outer cases of insects can be used to help identify the exact species of infestant, and the same also applies to rodent hairs. It must, however, be pointed out that the interpretation of SEM images require excellent entomological knowledge and because of this and the cost, the technique is only used in special cases by experts.

Packaging and miscellaneous materials

Clearly, the inspection technique must be used to see whether insects or traces of rodents are present, but once again in the case of packaging materials, inspection of the suppliers' factory should not be neglected – particularly when a serious infestation of supplies is discovered.

The items to inspect upon arrival of packaging materials are basically the same as those indicated in the previous section, i.e.:

● Vehicle used to deliver goods
● Outer packaging of materials
● The materials themselves

Special but not exclusive attention must be paid to detecting traces of rats and mice (hairs, excrement, urine) since this type of infestant is found in packaging materials and trucks.

In addition, a careful examination must be made of the pallets supporting the materials – they are often a nesting place for Blattae and other insects. Also, the folds of sacks or bags must be carefully observed.

Always record the species identified and the number of individuals found, to permit an ongoing control, to evaluate the results of protective measures taken and, more generally, to allow early intervention in any apparently abnormal situations which suggest potential infestation.

Outside areas

Introduction

Monitoring of outside areas surrounding buildings which are a possible source of pests must be performed systematically on an ongoing basis, recording all findings, action taken, and results. This approach makes it possible to alert the managers concerned immediately so that they can take effective action, thus avoiding costly and hard-to-control infestation. In the case of outside areas, monitoring requires the use of equipment to attract and/or kill infestants.

Given that the factory and warehouse buildings are surrounded by a wide empty path or area covered in tarmac or gravel, with no weeds, bushes, or rubbish (see Section 4.D.2), the following types of devices can be positioned around the buildings for insects and mice.

Insect traps

These are of several types and include:
● Poisonous baits for air-borne insects. These must be placed at sufficient distance from buildings to create a 'ring' around them, thus attracting air-borne insects without letting them get too near to building entrances. The devices are simple plates containing a piece of sackcloth or jute on to which the poisonous bait is poured. The bait can be made from an aqueous sugar solution containing Propoxur (see Table 4.7, p. 101), a fast-acting insecticide. The sugar and water attract the insects looking for food and the Propoxur kills them off rapidly.
● Baited traps. Within the ring of baits described above, a second ring of baited traps can be used. These normally consist of a plate raised a few centimeters off the ground, with a hole in the middle on which an inverted net cone rests, with an opening in the top. An inverted net cup is placed on the edge of the plate, thus closing the open-ended cone mentioned above. In this way a cavity is created between the two nets and this is the chamber in which the bait is placed. The bait's smell attracts the insects which enter from the base of the trap. They are attracted upwards by the bait and, once inside the trap, are unable to escape.
● Light traps. These work in the same way as baited traps except for the fact that the 'bait' consists of a source of artificial light.

- Glass-jar traps. These can be set against walls and consist of a normal glass jar whose neck is covered by a piece of dark cloth with a small opening. A poisonous bait – Bendicarb, for example – is placed inside the jar. This attracts the insect which, once inside the jar, is unable to find the way out and is lethally poisoned.
- Electrocuters. It is a well-known fact that insects are attracted by ultra-violet light (optimal wavelength = 356 nanometers). Ultra-violet lights are placed inside electric grilles and the insects attracted to the light are electrocuted when they enter the electric field. These traps are very efficient, particularly in the case of nocturnal air-borne insects. Troller[1] makes three important points in their use:

  1. The electrocuter must be carefully placed so that it obtains optimal effectiveness and does not attract insects into the plant from the outside.
  2. The pan or drawer into which the electrocuted insect falls must be emptied regularly to prevent infestation by dermestid beetles and other 'cannibals' feeding on the dead insects. In fact, a careful appraisal of the types and numbers found should be made by the inspector or person servicing the electrocuter. These data should indicate the type of insect population present, and provide valuable control information.
  3. The UV source should be renewed each year because it loses effectiveness with age.

- Pheromone traps. These traps – which capture insects by means of a glue spread in the lower part of the trap after they have been attracted by pheromones – will be discussed in greater detail in the section concerning monitoring of 'inside areas'.
- Cockroach traps. These are boxes with openings covered by strips of aluminum allowing the cockroaches to enter but not to escape. The cockroaches are attracted by substances such as flour, sugar, algae, potatoes, sesame, cheese, and honey without any insecticides. This means that they can be safely used in any environment. The cockroaches captured can then be killed, e.g. by drowning.

Rodent traps

The main types are as follows:

- Mechanical traps – Various types of traps are available on the market. The common wood (or metal) base snap or guillotine trap is quite efficient and inexpensive. Two sizes are available and it should be noted that rat traps do not work with mice nor mouse traps with rats. Live-catch or multiple-catch mouse traps are also suitable for food plants.

A good example in Italy is the 'Ugglan' model mouse trap. The real effectiveness of the traps depends on how well they are positioned and on the type of bait used. The traps should be placed along what are presumed to be the rodents' tracks and close to probable nesting positions, harborages or known feeding places. Katsuyama and Strachan[2] recommend that the effectiveness can be increased by creating artificial runways with boxes or by leaning a board against a wall over a trap. They advise that snap traps should be left unset for 2 or 3 days until the rodents become accustomed to them. The use of traps is illustrated in Figures 4.35 and 4.36.

As mentioned earlier, baits must be frequently changed as they tend to deteriorate rapidly and become unappetizing for their prey.

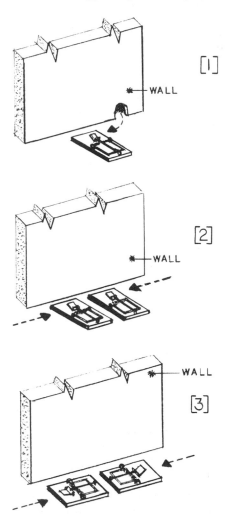

Figure 4.35 Placement of rat traps by a wall

CARDBOARD   EXTENSION
ON TRAP TRIGGER

Figure 4.36 Placement of rat traps on overheads

- Poisonous bait traps. These are containers holding edible substances liked by mice and mixed with rodenticides. Once again they should be placed along the route that mice are likely to follow to get to food and take it back to their nests. Consumption of the bait and the presence nearby of excrement, hairs, urine or other traces are useful as regards the monitoring of these infestants.

Inside areas

Introduction

Monitoring of areas inside the factory and warehouse is a task for the experienced. When monitoring techniques are correctly used, it is possible even to avoid insecticide applications or suspend them for considerable periods by:

- Carrying out appropriate cleaning procedures.
- Preventing the entry of infestants into food processing or storage facilities.

This insures that their presence in facilities is so low as to eliminate any real danger and will make large-scale disinfestation work unnecessary. Insecticide application is then undertaken only when the danger level of infestation is about to be exceeded. This section presents a detailed overview of:

- Examination of dust residues, e.g. after disinfestation.
- Analysis of traces of infestations.
- Use of food traps and baits.
- Use of light traps and of pheromone traps.

Examination of dust residues

The first, fundamental inspection must be performed *immediately after* disinfestation. In this way it is possible to discover exactly which species were present and how many dangerous types there were.

A simple technique that is recommended for the collection and counting of infestant insects – and that can be used at any time, not just after disinfestation – is examination of the dust residue. This is collected in the bags of the vacuum cleaners used in production premises, warehouses, etc. for cleaning floors, machinery, wireways, and so on. Inspection is done by sieving the material and inspecting it visually.

Analysis of traces of infestation

It is important to examine unused space and corners where unutilized equipment and various materials may be found, together with organic processing residue. In addition, a check must be made in all those areas where there may be deposits of organic materials attractive to infestants (for example, of upper parts of plant, wireways, and the interior of machinery). This inspection is for traces of:

- Coleoptera. Larvae (snakelike, continuous) and adults – a discontinuous trace left by the abdomen together with regularly spaced footprints.
- Lepidoptera. Larvae (silklike threads, agglomerating excrement, rubbish and exuviae) and adults – the adults will be seen since, as they are nocturnal, they lodge on walls and ceilings during the day.
- Flies. Brownish dot-shaped excrement on tiled surfaces, windows, lampshades, etc.
- Cockroaches. Dot-shaped or granular excrement near sinks, air-conditioning vents, electric panels, etc.
- Acarids. Small irregularly shaped masses with dusty appearance, recognizable using a magnifying glass, since the masses are in fact hundreds of individuals heaped together with their eggs, food residue and exuviae.
- Rodents. Along the routes regularly used by these infestants footprints will be noticed in the dust as well as signs left by tails (tail drag) and these can give a fair indication of the density of the population. Another sign is an oily track indicating a path currently used. If this track is covered with dust it means that the rodents' route has changed or that they are dead. Their excrement has a significant hair content. Making an accurate estimate of rodent numbers is not easy, but for practical purposes, their presence should trigger remedial action. For further indications of rodents, see Troller[1].
- Birds. Claw-marks in dust and easily recognizable excrement containing insect fragments.

- Bats. Excrement containing numerous fragments of the insects on which they feed.

The identification of traces of infestation is made easier by distributing talc, chalk or flour or 'tracking dust' on some areas of the floor to act as 'tracers'.

Use of food traps

There are a number of different types of food traps for the capture of insects typically infesting food-stuffs and for cockroaches, ants, flies and wasps.

- Pinninger traps. These are simple plastic netting bags (that rodents find difficult to gnaw) containing a mixture of seeds and bait substances. They are effective for catching Coleoptera adults and larvae, and the larvae of Lepidoptera, Psocoptera and Acari.

  Since they are small (200 × 100 mm with 1.5 mm mesh), Pinninger traps can be placed in all types of spaces, above machinery, near elevators and in cracks of walls and floors, i.e. where foodstuffs settle and where infestation usually starts. The traps must be checked weekly, by scattering the bait on a sheet of white paper on to which the insects fall. Classification, counting and recording of the insects thus collected is very useful for immediate identification in the various locations of:
  - the presence of dangerous species
  - their distribution
  - variation of presence during the year
  - the effectiveness of disinfestation
  - the onset of resistance to the disinfestation chemicals employed
- Cockroach traps. Several different types can be used. The most common are simple beakers containing a cone allowing the cockroach to enter and fall. Others, as already mentioned, are equipped with aluminum strips. There are types consisting simply of sticky surfaces that trap the insects and there are electrocuters consisting of live grids which kill the cockroach when they close the contact by walking over the device.

All these traps attract these insects using food baits, which consist of strong-smelling floury substances. More recently use has been made of the aggregation pheromone present in these insects' faeces and identified for the most common species of cockroaches. These traps must be placed in dark, moist and warm places, close to cracks, electric panels, sinks, etc. and where organic residues are present.

- Wasp traps. These are food traps in which the bait consists of sugar and protein mixtures combined with insecticides. Their use permits not only monitoring but also elimination of any wasps accidentally entering factories without having to spray insecticides inside the premises.
- Ant traps. The baits are made from sugar and proteins combined with aromatic substances and insecticides. The same baits are used with the addition of growth regulators, so that the ants transport the toxic substances to the ant nests – usually well hidden and protected – and kill or sterilize the queen ant or affect the development of new-born ants.

Light traps

Light traps for insects with positive phototropism can, if correctly positioned, be used not only for monitoring but also for mass-trapping. They are useful for environmental infestants and for revealing the presence of species colonizing stored foods. The source of light is ultra-violet and the traps must be carefully positioned in order not to attract insects from outside into the premises.

They can be designed to successfully electrocute Lepidoptera, Coleoptera, Diptera and, in some cases Hymenoptera. They are useful for controlling Anobidiae beetles, *Lasioderma* spp., and *Stegobium paniceum* and for destroying any bees and wasps accidentally entering buildings. They are also useful for capturing Ephemeroptera or Diptera capable of contaminating products being manufactured, but are not very effective for controlling flies and Lepidoptera infesting various types of foods.

Pheromone traps

Pheromones are substances secreted from various parts of insects' bodies and determine repeatable responses by individuals of the same species. The molecules dispersed in the environment are in fact picked up by other individuals of the same species by means of sensitive chemical receptors typically present in antennae and cerci.

Pheromones are classified according to the action they perform, i.e.

- Aggregation – secreted to increase the population when conditions are favorable (abundant food) or prior to reproduction (preceded by mating stimulated by sexual pheromones), or to indicate the way towards food or nests (track pheromones secreted by ants).
- Dispersion – for example, the pheromones used by ants to mark their colony's territory or the anti-aphrodisiac types secreted by male Coleoptera to stop their females mating.
- Aggression
- Recognition
- Sexual recall. These are the pheromones that have been most studied. They are designed to sexually excite and attract individuals of the same

species, or to inhibit attraction for different species even if these happen to be closely related. Sexual pheromones are useful for tracer traps (monitoring) and for mass-trapping when the substances concerned are available in synthetic form, as is the case of phyitidic Lepidoptera.

There are various types of traps, e.g. the pagoda version with a sticky base, used in the open; sticky strips hanging from the ceiling; or funnels with a collecting bag or jar underneath. The recall substance is normally contained in a synthetic dispenser.

To increase effectiveness, the sexual recall pheromones normally used are those that are common to several species and the anti-pheromonic elements are eliminated. The sexual pheromone can be combined with an aromatic substance, such as a foodstuff, to increase its power of attraction.

Coleoptera secrete sexual recall pheromones or those relating to pre-reproduction aggregation. Diptera, Hymenoptera and cockroaches mainly secrete aggregation pheromones that can be used for mass-trapping of cockroaches and against flies in combination with insecticides.

Male and female sexual pheromones have been identified for the three species of cockroaches most commonly found in Italy (*Perplaneta americana*, *Blattella germanica*, and *Blatta orientalis*).

Pheromones are used in two ways:

- To attract and capture insects – monitoring and/or control of infestations by means of mass-trapping using sexual and/or aggregation pheromones.
- To disorient and confuse insects, interfering with the mechanisms of distribution and reception of smells and thus with the laying of eggs. This, however, is an area that has yet to be studied fully and hence this 'confusion technique' is only rarely used today.

To summarize, the use of pheromones is of key importance in monitoring and makes possible, in conjunction with other monitoring techniques, highly focused and specific disinfestation when required. It can supersede conventional methods, with better results, lower costs and less risk of contamination of foodstuffs.

References

1. Troller, J. A. (1983) *Sanitation in Food Processing*. New York: Academic Press
2. Katsuyama, A. M. and Strachan, J. P. (eds) (1980) *Principles of Food Processing Sanitation*. Washington DC: The Food Processors Institute

4.D.4 Insect disinfestation

Introduction

Insect disinfestation is always important for the food manufacturer and is of special importance where dried foods and infant or dietetic products are made. For further reading see the relevant chapters in Katsuyama and Strachan[1] and Troller[2].

This type of disinfestation is performed using:

- Disinfestants in aqueous solutions and/or emulsions
- Gaseous disinfestants
- Powder disinfestants
- Disinfestation by means of fumigation ('fogging')
- Use of Juvenile Hormone, e.g. against Pharaoh's ants

The key guidelines to keep in mind when considering use of these insect disinfestants are that they:

- Can be used on a regular, programmed basis as part of systematic prevention
- *Must* be used when the results of monitoring and control make it likely that an area of the factory and/or warehouse facilities are infested
- Should be used in minimum quantities to prevent the toxic substances they contain from remaining present in premises, and in ways which minimize the risk of foods being contaminated.

Before going into the details of aqueous and gaseous disinfestants, the action that must be taken *before* they are used is outlined.

Preparations for disinfestation – the basic rules

Before disinfestation is undertaken, the premises to be disinfested must be thoroughly inspected to confirm that:

- They are perfectly clean.
- All materials and/or products have been positioned well away from the walls so that there is easy access to the latter. If this is not done, disinfestation should *not* be carried out.

Preparatory cleaning of premises is important since the disinfestants' degree of penetration is severely reduced by dirt. The presence of piles of materials or thick layers of dust provides an ideal hiding place for insects, enabling them to survive disinfestation.

There are two further rules that *must* be observed during disinfestation:

- Only the personnel directly responsible for disinfestation must be present in the departments when disinfestation is in progress.
- No production work must be carried out either in the area to be disinfested or in adjacent

departments. (Spray droplets are easily air-borne and can move fast from one area to another.)

Thus, disinfestation should ideally be undertaken during down-times, weekends, etc.

Disinfestants in aqueous solution

Introduction

Generally speaking, insecticides act in one of three ways, by:

- Contact
- Ingestion
- Inhalation

and in addition may or may not have a long-lasting residual activity. These insecticides are normally applied using sprayers. This equipment must:

- Produce a wide jet of spray so as to insure that the entire surface is treated.
- Feature a moderate and controlled degree of atomization – an unduly fine spray would cause the risk of the fine droplets remaining air-borne for too long and of consequent contamination of foods once production restarts. Where a limited area is to be treated for example, parts of plant, floors – particularly close to extremely critical points – a brush can be used to apply the solution.

There are a number of specific precautions that must be taken:

- The insecticide must be distributed only in the areas or points of plant and machinery that *do not come into direct contact with foods*.
- Machinery – when located in or close to the areas to be treated – must be protected with plastic sheeting.
- No open food products or ingredients must be present in the area. Products packed in hermetically sealed, gas-proof containers may be occasionally permitted provided that they can subsequently be used without risk of contamination. If they are not hermetically sealed, the products must be protected with heavy-duty plastic sheeting.
- Room temperature must be exactly measured before disinfestation. The temperature factor is *crucially important* for successful disinfestation since, if the temperature is too *low*:

Insects are *less mobile* – meaning that they will not come into contact with insecticides sprayed on walls and floors. In addition, their respiration rate will decrease, thus causing them to inhale insufficient amounts of the air-borne toxic substances.

Some chemicals *do not evaporate adequately*. Dichlorvos, for instance, gasifies very slowly below 18°C (64°F). Thus, if disinfestation is carried out during winter months, the premises concerned must

be kept heated to insure a constant temperature of 18 to 20°C (64 to 68°F).

Personnel responsible for carrying out disinfestation operations *must* wear:

- Waterproof overalls
- Heavy-duty rubber boots and gloves
- Gas mask with a filter appropriate for use with specific insecticide concerned.

The following sections describe some details of all the various elements concerning the use of insecticides in water-based solutions.

Preparation of solution and equipment

Preparation of the insecticide solution must be carefully done, following the steps listed below:

1. Fill the tank of the spray pump with the required amount of water and add the correct quantity of the concentrated insecticide to be used. If any of the concentrated substance or solution splashes on to the skin or near the eyes, immediately wash off splashes with abundant water.
2. Check that the spray nozzle is in perfect condition, without any blockages or faults.
3. Check that gas mask fits perfectly and is working efficiently. Once this has been done, it should be checked to see whether the filter has become exhausted, whether it has been correctly screwed into the mask, or whether it is damaged. (If the gas mask fails, it should be remembered that the air breathed in usually enters through cracks and gaps in other parts of the mask rather than through the filter.)

Spraying of solution

Once all preliminary operations have been completed, the actual spraying operation starts:

- Avoid spraying electrical wiring and insure that mains cables are not badly fixed or uncovered. This is to prevent short circuits or electric shocks during disinfestation.
- Spray at a close distance. This way it is possible to hit key points with great accuracy while avoiding contamination of other parts of premises or of food products. (If splashes should accidentally occur, the contaminated areas should immediately be washed very carefully.)
- When spraying, keep the jet of solution absolutely horizontal. This is important, since any sloping upwards of the jet of liquid may cause the person using the pump to be hit by drops of falling insecticide.

Warning! If persons involved in disinfestation experience breathing difficulties or note an unduly strong insecticide smell, they should *immediately switch off the pump and leave the premises concerned.*

Spraying targets

While keeping in mind that the insecticide solution must not come into direct contact with any foodstuffs or food products, there are some specific *targets* that must be covered by spraying:

- Walls. These must always be sprayed over as wide a surface as possible. Spraying should be very even, particularly if the insecticide concerned is of a type that continues to be active for a significant period of time (residual activity).
- Floor/wall joints. Spraying must be intensive in these points where food tends to deposit and so favors insects.
- Areas of floor under and surrounding machinery. These areas, due to the significant amount of edible residues, are particularly exposed to infestation.
- Powder-product packaging machinery. Even if kept reasonably clean this type of machinery is a high-risk area in terms of infestation. Any insects present here have only to travel a short distance to get inside the packs themselves – i.e. in a habitat where they can reproduce quite undisturbed. Once this happens, insects can easily migrate to other packs and quickly contaminate the whole product lot.

Operations subsequent to spraying

Once spraying has been completed, there are further actions that must be taken:

1. If the insecticides are of the volatile type (such as Dichlorvos) then the departments treated must be shut and kept closed for at least 24 hours. After this they must be opened and well aired for a further 24 hours before production personnel are allowed to return to them.
2. All equipment used (including clothing) must be thoroughly washed. Care must be taken that personnel do not get splashed once they have taken off their protective clothing.
3. Personnel should then take a shower to remove any insecticide residue accidentally coming into contact with the skin or hair.

Spraying equipment

Shoulder-carried equipment is recommended for many jobs. It has a small 10 to 15 liter (2 to 3 gallon) tank that can be adjusted in terms of jet length (6 to 8 m (6½ to 8½ yd) horizontally) and of nozzle width. As already mentioned, the jet must be well targeted and consist of droplets that are not too fine.

Many different types of disinfestation equipment are available. Generally speaking, the choice of one type rather than another is dictated by the following criteria:

- The device must be sufficiently capacious to do the job but not too heavy to make the work difficult.
- The valves must be easy to close fully.
- The nozzles must be of the right size, i.e. permitting a carefully adjusted jet or spray of liquid.
- The delivery line must be sufficient to insure full discharge of the right quantity of solution of active ingredient.
- Stainless steel equipment is preferable.

All sprayers have a number of common features, i.e.

- A tank
- A system for pressurizing the liquid
- A delivery line with a valve and a nozzle

Typical shoulder-carried equipment with tanks holding 10 to 15 liters (2 to 3 gallons) of liquid may be pressurized by a small two-stroke motor weighing about 8 kg (18 lbs.) able to run at two different speeds. This equipment measures approximately 600 × 400 × 300 mm (24 × 16 × 12 in). Small hand-operated pumps are useful for localized utilization on or around machines and for roach control. A sprayer having a tank capacity of either 0.5 or 1 liter (1 pint or 1 quart) is convenient and commonly used. In this case the operator squeezes a trigger to activate a plunger which forces the liquid out through the discharge nozzle. Depending on the nozzle used, this sprayer produces a fine jet or mist of pesticide. The advantage of these sprayers is that pressure can easily be controlled simply by using greater or lesser pressure when squeezing the trigger.

If large-capacity sprayers are chosen, a series of accessories must be provided, including:

- Pressure regulators and gauges
- Agitators for the liquid to be sprayed.

A key accessory is the *nozzle* as this determines the way in which the liquid is distributed and the size of spray particles. If there are no specific requirements, adjustable nozzles are recommended.

Another highly important accessory is the *valve* used to shut off the flow of liquid. It must be fast-acting and easy to use. The best type of valve to use depends on the density of the disinfestant. Rotary valves are often recommended and should be made in neoprene or teflon, i.e. of a type that cannot be corroded by chemical agents.

Finally, the flexible hose must be strongly built (once again neoprene is recommended) and of an appropriate diameter for the flow of liquid required.

Table 4.7 Main insecticides in aqueous solution recommended for disinfestation

Brand name	Primary ingredient	Type of action	Use (%)	Quantity	Activity
MAFU EC					
NUVAN 50E 50 EC DEDEVAP	Dichlorvos 50%	Ingestion Contact Inhalation	3	2 to 3 l/100 m^2	Non-residual
ACTELIC D	Pirimiphos Methyl 50%	Ingestion Contact Inhalation	3	1 l/10 m^2	Residual
BAYGON EC	Propoxur 20%	Ingestion Contact Inhalation	3	2 to 3 l/100 m^2	Residual
K-OTHRINE	Deltametrin 0.73%	Contact Ingestion	5	1 l/25 m^2	Residual

Note: Always confirm the local legislative position of *any* pesticide before using or giving approval for use.

For further information on these insecticides, reference may be made to Figure 3 in *The Agro Chemicals Handbook*, 2nd edition, 1987, published by the Royal Society of Chemistry [3].

List of insecticides in aqueous solution

Table 4.7 lists the main insecticides recommended for disinfestation of manufacturing facilities. The list indicates commercial brand names, type of action (contact, inhalation, etc.), percentage to be used, etc.

In Italy, experience is that these products are effective for disinfestation. However, every country has regulations and preferences in this matter, and the list is simply an example.

Depending on local availability and legislative status of these, alternatives may be used. However, in our experience, DDVP (dichlorvos 50%) has no substitute. It is a product which is of unequalled effectiveness because of its type of action – it gasifies strongly – its user safety and rapid hydrolyzation. It is a 'must' when infestation has occurred which requires fast and effective action.

It is important to remember that there is the possibility of some spray residues in the plant or at production start-up. For this reason, it is necessary that the possibility of residues in the product should be recognized and limits established and enforced.

Gaseous disinfestants

Warning of toxic properties

While liquid insecticides are only moderately toxic to man, gas insecticides are *highly toxic*.

Introduction

Disinfestation operations using these disinfestants need expert knowledge, experience and care. They should ideally be carried out by specialized companies specifically authorized to do this type of work by the national health authorities or agency having jurisdiction. Use only companies having a proven good record in disinfestation of food manufacturing premises.

The premises treated with gas insecticides *must* be capable of being hermetically sealed off. There are several reasons why this is an absolute requirement:

– In an open environment the gas disperses rapidly, not remaining in the premises long enough to kill pests.
– In an open environment the gas may well leak out from the premises being disinfested. This is *extremely dangerous* since the gas might *kill people* happening to pass by. Clearly, this risk must be given special consideration when the factory is located in an urban area or close to residential zones.

Gas insecticide disinfestation should never be undertaken if in the immediate vicinity – adjacent departments or premises – there are people working or, worse still, dwelling. Gas disinfestation means that not only must the premises to be disinfested be evacuated, but also adjacent departments throughout the entire period of disinfestation and subsequent ventilation. As with liquid insecticides, personnel using them must be properly equipped with waterproof overalls, heavy-duty rubber gloves and boots, and gas masks with appropriate filters.

Preparation for gassing

A series of actions must be performed prior to commencing actual disinfestation:

1. All openings (doors, windows, air vents, etc.) must be sealed with adhesive paper tape. Any crack visible to the eye must be sealed in the same way.
2. The area to be disinfested must be inspected thoroughly to identify those zones where gas would be dispelled with some difficulty during ventilation after disinfestation.
3. To check the effectiveness of gas disinfestation – which is a costly operation – place small cages containing insects at various stages of their life cycles (eggs, larvae, adults) in various points of the area to be treated, ideally in potentially protected or 'difficult' positions. They must be placed where they cannot readily be seen by disinfestation personnel.
4. The specific antidote, together with a method of measuring the escaping gas, e.g. a Draeger tube, should be available before gassing is undertaken.

Gassing

Gas (methyl bromide) is supplied as canisters and is carried into the premises to be disinfested. The quantity of gas to be distributed, and thus the exact number of canisters required for each room or area, is calculated on the base of cubic volume.

Phosphine is marketed as tablets which contain aluminum phosphide. For the production of gas, remove their outer wrapping: air humidity causes the hydrolization of the phosphide thus releasing phosphine. The tablets are distributed in relation to the volume of the area to be treated. Table 4.8 summarizes conditions of use for both of these gases.

It is *essential* for effective use of gas insecticides that there is *even distribution of gas* throughout the area. This can be achieved by:

1. Positioning the gas canisters at distances proportionate to the cubic volume to be disinfested.
2. Carefully orienting the nozzles of each canister.

Once this has been done, the various operators must:

1. Open the canister nozzles starting from the point farthest away from the exit and opening each subsequent nozzle as quickly as possible while moving towards the exit.
2. Seal the exit door on the outside using adhesive paper tape after confirming that all operators have left the room or area.
3. Place large, visible notices on the door and at other strategic places to warn everyone that gas disinfestation is underway.

The duration of disinfestation of the hermetically sealed departments is calculated according to insecticide concentration and environmental temperature. Keep as closely as possible to the indications provided in Table 4.8.

When using phosphine it is necessary to take the same precautions, positioning the tablets instead of canisters at distances proportionate to the cubic volume to be disinfested.

After gassing

At the end of the period of disinfestation the following actions must be carried out:

1. Open all doors and windows and allow *natural* renewal of air to dispel gas from the area.
2. Check that gas has in fact disappeared using specific gas detectors, particularly in those points where prior inspection indicated possible difficulties in gas dispersion and elimination. Only after this has been done can the premises be declared safe for a return to normal operations.
3. Remove the cages previously placed in the areas and check the survival rate of insects. If gas disfestation has been properly performed there should be no survivors.

Safety warning!

It is essential that *all* gassing operations are carried out by expert and well-trained personnel. *Always take full safety precautions*.

Table 4.8. Usage of main gas insecticides

Primary ingredient	Concentration (g/m^3)	Temperature (°C)	Time
Methyl bromide	27 to 30	Over 18	48 hours
Aluminum phosphide (producing phosphine)	6 to 15	About 15 From 15 to 20 Over 20	5 days 4 days 3 days

Note: Always confirm the local legislative position of *any* pesticide before using or giving approval for use.
For further information on these gas insecticides, refer to Figure 5 in *The AgroChemicals Handbook*, 2nd edition, 1987, published by the Royal Society of Chemistry [3].

Powder insecticides

Introduction

Wide use is made of powder disinfestants in industry generally, but in the specific case of the food-processing industry they must be used in *service areas only* (cellars, boiler rooms, etc. used as 'transit areas' by ants, cockroaches, etc.) and in protected places such as electrical control panels. This is because – clearly – the presence of air-borne powdered disinfestants in production departments must be avoided at all costs.

Types of duster

Powder disinfestants are applied using devices called 'dusters': these spread an even layer of powder and make it possible to get right into awkward corners, cracks, etc. There are three types of duster devices useful for the food-processing industry:

- Small hand-operated dusters. These have a capacity of 250 to 500 g (9 to 18 oz) and should be used to apply a thin layer of dust or to force dust into a small crack where insects may be hiding.
- Hand crank-operated dusters. These have a capacity of about 5 kg (11 lb) and can be used to apply powder to larger areas.
- Small electrically driven dusters. These produce a very fine layer of dust and can be used effectively to treat small cracks and wherever deep penetration is required.

The powder used is normally Propoxur-based.

Fumigation (fogging)

Introduction

Fumigation is used mainly in agriculture, but is occasionally used in the food-processing industry to disinfest premises. In the case of a closed environment, fumigation using oil-based fogs presents some risks and therefore product concentration must be meticulously calculated to avoid explosions. On no account must a concentration of over $3.78 \, l/1416 \, m^3$ (1 gall/2223 yd^3; 1 US gall/1852 yd^3) be used. Additionally, no flames must be lighted or electrical switches turned on.

In practice, fumigation is only really effective if strong currents of air are present to force the fumigation 'fog' into cracks and crevices and over the surface to be disinfested. These air currents must not in any case exceed a speed of 8 km/h (5 miles/h) as this would cause an unduly strong dispersion of product and hence imperfect disinfestation.

Types of fogger

The fumigation equipment described below – 'fog generators' – tend to become extremely hot and thus the people using them must take appropriate precautions. Thermal aerosol generators or 'foggers' disperse very small droplets by introducing an oil-based pesticide formulation into a chamber which is heated to a temperature sufficient to cause immediate vaporization of the oil. Heat is produced either by a flame in a heating chamber, by electricity, by steam or by an internal combustion engine. The key point to remember here is that the particles are *easily airborne* and hence extreme care should be taken during fumigation.

Larger foggers put out large volumes of fumigant fog or mist at a rate of some 120 to 240 l (26 to 53 gall; 32 to 63 US gall) per hour. Given this, in the food-processing industry they are only occasionally used for the treatment of buildings such as warehouses. Smaller foggers eject smaller quantities at a rate of some 20 to 60 l (4.4 to 13 gall; 5.3 to 16 US gall) per hour and these are the ones normally used for treatment inside buildings.

Aerosol generators are recently available which produce droplets of up to 6 microns by non-thermal means. They are economical and the aerosol penetrates cracks and cavities. However, small droplet size means that droplets remain suspended in air for a long time and extreme care must be taken during and after fumigation.

Use of 'Juvenile Hormone' (against Pharaoh's ants)

Pharaoh's ant (Monomorium pharaonis (L) probably originated in North Africa, but since the beginning of the nineteenth century has spread through international trade throughout the world. It has always been an exceedingly difficult pest to control and until recently the effective treatments relied on the relatively toxic and persistent organochlorine insecticides. A good account of the ant and its control is given in an ADAS leaflet[4].

The worker ant is about 2 mm (⁵⁄₆₄ in) long and yellow although the color is sometimes pale and hard to see until 'you know what to look for'. They may then be noticed in very large numbers! Heat-processed ants are a ginger color and therefore visible in a pale product.

The ideal temperature for development is 30°C (86°F) with a minimum of 18°C (64°F) although the colony can survive low temperatures for fairly long periods and the workers continue to forage. The colony spreads by 'budding' which means that new temporary and permanent nests are formed. Temporary nests have been found in laundry baskets, bakers' trays and other moveable equipment. 'Budding' means that the colony spreads rapidly. Both food and water are taken back to the nests and as these ants can carry pathogenic bacteria they must be considered a health hazard.

Before treatment is started, the extent of the infestation needs to be known. Baits of fresh liver cut into 1.5 cm (⅝ in) cubes should be laid along the skirting of walls, near sinks and where pipes pass through floors or walls. Leave overnight or undisturbed for several hours before examining. Start control measures from outside the area of infestation and work towards the center(s). It is important that treatment is done thoroughly and quickly so that *all* the colonies are exposed to the treatment at the same time.

Baits using the insect hormone analogue methoprene are highly effective. Methoprene is not poisonous to ants, man or animals. It acts by stopping the developing stages in the nest and by sterilizing the egg-laying queens. The colony therefore dies out. Baits of about 1 g of a mixture of honey, liver powder and sponge cake are used with a final percentage of 0.5% active material by weight. The baits are laid at about one per 3 m² (3.6 yd²) and repeated after one week. After about 20 weeks the infestation should have been cleared. To prove this, follow-up with liver bait. If any ants are found, re-treat the area immediately, remembering how colonies are spread and consider carefully whether the first treatment was sufficiently extensive.

References

1. Katsuyama, A. M. and Strachan, J. P. (eds) (1980) *Principles of Food Processing Sanitation.* Washington, DC: The Food Processors Institute
2. Troller, J. A. (1983) *Sanitation in Food Processing.* New York: Academic Press
3. RSC (1987) *The Agro Chemicals Handbook*, 2nd edn. The Royal Society of Chemistry, Nottingham, UK
4. ADAS (1985) *ADAS Leaflet 366. Ants Indoors*, Ministry of Agriculture, Fisheries and Foods (Publications), Lion House, Willowburn Estate, Alnwick, Northumberland NE66 2PF UK

4.D.5 Contract pest control

Introduction

Whether it is done 'in-house' or by contractors, pest control, if it is to be effective, needs operators who are not only competent but who also have a real interest in the work. The personal qualities required are those of careful, accurate and patient observation together with the ability and willingness to work with little direct supervision at times when the plant is quiet, e.g. at weekends.

Warning!
It must be clearly understood that the use of contractors for Pest Control does not diminish the accountability (responsibility) of plant management for its effectiveness.

Principles

Before a contractor starts work there must be a clear understanding by both parties as to what will be done, how and when. Failure to achieve this is a recipe for trouble. This understanding should lead to an agreement – in writing – which includes but is not necessarily limited to:

- A permitted list of pesticides. Remember that some legally acceptable materials may not be suitable for some of your packs for either technical or policy reasons.
- A list of quantities of active ingredients to be used in pesticide formulations. Be clear how the percentage is figured, e.g. W/W, V/V or W/V.
- A list of methods of application in sufficient detail to be sure of what should be done, where and when. This applies to routine and therefore 'on-going' work.
- On access to the premises. Names of security and technical personnel contacts who will be involved with the contractor should be provided.
- 'Emergency' pest control treatment should be agreed as needed. Records should be kept of why it was necessary, what was done and what 'follow-up' measures were taken.
- The method of reporting by the contractor of both findings and treatment. An inspection check-list is a valuable aid to a consistent approach. It should not be used in a restrictive or unthinking way. A good check-list should always have open-ended questions and include 'What else is of significance and should be reported?' The check-list conveniently includes an assessment or rating system that is agreed to be pertinent.

Katsuyama and Strachan[1] in their Chapter 11 give an example of a sanitation score sheet together with suggestions for developing standards or ratings. In addition, an audit system should be developed to provide independent confirmation that the contractor is delivering what has been agreed and also that effective pest control is being achieved.

The inspection and audit systems developed for your plant whether pest control is contracted out or done 'in-house' need to be based on a thorough HACCP analysis and reviewed periodically.

Reference

1. Katsuyama, A. M., and Strachan, J. P. (eds) (1980) *Principles of Food Processing Sanitation.* Washington, DC: The Food Processors Institute

4.E Utilities (services UK)

Introduction

The supply of utilities both to and within a department has sanitary implications especially for 'high-risk' or 'sensitive' products such as chilled foods or baby foods. These vary with the technology and the product, but an outline (overview) will be given of some of the sanitary considerations which apply to some important utilities.

4.E.1 Air and other process gases

Culinary air

This should be odorless and free from compressor lubricating oil and water. Troller[1] gives further information and there are Accepted Practices for supplying air under pressure in contact with milk, milk products and product contact surfaces published by IAMFES[2]. The same applies to other process gases, e.g. nitrogen or carbon dioxide, although these are usually supplied in cylinders or tanks. Compressed air usually has a very low microbial count but this may well not be true of 'Heating – ventilating – air conditioning' (HVAC) air. *Remember* that because air transports microorganisms, the air flow should always be from 'clean' to 'dirty' areas, *never* the reverse. Air used for transporting powders and in bagging or packaging zones will need HEPA filtration if 'high-risk' or 'sensitive' products are involved and numbers of pathogens, e.g. *Salmonella* spp., are to be minimized. This is done by passing air through pre-filters and then High Efficiency Particulate Filters (HEPA). When considering high-quality air, e.g. for aseptic work, it is prudent not to overlook the practical importance of the pre-filter system.

ABPMM requirements for high-quality air

The requirements for high-quality air are summarized in the Guidelines for Good Hygienic Practice in the Manufacture of Milk-Based Products, published by the ABPMM[3], as follows:

4.6. Air supplies
Whether it is used for primary drying, secondary drying, cooling, conveying or pressure equalization, air is a potential source of contamination. Particle loading in unfiltered air can be of the order of 10^8 particles per cubic meter of particles of 0.5 micron or greater.
    4.6.1. Air inlets should be positioned at least 1 meter (over 1 yd) above any floor or dust-collecting surface and away from any other

possible source of contamination, in order to avoid dust or water droplets being drawn into the system.
    4.6.2. All air which is in contact with finished product should be filtered to Class 1 of BS 5295, Part 1[4].

This will include air which is used post the main drying operation, such as fluid beds, fines transport, main transport and any air associated with storage and packing of finished product. This standard calls for a particle count not exceeding a total of 3000 particles m^3 of a size of 0.5 micron or greater. The greatest single particle present in any sample should not exceed 5 micron and the filter efficiency should be 99.995%.

General ventilation air into the high risk areas should be filtered to the lower standard of Class 3, with a filtration efficiency of 95%.

Air intake for the primary heating air of the spray drier only requires coarse filtration to protect the heater battery since it is normally heated to above 120°C (248°F). Precautions should be taken either by switching of the main fan or diversion of the air, to insure that the heater battery is up to temperature before air is sent to the drying chamber.

Where air sweeps are employed, these air inlets should be filtered to Class 1 or the powder generated separated into reject bags.

Where air is used for cooling banks, the final filter should be placed after the cooling coils which, due to their condensing action, are invariably sources of contamination and must be suitably designed to be readily cleaned.

Regular inspection is necessary to insure filters remain undamaged, dry and relatively clean. Choked filters can become a positive contamination hazard.
    4.6.3. Where sections of plant operate at a relative vacuum, contamination can be drawn in through any leaks. Every effort should be made to minimize air leaks through effective maintenance. Since achieving a completely hermetic seal is not always possible, the external surface of the plant which operates under vacuum should be kept scrupulously clean.
    4.6.4. Air filters should be fitted with pressure sensing devices measuring pressure drop across the filter as an indication of when replacement or cleaning becomes necessary. Filters should then be cleaned appropriately using steam or caustic solution at a high temperature (80°C (176°F) is recommended where possible, but some synthetic filters in use may not withstand such high temperatures) and chemically disinfected. The disinfectant used should also be effective against yeast and molds. Cleaning should be carried out in a manner which will not re-introduce contamination.

4.E.2 Vacuum

These lines can be a problem when associated with liquids, e.g. juices, since droplets are carried into the vacuum system. A vacuum tank designed to trap droplets and to be cleaned daily will, when installed near to the equipment, prevent mold growth along the lines.

4.E.3 Steam

When steam is used for culinary purposes, e.g. by direct injection into a product, it must be:

- Non-toxic – this restricts boiler feed water additives if steam is used directly from the mains.
- Non-tainting
- Filtered – Charlish and Warman[5] recommend a combined strainer and filter down to 2 microns.

Steam should be drawn from the top of a header to avoid rust or scale entering the take-off line. It is also worth remembering that 'dead-ends' in steam lines just like 'dead-ends' in water or product lines can cause problems. The most satisfactory rule is to disconnect lines not in use.

Boiler water additives

Steam which is in direct contact with ingredients or foods is properly regarded as a food ingredient and must meet regulatory requirements for potable water. When boiler feedwater is treated with corrosion inhibitors or water-conditioning compounds these may be 'carried over' by the steam, so their use should be controlled by having a *permitted list* for your plant and only using compounds on this list.

A 1981 list for the USA is given by Troller[1] as in Table 4.9.

Superheat

Steam for process purposes is normally used as saturated steam although in a large factory it will be generated with a few degrees of superheat. This is to reduce condensation in the distribution system (mains, UK). If it is essential that the steam is saturated, e.g. in a steam sterilizer, care must be taken to insure that it *is* fully saturated at the point of use. Temperature distribution problems can occur when steam has slight superheat. Water injection into the steam line may be used to insure saturation.

Steam pipes (and 'cold' lines) must be adequately insulated (lagged, UK) with a suitable material which does not pose a hazard to food products or personnel. See Chapter 5.B.

4.E.4 Process water

Whether water is used:

- as an ingredient
- for personal purposes, e.g. drinking or washing
- for the final rinse when cleaning food handling equipment
- in any way likely directly or indirectly bring it in contact with the product

it should be at least equal to the highest standard of drinking water required by local regulatory authorities. In the absence of such a standard use the latest edition of the World Health Organization's International Standards for Drinking Water.

Parasites

In some places, water-borne parasites can be a problem as they may be resistant to chlorination. Troller[1] states:

Giardia lamblia is the parasite most frequently involved in water-borne outbreaks. Like other helminths, this parasite is highly resistant to the chlorine concentrations normally employed in water supplies to kill bacteria. Additionally, cultural tests for coliform bacteria (the time-honored indicators of fecal contamination and its associated bacterial pathogens) will not predict the presence of this organism. *Giardia* contamination is most frequently encountered in private or semipublic water supplies, such as might be found in parks, in which water pretreatment (flocculation, settling, and filtration) is inadequate or nonexistent. Under most circumstances, pretreatment conditions are sufficient to remove cysts of *Giardia* from water supplies; however, in cases in which a threat may exist (such as a surface water supply subject to contamination from human or animal waste) most conventional chlorine treatments are neither of sufficient concentration nor adequate duration to kill this organism. As an indication of the prevalence of giardiasis, about 27% of all reported cases of water-borne diseases in 1977 were attributed to this disease.

Cryptosporidium has also caused outbreaks in the UK.

It should also be noted that parasites may be resistant to heat. If there is the possibility of their presence appropriate expert advice must be sought and taken.

Hardness

Remember that potable, i.e. 'safe to drink', water may not necessarily be suitable for all products, e.g. traces of iron may make a supply unsuitable for

Table 4.9 Approved boiler water additives for steam generation – food contact situations[a] (1981 data from Troller[1])

Substance	Limitation
Acrylamide – sodium acrylate resin	Contains less than 0.05% acrylamide monomer
Ammonium alginate	
Cobalt sulfate	
Lignosulfonic acid	
Monobutylethers of polyethylene – propylene glycol	See specific regulations[a]
Polyethylene glycol	See specific regulation[a]
Polyoxypropylene glycol	
Potassium carbonate	
Potassium tripolyphosphate	
Sodium acetate	
Sodium alginate	
Sodium aluminate	
Sodium carbonate	
Sodium carboxymethylcellulose	
Sodium glucoheptonate	Less than 1 ppm cyanide impurity
Sodium hexametaphosphate	
Sodium humate	
Sodium hydroxide	
Sodium lignosulfonate	
Sodium metasilicate	
Sodium metabisulfite	
Sodium nitrate	
Sodium phosphate (mono-, di-, tri-)	
Sodium polyacrylate	
Sodium polymethacrylate	
Sodium silicate	
Sodium sulfate	
Sodium sulfite	
Sodium tripolyphosphate	
Tannin	
Tetrasodium EDTA	
Tetrasodium pyrophosphate	
Cyclohexylamine[b]	Less than or equal to 10 ppm
Diethylaminoethanol[b]	Less than or equal to 15 ppm
Hydrazine[b]	0 ppm in steam
Morpholine[b]	10 ppm in steam, exclude contact with milk
Octadecylamine[b]	3 ppm in steam, exclude contact with milk
Trisodium nitrilotriacetate	5 ppm in boiler feedwater, exclude where steam contacts milk

[a] Code of Federal Regulations, 21.173. 310
[b] May be alone or in combination. See specific regulation.

Warning Permitted lists of additives may change rapidly. Check any proposed usage with the local regulatory authority.

some soft drinks; calcium levels in hard water may affect the texture of blanched navy beans, etc.

Water quality is also important in plant cleaning. As Tamplin[6] remarks:

> To avoid water scale deposition detergent formulations must be selected for the specific water hardness existing on site. In nearly all cases, detergent costs will increase in proportion to the mineral salt content, because expensive detergent ingredients will be required to combat water hardness.
>
> The water hardness is a very important factor in detergent formulations and must be known before

detailed detergent recommendations can be given.

Calcium and magnesium salts, dissolved by rain falling on certain rock formations, produce 'hard' water. Tamplin[6] quotes the US Geological Survey (Table 4.10).

UK values are similar to those of the US, but German values are higher (See Kessler[7]).

Water hardness is divided into two types – temporary and permanent. Temporary hardness is due to the presence of calcium and magnesium *bicarbonates*. It can be removed by boiling, although heat is only one method of decomposing bicarbonates into insoluble carbonate salts. Some alkalis precipitate hardness constituents from water, unless other additives are present. Kettle 'fur' is a typical carbonate scale which can occur in heat exchangers, boiler tubes, etc. and reduces rates of heat transfer. Scale may also harbor microorganisms. Permanent hardness is caused by calcium and magnesium *sulphates* and other salts and is not removed by boiling. Hardness is sometimes given in other units than those used in Table 4.10; see the Appendix for conversion tables at the end of this section.

Table 4.10 US Geological Survey classification of water hardness

Hardness	Total calcium + magnesium ions expressed as ppm calcium carbonate
Soft	0–60
Moderately hard	60–120
Hard	120–180
Very hard	Over 180

Soft water is preferred for cleaning processes involving alkaline detergents. It gives greater detergent economy and more efficient detergent action, and it prevents scale formation. However, soft water may be corrosive. Water used for Cleaning In Place (CIP) may need to be softened to avoid blockage of heat exchangers and delivery pipework, which can result in considerable reduction of cleaning efficiency.

Storage

In any plant it is necessary to store some potable water, and in a large plant with very variable demands this can be of hundreds of thousands of liters. Imholte[8] states that: 'The storage system should be designed with no dead ends so that water may circulate or flow freely. Water should not be allowed to stagnate.'

This is amplified in DHSS Codes of Practice No. 10[9] as follows:

5.5 Storage of Water

5.5.1. All potable water storage tanks should be fitted with closely fitting covers to exclude contamination from dust, insects, birds or rodents.

5.5.2. All storage tanks should be fitted with a means of access to permit adequate cleaning as often as necessary and, in any case, not less than once a year. Water in the tanks should be sampled for microbiological analysis at suitable intervals.

5.5.3. There is evidence that potable sources of supply containing small numbers of microorganisms can cause trouble because of proliferation of bacteria in the distribution system, particularly in storage and header tanks. This is particularly true when the water remains static over holiday periods and weekends.

Cross-connections and back-siphonage

After storage, the water will enter the internal (plant) distribution system, where problems can and do occur. Imholte[8] states:

Water supply lines should be installed to eliminate dead ends. Water lines no longer in use should be removed. Water that is not allowed to move freely stagnates, and a buildup of microorganisms can occur, leading to contamination of the water supply. Contamination from rust, scale, and grease can occur when repairs are made on old piping and pumps. New water piping installations should be made of corrosion-resistant materials.

It is very easy to take the safety of a water supply for granted. Many problems can be avoided with minor attention to one of the most important ingredients used in the typical food manufacturing plant, water.

One aspect of safe water which should not be overlooked is the possible growth of *Legionella* spp. (see also Section 4.B.2). These occur in many waters and will multiply in dead-ends. To maintain cold water lines below the recommended 20°C (68°F) insulation (lagging, UK) may be needed.

In a large plant, the 'potable water' system may include heat-regenerating units and heat exchangers to provide a piped hot water system, as well as high-pressure hot and/or cold hose lines for cleaning purposes. In an old plant with extensive pipe runs, 'dead-ends' or 'dead-sections' can present real problems and have been known to be the explanation of sudden and unacceptably high numbers of microorganisms and taints. This is more likely to happen when the person responsible for installation has retired or moved away and no reference drawings were made.

Remember, 'dead-ends' are rarely the result of deliberate design. They are usually caused by a 'temporary' or rushed installation job which is not corrected. Where no reference drawings can be found, long and complex pipe-runs may be 'lost', especially in an old plant. Tracing complex pipe runs, which may need to be done as part of an investigation of a problem, has been known to take many man days – a costly and frustrating procedure.

In principle, hot water for 'process' and 'domestic' supplies should be piped into a recirculating system to avoid dead-ends. It is good practice to keep the length of 'spurs' or 'branches' to less than 6 m (20 ft), and to lag hot water lines. This will mean that they can deliver hot water at 50°C (122°F) within one minute at the draw-off point.

The best cure for these problems is to 'do it right – the first time', and make *and keep* reference drawings or sketches of installations.

Both cross-connections and back-flows can cause unexpected problems in a processing plant. A cross-connection is a physical connection, either temporary or permanent, between systems such that water can flow between them. A back-flow or back-siphonage occurs when contaminated water is drawn back by reduced pressure into a potable water line. Imholte[8] illustrates how this may happen (his Figure 3–57). Additionally, Katsuyama and Strachan[10] deal with this subject in their Chapter 8. Lack of forethought, supervision or care are the likely root causes of these happenings. Remember that HACCP was derived from engineering principles.

Vacuum breakers

Backflow preventers or vacuum-breakers should be used, where required, in the water system. Where applicable, they should be approved by the local regulatory agency. If there is no local regulatory agency to follow, Katsuyama and Strachan[10] give clear, useful diagrams.

Remember, back-flow can also be the result of back-pressure – even coming from a food manufacturing process. This can happen, e.g. through valve leakage or control malfunction.

Re-use of potable water

Imholte[8] states that

water may be safely re-used to perform cooling functions. However, care must be taken with such systems.

All pipelines, tanks, cooling towers and other such equipment used in the re-used water system must be installed so that they can be readily inspected and cleaned. The water supply line to this system must be installed with a backflow preventer, or in situations where the plumbing code requires it, an air gap. The piping assembly should also include the necessary valves to completely drain the system down for cleaning and fresh water replacement.

Remember, not only cleaning but disinfection may be needed in such systems to avoid microbial hazards, e.g. *Legionella* spp.

For particular purposes, special types of potable water may be needed, e.g. de-ionized water, sterile water or condensate. These will be in dedicated and isolated lines for the purpose which may require them to be made of specific materials.

Non-potable water

Supplies of non-potable water include: untreated surface water, e.g. river water; re-used water; sea water; chlorinated cannery cooling water and any water of questionable quality. Suitable uses for non-potable water include cooling of refrigeration units or compressors, or use in some fire protection lines. It is very important that non-potable supplies should always be recognized as a source of potential danger.

Warning do not use non-potable water:

- In any part of a fire protection system, including sprinklers, for any area that prepares, handles or stores food.
- For any sanitation operation, e.g. washing floors, walls, ceilings, or any food equipment.

Remember that it is always good practice and may well be locally mandatory that potable and non-potable water should be in separate, independent distribution systems. It is not acceptable to rely on separation or isolation by valve arrangements. Valves can and do leak and pressure differentials may be momentarily altered or reversed.

Potable water used to supplement any non-potable supply *must always be positively protected* against contamination from back-pressure or siphonage, e.g. by an adequate air-gap.

Non-potable lines must be clearly and readily identified as such. Katsuyama and Strachan[10] quote the USDA suggested identification system given in Table 4.11.

Table 4.11 USDA pipeline color codes

Fire lines	Red
Sewer lines	Black
Edible brine lines	Green plus name
Inedible brine lines	Black
Air lines	White
Potable water lines	Green
Non-potable water lines	Black
Inedible product lines	Black plus name
Ammonia lines	Blue
Edible product lines	Green plus name
Curing pickle lines	Green plus name

It is not necessary to paint the whole line although it may be convenient to do so. A band of the appropriate color about 300 mm (12 in) placed at suitable intervals along the pipe is adequate.

Always remember that non-potable supplies anywhere in a food plant are a potential danger. Management is accountable for the proper control of this danger.

Water re-use

Water re-use may be required because of either availability or cost. Remember that water conservation may also be achieved by the development of more efficient ways of use in production, e.g. in better designs of vegetable washers; or in cleaning, e.g. the use of High Pressure Low Volume (HPLV) hoses.

Water that is re-used must be completely fit for its intended purpose after any treatment which may be given. It should not be regarded as *potable*. Imholte[8] recommends that re-use of potable water should be limited primarily to non-food and

Table 4.12 Water-economy checklist (from Katsuyama[11])

Operation or equipment	May recovered water be used?	May water from this equipment be re-used elsewhere in plant?	Source of water for re-use in equipment[a]
1. Acid dip for fruit	Yes	No	Can coolers
2. Washing of product			
A. First wash followed by second wash	Yes	Yes[a]	Can coolers
B. Final wash of product	No	Yes[a]	
3. Flumes			
A. Fluming of unwashed or unprepared product (peas, pumpkin, etc.)	Yes	Yes[a]	Can coolers
B. Fluming partially prepared product	Yes	Yes[a]	
C. Fluming fully prepared product	No	Yes	
D. Fluming of wastes	Yes	No	Any waste water
4. Lye peeling	Yes	No	Can coolers
5. Product-holding vats; product covered with water or brine	No	No	
6. Blanchers – all types			
A. Original filling water	No	No	
B. Replacement or make-up water	No	No	
7. Salt brine quality graders followed by a freshwater wash	Yes	Only in this equipment	
8. Washing pans, trays, etc.			
A. Tank washers – original water	No	No	
B. Spray or make-up water	No	No	
9. Lubrication of product in machines such as pear peelers, fruit size graders, etc.	No	Yes[a]	Can coolers
10. Vacuum concentrators	Yes	In this equipment after cooling and chlorination	
11. Washing empty cans	No	No	
12. Washing cans after closing	Yes	Yes[a]	Can coolers
13. Brine and syrup	No		
14. Processing jars under water	Yes	For processing	Can coolers and processing water
15. Can coolers		Water from these coolers may be	
A. Cooling canals		re-used satisfactorily for cooling	
1. Original water	No	cans after circulating over cooling	
2. Make-up water	Yes	towers, if careful attention is paid to	
B. Continuous cookers where cans are partially immersed in water		proper control of replacement water, and to keeping down	
1. Original water	No	bacterial count by chlorination and	
2. Make-up water	Yes	frequent cleaning	
C. Spray coolers with cans not immersed in water	Yes	This water may be re-used in other places as indicated	
D. Batch cooling in retorts	Yes		
16. Cleanup purposes			
A. Preliminary wash	Yes	Yes[a]	Can coolers
B. Final wash	No	No	
17. Box washers	Yes	No	Can coolers

[a] A certain amount of water may be re-used for make-up water and in preceding operations if the counterflow principle is used with the recommended precautions.

non-cleaning uses. He suggests, in his Chapter 3, ways in which this could be done.

Katsuyama[11], in his excellent Chapter 2.1 on water conservation, states:

Water re-used for contact with food products must:

- be free of microorganisms of public health significance
- contain no chemicals in toxic or otherwise harmful concentrations
- be free of materials or compounds which could impart discoloration, off-flavor, or off-odor to the product or otherwise adversely affect its quality and
- be chemically and physically acceptable for its intended use,

and adds:

Reclaimed water can be used in one of two ways; it can either be re-used within the system from which it was recovered (i.e. recirculated) or it can be used in some other operation. Its suitability for use in any operation is dictated by the quality of water required in that operation. The final operations will require water of high quality, while water quality requirements are less stringent for intermediate and preliminary steps. Water used to convey waste materials can virtually be of any quality aesthetically acceptable. A checklist indicating some potential uses of water from various unit operations is provided in the table (Table 4.12).

Water that is re-used should be run in a system (circuit, UK) separate from that used for potable water and with *no* cross-connections. Usually some topping up (make-up, UK) with potable water will be needed. The potable water supplied *must be positively protected* against contamination from back-pressure or siphonage, e.g. by an adequate air-gap.

Appendix: Water Conversion Tables

Water hardness is expressed in a variety of units other than ppm calcium + magnesium given as calcium carbonate. For convenience, the following is taken from *The Examination of Waters and Water Supplies* (Thresh, Beale and Suckling,) 1949, 6th edn, E. W. Taylor, J. & A. Churchill, London.

To convert:
Grams into grains multiply by 15.432
Grains into grams multiply by 0.0648

US gallon = 231 cubic inches = 3.7854 liters
= 0.833 British gallon
British gallon = 1.2 US gallon = 4.546 liters

	Grains per US gallon	Grains per British gallon	Parts per 100 000	Parts per million
1 grain per US gallon	1.000	1.20	1.71	17.1
1 grain per British gallon	0.830	1.00	1.43	14.3
1 part per 100 000	0.580	0.70	1.00	10.0
1 part per million	0.058	0.07	0.10	1.0

1 grain per British gallon = 143 pounds per million gallons.

Hardness conversion table

Hardness	Parts per million	Grains per British gallon	Parts per 100 000	German degree
Parts per million, as $CaCO_3$	1.0	0.07	0.10	0.056
Grains per gallon (degrees Clark) as $CaCO_3$	14.3	1.00	1.43	0.80
Parts per 100 000 as $CaCO_3$	10.0	0.70	1.00	0.56
German degrees = parts per 100 000 as CaO	17.8	1.24	1.78	1.00

Note. It is usual practice to express the total calcium and magnesium ions as 'calcium' ions.

Further reading

World Health Organization. Check the latest edition of *International Standards for Drinking Water*, World Health Organization, Geneva, Switzerland
Apart from *National Standard Texts* reference may be made to a symposium report published 1977 in *Water Quality*, edited by F. Coulston and E. Mrak. New York: Academic Press
Poretti, M. (1990) Quality control of water as raw material in the food industry. *Food Control*, **1**, (2), 79–83
This paper also deals with equipment.

4.E.5 Liquid waste

Two categories of liquid waste may be recognized in a food plant which require separate and distinct drainage systems. The first is the 'domestic' sanitary waste, e.g. from lavatories, urinals and toilet facilities. The second category is general plant drainage, e.g. from process equipment or floor drains. Any operation which produces very large volumes of waste, or waste which requires special treatment, should be considered for a separate system.

Sanitary waste lines are an obvious microbiological contamination hazard and should be kept separate from general plant drainage lines. Imholte[8] makes the points that sanitary lines must not discharge into grease interceptors or other waste handling equipment and neither should they enter the main sewer lines in any way which allows the possibility of sanitary sewage 'backing-up' through the general plant drainage system since this would be a serious contamination hazard.

It is important therefore that sanitary and all other liquid waste systems have sufficient capacity under maximum loading conditions to remove liquid wastes quickly and efficiently. Remember – if rain water or 'storm water' is run into the general plant drainage lines these lines must have the capacity to handle the worst storm conditions that can be foreseen.

The routing of sanitary waste lines should, as far as possible, be around storage and processing areas. Sanitary lines should *never* be installed over potable water storage tanks or over exposed ingredients, food processing equipment or finished products because leakages will cause sewage contamination. Remember that liquid waste lines are sloped, so leakage at one point may contaminate an area at some distance from the point of leakage. Sanitary lines like any others, must be expected to block occasionally. When this happens, it is important that cleanouts ('rodding access points', UK) should be located outside manufacturing areas.

Vents must be provided for sanitary waste lines and it is necessary that these should not be located near ventilation or process air intakes.

It is prudent to have and keep updated a complete set of sanitary waste piping drawings. These may be very useful when changes in use of areas are proposed.

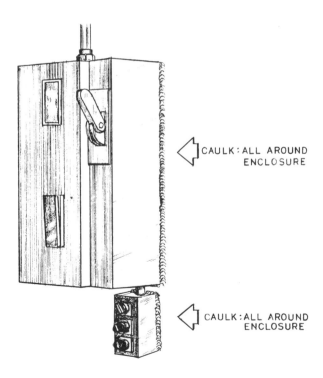

CAULK: ALL AROUND ENCLOSURE

CAULK: ALL AROUND ENCLOSURE

IF WALL MOUNTING ELECTRICAL APPARATUS IS AN ONLY OPTION. A COMPLETE SEAL WITH A PLIABLE CAULKING MATERIAL SHOULD BE REQUIRED.

Figure 4.37 Wall mounted electrical apparatus

Overhead general plant drainage lines may be unavoidable in ingredient and packaging storage areas, processing areas and finished product storage areas. It is obviously preferable to locate them along walls. Katsuyama and Strachan[10] recommend installation and regular cleaning of shields or deflectors whenever potential leakage constitutes a contamination hazard. This is *not acceptable* practice for sanitary lines.

4.E.6 Electricity

The sanitary design of electrical systems presents difficult and complex problems. The sanitary design ideal is either to have systems and components completely sealed or exposed so that they can be easily and thoroughly cleaned. In 'dry' areas, 'dust-tight' or 'explosion-proof' features exclude not only dust but also insects. In 'wet' areas 'water-proofing' is essential because water and electricity do not 'mix'. However, there are varying levels of 'water-proofing', and 'splash-proof' may not be sufficient. In many 'wet' area installations, hoses will be freely used and electrical equipment needs to be 'hose-proof'. Electrical equipment and components must also be mechanically strong, to withstand vibration and the type of impact damage that is inevitable in a food plant.

Some features of electrical distribution systems, e.g. trunking for cables, may actually assist the distribution of pests. It has been known for this to be discovered when the power cables inside trunking showed signs of rodent attack.

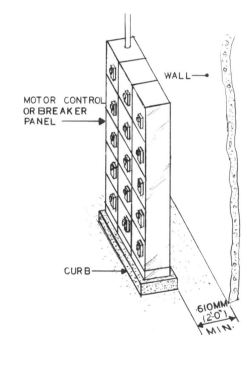

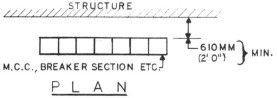

MOTOR CONTROL CENTER, SWITCH GEAR AND CONTROL PANELS SHOULD BE POSITIONED AWAY FROM WALLS OR OTHERS STRUCTURES. THIS PERMITS CLEANING AND MAINTENANCE ALL AROUND EQUIPMENT

Figure 4.38 Free-standing electrical gear

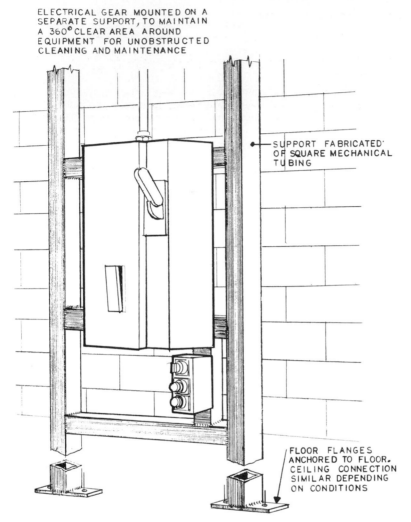

ELECTRICAL GEAR MOUNTED ON A
SEPARATE SUPPORT, TO MAINTAIN
A 360° CLEAR AREA AROUND
EQUIPMENT FOR UNOBSTRUCTED
CLEANING AND MAINTENANCE

SUPPORT FABRICATED
OF SQUARE MECHANICAL
TUBING

FLOOR FLANGES
ANCHORED TO FLOOR.
CEILING CONNECTION
SIMILAR DEPENDING
ON CONDITIONS

Figure 4.39 Electrical gear mounted on a frame

Electrical trunking should be kept lidded and there are sanitary advantages for side-lid mounting. This is first, because if the lid is left off, it is more easily visible and second, it reduces the ingress of dust or food material from overhead spillages.

It is difficult to reconcile the differing requirements of safety, efficiency and hygiene, and careful attention to detail is needed. Imholte[8] has an excellent chapter on electrical systems which illustrates a practical concern with sanitary design. For example, he advises that wall-mounted push-button or other stop–start stations are either sealed to the wall with a pliable caulking as shown in Figure 4.37. (which is Figure 5.6 in Imholte[8]) or extended from the wall by at least 50 mm (2 in) to allow for cleaning. Although extension of such stations from equipment framework may increase the probability

of mechanical damage, Figures 4.38 and 4.39 show how free-standing electrical equipment may be installed.

The installation ('running', UK) of conduit pipework for cables needs careful thought if it is to be both effective and sanitary. It is not unknown for this pipework to form an uncleanable dirt trap. The possibility of other uses (or misuses) should be kept in mind, e.g. an electrical conduit may be wrongly used as a convenient foothold or platform by those who are cleaning, adjusting or maintaining equipment.

References

1. Troller, J. A. (1983) *Sanitation in Food Processing.* New York: Academic Press

2. IAMFES (1972) 3 – A Accepted practices for supplying air under pressure in contact with milk, milk products, and product contact surfaces. *Journal of Milk and Food Technology*, **42**, 172–179

3. ABPMM (1987) *Guidelines for Good Hygienic Practice in the Manufacture of Milk-Based Powders*, The Association of British Preserved Milk Manufacturers, 19 Cornwall Terrace, London NW1 4QP

4. BS1 (1976) *BS5295: Environmental Cleanliness in Enclosed Spaces. Part 1. Specification for Controlled Environment Clean Rooms, Work Stations and Clean Air Devices*, London: British Standards Institution

5. Charlish V. R. and Warman, K. G. (1980) Solving hygiene problems by process design. In *Hygienic Design and Operation of Food Plant*, edited by R. Jowitt, pp. 35–51. Chichester: Ellis Horwood

6. Tamplin, T. C. (1980) CIP technology, detergents and sanitizers. In *Hygienic Design and Operation of Food Plant*, edited by R. Jowitt, pp. 183–225. Chichester: Ellis Horwood

7. Kessler, H. G. (1981) *Food Engineering and Dairy Technology*, English edn, Vertag, A. Kessler, PO Box 1721, D-8050 Freising, Federal Republic of Germany

8. Imholte, T. J. (1984) *Engineering for Food Safety and Sanitation*. Crystal, MINN: Technical Institute of Food Safety

9. DHSS (1981) *Food Hygiene Codes of Practice No. 10. The Canning of Low Acid Foods – A Guide to Good Manufacturing Practice*. London: HMSO. (Note, this is out of print and currently being revised)

10. Katsuyama, A. M. and Strachan J. P. (eds) (1980) Water supply. In *Principles of Food Processing Sanitation*, pp. 155–166. Washington, DC: The Food Processors Institute

11. Katsuyama, A. M. (1979) *A Guide for Waste Management in the Food Processing Industry*. Washington, DC: The Food Processors Institute

Chapter 5
Sanitation

5.A. Sanitary (hygienic) design

Good sanitary (hygienic) design is essential if:

- The product is to be protected from microbial or other contamination
- Cleaning effectiveness is to be maximized and cleaning costs reduced.

It is important, however, to realize that good work in hygienic design is complementary to good plant layout and good departmental layout and management.

5.A.1 Plant (factory UK) layout

Alternative outline layouts

The ideal plant layout integrates the departmental areas in a logical way to provide for a smooth flow of materials and services and Hayes[1] discusses this in

his Chapter 6. He gives alternative layout suggestions, two of which are shown here as Figures 5.1. and 5.2. He wisely remarks that much depends on the amount of land space available. It may also depend on the shape of the site, particularly in respect of road and rail access.

In practice, a straight line product flow as in Figure 5.2. has many advantages if it can be achieved. However, it must be stressed that good layout does not require lavish or luxurious buildings – it does mean spending so as to maximize the probability of good management producing safe, wholesome food economically.

5.A.2 Manufacturing department layout

Good department layout is crucially important in order to minimize the risk of cross contamination,

119

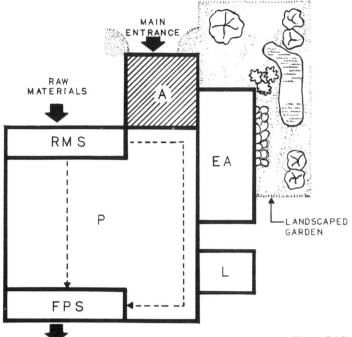

Figure 5.1 Layout with support departments attached to main envelope (from Hayes[1])

LEGEND

A	ADMINISTRATION
EA	EMPLOYEE AMENITIES
FPS	FINAL PRODUCTS STORAGE
RMS	RAW MATERIALS STORAGE
L	LABORATORY
P	PRODUCTION AREA

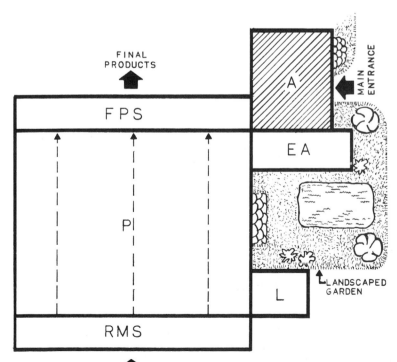

NOTE :
REFER TO LEGEND ON
FIGURE 5.1

Figure 5.2 An alternative layout to that in Figure 5.1 (from Hayes[1])

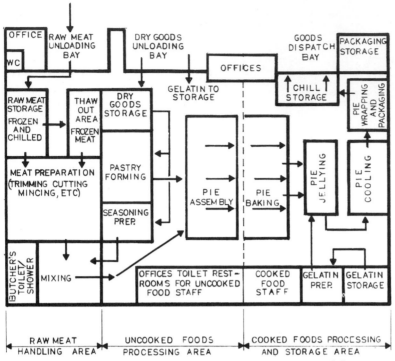

Figure 5.3 Production flow in a pork pie processing factory (from Hayes[1])

which has a high risk for public health and product spoilage. It is important to separate:

- Microbiologically 'high' and 'low' risk materials and processes
- Personnel working with 'high' and 'low' risk materials and processes

and to prevent cross traffic between 'high' and 'low' risk materials and processes.

Production flow in a pork pie department

Hayes[1] makes these points very clearly in his Chapter 8, part of which is summarized here. The British pork pie has a meat, fat, cereal and/or rusk center with spices etc. set in gelatin inside a pastry crust. It may be eaten either hot or cold. Pork pies are often regarded as a 'snack' food which is stable if properly chilled although it may be exposed for sale at 'cool' rather than 'chill' temperatures.

Hayes[1] uses the layout shown here as Figure 5.3 to give an illustration of the application of microbiological principles to the layout design as well as the operation of a food plant and processes. You will find his discussion on pages 257 to 260. His figure shows clearly the separation of 'high' and 'low' risk materials, processes and personnel. The need for this is shown by considering the three

components of the pie; raw meat, pastry and gelatin along with the processing steps.

For cost reasons, meat of 'manufacturing grade' is used. There is the possibility that it may carry low numbers of pathogens, e.g. *Salmonella* spp., so separation of raw meat minimizes the risk of cross contamination of the cooked product. After thawing and size reduction, the meat is mixed with other ingredients, e.g. curing salts and spices, before being filled into pastry cases and cooked. Cooking is by baking in ovens which are 'double ended' so that they can be unloaded on the 'clean side'.

Pastry is a mixture of flour, fat and water which is formed into a casing. After filling with meat, pies are lidded and baked before gelatin solution is injected into either hot or cool pies. The clear risk of cross contamination is the reason for separate cooked and uncooked processing areas.

Gelatin is a good growth medium, still liquid at about 40°C (104°F) which is near optimum for *Salmonella* spp. Unless the gelatin is properly preheated, *Salmonella* spp. may survive and multiply before the pies are completely chilled.

To minimize cross contamination and re-contamination of the pies, baking, jellying, cooling, wrapping and packaging all take place in the cooked foods processing and storage area.

It is important to realize that the Hazard Analysis

Critical Control Point (HACCP) system can and should be applied to plant layout as well as to the processing stages.

The discussion by Hayes[1] underlines the need to design as well as construct, maintain, and operate hygienic food handling equipment. It is an excellent illustration of the application of microbiological knowledge to the design and operation of food plant and processes.

Reference

1. Hayes, P. R. (1985) *Food Microbiology and Hygiene*, London/New York: Elsevier Applied Science

Further reading

Vinson, H. G. (1990) Hygienic design of a dairy. *Journal of the Society of Dairy Technology*, **43**, (2), 39–41

5.A.3 Sanitary (hygienic) design of equipment

Introduction

Sanitary design is sometimes regarded by management as an obviously 'Good Thing' but of real concern to someone else. Because of Public Health and cost implications this is an inadequate response. Examples of instances where food handling equipment has been implicated as the cause of food-borne illness or major spoilage are given by ICMSF[1], and are shown in Table 5.1.

Honest reflection on experience within any company that has been operating for a few years will identify spoilage incidents which in part at least were caused by poor sanitary design – even if the true full costs of these incidents are not accurately known. Of course, choice of equipment is often determined by delivery as well as price. However, always remember that purchase price is not the full true cost, which is the total expenditure during the working life of the equipment. Katsuyama and Strachan[2] make the point that there may be little relationship between the initial cost of a piece of equipment and cost per unit of product that is handled. This latter cost is made up of the cost of performing the operation plus the depreciation (or replacement), maintenance and cleanup costs figured on a 'per unit of product' basis.

It can be a worthwhile question to ask whether your current costing system actually reflects *all* the true costs incurred. Remember, *costs* are facts but *costings* include opinion.

What sanitary design must deliver

It should be appreciated that sanitary design is an intrinsic property of equipment, and later modifications to correct deficiencies are rarely simple, cheap or wholly successful. Le Corbusier's words, quoted in Campden Technical Manual 17[3], 'Good Design is Intelligence made Visible' apply fully to sanitary

Table 5.1 Examples of equipment-related spoilage or food-borne illness

Equipment	Problem	Consequences	Correction
Grain silo	Areas of high moisture	Moldy grain[a]	Proper ventilation and grain turnover
Can reformer	Holes in cans of salmon	Botulism	Proper maintenance of equipment
Gelatin injector	Welds difficult to clean	Salmonellosis from meat pies	Smooth weld
Wood smoke sticks	Bacteria surviving cleaning	Spoilage of sausage	Replace wood with metal
Heat exchanger (cooling side)	Cracked cooling unit permitting entrance of contaminated water	Salmonellosis from milk powder	Replace heat exchanger
Pump	Worn gasket	Spoilage of mayonnaise	Replace gaskets more frequently
Deaerator	Not properly cleaned or located in processing scheme	Contamination of pasteurized milk, enterotoxigenic cheese	Properly clean deaerator and move upstream of pasteurizer
Commercial oven	Poor heat distribution	Areas of under-cooking, rapid spoilage, potential food-borne illness	Correct heat distribution in oven, monitor temperature to detect failure

[a] Molds can produce a range of aflatoxins.

design of equipment. This is because good sanitary design must deliver:

- Efficient performance of the intended task, e.g. mixing, heating, chopping, etc. Where pertinent, this includes monitoring, control and recording of its function or performance.
- No additional contamination of the product, chemically, physically or microbiologically. *Note*: microbial count may increase because of normal operation of the equipment, e.g. a dicer will break and spread microbial clumps thus increasing the colony-forming unit (cfu) count.
- Maximum protection of the product from external contamination *and* minimum contamination by the product of other equipment and plant structures.
- Cost effectiveness as installed – for production, cleaning and maintenance over the whole of its intended life when compared with alternative equipment.

Remember, even the 'best design' needs to be checked out. HACCP is a powerful tool to insure that this is done. The principles given in Chapter 3 apply and need to be used both for production aspects of design as well as for cleaning and sanitation.

Imholte[4] has wise words on costs and on the principles of sanitary design. On costs:

Without good (sanitary) engineering, food safety and sanitation programs can become significantly more costly. In addition, the effectiveness of these programs is reduced, and problems such as food poisoning, spoilage, insect infestations, and injurious foreign material contaminations are more likely to occur.

On principles:

The principles of food safety and sanitation are sometimes perceived as common knowledge, which any sensible, responsible person should recognize. On the contrary, food safety and sanitation is a complex subject. The more that engineers, in particular, know about it, the better their judgments and decisions are likely to be.

His comments about engineers apply equally to management and technologists. Remember, the principles include design, construction, installation, use (for both production and cleaning) and maintenance.

Design of exterior

What good sanitary design must deliver applies equally to the externals of equipment. As Ridgeway and Coulthard[5] state:

The exterior of the machine is not something that can be tackled in isolation. It is an integral part of

the complete design. It is no solution to put easy-to-clean boxes around dirty machines.

However, this paper concentrates on the various hygiene aspects of a machine's exterior.

First a definition of what we mean by exterior. The exterior of a machine, or plant, is those parts of the equipment for which the operator, or cleaning staff, would normally be responsible. This includes not only the outer surfaces of the machine, but also the product areas, for which the operator has responsibility.

The general philosophy in tackling hygiene is three-fold:
1. If a basic process is, in itself, one which creates dirt, dust and unhygienic conditions, attempt to replace that process with a cleaner one.
2. If a basic process change cannot be made, then separate the product zones from the non-product areas.
3. Position operator controls so that the operator does not have to touch the product.

Remember, whatever the process, exteriors need to be clean. This is because employee morale and pride in work is enhanced if cleaning is easy and effective.

Effect on contamination (complaint) rates

The need to take a comprehensive view of sanitary design is illustrated by Robson and Barnes[6] in a paper dated 1980. They worked for Marks and Spencer Ltd in the UK, one of the best-known retail groups with a high reputation for 'Quality Foods' and other non-food items. Their paper is concerned with non-microbial contamination and although they are careful to avoid giving rates of complaints, the numbers are meaningful. They state:

Firstly, some facts: in 1977 our company received 13,477 customer complaints of foreign bodies in foods. This was a marked increased over the previous year, even allowing for the greater number of units sold. Of this total, 6289 or 47%, were in categories that could have been due, in whole or in part, to defects in machinery or its design. These complaint types are set out in the table below.

Foreign body type	Number	Percentage of total foreign body complaints
1. Dirt	1303	9.7
2. Metal (detectable)	966	7.2
3. Plastic/rubber	963	7.1
4. Oil/grease	762	5.6
5. String	587	4.4
6. Wood	571	4.2
7. Fibers	525	3.9
8. Glass	257	1.9
9. Bristles	202	1.5
10. Metal (non-detectable)	153	1.1

In addition there were 1166 (8.7%) complaints due to insects and a proportion of these were undoubtedly derived from poor factory and machinery design.

These figures are facts. What follows is conjecture.

Marks and Spencer represents about 2% of the food trade of the UK. If their experience is reasonably representative of the whole, then foreign body complaints on foods sold throughout the UK in 1977 should have been of the order of 700,000. Since experience shows that only between 10% and 20% of the actual foreign body complaints find their way back to the company, then it is possible that the actual number of incidents nationally could be of the order of 3 to 4 million. Perhaps the food industry should be thankful that only 13,000 foreign body complaints were received by Public Health Departments during 1976 (the last year of published figures).

Now, if poor plant and factory design is responsible for a third or a half of this figure, then this today is a massive and urgent subject, especially when many highly emotive foreign body incidents are 'unseen' by the consumer and hence are not the subject of complaints, such as contamination due to rodents, S.P.I. (stored product insects) etc.

Some of their categorization may not be immediately obvious – string was not put in by a machine, but it should have been removed. Even allowing for all possible doubts, these figures illustrate the clear need to use management and technological resources to achieve good sanitary design.

Where to get help

This section cannot meet the readers' specific needs because each reader's likely concern is for a particular item of equipment – and decisions are wanted quickly.

Troller[7] writing in 1983 stated that there were over 2000 companies in the USA fabricating food processing and service equipment. This great variety of equipment means that no book can give detailed consideration to the exact item of equipment that you want to buy even if it is not custom-built or extensively modified for your purposes.

Ideally, fully adequate sanitary design skills and resources should be available in-house. Imholte[4] gives his opinion that:

It takes a design engineer about two years to develop his sanitary design skills. The engineering criteria are not particularly complex, and they can be mastered fairly quickly. Nonetheless, more time is needed to acquire a better understanding of the different manufacturing processes and their particular design needs.

However, those who are not experienced design engineers must be expected to take significantly longer than two years to acquire the requisite skills.

Alternatively, there may be a Research Association or similar center of expertise familiar both with your processes and range of equipment available. The conscientious equipment manufacturer will set out to achieve high standards of sanitary design but this may be for the most usual customer application, and your process may have special needs. Over the last three decades, standards have improved considerably but they are not universally adequate.

What follows here is a guide to principles which, together with the summary of the rather thin and scattered technical literature, may assist the user in asking pertinent questions about proposed or existing equipment. These will be considered under the following headings:

- Basic principles
- Materials of construction
- General design features
- Details of construction (fabrication, UK)

Basic principles of sanitary design

These are well expressed in the seven points of the USA document 'Sanitary Design' (taken from *Food Processing* October, 1964 and April, 1965 and *Food Processing Catalog*, 1964/5) and quoted in the Appendix to Jowitt[8]:

- All surfaces in contact with food must be inert to the food under the conditions of use and must not migrate to or be absorbed by the food.
- All surfaces in contact with food must be smooth and non-porous so that tiny particles of food, bacteria, or insect eggs are not caught in microscopic surface crevices and become difficult to dislodge, thus becoming a potential source of contamination.
- All surfaces in contact with the food must be visible for inspection, or the equipment must be readily disassembled for inspection, or it must be demonstrated that routine cleaning procedures eliminate possibility of contamination from bacteria or insects.
- All surfaces in contact with food must be readily accessible for manual cleaning, or if not readily accessible, then readily disassembled for manual cleaning, or if clean-in-place techniques are used, it must be demonstrated that the results achieved without disassembly are the equivalent of those obtained with disassembly and manual cleaning.
- All interior surfaces in contact with food must be so arranged that the equipment is self emptying or self draining.

- Equipment must be so designed as to protect the contents from external contamination.
- The exterior or non-product contact surfaces should be arranged to prevent harboring of soils, bacteria or pests in and on the equipment itself as well as in its contact with other equipment, floors, walls or hanging supports.

Additions

Four additional points are made in The Campden Technical Memorandum 289[9]:

- In design, construction, installation and maintenance it is important to avoid dead space or other conditions which trap food, prevent effective cleaning and may allow microbial growth to take place.
- The requirement of guarding machinery to insure safety in operation may easily conflict with sanitary (hygiene) requirements unless considerable care is taken in design, construction, installation and maintenance.
- Noise suppression is important in providing acceptable working conditions. However, many noise reducing materials can give rise to microbiological or infestation problems unless care is taken in their selection, installation and maintenance.
- It is important that the equipment itself is so designed, installed and maintained that it does not cause product contamination. Examples of possible contamination are: lagging which may break up or insufficiently secured nuts and bolts. Such hazards should be designed out of the system.

The Campden group endorsed the comments of the FMF/FMA Committee quoted in Jowitt[8]:

There is no substitute for common sense, and no specification can be complete enough to insure a sanitary (hygienic) design. Good design requires much attention to detail. It is seldom that a single plant item can be considered in isolation; one must consider the process itself, the environment in which the equipment is placed and, of course, the maintenance which will be provided throughout the life of the equipment. The compatibility of the equipment with the product, the environment and also the cleaning fluids are of vital importance.

It must be recognized that there are degrees of sanitary (hygienic) design generally related to the permitted tolerance of microbial infection in the final product. Designs must allow for easy dismantling of components for rapid cleaning, or in-place cleaning methods must be incorporated as part of the basic design. It is suggested that the latter will give more reliable microbiological results, particularly on large installations and where an automated cleaning cycle is justified.

The Campden group added that the frequency of cleaning depends on the nature of the product and processes. Cleaning must prevent microbial or other contamination of subsequent production.

The most common design faults which cause poor cleanability are:

- Poor (or nil) accessibility
- Inadequately rounded corners (minimum radius should be 6.4 mm (¼ in)
- Sharp angles
- Dead ends (including poorly designed seals).

There is a useful discussion of Hygienic Design by Milledge[10] which has excellent references. Further 'principles', together with some details are given by Stinson[11], by Katsuyama and Strachan[2] pages 111–112, and also in the USDA accepted 'Meat and Poultry Equipment' MPI – 2 dated 1990[12]. This document also gives definitions of terms which are often used by processors. These are given below.

Some USDA definitions

Accessible: Easily exposed for regular cleaning and inspecting with simple tools such as those normally carried by cleaning personnel.

Readily accessible: Easily exposed to sight and touch for regular cleaning and inspecting without the use of tools.

Removable: A component part can be separated from the principal part with simple tools such as a screwdriver, pliers, or open-end wrench.

Readily cleanable: The equipment can be cleaned with hot water, cleaning agents, and scrubbing implements normally used by cleaning personnel.

Corrosion resistant material: A material which maintains its original surface characteristics under prolonged exposure to the normal environment including product, ingredients, ambient conditions, and cleaning and sanitizing materials.

Acceptable materials: Materials which have been found to be chemically acceptable and physically suitable for the purpose intended.

Sealed: Having no openings that permit the entry of product, dirt, or moisture.

Product zone: All surfaces of the equipment which may normally be directly or indirectly exposed to product or ingredients.

Non-product zone: All surfaces of the equipment outside the product zone.

The reader should be aware of a limitation of the usefulness of such excellent definitions and the general principles given above in that they are open to a variety of interpretations. They need to be made more detailed and specific when they come to be applied to specific product(s) and process(es).

Materials of construction

Campden Technical Manual 17[3] sets forth the qualities of these materials as follows:

> Good sanitary (hygienic) design of equipment used in the manufacture of foods and beverages requires that all surfaces in contact with the product must be non-toxic, inert to the product under the conditions of use, must not have constituents which migrate or are absorbed by the product and, in addition, must be resistant to (i.e. be inert to) cleaning and disinfecting agents under normal (or expected) conditions of use.
>
> Food contact surfaces should permit the easy removal of food residues and soils during cleaning so that microbial growth is prevented. Surfaces should be smooth, hard, continuous and free from cracks, crevices and pits. Ideally, materials should therefore be such that the original surface finish is maintained during the working life* of the equipment: furthermore, the material should not deform and should be resistant to denting, chipping, flaking and delamination.
>
> Equipment surfaces not directly in contact with foods, for example machinery framework and exterior cladding, should also be smooth to facilitate cleaning. In addition, they should be made of corrosion resistant material or protected against corrosion. Painting of equipment surfaces must be restricted to non-product contact surfaces and precautions should be taken to insure they are not above exposed food materials.

A detailed discussion of commonly used materials then follows on in Technical Manual 17.

The HACCP approach applies to the choice of materials, as it is not just the mechanical properties of the material which are important. Questions need to be asked about the properties of the foods in contact with the material and about contact times and temperatures of cleaning and sanitizing materials. When considering sanitizing materials, remember to include pesticides if these may be involved. Also important are up-stream and down-stream operations. As an example, if a metal screen has magnetic properties, this can simplify down-stream anti-contamination measures which guard against the effects of possible breakage.

* The term 'working life' should be understood to include the expected abuse as well as normal wear caused by production.

In practice, some compromises have to be made. As an example, cast iron is not the ideal material for contact with wet foods. However, the bowls of bowl choppers used for meat are usually made of cast iron. If, after cleaning and drying they are lightly oiled with an edible vegetable oil, cast iron bowls have been found to be satisfactory in use.

Warning: Be prepared to answer these two questions

1. What can go wrong?
2. When it does go wrong, what is the planned action to protect consumers?

It is not easy for the technologist to get the information that is needed to make decisions. For example, what grade and finish of stainless steel is needed for a specific surface?

Stainless steels

Both Katsuyama and Strachan[2] and Campden Technical Manual 17[3] discuss the properties and grades of stainless steels. As with equipment, it is worth remembering that the cheapest first cost is not necessarily the least costly over the intended life of the material (see Table 5.2).

Table 5.2 Commonly used grades of stainless steels

AISI Code	Properties
304	Moderate corrosion resistance
304L	Variant for welding without subsequent annealing
316	Good corrosion resistance
316L	Variant for welding without subsequent annealing

AISI – American Iron and Steel Institute

Campden Technical Manual 17[3] has a detailed review of stainless steels and also gives a useful international comparison of grade designations shown here as Table 5.3.

The finish of stainless steel is not just a matter of appearance; on product contact surfaces it has an effect on sanitation. Katsuyama and Strachan[2] discuss this in the context of US practice and Campden Technical Manual 17 deals with both US and European conventions.

Campden Technical Manual 17[3] gives a detailed description of surface 'roughness' as well as corrosion and explains the differences between US root mean square (RMS) units and the European unit R_a. A table of finishes is also given (see Table 5.4). Additionally, this manual has useful material on corrosion of stainless steels.

In the UK, specifications for the finishes of tubing for the food industry are usually as given in ISO 2037 and BS 4825 Part 1. (See Further Reading.)

Table 5.3 Nearest equivalent specifications to those given in BS 1449

British BS 1449 Part 2 (1983) 284S16	France AFNOR	W. Germany Deutsche Stoff no. 1,4371	Italy UNI	Japan JIS	Sweden SIS 2357	United States AISI 284
301S21	Z12CN17–08	1,4324	X12CrNi1707	SUS 39	2331	301
		1,4301	X5CrNi1810	SUS 27	2332	302
		1,4300	X10CrNi1809	SUS 27	2331	302
304S11	Z2CN18–10	1,4306	X2CrNi1811	SUS 28	2352	304L
304S15	Z6CN18–09	1,4301	X5CrNi1810	SUS 27	2332	304
304S16	Z6CN18–09	1,4301	X5CrNi1810	SUS 27	2332	304
305S19	Z8CN18–12	1,3956	X8CrNi1812	SUS 62		305
309S24	Z15CN24–13	1,3828	X16CrNi2314	SUS 41		309
310S24	Z12CN25–20	1,4842	X6CrNi2520	SUS 42	2361	310
315S16					2340	
316S11	Z2CND17–12	1,4404	X2CrNiMo1713	SUS 23	2353	316L
316S13	Z6CND17–11	1,4401	X5CrNiMo1713	SUS 32	2343	316
317S12	Z8CND19–15	1,4438	X2CrNiMo1816			317L
317S16					2366	317
320S31	Z8CNDT17–13	1,4573	X6CrNiMoTi1713		2350	
321S31	Z6CNT18–11	1,4541	X6CrNiTi1811	SUS 29	2337	321
347S31	Z6CNNb18–11	1,4544	X6CrNiNb1811	SUS 43	2338	347
504S17	Z6C13	1,4000	X6Cr13		2301	403
405S17	Z6CA13	1,4002	X6CrA113	SUS 28		405
410S21	Z12C13	1,4006	X12Cr13	SUS 51	2302	410
420S45	Z30C13	1,4028	X30Cr13	SUS 53	2304	
430S17	Z8C17	1,4016	X8Cr17	SUS 24	2320	430
434S17	Z8CD17–01	1,4078	X8CrMo17			
442S19						442

Taken from Campden Technical Manual 17.

The effect of surface finish on cleaning is discussed by Milledge and Jowitt[13], who conclude that it may be the nature of the surface irregularities which is important rather than their magnitude as indicated by the R_a value. They also point out the cost-effectiveness of electropolishing. A more recent paper by Timperley[14] states that for a hard surface finish:

> There is no evidence from the tests conducted that the surface finish of stainless steel, provided the R_a value is less than 1 micro meter, has any effect on cleanability using direct impingement from a spray jet irrespective of the finishing method. These findings are contrary to those for pipelines cleaned-in-place for which there was a correlation between surface finish and cleanability. It would, therefore, seem unnecessary to invest in highly polished finishes on the grounds of cleanability when spray jets are used. However, the initial assessment of the cleanliness of equipment, when hand-held lances are used, is on bright surfaces; this should be considered when specifying a finish. Electropolishing is a cost effective method of obtaining a bright finish and it is reported that, in practice, operating periods may be extended, cleaning times reduced and anti-stick properties improved.

The smoothness of internal pipe surfaces is discussed by Timperley[15]. He comments that in the UK longitudinally welded tube has replaced seamless tube but that the surface finish can vary. The relationship of surface finish to residual microbial count and the effect of surface finish on cleaning time is discussed in his paper. Timperley concludes that 'It is important when ordering tube to specify that the internal finish should be less than 1.0 μmRa' as required by ISO and BSI standards.

All the above sources reinforce the fact that choice of grade and finish is not necessarily simple, since the real question is 'How good is the evidence that the proposed grade and finish will be suitable for the equipment, process(es), products, cleaning and sanitizing materials and temperatures throughout the expected life of this equipment?' There is, of course, no better evidence than first-hand practical experience over a period of time. The prudent will make sure that initial costs or delivery dates are not the only consideration in the choice.

Table 5.4 Stainless steel surface finishes

Finish no.	Description	Roughness – R_a (typical value=/μm)	Notes
Mill finishes			
0	Hot rolled and softened but not descaled	–	Suitable only for certain heat resisting applications, as the presence of oxide scale impairs resistance to corrosion. Surface inspection is not practicable
1	Hot rolled, softened and descaled	3.75–6.25 (5.0)	Generally used when smoothness and uniformity of finish are not important. It is permissible for grinding marks to be present
2D	Cold rolled, softened and descaled	0.375–1.0 (0.50)	A uniform matt finish
2B	Cold rolled, softened, descaled and lightly rolled on polished rolls	0.125–0.50 (0.275)	A smooth finish for general applications brighter than finish no. 2D
2A	Bright annealed (BA)	0.05–0.125 (0.088)	A cold rolled reflective finish retained through annealing
Polished finishes:			
3.A	Ground	1.0–2.0 (1.5)	A uniform coarse ground finish generally obtained by treatment with abrasives of 80 to 100 grit size
3.B	Dull buffed	–	A uniform straight grained finish, generally achieved in one pass, applied to a 2B or 2A mill finish; produced by polishing with 180 to 220 grit size abrasive belts
4	Polished with fine grit	0.20–1.5 (0.8)	The final polishing generally produced by treatment with successively finer abrasives of 180 to 240 grit size, gives an undirectional texture, not highly reflective
7	Bright buffed	–	A bright finish, generally applied to a 2B or 2A mill finish, by treatment with a fiber and/or cloth mop, as well as with a suitable polishing compound
8	Bright polished (mirror)	–	A bright reflective finish with a high degree of image clarity. Produced by polishing with successively finer grits followed by buffing with very fine polishing compounds

Taken from Campden Technical Manual 17. Compiled from BS 1449: Part 2, 1975 and 1983.

Other materials of construction

Table 5.5 is an index to the pages in the documents quoted where reference may be found to materials of construction commonly in use. Whether using 'traditional' or 'new' materials they must be suitable for intended use with product, processes, cleaning and sanitizing agents.

Good sanitary design features

These may be summarized as those which:

- Avoid chemical, physical or microbial contamination of the product. This means, for example, that there are no places where product lodges, builds up and then falls back into the product stream. It

also means that component parts of the equipment are either accessible for cleaning and maintenance or are completely sealed.
- Give maximum protection to the product from external contamination, e.g. tank lids.
- Simplify cleaning, thus promoting cost-effectiveness. To illustrate how this may be done, consider the following outline examples.

Remember: good sanitary design is largely about careful attention to detail

Avoidance of contamination

One of common features of poor sanitary design is the existence of inherent 'dead spots' or 'dead ends'. They are better described as 'filth traps'.

Table 5.5 Materials of construction

Item	Campden Technical Manual 17[3]	Hayes [6]	Imholte [4]	Jowitt [8]	Katsuyama and Strachan [2]	Troller [7]	'Other'( )	Notes
								Reference to page(s) in:
Stainless steel	2.1 to 2.3	204, 205	114	20, 247, 248	114–116	45, 46		
S/S surface finish	2.5 to 2.9				115–116	46	(13)57–62* (14)31–42+	*Cleanability and finish. +Cost effectiveness of finish.
S/S corrosion	2.9 to 2.17	205, 206			116			
Stellite					116			
Titanium	2.20	207	114		116			
Iconel			114					
Mild steel	2.19	206	114		114			
Black/cast iron	2.18	206	114		114			
Aluminum	2.20		114		117	46		When friction occurs between meat or fat and aluminum a black oxide is produced which discolors the meat. Anodizing the aluminum does not eliminate this problem.
Brasses (copper Brass, bronze)	2.20	206, 207	21	117	46			
Monel					116			
Lead		207						
Solder		207						
Galvanized iron	2.19		115					
Tin plate			115					
Nickel plate			116					*Warning* – Hypochlorite is not compatible with nickel.
Chrome/nickel plate			116					
Cadmium (plate)		207	115			46		*Toxic!* Do not use.
Wire			116					
Enamel					117			
Wood	2.22	208, 209	116		113			
Rubber	2.21	208, 209	116	248	117			
Glass	2.21	208, 209	116		117			*Beware!* Contamination risk to product.
Fiberglass			116					
Plastics	2.21	207, 208	116	21	117, 118			
Cloth/fabric fibers			116		17			
Teflon			118					
Caulking compounds			118					
Asbestos								*Warning!* Do not use.
Paint			129, 130					*Beware!* Contamination risk to product.
Jointing materials				247				

It needs to be remembered that microorganisms are very small, and that what is visually small to a human are to microbes the 'wide-open spaces' where they can multiply. Furthermore, if the cleaning does not remove material because design features shield it, then this becomes a 'designed in' source of contamination.

It is therefore axiomatic that sharp corners, ledges, crevices and dead ends anywhere in the product area are *unsanitary*.

To produce a sanitary finish both good design and good workmanship are required, as can be seen in the simple example of welded joints in Figure 5.4. However, good workmanship alone cannot compensate for poor design.

An example of care needed in design detail is taken from Campden Technical Manual 17[3]. Figure 5.5 shows an unsatisfactory design for an instrument probe inserted into a tank. The probe can function, but the poor design prevents effective cleaning without disassembly.

As an illustration of the influence of design, the

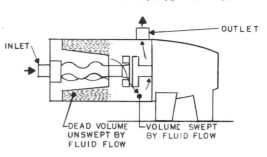

A. INITIAL DESIGN

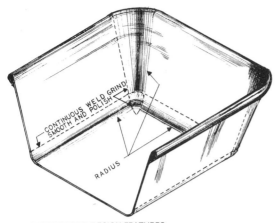

USE RECOMMENDED DESIGN FEATURES
FOR TOP RIM

Figure 5.4 Properly welded joints provide a sanitary finish

joint Technical Committee, FMF/FMA[8], show how 'dead' areas may be avoided by a change of position in the outlet of a pump (see Figure 5.6.).

In summary, while good fabrication techniques make good design effective, bad design contributes to contamination.

Maximum product protection

As a general rule, tanks should be lidded (fitted with lids). The design of the lid, however, may be unsatisfactory and need improvement. Campden Technical Manual 17[3] shows how this may be done in progressively better ways on pages 3.53 to 3.58 of that manual.

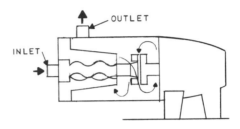

B. IMPROVED DESIGN WITH FULL VOLUME SWEPT BY FLUID FLOW

Figure 5.6 Improvements in pump design (from FMF/FMA (now FDF) Joint Technical Committee[8])

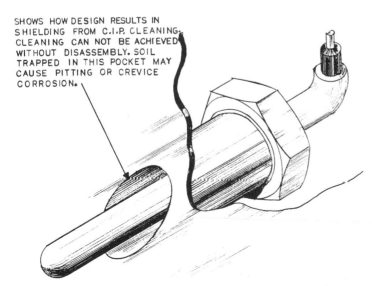

SHOWS HOW DESIGN RESULTS IN SHIELDING FROM C.I.P. CLEANING. CLEANING CAN NOT BE ACHIEVED WITHOUT DISASSEMBLY. SOIL TRAPPED IN THIS POCKET MAY CAUSE PITTING OR CREVICE CORROSION.

Figure 5.5 Unsatisfactory design of instrument pocket (from Campden Technical Manual 17[3])

130

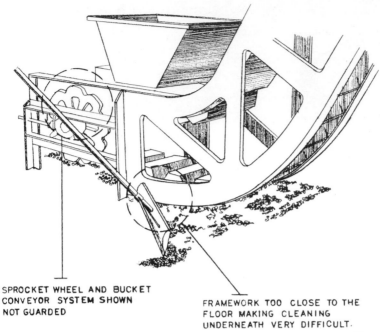

SPROCKET WHEEL AND BUCKET
CONVEYOR SYSTEM SHOWN
NOT GUARDED

FRAMEWORK TOO CLOSE TO THE
FLOOR MAKING CLEANING
UNDERNEATH VERY DIFFICULT.

Figure 5.7 Bad bucket elevator design (from Campden Technical Manual 7[16])

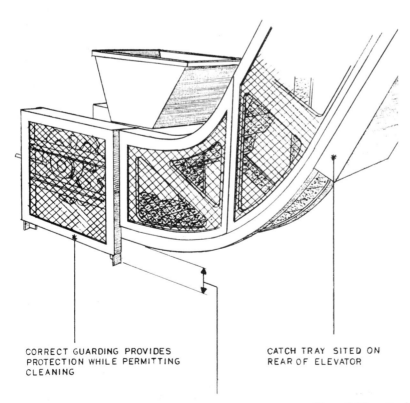

CORRECT GUARDING PROVIDES
PROTECTION WHILE PERMITTING
CLEANING

CATCH TRAY SITED ON
REAR OF ELEVATOR

MINIMUM CLEARANCE FROM
FLOOR SHOULD BE 200MM(8")

Figure 5.8 Improved bucket elevator design (from Campden Technical Manual 7[16])

Simplify cleaning

Both design and installation are important to achieve this aim. The way in which good sanitary design helps is illustrated in Campden Technical Manual 7[16] on pages 72 and 73, with bucket elevator design. These are shown as Figures 5.7 and 5.8.

Remember, good sanitary design plus good installation, maintenance and cleaning makes possible maximum efficiency and cost effectiveness. Table 5.6. gives further references to sanitary equipment design which may help with your problem. However, the technical literature must be expected to offer ideas for consideration rather than detailed rules. The semi-monthly *Food Processing* by Putman Publishing Co., Chicago, Illinois is a useful source of ideas and US trade information.

Equipment construction (fabrication) – aims

As with other aspects of sanitary design, important aims are to:

- Give maximum protection to the product. This means, for example, that fastenings should not break or work loose with consequent product contamination.
- Provide product contact surfaces adequate for the design function which will not contaminate the product and are readily cleanable.
- Provide junctures which minimize 'dead' areas where chemical or microbial contamination may occur. Remember, ledges are *unsanitary*.
- Give ready access for cleaning and maintenance. This means, e.g. that removable panels should not require the use of tools.

Once again, attention to detail is the essence of sanitary design.

Details – plate fabrication

This is illustrated in Campden Technical Manual 7[16], examples of good construction. A sample of this material is given as the following list of DOs and DON'Ts and as Figure 5.9 which shows a good design for the fabrication of a hopper rim (reprinted from Campden R.A. Technical Manual 7[16] recommendations for the bending of sheet plate):

Do

1. *Do* use welded joints. For ease of construction these could be made at the positions shown by the dotted lines on Figure 14. [Our Figure 5.9]
2. *Do* grind and polish the weld joints to give a smooth finish so that debris does not lodge and microbial slime develop.
3. *Do* radius corners. The optimum radius for cleaning is 20 mm (0.75 in) Ref. 3-A Sanitary Standards.

4. *Do* have vertical or steep sides to minimize the lodging of product.
5. *Do* use the recommended design features for top rim – see Figures 18–20 (page 15) which show design faults and recommended alternatives.

Don't

1. *Don't* use overlap joints fixed by nuts and bolts or rivets because:
   (a) The overlap provides an area where product debris becomes trapped and encourages microorganisms to grow.
   (b) The overlap area is impossible to clean and disinfect efficiently.
   (c) The fixing system collects debris and presents a potential customer complaint hazard.

Details – fastenings

One of the most usual failures of sanitary design is the use of unsuitable fastenings. Imholte[4] wisely observes that 'despite efforts to eliminate nuts, bolts, and machine screws there will always be some need for them'. He advises that when needed they should have magnetic properties so that downstream magnets as well as metal detectors have a chance to remove them. However, some types of stainless steel are less magnetic than others, and it may not always be possible to follow this excellent advice.

One important detail is to make sure that the threads are 'blind', i.e. covered, in the product region. This is because threads are very effective at trapping material and are hard to clean. Imholte's pages 126–129 on assembly detail are worth careful study.

Campden Technical Manual 7[16] has a useful section on 'Recurring Faults' which clearly illustrates the need to avoid:

- Product contamination because of interference with product flow or the formation of 'dead' areas.
- Nuts, bolts, washers etc. working loose and falling into the product flow.

Table 5.7 gives references to sanitary design features which may help with your problem. However, the technical literature normally offers ideas for consideration rather than detailed design rules.

Closed systems

These may be defined as product pipelines and tanks. In principle, a pipeline should have an interior surface which is smooth (to allow efficient cleaning) and be of the same internal bore throughout. Additionally, pipelines should be filled

Table 5.6 Equipment design

Item	Campden	Hayes [18]	Imholte [4]	Jowitt [20]	Katsuyama and Strachan [2]	Troller [7]	Notes
					Reference to page(s) in:		
Bulk storage and use bins	(TM) 7: 12–13		143–147		121		
Carts						63[a]	[a]General comment not detailed information.
Catwalks	TM7: 88–91		32–3				
Chutes	TM7: 26–31				118		
Coolers						72–73[a]	[a]General comment not detailed information.
Conveying systems							
Belt conveyors	TM7: 34–69	234–237 refers to Campden	126–127: 133–134, 135		120	60–62	
Bucket elevators, lifts or conveyors	TM7: 72–85	237–238 refers to Campden	136–139			62–63	
Multi-lane conveyors	TM8: 65–79						Primarily for can handling but may have wider application.
Pneumatic systems			139–143				
Screw conveyors		211, 234	134–137			59–60	
Cutters, choppers, slicers and grinders		230–231				63[a]	[a]However, his statement that 'sanitary requirements are rather minimal' does not always apply.
Dehydrators and driers			44[a]		64[a]		[a]General comment not detailed information
Direct steam cooking			110–111[a]	41,46[a]	122[a]		[a]General comment not detailed information.
Drives, drive coupling guards and motor mountings	TM17: 3.61–3.62		121–126		120–122		
Drives and drive motors		228–230 refers to Jowitt	108–110	272–273			
Dust collection systems			151–154				
Food forming and assembly equipment		232–233					
Supporting framework		108,120		118,119		49–53	
Hand utensils						75[a]	[a]General comment not detailed information.
Heaters		239–242[a]				70–72[a]	[a]General comment not detailed information.
Hoods			110[a]		122–123[a]		[a]General comment not detailed information.
Hoppers and flumes	TM7: 8–11				118		

Table 5.6 Continued

Item	Campden	Hayes [18]	Imholte [4]	Jowitt [20]	Katsuyama and Strachan [2]	Troller [7]	Notes
Kettles	TM17 Index[b], 3.2, 3.37–3.42, 3.53–3.62		110–111				[b]Includes ancillary equipment
Mixers	TM17: 3.37–3.48	231–232[a]				65[a]	[a]General comment not detailed information.
Pipework and couplings	TM17: Section 5	222–228	150–151	265–271	39,87–88		
Plant exterior		213–214		53–77			
Process controls			130–132			48	
Pumps	Technical Memorandum No. 285[b], 28 pp	216–221	151	252–256	123	69–70	[b]Guide on types, performance, selection and design of sanitary pumps.
Sifters and separators		233–234	265, 266			74	
Recurring faults	TM7: 1–4, 18–23	209–210					Construction or fabrication faults
Tanks	TM17: Section 3[b] 64 pp	214–216	147–151	39, 248–251	119,121	65–69	[b]Includes Kettles
Valves		217–224		91–92, 257–264	119,123	56–58[b]	[b]States Butterfly valves need inspection and cleaning.
Washers and peelers		223[a]				73[a]	[a]General comment not detailed information.

with product during use. It has been known for a horizontal pipe of large diameter not to be filled. This was shown by the appearance of soil after cleaning and resulted from a combination of design and installation faults.

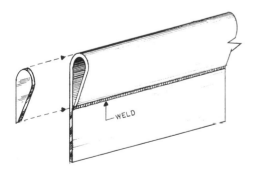

WHERE THIS METHOD OF CONSTRUCTION IS USED, OPEN ENDS SHOULD BE "CAPPED" TO PREVENT DEBRIS FROM BEING TRAPPED. THE WELD MUST BE GROUND AND POLISHED TO GIVE A SMOOTH SURFACE

Figure 5.9 A good design for the fabrication of a hopper rim (from Campden Technical Manual 7[16])

Timperley[15] discusses the difference between the welded and expanded versions of the IDF coupling designed for cleaning in place. He points out that the expanded version is not as cleanable as the welded version because the joint ring protrudes into the base as a result of the increase in diameter of the tube from the expanding operation.

With regard to welding operations, Timperley and Lawson[17] discuss orbital welding and show on page 90 of their paper a cross section of a weld made in this way. They point out that 'orbital welding of pipes is an automatic process, and if properly programmed, it gives consistently good welds which do not require polishing'. Further detailed information is to be found in Section 5 of Campden Technical Manual 17[3], which deals with pipelines and couplings.

With regard to tanks, provided that the design allows for free draining, that corners are adequately radiused and the surface finish and welds are smooth; then design and fabrication problems are usually associated with ancillary equipment and fittings.

Further information is found in section 3 of Campden Technical Manual 17[3] as well as in 3-A

Table 5.7 Equipment construction (fabrication)

Item	Campden	Hayes [18]	Imholte [4]	Jowitt [20]	Katsuyama and Strachan [2]	Troller [7]	Notes
			Reference to page(s) in:				
Agitator blade attachment	TM17: 3.39–3.40						
Baffles	TM17: 3.47–3.48						
Bearings						47	
Cabinets or enclosures			121–122				
Coves or fillets			119–120		119–120		
Cracks and crevices			120				
Dead ends		209–210	119	92–94			
Fastenings (nuts, bolts, rivets, etc.)	TM7: 18–21 TM17: 3.45–3.48		128–129				
Framing			120–121				See also Tubing
Inspection windows	TM7: 22–23						
Kettle lid hinge assembly	TM17: 3.59–3.60			251[a]			[a]Sketch for 'Tank Cover' Page 251.
Lap seams			117–118				
Ledges			117–118				
Leveling pads			121				
Motor mounts			121				
Motor take-up rails			121				
Mounting pads			120–122				See also Tank Supports (feet)
Pittsburgh seams			84, 120				
Rolled edges	TM7: 14–15		119		120	76	
Rounded corners			119		118–121[a]		[a]Text 119–120
Sheet plate	TM7: 12[a], 13, 15						[a]Bending and Fabrication
Smooth surfaces	TM17: 2.7–2.9		120				
Soldering						47	
Tank supports (feet)	TM17: 3.21–3.24						See also Mounting Pads
Tubing (construction)				118			See also Framing.
Void areas (hollow UK)			118–119				
Welds	TM17: 2.9, 2.17–2.18, 2.27–2.28		126–129[a]	89–91		45–46	[a]Note good advice on weld specification – get a sample from fabricator

Table 5.8 List of published sanitary equipment standards (Adapted from Table 5.2 ICMSF[1]

Product(s)	Organization	Address	Publication(s)	Date
Food and beverages	American Society of Mechanical Engineers	American Society of Mechanical Engineers, 345 East 47th Street New York, New York 10017, USA	Food, Drug and Beverage Equipment (ANSI – ASME F2.1–1982)	1982
Bakery	Baking Industry Committee Sanitation Standards	American Institute of Baking, 1213 Bakers Way, Manhattan, Kansas 66502, USA	Sanitation Standards for the Design and Construction of Bakery Equipment and Machinery	1986
Milk	International Association of Milk, Food and Environmental Sanitarians. Sponsoring organizations include IAMFES, Dairy Industry Committee, and US Public Health Service	International Association of Milk, Food, and Environmental Sanitarians Inc., Ames, Iowa 50010, USA	3-A Sanitary Standards and Accepted Practices. *N.B.* New standards and revisions to existing standards are published in the *Journal of Food Protection* as they become available	1986 onwards
Eggs	International Association of Milk, Food and Environmental Sanitarians. Sponsoring organization include IAMFES, Dairy Industry Committee, and US Public Health Service	International Association of Milk, Food and Environmental Sanitarians Inc. Ames, Iowa 50010, USA	E-3-A Sanitary Standards and Accepted Practices *N.B.* New Standards and revisions to existing standards are published in the *Journal of Food Protection* as they become available	1986 onwards
Vended foods	National Automatic Merchandising Association	National Automatic Merchandising Association, 20 North Wacker Drive, Chicago, Illinois 60606, USA	Standards for the Sanitary Design and Construction of Food and Beverage Vending Machines	1984
Food and beverages	National Sanitation Foundation	National Sanitation Foundation, PO Box 1468, Ann Arbor, Michigan 48106, USA	21 Food Service Equipment Standards. (Each is for a different type of equipment and is updated every 5 years)	Current editions
Meat and poultry	US Department of Agriculture	Superintendent of Documents, Government Printing Office, Washington, DC 20402, USA	United States Inspected Meat and Poultry Packing Plants: A Guide to Construction and Layout. Agriculture Handbook No. 570	1984
Meat and poultry	US Department of Agriculture	Superintendent of Documents, Government Printing Office, Washington, DC 20402, USA	Accepted Meat and Poultry Equipment MPI-2	1987

Sanitary Standards and Accepted Practices (for the 3-A address, see Table 5.8).

Tanks are usually designed to be cleaned-in-place (CIP) and while there is a considerable range of standard CIP packages, Timperley[15] wisely observes: 'It does not necessarily follow that standard units are suitable for all plants and in certain cases purpose-designed CIP units may be required.' He also warns that if equipment is of poor sanitary design then it will never be cleaned satisfactorily.

Further information

In practice, concern about sanitary design is for one specific machine or item of equipment which is to be bought, modified or custom built. There are many different types of equipment used in the food industry and for this reason, the technical literature is both scattered and usually concerned with one aspect only. The continuing series of Campden Research Association publications on Hygienic (Sanitary, USA) Design is aimed at improving this situation so that the user can ask for, and get, equipment of better sanitary design. General considerations are listed in:

- Katsuyama and Strachan[2] on pages 111 and 112.
- W.S. Stinson[11] on pages 98 to 107.
- Campden Technical Memorandum 289[9] on pages 4 to 7.

Useful guidance is also contained in the joint FMF/FMA 1967 publication on tanks, pumps and pipework given as an appendix to Jowitt[8]. Principles are dealt with on pages 244 and 245; tanks on pages 249 and 250; pumps on pages 253 and 254; valves on page 259; and product on pages 265 and 266. The full appendix covers pages 241 to 286.

References

1. ICMSF (1988) *Microorganisms in Foods 4 – Application of the Hazard Analysis Critical Control Point (HACCP) System to Ensure Microbiological Safety and Quality,* International Commission on Microbiological Specifications for Foods, Oxford: Blackwell Scientific
2. Katsuyama, A. M. and Strachan, J. P. (eds) (1980) *Principles of Food Processing Sanitation,* Washington, DC: The Food Processors Institute
3. Thorpe, R. H. and Barker, P. M. (1987; publication continuing) *Technical Manual 17. Hygienic Design of Liquid Handling Equipment for the Food Industry,* The Campden Food Preservation Research Association, Chipping Campden, Gloucestershire GL55 6LD
4. Imholte, T. J. (1984) *Engineering for Food Safety and Sanitation.* Crystal, MINN: The Technical Institute of Food Safety
5. Ridgeway, R. R. and Coulthard, R. (1980) Hygienic design of plant exteriors. In *Hygienic Design & Operation of Food Plant,* (edited by R. Jowitt), pp. 53–77. Chichester: Ellis Horwood
6. Robson, J. N. and Barnes, G. (1980) Plant design to minimize non-microbial contamination. In *Hygienic Design & Operation of Food Plant,* (edited by R. Jowitt), pp. 121–135. Chichester: Ellis Horwood
7. Troller, J. A. (1983) *Sanitation in Food Processing,* New York: Academic Press
8. Joint Technical Committee, FMF/FMA. (1980) Appendix – Hygienic design of food plant (1967), Joint Technical Committee, FMF/FMA. In *Hygienic Design and Operation of Food Plant,* (edited by R. Jowitt), pp. 241–286. Chichester: Ellis Horwood
9. Campden R. A. (1982) *Technical Memorandum 289. The Principles of Design for Hygienic Food Processing Machinery,* Campden Food Preservation Research Association, Chipping Campden, Gloucestershire GL55 6LD.

10. Milledge, J. J. (1981) The hygienic design of food plant. *Proceedings of the Institute of Food Science and Technology (UK),* **14,** 74–86.
11. Stinson, W. S. (1978) Sanitary design principles for food processing plants. *Food Processing,* **39,** (8), (mid – July Guide and Directory) 98–108
12. USDA (1990) *Accepted Meat and Poultry Equipment,* MPI–2 United States Department of Agriculture, Washington, DC
13. Milledge, J. J. and Jowitt, R. (1980) The cleanability of stainless steel used as a food contact surface. *Proceedings of the Institute of Food Science and Technology (U.K.),* **13,** 57–63
14. Timperley, D. A. (1984) Surface finish and spray cleaning of stainless steel. In *Profitability of Food Processing,* Institution of Chemical Engineers Symposium Series No. 84, Institution of Chemical Engineers, Rugby CV21 3HQ, UK, p. 31 onwards
15. Timperley, D. A. (1981) Modern cleaning and recovery systems and techniques. *Journal of the Society of Dairy Technology,* **34,** (1), 6–14
16. Campden R. A. (1983) Technical Manual No. 7. *Hygienic Design of Food Processing Equipment,* Campden Food Preservation Research Association, Chipping Campden, Gloucestershire GL55 6LD
17. Timperley, D. A. and Lawson, G. B. (1980) Test rigs for evaluation of hygiene in plant design. In *Hygienic Design and Operation of Food Plant,* (edited by R. Jowitt), pp. 79–108. Chichester: Ellis Horwood
18. Hayes, P. R. (1985) *Food Microbiology and Hygiene,* London/New York: Elsevier Applied Science
19. Thorpe, R. H. and Barker, P. M. (1985) Technical Manual 8. *Hygienic Design of Post Process Can Handling Equipment,* Campden Food Preservation Research Association, Chipping Campden, Gloucestershire GL55 6LD
20. Jowitt, R. (ed) (1980) *Hygienic Design and Operation of Food Plant.* Chichester: Ellis Horwood

Further reading

Additional useful material is to be found in the following textbooks, manuals or standards:

- Hayes[18] – Chapter 7, Design of Food Processing Equipment
- Imholte[4] – Chapter 4, Equipment
- Troller[7] – Chapter 4, Process Equipment
- Campden Technical Manual 7[16] – Hoppers, Recurring Faults, Chutes, Product Transfer, Inspection Tables, Elevators, Catwalks.
- Campden Technical Manual 8[19] – Multi-lane Conveyors
- ICMSF Microorganisms in Foods 4[1] – Chapter 5, Hygienic consideration in the design and use of equipment. This chapter has a 'list of published sanitary equipment standards' (their Table 5.2). A modification of this table, with only US information, is shown as our Table 5.8. However, the UK reference in the ICMSF Table 5.2 (page 83) appears to be a misprint and probably refers either to Technical Manual 7[16] – Hygienic Design of Food Processing Equipment 1983 or more likely to Technical Manual 17[3] – Hygienic Design of Liquid Handling Equipment 1987. This is

being published in parts or sections, and sections 17–1 Introduction; 17–2 Materials of Construction and 17–3 Vessels (Tanks and Kettles) are already published. Sections 4 – Stirrers and Agitators and 5 – Pipelines and Couplings were published 1990.

• Barker. P. (1982) *Technical Memorandum No. 285. Pumps for the Food Industry*, Campden Food Preservation Research Association, Chipping Campden, Gloucestershire GL55 6LD
• ASME/ANSI. (1987) An American National Standard Food, Drug and Beverage Equipment. ASME/ANSI F2.1 – 1986. The American Society of Mechanical Engineers, United Engineering Center, 345 East 47th Street, New York, NY 10017. This gives helpful definitions as well as information on materials of construction, design and construction and specific criteria for some types of equipment.
• ISO (1980) ISO 2037 Metal Pipes and Fittings – Stainless Steel Tubes for the Food Industry, International Standards Organization, Central Secretariat, Geneva, Switzerland, (Note: National Standards Organization usually act as ISO sales agents.)
• BSI (1984) BS 4285: Part 1: 1972, Pipes, British Standards Institution, London

5.B. Installation and alterations

5.B.1 Installation

Introduction

Plant management has the clear accountability (or responsibility) for the sanitary status of installations. Because machine installation may be done by those having little or no experience of working in a food plant it is essential that contractors understand what should be done and how it should be done to achieve the required standards. This applies equally to building work and equipment installation. See Chapter 4.C.2 for working with contractors.

Principles

Sanitary design concerns not only individual items of equipment but also the complete installed line or plant.

Campden Technical Manual 17[1] makes the following eight points to which some comments have been added. Remember, in the location and installation of any items of process equipment:

1. There should be sufficient height to allow adequate access for inspection, cleaning and maintenance of the equipment and for the cleaning of floors.

Hayes[2] states that this means at least 200 mm (8 in) off the floor. This is satisfactory for narrow equipment, but for large items of a meter (yard) or more it is better to follow Ridgway and Coulthard[3]. They state that ' sufficient height to enable a brush, held at 30° to the horizontal, to reach the machine's center line allows the complete floor to be cleaned, without stooping, from both sides of the plant'.

2. All parts of the equipment should be installed at a sufficient distance from walls, ceilings and adjacent equipment to allow easy access for inspection, cleaning and maintenance, especially if lifting is involved.

Hayes[2] interprets this as being about a meter (yard) from the nearest ceiling, wall and adjacent equipment.

3. Ancillary equipment, control systems and services connected to the process equipment should be located so as to allow access for maintenance and cleaning.

To this the ICMSF[4] add:

It is worth repeating an earlier observation about the need properly to encase and waterproof all electrical and other service or control facilities connections; leakage from conduits can be a potent source of contamination. Similarly it is important to install liquid flow connections, such as vacuum breakers, that do not permit back-siphonage of liquids, as this can lead to microbiological contamination of processed product by unprocessed product or even by waste water.

Additionally, the possibility of condensation and its cure needs consideration before and during installation.

4. Supporting framework, wall mountings and legs should be kept to a minimum. They should be constructed from tubular or box section material which should be sealed to prevent ingress of water or soil. Angle or channel section material should not be used.

Katsuyama and Strachan[5] advise rotating horizontal square tubing by 45° to eliminate flat 'soil' collecting surfaces.

5. Base plates used to support and fix equipment should have smooth, continuous and sloping surfaces to aid drainage. They should be coved at the floor junction. Alternatively, ball feet should be fitted.
6. Pipework and valves should be supported independently of other equipment to reduce the chance of strain and damage to equipment, pipework and joints.
7. Avoid draining equipment directly onto the floor.
8. Avoid as far as possible installation practices that introduce for example ledges and soil traps, recessed corners, spot or tack welds which result in incompletely sealed seams and projecting bolt threads.

Katsuyama and Strachan[5] advise the welding to structural members of tubular supporting braces, together with the capping of open tubes and the avoidance of drilling into tubes.

In the Campden Technical memorandum No. 289[6] a series of *do's and don'ts* are given for installation (as well as maintenance) which include:

Do Protect food from external contamination.

Do Check that denting, chipping, flaking and delamination of construction materials (materials of construction) has not occurred.

Do Insure that hot or cold water is available in sufficient quantity and at the right temperature.

Do Provide adequate drainage holes for wash-down water.

Do Hang pipework carefully to insure a natural fall to a designed drainage point (minimum gradient of 1 in 240). Particular care must be taken with the installation of plastic pipelines.

Do Make sure drainage and curbing are fully adequate and that floors are hard surface and slip-resistant.

Do Check clean in place (CIP) systems for scale build up, corrosion and airlocks.

To which may be added:

Do Take care to align pipe couplings correctly and apply the proper torque when tightening.

The Technical memo 289[6] also includes the following *don'ts*.

Don't Use lagging (insulation) which is insufficiently protected and where possible misuse could result in product contamination. (As a comment, it is known that damage to lagging has occurred by careless movement of scaffolding poles, by ladders being propped against it and by being walked upon.)

Don't Install pipes with blanked-off 'T' joints and dead-ends.

Don't Create dirt traps by installing pipes and conduits hard against machinery or structural surfaces.

Campden Technical Manual 17[1] points out that trap points or areas where dirt, product, etc. ('soil') accumulate may occur because of either bad design or poor installation. These are likely to have the consequences that:

1. Cleaning operations will involve more time and increased costs.
2. Clean-in-place (CIP) procedures may not remove all soil deposits.
3. Failure to remove soil will result in microbial growth and contamination of product.
4. The risk of a corrosion attack associated with retained soil is increased.

Remember The wise words in this Campden Manual 17[1]: 'Sanitary (hygienic) design goes beyond product contact surfaces and involves consideration of any area where there is a risk of build-up of material or microorganisms which could affect product quality or provide an opportunity for aerial contamination of either product or product contact surfaces.'

Warning Be clear as to the full cost-in-use implications of lower specification of materials and workmanship. The 'cheaper' proposals may not be the most cost-effective.

Access to equipment

This, particularly if the equipment hook-up is large and complex, can present real problems, but it is essential for cleaning and maintenance. It is usually done by providing ladders or stairways and catwalks. Hayes[2] states:

It is important to insure that catwalks are constructed so that debris cannot be transferred from footwear to underlying equipment or food; thus decking should be constructed from suitable

non-slip (slip resistant) solid plate rather than mesh and the plate should be angled to incorporate a kick-stop.

This applies equally to stairs (staircases, UK) near or above food lines (see his Chapter 4 'Stairs').

Pipework

This requires good sanitary design and installation, because apart from conveying material it is part of the 'overheads' which collect dust and dirt. Badly installed or maintained pipework can be expected to leak, and may be a source of direct contamination or, if the material is suitable, e.g. liquid sugar, be a pest control problem. There is additional and valuable information on pipelines in Campden Technical Manual 17[1] Chapter 5 – Pipelines and Couplings.

The type of support used is important to minimize dirt accumulation and make cleaning quicker. Troller[7] states: 'Piping should be installed at least 150 mm (6 in) from walls and floor to provide for thorough cleaning around it.'

The design of hanger suspension rods and braces requires some thought. What is wanted is a design which is easy to clean and minimizes the accumulation of dust. Round tubing and installing angle iron as an inverted V are design details which are often used. Campden Technical Manual 17[1] shows a good design which positively locates the pipes and is shown here as Figure 5.10.

Imholte[8] deals with pipework on pages 96–101 and should be consulted. He includes hangers, wall brackets and floor supports and highlights insulation (lagging, UK) problems. These include the following:

- Liability of insulation to insect and mold attack. (*Note*: it also provides a runway for rodents.)
- Fiberglass insulation does *not* scratch the cuticle (wax-like surface) of insects, which would cause them to dehydrate. On the contrary, stored pest insects get along well in it.
- Loose canvas jackets over insulation harbor insects.
- Provision of mechanical protection for insulation may be needed.

When a pipeline is being installed which has couplings that use gaskets, be sure that the correct gasket is used and that the pipework and coupling are both correctly aligned and tightened. Failure to do this can be expensive, especially with aseptic lines.

Remember that the gasket must not only be of the right size and shape, it must be of a suitable composition. What is suitable depends on the food, detergent and sanitizer with which it will be in contact and the maximum temperature to which the gasket is exposed.

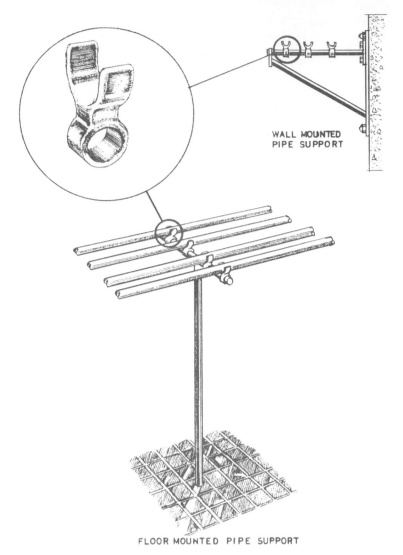

WALL MOUNTED
PIPE SUPPORT

FLOOR MOUNTED PIPE SUPPORT

Figure 5.10 Diagrams showing pipe hangers with locators (from Campden Technical Manual 17[1])

Be aware that changes of contact materials or temperatures may reduce the suitability of the gasket material in use. Information on properties of gasket materials is not easily available, but Alfa-Laval has collected data which is reproduced here, with permission, as Table 5.9. Note that all technical specifications contained in this table are given only for guidance, and Lavrids Knudsen Maskinfabrik A/S shall not be responsible for any errors in this material. However, manufacturers will be able to supply technical information and specifications of their particular products and materials.

Electrical systems

Apart from the pure engineering expertise required, these present, as Imholte[8] states, 'special sanitary design challenges'. He deals with them in his Chapter 5 which should be consulted by the technologist. These involve the need to protect electrical equipment from adverse effects of water, steam, dust and wet product, vibration, etc. and also the need to provide easily cleanable hardware which will function efficiently in a particular local environment.

Table 5.9 Product and chemical resistance of flexible rubber materials

Rubber may in principle be attacked by chemicals in two ways:

1. Swelling (partly reversible) or dissolution.
2. Chemical attack.

Product or process		ISO R 1629 Designation of flexible rubber material				
		A	B	C	D	E
Dairy products (milk, cream)		3	3–4	3–4	–	–
Dairy products (sour milk products)		3	3–4	3–4	–	–
Brewery products (beer, hops etc.)		3	3–4	1–2	2–3	–
Wine and yeast		3–4	4	4	2–3	–
Animal and vegetable fats	100°C	3–4	1	3–4	3–4	–
Water and water solutions	<70°C	3–4	4	3–4	3–4	4
Hot water and steam	<120°C	2	3–4	2–3	2	3–4
Concentrated fruit juices and etheral oils	<100°C	1	1	1	3	1
Non-oxidizing acids	<80°C	1–2	2–3	1–2	2	4
Oxidizing acids	<80°C	1	2	1	2	2–3
Weak concentrate of lye	<100°C	2	4	2	2	4
Strong concentrate of lye	<100°C	1	3	1	1	3–4
Mineral oils	<110°C	3	1	2	3–4	1
Aliphatic carburetted hydrogen (hexane)		3	1	1	4	1
Aromatic carburetted hydrogen (benzole)		1	1	1	3	1
Alcohols		1–3	3–4	3–4	1–3	3–4
Esters and ketones		1–2	2–3	1–2	1–2	1–3
Ether		1	2–3	1	1–3	2–3
Methylene chloride		1	1–2	1	2–3	1
Ozone and atmospheric conditions		1–2	4	4		3

Key to designation:

A = NBR Nitrile rubber
B = EPDM Ethylene Propylene rubber
C = Q Silicone rubber
D = FPM Fluorinated rubber
E = HV–IIR Butyl rubber

The attached schedule contains data which have been compiled from the results of our own tests and applications and from the recommendations of our raw material suppliers.

The data should be considered as recommended norms and will be brought up-to-date from time to time.

In case of doubt or lack of information it would be advisable to contact us direct. This will enable us to investigate specific applications.

All norms are on the basis of constant contact with the specified products.

Basis of judgment:

1 = Unsuitable
2 = Limited suitability
3 = Normal suitability
4 = Highly suitable
– = Not recommended for other reasons

Information supplied by Alfa-Laval and Lavrids Knudsen Maskinfabrik A/S.

It should be added that switch and relay boxes need to be well sealed so that they are either waterproof for wet areas or dust and insect proof in dry areas. In these electrical matters, the advice of an expert with proven sanitary design experience should be sought.

References

1. Thorpe, R. H. and Barker, P. M. (1987); publication continuing). Technical Manual No. 17. *Hygienic Design of Liquid Handling Equipment for the Food Industry*, The Campden Food Preservation Research Association, Chipping Campden, Gloucestershire GL55 6LD

2. Hayes, P. R. (1985) *Food Microbiology and Hygiene*, London/New York: Elsevier Applied Science
3. Ridgway, G. R. and Coulthard, R. (1980) Hygienic design of plant exterior. In *Hygienic Design & Operation of Food Plant*, edited by R. Jowitt, pp. 53–77. Chichester: Ellis Horwood
4. ICMSF (1988) *Microorganisms in Foods 4 – Application of the Hazard Analysis Critical Control Point (HACCP) System to Ensure Microbiological Safety and Quality*. International Commission on Microbiological Specifications for Foods, Oxford: Blackwell
5. Katsuyama, A. M. and Strachan, J. P., (eds) (1980) *Principles of Food Processing Sanitation*, Washington, DC: The Food Processors Institute
6. Campden R. A. (1982) Technical Memorandum 289. *The Principles of Design for Hygienic Food Processing Machinery*, Campden Food Preservation Research Association, Chipping Campden, Gloucestershire GL556LD
7. Troller, J. A. (1983) *Sanitation in Food Processing*, New York: Academic Press
8. Imholte, T. J. (1984) *Engineering for Food Safety and Sanitation, Crystal, MINN: The Technical Institute for Food Safety*

5.B.2 Alterations

Principles

Alterations or modifications to equipment are made either as part of a long-term plan – when sanitary design principles can be applied at the planning stage – or in response to an immediate need. The latter is the hazardous situation and causes most of the contamination problems, e.g. with dead areas, unsuitable fastenings, poor weld finish, etc. The usual and dangerous fallacy is 'We'll fix it properly later' – when in practice 'it' gets forgotten because of other, more pressing, priorities. The question to ask is 'Why can't the job be done right (correctly) the first time?' Is it because of a lack of foresight and planning or is it a lack of resources, e.g. in skills, parts, specialist equipment, etc.? A further question is 'What, if anything, has been learned from the experience?'

Major alterations to equipment may also involve re-location and some building work. Sanitary design principles apply to all of this work and where contractors are involved so do the considerations given in Chapter 4.C.2.

Management is accountable for its plant and should be fully aware of the true costs and benefits of alternative strategies or options when alterations need to be made. Management is also accountable for the food sanitation training and the supervision of personnel. Top management finally makes the decision 'This is important – we will do it' and so must insist that account is taken of *all* the technical considerations on which product safety and quality and cost-effectiveness depend.

5.C Maintenance

Introduction

Maintenance is done to insure that equipment continues to work within design tolerances or specifications. It must be carried out in a way which avoids contamination of ingredient materials, products or packaging materials and also permits effective cleaning before production re-starts.

Remember Maintenance operations should always be used as an opportunity to check for machine part wear or other features which mean that product contamination may have occurred.

5.C.1 Preventative or breakdown maintenance

Principles

There is general agreement with the sentiments of Imholte[1] on preventive maintenance:

> Preventive maintenance is commonly regarded as a necessary element of high productivity. While its contribution to productivity is undisputed, it is also paramount for product safety. Preventive maintenance avoids costly and painstaking product recalls or product defects. Its value cannot be overemphasized.

When a situation develops in which preventative maintenance is the theory but breakdown maintenance is the practice, it must be questioned as to whether the real cost of breakdown is known. Budgetary systems may not highlight these real costs especially if a down-time allowance is built into the system. Experience with running aseptic lines, where very high utilization rates are needed for satisfactory running, underlines real costs of equipment failure. Less obvious but no less real are, for example, the costs of a conveyor drive failure which stops a line for around two hours.

Clearly, although it costs money, high quality preventive maintenance means that breakdown or 'emergency' maintenance is rarely needed. 'Emergency' maintenance has hazards which may be overlooked:

- There is pressure to work fast, which may lead to lack of care.
- Under the pressure, unsuitable materials may be used.
- 'Temporary' repairs once made tend to be forgotten. Imholte [1] states:

> Temporary repairs to prevent leaks, dust emissions, and product spills, and to prevent metal-to-metal contact will probably always be necessary. However, temporary repairs have a way of becoming permanent. This tendency is of concern because many of the materials used to make temporary repairs, such as tape, cardboard, and wood represent a serious food contamination threat.
>
> Temporary metal baffles or diverters are rarely secured adequately and, as a result, they eventually make contact with moving metal components. Caulking materials used to plug small cracks end up in the product. Baling wire is sometimes subject to vibration or constant flexing which causes it to break. String, of course, is also easily broken.
>
> To prevent contamination, make permanent repairs as soon as possible. Dating (temporary) repairs will keep them from becoming permanent.

- Under pressure to work fast, cleaning-up after the job is done may be incomplete and small items such as nuts, bolts, washers, etc. may be left in places where vibration, etc. may move them so that they contaminate the product. Complete reliance should *never* be placed on down-stream equipment such as magnets or metal detectors to remove or reveal *all* such contamination. Remember, the true cost and disruption to the business of a voluntary (or worse still, compulsory) product recall due to avoidable contamination.

Pipelines

Pipeline maintenance is an operation which may well result in ingredient, product, or packaging contamination. There are a number of straightforward ways to minimize this contamination during maintenance of pipelines. As a convenience to the reader these are given in the Appendix to this section.

Reference

1. Imholte, T. J. (1984) *Engineering for Food Safety and Sanitation*, Crystal, MINN: The Technical Institute of Food Safety

Appendix: Minimizing contamination from pipeline maintenance

Product lines:

- Remove all product before starting work.
- Take care to avoid splashing ingredients, product or packaging materials.
- Do not empty waste into ingredient or product containers.
- Cap or protect open ends on lines while working in an area.

- Be sure to clean repaired pipe-lines thoroughly to avoid damage to equipment, e.g. pumps, and product contamination. Remember, a thick product will 'pick-up' metallic contamination, e.g. swarf, which inadequate cleaning with 'thin' detergent solutions may leave behind.

Product and utility lines:

- Fabricate as much as practicable in the engineers' work-shop and weld, thread or cut in an area screened (isolated) from ingredient or packaging material storage and product handling. 'Magnetic mats' at the work-shop exit help to limit the spread of steel shavings (swarf, UK) and other magnetic small pieces of metal from the work-shop.

Utility lines:

- Be sure that lines have been carefully emptied before starting work.
- Use PTFE or graphite compound on male threads only.
- Wipe off excess jointing compound immediately.
- Inspect 'trouble points' at pre-determined frequencies. These include:
- Leaking valve stems – worn or defective packing.
- Leaking flange joints – 'dried out' (permanently distorted, UK) gaskets.
- Leaking flange joints – stripped nuts or bolts.
- Leaking product line couplings – misalignment of parts or wrong or missing gasket.
- Leaking threaded joints – replace or weld line.
- Check that lines passing through floors are correctly sleeved and that sleeves are correctly installed.
- Check that 'cold' or 'sweating' pipes are insulated.
- Check (audit) that replacement fittings are of the correct type, of appropriate materials, and are correctly installed.

5.C.2 Maintenance operations

Introduction – stages of operation

Whether simple or complex, the operations for product safety may be considered as a number of stages:

- Preparation – personnel, planning, isolation of worksite
- The task or job
- Clearing-up (cleanout)
- Plant cleaning
- Inspection and start-up

Note: For information on Contractors' Personnel see Chapter 4.C.

Preparation – personnel

It may seem strange to consider this item, since those involved will be adequately skilled for their tasks. However, maintenance engineers and mechanics (fitters, UK) may, without realizing it, be the cause of sanitation or contamination problems. Remember that they sometimes have to work under difficult, cramped and/or hot, wet conditions. This is a partial explanation but is not acceptable as an excuse. Improvements should be sought by management especially if dirty overalls are worn, greasy tools and rags are brought into ingredient storage or manufacturing areas and if the area around the work is not left tidy when the maintenance job is finished.

Specific training should always be given so that the need to avoid product contamination and the reasons and importance of particular working practices in food plants, e.g. the use of food grade lubricants, are fully understood.

Personal hygiene rules appropriate for the plant apply to all management and workers at all times, e.g. the ban on smoking, absence of jewelry and wearing of hats and, where appropriate, beard snoods as protection against contamination by hair. As Robson and Barnes[1] state:

> It is interesting how an old and badly-designed plant can be kept clean and relatively free from foreign matter by careful attention and vigilance. It is equally significant that modern, well-equipped factories can rapidly deteriorate and cause hazards to the food product if they are not operated and maintained correctly. The difference between these two extremes is the attitude of mind of the people working in the factory. Without proper guidance this attitude can prejudice the care and thought that went into the original design of a safe and hygienic food production unit.

Preparation – planning

Apart from the engineering aspects, there is the need to inform production departments and, where appropriate, quality assurance, of the nature of the work to be done on equipment or facilities together with its timing.

Planned maintenance should obviously not be scheduled during production periods or in areas where ingredients, products or packaging materials could be contaminated by materials, odors, dust, etc. resulting from maintenance operations.

Emergency or breakdown maintenance or repair in or above an area where ingredients, food or packaging materials are open to contamination requires immediate action. Shut down and remove ingredients, food or packaging materials to a place where they may be kept safely. As necessary, processing equipment in and immediately around the work area should be covered with clean drop cloths or plastic sheets to protect the equipment from contamination. Remember that such things as metal shavings, filings (swarf, UK) and hot flying particles of weld metal *must* be contained in the work area.

Preparation – Isolation of work site

Where it is practical to do so, the equipment to be maintained should be taken to the maintenance area. If this is impractical, then, as necessary, the work area should be isolated using appropriate barriers, shields, drop cloths or plastic sheets to contain possible contamination within the work area.

Warning Equipment including storage silos, conveyors and tanks must preferably be emptied of food material *before* maintenance commences, *or* be isolated and *all* material dumped afterwards. A full (scheduled) clean is required before further use with food materials. Be sure the cleaning and inspection are thorough.

The task or job – care of tools

Mechanics (fitters, UK) should be provided with adequate facilities for their tools and equipment when working in production or storage areas. This includes not only tool boxes (or chests) and belt-worn holsters but also, as appropriate, other facilities such as trolleys for tools or parts such as filler valves, magnetic trays, etc.

It is good practice to keep tools as clean and free from dirt, oil and grease as is practical. When the job is finished or at the end of the day, tools should be returned clean to the appropriate lockable work-shop area or tool room. They, like parts or materials, must *not* lie around on food contact areas *nor* be stored in food or ingredient containers.

Experience has shown that it is necessary for critical operations, e.g. milk dryer rooms, to have dedicated sets of tools and equipment (e.g. a work bench) which do not leave the high-care or 'clean' area. No tools may enter this area unless they have been disinfected.

The task or job – use of materials

As Imholte[2] states:

The use of proper repair materials is as important as promptness in making equipment repairs. Use only foodgrade materials which have been approved for that specific use. If the original components are not available, make sure that the materials substituted are of a satisfactory food-grade quality.

Some materials are likely to cause contamination incidents if they are misused. Apart from string and wire misused for 'temporary' repairs, these include:

- Wood. Because this cracks and splinters, avoid its use if possible – even for shield frames, packing (jacking) equipment into position, etc. Remember wooden tool handles can break and splinter and the mechanic must also be aware of this risk. Wood tanks, however, have traditionally been used for brining vegetables. An account of maintenance care at the end of a season is given by Katsuyama and Strachan[3].
- Wire brushes and scouring pads. These should not be used on food contact surfaces nor above or near open ingredients, food or packaging materials. Wire brushes may legitimately be used by responsible maintenance personnel to prepare surfaces for painting but only when there is no production in the area. Broken or loose wires mean that the wire brush should be replaced and that contamination may have already occurred.
- Wipe rags are essential for maintenance work but care must be taken to avoid threads or pieces of worn rag getting into product or equipment. If this should happen, the mechanic should inform the Production Department immediately so that action can be taken to prevent affected food being shipped. Soiled, oily and greasy rags should be promptly discarded into containers designated for this purpose. Paper wipes are acceptable provided they are of a non-shredding type and proper anti-contamination precautions are taken. Remember, paper towels can frequently be found as litter in a department unless supervision is alert and effective.
- Solvents are usually toxic, flammable and volatile and may cause taints in foods. Be sure that only authorized solvents are used for specific purposes. Avoid their use in areas where ingredients, product or packaging materials could be affected. If they have to be used, e.g. cleaning or thinning inks used in printers or coders, have and enforce a written procedure for this use which includes proper disposal of solvent wipes, spilled and spent (used) solvent.
- Caulking materials should be used only when absolutely necessary and not in or near food product zones if there is the possibility of contamination. Food grade, i.e. those containing no known toxic substance, caulking materials should be used.
- Release (silicone) sprays are sometimes used to free or stop squealing (squeaking, UK) of belts. They should be carefully selected and used as they may contaminate and affect the product. Imholte[2] states that silicones affect the leavening (raising, UK) action of cakes.
- Boiler feedwater additives. These materials may

be toxic or tainting. Where steam is used for culinary purposes, e.g. in direct steam injection, or where condensate is used, e.g. in some aseptic systems, only food grade additives may be used.

- Lubricants (oils and greases and hydraulic fluids). All of these materials used on the plant site should be classified as:

Food grade (edible)
Non-food grade (non-toxic)

or

Non-food grade (toxic)

Each class should be defined. Products should be listed individually and their use unambiguously prescribed.

Only food grade (edible) lubricants should be used in food product zones. Currently (1990), the USDA list of Chemical compounds classifies these as H-1. Mineral and vegetable oils usually qualify, and the FDA tolerance is 10 parts per million for mineral oil in food products.

Non-food grade (non-toxic) lubricants are considered in the USA as contaminants and have a zero tolerance in foods. To be included in this category the oral LD_{50} should be 5 grams per kilogram or greater. Many widely used lubricants come into this category.

Non-food grade (toxic) lubricants. These should preferably be *banned* from the plant site. If there is truly no alternative to their use then a technically sound, unambiguous, written approval system giving full details of materials and uses should be established and enforced. Only materials in full compliance with this approval system should be permitted or used on-site.

The task or job – use of components

Some equipment components are more likely than others to cause problems associated with contamination. Imholte[2] details thirteen (on pp. 175–177) to which reference should be made.

Imholte's list ranges from broken bearings and seals to peeling paint and conveyor misalignment resulting in wear contamination of food. Other components are likely to cause contamination because they are easily (and carelessly) left where they may cause harm. Examples of these are nuts, bolts, small machine parts, wire, string, sleeving, etc. Oils and grease become contamination risks usually because of over-use but sometimes because of disregard of the basic rule: *never* use food containers for any other purpose.

The task or job – testing

After the work has been completed, thorough test-running of the equipment *must* be done before

cleaning and putting it back into service. Be sure that it is running properly and that there is nothing else wrong with it.

Clearing up (cleanout)

When the maintenance and repair work is comapleted, or at the end of the work day, the area should be left clean and tidy. Imholte[1] directs attention to the following causes of contamination on completion of maintenance work:

1. Drill shavings/metal filings (swarf, UK). All such materials must be removed.
2. Weld slag. Chip and brush welds, and remove all loose slag. Pick up welding rods, including the small ends of spent rods.
3. Repair materials. Remove nuts, bolts, screws, gaskets, and repair plates that have not been used, and return to the storeroom. Often a handful of nuts and bolts is left lying on the equipment or on a nearby ledge. Loose objects of this nature are sure to find their way into the product.
4. Electrical components. Electrical materials such as wire and nuts fall into the same category as the repair materials mentioned above, and the same precautions apply.
5. Escessive lubricants. Many jobs require greasing and oiling. Make certain that all excess lubricant is wiped off and that the grease rags themselves are not left behind.
6. Tools/check out instruments. Remove all tools and instruments used in the repairs and return them to their appropriate storage areas. Provide trouble lights [inspection lights, UK] with plastic-coated, shatterproof bulbs, and be sure to remove them from the repaired equipment.

It is very important to make certain that the surrounding work area as well as the repaired equipment is clean. Sweep up metal filings and drill shavings, in particular. Pick up all leftover material and return to storage.

Remember the need to sweep the surrounding area free from metal filings and drill shavings (swarf, UK). This is because sharp metal particles easily become attached to the shoes or boots and can be taken to other parts of the plant and left there, thus causing unexpected contamination.

Plant cleaning

After all maintenance work, the equipment will need to be cleaned to the necessary pre-production standard. Depending on individual circumstances, a full scheduled cleaning may or may not be needed. If in doubt, give the full scheduled clean. If the environment might have been affected by the maintenance work, it too will need cleaning.

Particular care needs to be taken with kettles and other open vessels. It is wise to assume that contamination has occurred and clean accordingly. When cleaning associated pipework, remember that 'thick' product will push out heavy debris, e.g. nuts or washers, which have been left in places where 'thin' detergent may not have dislodged them.

Inspection and start-up

The Production Departments must be sure by checks and inspection that the equipment is ready for production use. Dismountable equipment should be opened for inspection. Depending on the organization within a company, Quality Assurance may be asked to inspect. This inspection should be regarded as a 'double check' as the prime responsibility (and accountability) is with the user, i.e. the Production Department.

Only when the Production Department representative is satisfied that maintenance and cleaning are satisfactory should the start-up begin.

References

1. Robson, J. N. and Barnes, G. (1980) Plant design to minimise non-micropbial contamination. In *Hygienic Design & Operation of Food Plant*, edited by R. Jowitt, pp. 121–135. Chichester: Ellis Horwood
2. Imholte, T. J. (1984) *Engineering for Food Safety and Sanitation*. Crystal, MINN: The Technical Institute of Food Safety
3. Katsuyama, A. M. and Strachan, J. P. (eds) (1980) *Principles of Food Processing Sanitation*. Washington, DC: The Food Processors Institute

5.D Cleaning and disinfection

5.D.1 Introduction to cleaning systems

Why we clean

Cleaning and disinfection of food contact surfaces and the environment is done:

- As part of the achievement of overall control (chemical, physical, microbiological) of a specific ingredient or food product handled by a particular item or line of equipment.
- To maintain the performance of equipment, e.g. a filler, within design limits.
- Because it is part of good plant management which has an important effect on employee morale.

These apparently simple and obvious reasons lead to the development of cleaning systems which require careful technological study and may become complex.

As an example of the need for study, the ICMSF HACCP book[1] states:

> Under certain circumstances it may be best not to attempt to make an environment aesthetically clean, because more harm than good can be done. For example, dust which may accumulate on an overhead pipe in a food handling area might best be left untouched except at special cleaning intervals. If the dust removal cannot be done effectively in the normal time available for cleaning, the food handling equipment below may become contaminated. The results of microbiological analysis of apparently unclean materials such as dust or condensing moisture should be interpreted in terms of the hazard the accumulations pose to the microbiological status of the product, thereby dictating the frequency and necessity for cleaning.

Definitions and terms

Cleaning systems

It is important to be clear about the meaning of terms in order that issues may be addressed clearly and rationally. This is particularly so when terms in common use, e.g. 'clean', must be given a particular meaning before progress can be made.

For the purpose of this chapter, terms are defined in a way which is intended to be helpful for users internationally. Some terms which are commonly used are discussed in the following text, others appear in the 'annex' to this section.

It is emphasized that Plant Management has the responsibility and accountability for ensuring that fully adequate and effective cleaning systems are developed, used and monitored. Following the definition used by Dunsmore *et al.*[2], this is the sum of those practices which maintain product contact surfaces and the environment in a condition which ensures that they do not impair the safety or the quality of the food.

This may be done either 'in-house' or by the use of specialist advisers and contractors. The use of these outside specialists does not, however, remove responsibility from Plant Management. As Troller [3] states:

> It is often cost-effective for companies to rely on contract cleaning services, in the same manner as they employ contract pest control services. As with the latter, it is the food company's responsibility to describe carefully what is expected of the contractor, to determine when and how frequently cleaning should be carried out, and to monitor the effectiveness of the cleaning operations that are performed. Written contracts are absolutely essential to insure that these points are thoroughly understood by both parties. Naturally, the contractor cannot do the job thoroughly if they are not provided with essential information and adequate access to the areas to be cleaned.

Clean

The word 'clean' means that a surface is clean:

Chemically – when materials used in plant cleaning and/or sanitizing are removed by rinsing and when scale and other deposits have been removed.

Physically – when all visible 'soil' or residues are removed by the cleaning operations.

Microbiologically – when numbers and types of microorganisms are reduced to an acceptable level. See the annex for further discussion.

Remember the ICMSF[1] comment:

> Surfaces may appear clean and still be microbiologically unacceptable. Conversely, surfaces may not appear clean but in fact be quite acceptable for certain operations. Microbiological samples of food contact surfaces might have to be taken to verify what the human senses perceive as clean or unclean.

Other terminology –

There is some difference in terminology between countries which may be confusing.

Cleaning and sanitizing (or disinfection) are technologically distinct operations. However, since the objective is to achieve chemical, physical and microbiological cleanliness the combined operations are sometimes called *cleansing*, a term which is used in some milk and dairy legislation. Note that cleaning system components have both cleaning and sanitizing functions.

Sanitizing is the usual term in North America for what is known as *disinfection* in Europe. It is incorrect to use *sterilizing*, which properly means killing all organisms, when *sanitizing* is meant.

Sanitation is the usual term in North America for what is known as *Hygiene* in Europe.

Definition of these and other terms are given in the annex to this section.

Annex – A glossary of definitions and terms

The following have been obtained from various sources, principally from Elliott[4] and Tamplin[5]. The list is not exhaustive but is intended as a guide.

Anionic Possessing a negative electrical charge.

Bactericide A chemical agent which under defined conditions of use is capable of killing bacterial cells, but not necessarily bacterial spores. It is important to understand that the use of a bactericide will not effectively remedy poor cleaning. It should not be expected to be effective outside the defined (specified) conditions of use.

Bacteriostat A chemical agent which under defined conditions of use inhibits the increase in numbers of a bacterial population. It should not be expected to be effective outside the defined conditions of use.

Break point chlorination The point at which the chlorine demand of a water has been reached, after which the free residual chlorine concentration increases in nearly direct proportion to additional amounts of chlorine added.

Buffer A chemical agent or agents in solution which stabilize the pH. This is a measure of the acidity or alkalinity of the solution. During cleaning the pH of a buffered cleaner is not much changed by dilution or small amounts of soil, e.g. acidic tomato paste.

Cationic Possessing a positive electrical charge.

Chelating power This enables a cleaning solution to hold unwanted ions, such as calcium and magnesium, in solution or to re-dissolve precipitated salts. This property is important with hard waters, as calcium and magnesium can interfere with cleaning or rinsing. Chelating agent(s) are organic compounds which do the same job as inorganic sequestering agents.

Clean A surface is clean:

Chemically when materials used in cleaning and/or sanitizing are removed by rinsing. Water used for rinsing *must* be potable, i.e. of such chemical and microbiological quality that it is wholesome and fit for human consumption. There can be other chemical residues on product contact surfaces, e.g. calcium phosphate on dairy heat exchanger plates, which must be removed by suitable treatment.

Physically when all 'soil', scale, or residue has been removed by cleaning. This is usually judged visually and it is important to remember that a surface which looks clean physically is not necessarily microbiologically clean.

Microbiologically when the numbers and kinds of microorganisms are reduced to an acceptable level. It is important to understand that in principle the requirement depends on:

- The equipment and the way in which it is used
- The product(s) handled by the equipment

In practice, within any one plant a single microbiological standard is used which is acceptable for most or all of the equipment. Strictly limited exceptions may be made from the single standard after careful justification. This is a management need rather than a technological requirement.

A simpler approach is that cleaning done properly should result in only organisms derived from the final rinse water being present. This is consistent with results actually achieved.

For use in UK Canneries, the following standard given in Campden Technical Manual 17[6] is generally accepted:

Total number of microorganisms (colony-forming units) following swabbing

Grade	Per square foot	Per square decimeter
Satisfactory	0–5000	0–540
Fairly satisfactory	5000–25 000	540–2700
Unsatisfactory	over 25 000	over 2700

Other levels suggested for microorganisms are:

Numbers per square decimeter	Application	Reference
100	Dairy	Davis[7]
100	Multi-use and single-service containers for pasteurized milk and milk products; also for bottled water	Favero, Gabis and Vesley[8]
800	Meat industry	Goldenberg and Relf[9]
1000	Food service equipment	Favero *et al.* [8]
1000	Not given, but 'reasonable for general manufacturing or food service plant'	Timperley and Lawson[10]
≤7.5	Yeast on soft drink manufacturing equipment	Tamplin[11]

Additionally, Favero *et al.*[8] state that for floors, a 4 to 5 log cycle reduction in count is to be expected following cleaning and sanitizing.

Cleaner A mixture of chemicals designed to remove soil from equipment, environment or hands. Also used (UK) to describe the person employed to do a cleaning job.

Cleaning This is the removal of 'soil', slime and microorganisms by physical, chemical or mechanical means from surfaces. It may be followed by sanitizing (disinfection).

Cleansing This is a combination of the process of cleaning and sanitizing (disinfection) and is used in some milk and dairy legislation.

Contamination Objectionable material in a food product.

Corrosion inhibitors These are chemicals used in defined conditions to minimize corrosion of certain metals.

Deflocculation (dispersion) The action of breaking up aggregates or flocs into individual parts.

Detergents Substances that either alone or in a mixture assist cleaning. They include soaps, organic surface-active agents (such as synthetic detergents), alkaline materials and also acids in certain instances.

Detergent–Sanitizer A mixture of chemicals formulated to clean and disinfect food contact surface simultaneously.

Disinfection (sanitization) or disinfecting (sanitizing) The application of effective chemical or physical agents or processes to a cleaned surface or of an effective chemical or physical agent to a water supply with the intention of reducing the numbers of microorganisms to a level at which they can be properly assumed to present no risk to health. Sanitization is also used to control spoilage organisms. Sanitizing is generally understood to mean the destruction of vegetative microorganisms but not necessarily spores.

Dispersing and suspending power This is the ability to first bring and then keep soil in suspension. Once dispersed, soil must be kept in suspension so that particles do not re-deposit and foul the clean surface.

Emulsification The suspending and keeping in suspension of small globules of oil or fat in water.

Free residual chlorine That part of the total chlorine content that is in the form of hypochlorous acid (HOCl) and which will therefore react quickly and readily.

Fungicide A chemical agent which under defined conditions kills fungi, including their spores. It should not be expected to be effective outside the defined conditions of use.

Hydrophilic Having an affinity for, or being capable of dissolving in, water.

Hydrophobic Antagonistic to or incapable of dissolving in water; having an affinity for oils and fats.

Hygiene A European term meaning the same as and being interchangeable with the North American term 'sanitation'. It is the science and technology (1) of the establishment and maintenance of health; (2) of conditions or practices (such as cleanliness) which are conducive to health.

In-plant chlorination The addition of chlorine to the entire water supply of a plant beyond (in excess of) the chlorine demand to establish a low level, e.g. a few mg per liter (ppm), of free residual chlorine to be used for continuous application to food contact surfaces.

Nonionic Lacking an electric charge. Nonionic wetting agents consist of a balance of negatively (anionic) and positively (cationic) charged components resulting in a net neutral state.

Organic dissolving power The ability to solubilize proteins and fats.

Pathogen A specific causative agent of disease, e.g. bacterium, or protozoa or virus.

Peptizing The formation of a colloidal solution from protein soils that are only partially soluble. Note that a cleaner peptizes proteins by chemically breaking the peptide bonds.

Sanitary Adequately hygienic to insure a safe, sound, wholesome product fit for human consumption.

Sanitation A North American term with the same meaning as the European term 'hygiene'.

Sanitizer A chemical agent used as a disinfectant (see 'Disinfection').

Saponification The chemical reaction between an alkali, e.g. sodium, and an animal or vegetable fat, i.e. long-chain fatty acids, to form a soap.

Sequestering agent Used to describe an inorganic material with chelating power. It will therefore hold unwanted ions, e.g. calcium or magnesium, in solution so that they cannot form precipitates or interfere with cleaning or rinsing.

Soil Any undesirable material, including food residues, scale, atmospheric dirt, dust, etc., which should be removed by cleaning.

Sporicide A chemical agent which under defined conditions is capable of killing bacterial spores. It should not be expected to be effective outside the defined conditions of use.

Sterile Free from *all* living organisms.

Sterilizing agent (sterilant) A chemical agent capable of destroying any form of life. *Note*: The term detergent-sterilizer is a misnomer; it is in fact a detergent-sanitizer.

Surfactant See 'Wetting Agent'.

Suspension A dispersion of solid particles in a liquid.

Water hardness A characteristic of water imparted by salts of calcium, magnesium and iron that causes curdling of soap, increased consumption of soap or detergent and deposition of 'lime' scale.

Water softening The removal or inactivation of

water hardness. This may be done by the 'lime-soda' process, ion-exchange, chelation/sequestration and, for 'temporary hardness', by heating. *Note*: It may be cheaper to soften water rather than use a sequestering agent.

Wetting agent (surface-active agent, surfactant) This is a substance that lowers the surface tension of water or cleaning solution thus increasing the ability of the solution to contact all surfaces and penetrate the soil.

Note: Further terms are defined in the Glossary of ICMSF (1980) *Microbial Ecology of Foods, Volume I Factors Affecting Life and Death of Microorganisms*. New York: Academic Press

References

1. ICMSF (1988) *Microorganisms in Foods – 4, Application of the Hazard Analysis Critical Control Point (HACCP) System to Ensure Microbiological Safety and Quality*. International Commission on Microbiological Specifications for Foods, Oxford: Blackwell Scientific
2. Dunsmore, D. G., Twomey, A., Whittlestone, W. G. and Morgan, H. W. (1981) Design and performance of systems for cleaning product contact surfaces of food equipment: A review. *Journal of Food Protection*, **44**, (3), 220–240
3. Troller, J. A. (1983) *Sanitation in Food Processing*, New York: Academic Press
4. Elliott, R. P. (1980) Cleaning and sanitizing. In *Principles of Food Processing Sanitation*, edited by A. M. Katsuyama and J. P. Strachan, pp. 61–89. Washington, DC: The Food Processors Institute
5. Tamplin, T. C. (1980) CIP technology, detergents and sanitizers. In *Hygienic Design & Operation of Food Plant*, edited by R. Jowitt, pp. 183–225. Chichester: Ellis Horwood
6. Thorpe, R. H. and Barker, P. M. (1987 – continuing publication). Technical Manual No. 17. *Hygienic Design of Liquid Handling Equipment for the Food Industry*, Campden Food Preservation Research Association, Chipping Campden, Gloucestershire GL55 6LD, 1.1–1.2
7. Davis, J. G. (1956) *Laboratory Control of Dairy Plant*. London: Dairy Industries
8. Favero, M. S., Gabis, D. A., and Vesley, D. (1984) Environmental monitoring procedures. In *Compendium of Methods for the Microbiological Examination of Foods*, edited by M. L. Speck, pp. 49–54. Washington, DC: American Public Health Association
9. Goldenberg, N. E. and Relf, C. J. (1967) Use of disinfectants in the food industry. *Journal of Applied Bacteriology*, **30**, (1), 141–147
10. Timperley, D. A. and Lawson, G. B. (1980) Test rigs for evaluation of hygiene in plant design. In *Hygienic Design and Operation of Food Plant*, edited by R. Jowitt, pp. 85–86. Chichester: Ellis Horwood
11. Tamplin, T. C. (1981) Cleaning in Place (CIP) systems and associated technology. In Developments in Soft Drinks Technology – 2, edited by H. W. Houghton, pp. 174–175. Barking: Applied Science Publishers

5.D.2 Overview of cleaning systems

Introduction

Responsible management accepts that cleaning is an integral part of production. It is therefore necessary to develop, implement and maintain cleaning schemes that can be shown to be fully adequate and cost-effective. This requires:

- Definition of the cleaning task or job
- Selection of the optimal cleaning system
- Application of managerial skills and procedures to define, assess and change, as necessary, the chosen cleaning system. This will include written procedures (schedules) and on-going assessment (audit).

Remember Use clean-up to check for machine wear or other features, such as missing nuts, etc., which mean product contamination may have occurred.

Be clear that management is accountable for the sanitary status of plant and equipment. Good sanitary design together with the proper application of technically effective cleaning systems are needed to achieve and maintain a satisfactory sanitary status. Poor sanitary status is caused by management's lack of knowledge together with its attitudes and policies.

HACCP application

The Hazard Analysis and Critical Control Point (HACCP) System as a way of working is as applicable to cleaning systems as it is to production systems (see Chapter 3). Since HACCP is developed for a specific situation in an actual plant, the examples given are for illustration only. They do, however, show how:

- Its use raises open-ended questions
- It links to related analyses
- It can be extended to other issues, e.g. safety issues, *but beware when making the analysis not to lose sight of its primary purpose.*

Two examples, both hypothetical, are given below. The first is of cleaning guard panels on a cannery filling machine, the second is part of an analysis which could be made using the hypothetical plant cleaning schedule from Shapton[1].

Example 1 – guard panels

Guard panels are made from stainless steel sheet about a meter square. One side (the top) is curved to hook over a supporting tubular rail when the filler is in use. Handles are provided to lift the panel on and off this rail. The base of the panel is held in place by a channel section. The panel is not interlocked with the filler which can therefore be run with the panel removed during cleaning.

When cleaning of the filler starts, the panels are removed for separate cleaning. After the line and filler are cleaned, the panels are replaced.

The purpose of cleaning is to remove spillage which if left could cause an offensive smell and corrode the metal. Visual cleanliness is sufficient.

An analysis is given in Table 5.10. which may raise some unexpected points. It is worth noting that a real advantage of the HACCP approach is that it promotes open-ended lateral (wide-ranging) thinking in a multi-disciplinary team as well as producing a useful analysis. It is also worth noting that the analysis must be in the detail needed to deliver the required results.

Example 2 – Part of manufacturing line

This analysis is of the portion of the line shown in Figure 5.11., including and downstream of the reheat kettle. Assume that the schedule is being reviewed and amended using a HACCP approach. Visual monitoring and microbiological verification must be satisfactory when procedures are followed.

There are a number of background questions which are assumed to have been asked before consideration of this part of the line, e.g.:

- Should 'operators' be listed? Is the demarcation between 'cleaners' and 'fitters' (mechanics) valid?
- Equipment is on three floors, ground (or first floor, USA), mezzanine and first (or second floor, USA) – how are equipment controls and communications, e.g. intercom, arranged? Is time wasted or are additional staff needed because of deficiencies?
- Is it better to make up detergent in the kettles or would a supply of hot ready-to-use detergent be more cost-effective? (*Note* – heating metal

Table 5.10 HACCP Analysis of guard panel cleaning

Step number	Concern	Control – by training and supervision at each step plus:	Notes/comment
1. At start of filler cleaning, switch off (switch out) filler.	Safety of cleaners.		
2. Remove guards and *either* place on clean floor away from filler.	Mechanical damage to guard – it may not fit securely back in place thus affecting safety of operators.	Written schedule to specify correct practice. (This might be one or both options.)	Consider improved guard design to increase strength within acceptable weight limits.
or place against clean wall.	Mechanical damage to wall.		Consider improved protection for wall.
3. Use of steam/water hose to clean guards.	1. Safety of cleaners. 2. Hose water pressure or temperature may be low thus (a) making cleaning more difficult or less effective or (b) increasing time of cleaning to be unacceptable.	Rules for correct operation to be posted at steam/water mixer point. Sufficient steam and water must be available when needed.	This may mean that the main boilers are used outside of normal production hours. Consider provision of auxiliary steam generator for cleaning. Alternatively, consider a high pressure low volume system or foam cleaning.
4. If guard cannot be cleaned using hose, scrub with standard hand scrubbing alkali-hypochlorite mixture. Rinse thoroughly, using hose.	Use of incorrect brush (e.g. wire brush) could damage surface and start corrosion.	Issue or have readily available in the department the correct brush. Restrict availability and use of non-food contact brushes.	Be sure that brushes are bought against a clear written specification.
	Use of CIP alkali-hypochlorite mixture which could injure cleaner.	Issue or have readily available in the department the correct mixture in food grade utensils.	
5. After filler cleaning is finished, check filler is switched off (switched out) and replace guards.	Guards must fit correctly in place for operator safety.		Modify channel design, e.g. use bar guides, if channel cleaning or jamming of guard is a problem.

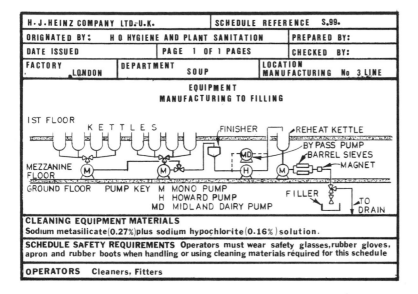

METHOD –after production use		
OPERATORS		
Cleaners	1	Remove finisher and barrel sieves
Cleaners	2	Hose out kettles with hot water 60°C then flush through complete system to waste at filler. Hose out finisher and reheat kettle.
Cleaners	3	Half fill kettles with hot water, start stirrers, add detergent, top-up with hot water then raise solution temperature to 80°- 85°C.
Cleaners	4	Pump detergent soluton to fill re-heat kettle then stop pumps.
Fitters	5	Disconnect finisher pump and connect in by-pass pump.
Cleaners	6	Pump remaining detergent via by pass pump through system to waste at filler.
Cleaners	7	Repeat operation 2 with cold water.
Fitters	8	Dismantle finisher pump and remove magnet and plug cocks.
Cleaners	9	Inspect kettles and finisher and if necessary, together with pump components, magnets, barrel and finisher sieves and plug cocks, scrub with detergent solution (half strength of C I P solution) then rinse in cold water.
Fitters	10	Reassemble pump, replace sieve magnets and plug cocks, disconnect by-pass pump and reconnect product pump into line.

KEY INSPECTION POINTS   Finisher, Barrel Sieves. Plug cocks and seatings

Figure 5.11 Hypothetical plant cleaning schedule (from Shapton[1])

equipment downstream causes loss of detergent temperature at point of use.)

- How does the cleaning of the manufacturing line relate – especially in time – to filler cleaning? Are rinses and detergent used in filler cleaning? If so, how is the filler by-pass cleaned – if not, could the line cleaning be adapted?

The HACCP analysis is given in Table 5.11.

Patterns of cleaning

These patterns will vary according to the environment, equipment and products. In a complex plant, skillful management is needed to minimize downtime for cleaning.

Because of the variation of plant and equipment, little that is useful can be written here apart from the obvious points that cleaning usually follows the

Table 5.11 HACCP Analysis of Line Cleaning – see figure 5.11

Step number	Concern	Control – by training and supervision at each step plus:	Notes/comment
Note			
No reference to wear checks on mono-pump stators or to integrity checks of finish and barrel sieves.	Product contamination	Amend to specify action to be taken if defects found. Cross reference to other procedures as needed.	Coherent, integrated procedures are a basic requirement for TQM. *Note*. 'Red–amber–green' concepts apply to cleaning activities.
1. Remove sieves.	No clear reference to: 1. Cleaning, e.g. by hose or brushing, unless step 9 refers. 2. Inspection – for cleanliness or damage. Storage until re-assembly. *Note*: damage, e.g. by dropping on floor could weaken metal.	Clarify text. Amend text to include how and when this should be done. Resolve, then specify what is wanted in the text.	This illustrates need for clear, unambiguous text. If brevity is needed then reference may be made to other documents, e.g. Operating Instructions or Training Modules.
2. Hose out kettles, etc.	Source of hot water. A piped supply is implied but no temperature limits are given.	Clarify text. Develop and state limits.	
	If a steam/water hose is to be used, how is temperature known and controlled?	Clarify text. Develop and state limits.	
3. Detergent make-up.	Is hot water at 80–85°C?	Specify what is wanted in text.	Comment as at Step 1.
	Is sodium metasilicate supplied as solid – if so does undissolved material cause kettle valve wear? If not, how is material supplied?	Resolve, then specify what is wanted in text.	If time needs to be cut, consider other ways of making required volume of hot detergent.
	How are amounts of detergents measured?	Specify what is wanted in text.	
	Should not cleaners be told to save quantity, e.g. 5 liters of mix, for Step 9?	Specify source of detergent used at Step 9.	
	Safe use of hypochlorite.	Specify in text, e.g. Distribute contents of closed containers between 4 kettles.	
4. Fill reheat kettle.	If finisher or kettle is over-filled is there danger to other operators? If finisher or kettle overflows, what is maximum loss of liquid before cleaning is ineffective downstream?	Resolve, then specify what is wanted in text.	
	Are pumps speeds matched to deliver correct cleaning velocities in pipework?	Specify here or cross reference to data or information document.	
5. Finisher pump.	Possible ambiguity – it is Howard pump.	Amend text.	This is minor point but it may as well be right.

Table 5.11 Continued

Step number	Concern	Control – by training and supervision at each step plus:	Notes/comment
6. Pump to waste at filler.	Is this to be done by flooding the filler or through drain by-pass at filler, or through both?	Specify what is wanted in text.	If flow is to be sent through filler by-pass for *x* minutes, how is this to be timed? *Note*: Contamination risk if cleaners wear watches.
7. Cold water final rinse.	What is minimum volume of cold water used? Flooding (full bore) is needed if detergent is to be rinsed from upper part of pipelines.	Resolve, then specify what is wanted in the text.	
8. Dismantle finisher pump, etc.	Mechanical damage. No clear reference to cleaning or inspection.	Resolve what is wanted and amend text.	See comments on Step 1.
9. Inspect, brush components.	Damage to components during handling.	Resolve, then specify what is wanted in the text.	
	Contamination risks because of unsuitable brushes or detergent containers.	Issue or have readily available in the department the correct brushes and containers. Restrict availability and use of non-food contact brushes.	Be sure that brushes are bought against a clear written specification. Note that this is the same 'control' and 'comment' as in Example 1. Once a careful and detailed analysis has been made the information can be used wherever it is pertinent.
10. Reassemble line.	Mechanical damage.		Be sure that the pipeline downstream of the Howard pump to the by-pass link is properly cleaned. If not, the linking pipework may need to be moved immediately down-stream of the Howard pump outlet.

product flow along a line and that the timing of equipment availability for cleaning is a key factor.

Environment

It is a mistake to think only of equipment cleaning, since the environment may be the source of pathogens or serious spoilage organisms. Drain channels, drains, floors, roofs near air intakes have all been implicated as significant sources of micro-organisms. Environmental clean-up is usually only done during a shut-down period and can be difficult and expensive. Sanitary design, installation and maintenance all have a part to play in minimizing the risks and simplifying environmental cleaning. The usual pattern is to use dry and wet cleaning as appropriate, make good the deficiencies, reclean

and sanitize until tests prove that the contamination source has been eradicated.

Dry cleaning

Dry cleaning methods are used where the products are hygroscopic or where water can react to form hard deposits which are difficult to remove. The principal Public Health risk is that failure to control moisture can permit the growth of pathogens, e.g. *Salmonella* spp., in the processing environment which then contaminate any food being processed. Environments usually dry cleaned include plants producing flour, chocolate, peanut butter, dry milk products, dry soup and snack mixes, and dry infant formulae.

Dry cleaning is essentially the mechanical removal

of soils using sweeping, brushing, wiping, and vacuuming. Vacuuming does not spread dust and dirt, but picks it up. It is therefore, in principle, the desirable system to use. Disinfection following dry cleaning is not easy, although 70% ethanol may be used and allowed to dry off before equipment is reassembled.

Wet cleaning in a dry area needs great care and it is usually best to remove equipment or parts needing wet cleaning to a separate area. Make sure that wet-cleaned parts are thoroughly dry before returning to the dry area for reassembly.

Wet cleaning

This may be done out of place or in place or, on large equipment, by a mixture of both methods as appropriate.

Clean out of place (COP)

This may be summarized as a series of steps:

- Remove food products and open containers from the area surrounding the equipment to be cleaned
- Dismantle the equipment to expose the surfaces to be cleaned. Small parts and components may be removed for cleaning in a wash area equipped with sinks, hoses, storage racks, ultrasonic baths, recirculating tank washers, etc.
- Clear the line of product using shovels, scrapers, brushes, squeegees, hot or cold, low- or high-pressure hoses, or flushing with water, as appropriate
- Apply detergent using, as mechanical aids to soil removal, brushes, 'boiling', high- or low-pressure hoses as appropriate
- Thoroughly rinse all food-handling parts of the equipment with cold potable water of suitable quality
- Sanitizing is best done after cleaning and, if chemicals are used, should normally be followed by a final rinse to remove the sanitizing agent. A final rinse is not always necessary, e.g. if Quaternary ammonium compounds have been used at the permitted, appropriate level
- After the final rinse, complete the reassembly of equipment and allow to drain and dry. Note – pipelines and closed equipment cannot be expected to dry. For this reason, it is considered good practice to rinse this type of equipment when it has stood for some hours between cleaning and use. For example, where hot product is handled the equipment should be rinsed with hot water (82°C, 180°F) before re-use. Where cold product is handled, equipment should be rinsed with cold water containing 20 mg per liter (20 ppm) available chlorine before re-use

Clean in place (CIP)

CIP is mainly used for cleaning liquid handling systems. It comprises a number of steps:

- Drain the system of product.
- Prepare the circuit(s), e.g. by switching controls to 'clean', installing key pieces or flowplates, confirming availability of detergents, etc.
- Pre-rinse to remove product residues (gross soil)
- Circulate hot detergent to remove residual soil
- Rinse with potable water of suitable quality. This rinse is the intermediate rinse if chemical sanitizing and/or acid scale removal is included in the CIP cleaning. It is the final rinse if there is no further step in the CIP cleaning
- Sanitize to reduce microbial numbers to an acceptable level
- Final rinse with potable water of suitable quality

Cleaning complex systems

In complex cleaning systems which are often large, a mixture of COP and CIP methods outlined above are usually used. Such systems are likely to be programmable logic controller (PLC) or computer controlled with manual interfacing. It is critically important that the control system can be proved to operate as designed and that the system is capable of delivering what is intended. This may not be the simple matter it seems. Consider a pump operation; it is easy to confirm that electricity is supplied to the motor, more difficult to show that the shaft is turning, and still more difficult to show that pumping is in accordance with the design criteria. The only sure way to demonstrate this operation may be to measure the suction or discharge pressures or incorporate a no-flow probe.

Assessment of cleaning – what must cleaning deliver?

There is a great deal more to assessment than simple visual inspection although this plays an essential part. The first issue to address is 'what must cleaning deliver?

When cleaning is completed, what assessment must be made to show that it is satisfactory? Referring to the Glossary in the Annex of Section 5.D.1. a surface is 'clean':

Chemically – When cleaning and sanitizing materials have been removed by rinsing.

Physically – When all 'soil', scale or residue has been removed.

Microbiologically – when the numbers and kinds of organisms are reduced to an acceptable level, e.g. for cannery purposes, the following standards apply:

Grade	Total numbers of microorganisms (c.f.u. after swabbing)	
	Per square foot	Per square decimeter
Satisfactory	0–5000	0–540
Fairly Satisfactory	5000–25 000	540–2700
Unsatisfactory	Over 25 000	Over 2700

Note: Other situations may require different (and tighter) standards (see 5.D.1, Annex)

Considering the assessment of each of these in turn:

Chemical assessment

If there is confirmation that the final rinse has been done, no further tests or assessment are usually made. If there is any doubt about the final rinse, it is prudent to repeat and/or lengthen that part of the cleaning cycle. If tainting of products, e.g. by chlorophenols, has occurred, appropriate chemical tests for residual in the rinse water may be made but they are unlikely to show positive if the cause of taint is in extremely low concentration.

Physical assessment

This is usually assessed by visual inspection. For pipelines or difficult-to-see parts of equipment, e.g. tanks, special lighting sets and mirrors or more sophisticated fiber optics may be needed. With complex pipework systems it is well worth while to 'design in' inspection points where 'worst case' conditions are likely to occur. These consist of a short length of pipe – about 30 cm (1 ft) long which can be readily disassembled for inspection. It is also useful to remember that when closed equipment, e.g. pumps, tanks, pipelines, has stood unused for a day or two, the smell when opened can give an indication of physical and microbiological cleanliness to the experienced sanitarian. Ultra-violet (UV) light may also be useful if the soil deposits fluoresce.

Microbiological assessment

This is assessed by one of the standard microbiological techniques to determine numbers and kinds of organisms. It is often sufficient to know the total numbers ('count'), but for some purposes, it is important to know what kinds of organisms, e.g. fecal coliforms, yeasts, etc., are present.

The methods usually used are:

- Swab. This is the most usual technique. The exact area swabbed is less important than the ability to consistently swab the same area. Avoid concentrating on easy-to-swab areas and remember that it is the 'worst case' area which is the most important. After suspension in a diluent the organisms are counted, usually by a standard plate count.
- Rinse. This is usually applied when it is convenient to use a sterile rinse to wash off the organisms, which are then counted.
- Direct contact. In these methods, RODAC (Replicate Organism Direct Agar Contact) plates or sticky surface sampling tapes are applied to the surface. Organisms stick to the surface and may then be cultured and counted.

It is important to appreciate that, irrespective of the method(s) used, their purpose is:

- To confirm that the cleaning system being developed achieves acceptable microbiological results
- To provide surveillance or audit information which shows that acceptable microbiological results continue to be obtained

To understand the significance of the results both microbiological and statistical sampling aspects need to be considered.

Who monitors cleaning?

Cleaners should inspect their own work and report to their supervisor if the method does not produce the expected result. Line supervision should checkout the cleaning after completion but before the cleaning crew leaves the area. The checks are visual and reported by exception, i.e. when something is wrong. It may be that there is a problem with the method, in which case urgent technical assistance is necessary, since the method was originally satisfactory. Alternatively, the method may not have been applied correctly. Training or retraining together with additional supervision of the cleaning process may then be appropriate.

Technical (often laboratory) personnel will usually be involved in microbiological assessment, as well as in developing cleaning systems. It may be worth exploring alternative ways of getting the assurance required, e.g. measuring the chlorine loss of an intermediate rinse water as a measure of cleaning efficiency. It may also be worth considering what tests could be done by production (and cleaning) operators and supervisors.

Periodic sanitation reviews ('surveys' or 'audits') are required in addition to more frequent but less extensive inspections made by plant management. To be effective, these periodic reviews need careful organization. Just 'walking through' a plant is useful in that it shows management interest, but it is not a review. The starting point of a review is an agreement on specific criteria, e.g. what is 'acceptable' or 'satisfactory' in the specific context of the area or equipment under review. It is important that the committee or team making the review has sufficient seniority and expertise. This insures that the reviews will be thorough, since what may be missed by one person may well be noticed by another, and also that the final report will be taken seriously so that effective 'root cause corrective (remedial) action' will be taken quickly.

In order to compare reviews over a period of time, some form of report must be made and kept on file. The inspection evaluation or rating form may use numerical ratings or descriptions such as 'unacceptable', 'needs improvement', 'acceptable', 'very good', which have a previously agreed meaning. It is important that whatever review and report system is chosen should include 'open-ended' questions, allow for 'observations' or 'comments' and provide an effective basis for comparison between past and present reviews.

The size of the plant and the complexity of the processes or the number of products influence the layout of the inspection form, the composition of the committee or team and the appropriate frequency of the reviews. Reference to microbiological cleanliness should be included in these reviews.

The 'checklist' approach has advantages and disadvantages which need to be understood. The advantages include:

- It provides a common list of things to be inspected which every inspector can use
- It helps to insure that other items are not forgotten when a deficiency is found
- It provides a consensus view of what is important
- It saves the inspector having to think out the list of checks each time, or rely on previous experience or opinion as to what should be checked
- It can be used as the basis for inspections but not as defining the inspection. In a well-designed checklist, questions should be 'open-ended' and the last question is always 'What has been missed?'

The main disadvantage is that it may be used in a mechanical, unthinking or 'wooden' manner and inhibit the inspector from making an intelligent, thoughtful and balanced assessment.

Shapton[1] gives an example of a plant sanitation checklist and the ideas can be adopted for cleaning assessment of environment and equipment.

Although it may not appear in standard textbooks, it is important that a 'Quality Audit' of cleaning systems should be made at suitable intervals. It should be understood that this includes a 'Systems Audit' which addresses the question 'Can the system of cleaning and its monitoring demonstrate that it delivers what is properly expected of it?' It is essential that this audit is conducted by those who have sufficient technical expertise and

managerial authority to insure that appropriate corrective (remedial) action is taken promptly. Simply recording deficiencies without taking effective action is a waste of time and money. It is also potentially dangerous since senior management think that the system is working when it is not.

Keeping and use of records

The key factor in the making of records is their intended use. There is considerable value in knowing that a cleaning system was proved to be effective when it was established. There is also value in knowing the trends of inspections, tests and surveys. The objective in studying trends is to take corrective (remedial) action before loss of control of the product(s) or process(es) occurs. The optimal way of doing this depends on individual plants but apply the rule 'the simpler the better', otherwise significant details get lost in a mass of data.

Cleaning in practice – management role and accountability

Successful cleaning results from using management skills to apply technical knowledge. To do this, personnel must be selected, trained, organized, motivated and supervised. ICMSF[2] put this as the foremost factor. They state:

Management must resolve at least nine issues to maintain an effective cleaning staff.

1. It is common for new employees to be placed on the cleaning staff until a more desirable job becomes available. New employees are frequently not familiar with the overall production process and do not comprehend the impact the quality of their work may have upon the products.
2. Inadequate training of personnel in proper procedures for cleaning and disinfection may lead to microbiological problems.
3. As cleaning and disinfection often takes place after production is completed and frequently at night, there may be inadequate supervision to monitor the quality of work as it is being performed and to make corrections as needed.
4. Inadequately cleaned equipment may be put into use rather than being recleaned.
5. The cleaning staff commonly hold the lowest paying positions in the food establishment.
6. Management uses the cleaning staff as a pool from which workers are drawn for use by the production department.
7. For various reasons (e.g. the undesirable task of cleaning, working at night, low pay), the cleaning staff has a rapid rate of turnover and a high level of absenteeism.
8. Often there is no acknowledgment of a job well done or incentive for better performance

Table 5.12 Some causes and control of poor sanitation[a]

Cause	Effect	Detection	Control
Water too hard.	'Stone' or 'fur' formation or thin white deposit.	Visual.	Use periodic acid wash, more water conditioning materials in detergent mix or softened water.
Unsanitary cleaning equipment, e.g. brushes, mops, cloths.	Spread of microorganisms.	Microbiological tests. Smell or visual in bad instances.	Use cleaning equipment of sanitary design or include a cleaning and sanitation step for this equipment.
Insufficient frequency of cleaning.	Build up of soil. Soil becomes hard and difficult to remove.	Visual. Microbiological tests.	Shorten cleaning intervals. Include partial cleaning between existing cleaning periods.
Unsatisfactory cleaning.	Soil (food residues) remain after cleaning. Reduced sanitizer (disinfectant) efficiency.	Visual. Microbiological tests.	Reclean and then: (a) Check that written procedures exist, and (b) That they are understood and are followed. Repeat clean under comparable conditions of soiling. If still inadequate, review and improve procedures to give proven satisfactory results. Alternatively, if correct procedures are *not* followed, investigate why. If there is no good reason why they cannot be followed, disciplinary action should be taken.
Incorrect water temperature Too hot Too cold	 Coagulation of protein soils. Fat not removed.	 Visual. Visual.	 Provide an adequate supply of water at appropriate temperature(s).
Inadequate rinsing (pre-rinse).	Detergents less effective leaving residual soil.	Visual. Microbiological tests.	Apply correct rinse procedure or increase flow or rinse time.
Sanitizer (disinfectant) contact time too short.	Sanitizer less effective.	Microbiological tests.	Check procedure is correctly applied. Increase contact time if necessary.
Sanitizer too dilute.	Sanitizer less effective. Possibility that strains of organisms may be selected, or become adapted to sanitizer, e.g. pseudomonads.	Microbiological tests.	Check procedure for dispensing sanitizer and way in which use dilution is made. Change procedure if necessary.
Wrong sanitizer (disinfectant) used.	Sanitizer less effective.	Microbiological tests.	Revise procedure for issue of sanitizer (disinfectant), e.g. use color code or symbol.
Wrong detergent used.	Detergent less effective leaving residual soil.	Visual. Microbiological tests.	Revise procedure for issue of detergent, e.g. use color code or symbol.
Residual moisture.	Local multiplication of microorganisms, e.g. in 'pools'.	Microbiological tests. Visual.	Check slope of pipework and/or design and installation of equipment.

[a] Adapted from Table 6.3 in *Microorganisms in Food 4*, Application of the hazard analysis critical control point (HACCP) system to insure microbiological safety and quality ICMSF [2]; Some common causes of inadequate cleaning and disinfection.

on the part of either the cleaning staff or supervisory personnel.
9. There may be unwillingness or inability to pay for the equipment or personnel to maintain an effective cleaning and disinfection program.

Warning If these issues are *not* effectively resolved, personnel will rapidly conclude that sanitation is of low priority to the management and so, quite properly, it should be of low priority to them. Once this idea gets accepted, then it will prove very difficult to change.

Among other common causes of poor sanitation given in ICMSF[2] are:

• Insufficient time to implement established procedures.
• Poor sanitary design of equipment or poor installation – this includes 'in-house' modifications which make parts less accessible or create filth traps in equipment.
• Use of inadequate, worn out or incorrect equipment and/or materials for cleaning.
• Lack of hot water or steam during cleaning time.

Further obvious causes are given in Table 5.12. which has been based on ICMSF[2] material.

Cleaning systems (procedures) and schedules

It is not sufficient to develop adequate cleaning systems or procedures. Unless they are written into a schedule there is no real point of reference from which to make changes or to use for purposes of audit or discipline. The UK Codes of Practice No. 10 for canneries, Section 6.1.3.[3], applies equally to any food production:

> Cleaning should be a properly planned and managed function. It is important that all cleaning procedures used in a plant have been properly evaluated and are written down in a cleaning schedule. This schedule should set out the frequency of the cleaning procedures throughout the plant and should specify material, equipment and procedures needed for each item of processing equipment and each area of the premises. All operatives responsible for cleaning should be trained in the proper use of equipment and the proper methods of cleaning required in the plant as set out in the schedule.

The form of the schedule should be appropriate for the operation. What is required is that it should be sufficiently detailed and unambiguous so that when followed, it delivers only the desired results. If, for brevity, it is written so that it may only be performed by personnel with specific training, then the training manual must be sufficiently detailed and unambiguous to be fully adequate as a logical supporting document.

Equally, with complex CIP systems, the cleaning schedule should address the routine operation, making adequate cross references to the engineering or technical manuals (in the local language) which should be developed and supplied with the equipment.

This may sound complicated but it need not be so. The simplest possible, clear unambiguous drawings and text are what is needed. The question to ask is 'what is needed to properly manage the cleaning operation?'

References

1. Shapton, D. A. (1986) Canned and bottled food products (soups, mayonnaise and sauces). In *Quality Control in the Food Industry*, 2nd edn, edited by S. M. Herschdoerfer, Vol 3, pp 293–296. London, Orlando, FLA: Academic Press
2. ICMSF (1988) Cleaning and disinfecting. In *Micro-organisms in Foods – 4, Application of the Hazard Analysis Critical Control Point (HACCP) System to Ensure Microbiological Safety and Quality*, pp. 93–116. International Commission on Microbiological Specifications for Foods, Blackwell Scientific
3. DHSS (1981) *Food Hygiene Codes of Practice No. 10 – The Canning of Low Acid Foods – A Guide to Good Manufacturing Practice*, London: HMSO (out of print, but a second edition is in preparation)

5.D.3 Development of cleaning systems

Introduction

Since cleaning is an integral part of production, the design of the cleaning system must be consistent with the proposed use of the line. In particular, the time available for cleaning may well determine the design of the system. Imholte[1] states:

> A good design can greatly reduce cleaning costs and at the same time vastly improve the quality of cleaning. To accomplish this, it is necessary to define the cleaning job and then select the right cleaning system to get it done. Because it sometimes seems difficult to justify the capital investment required for various [sophisticated] cleaning systems, it is important not to overlook possible productivity improvements. Properly applied, [these] cleaning systems get the cleaning job done faster, putting the manufacturing system back on stream more quickly. This results in optimum case outputs and reduces the need for overtime wages. Cleaning functions can be productive!

Development of cleaning systems will be considered under the following headings:

• Soils
• Dry cleaning

- Water
- Significance of equipment design and construction
- Choice of detergent: safety; properties; classification; identification; delivery to point of use; typical formulations
- Application of detergent: safety; cleaning out of place (COP); LPHV, HPLV, foams and gels, power brushing, manual cleaning; cleaning in place (CIP)
- Note on commissioning cleaning systems
- Evaluation of cleaning systems

Soils

Although often underestimated, a study of the soil or soils is the foundation of the development of a cleaning system. Both Watrous[2] and Tamplin 1980[3] underline the importance of soils and their study as a basis for the design of cleaning systems. Tamplin[3] gives a table of soil characteristics including the effect of heating various components. Consideration of the soil should include at least the following questions.

What is the soil?

Its chemistry and physical condition are important in deciding what type of detergent is likely to be successful. For recipe products, it is likely that:

- Acidic material, e.g. tomato products and pickles, will be fairly easy to remove
- Starch/protein/fat/salt soils with vegetable material will be moderate or difficult to remove
- Starch/protein/fat/salt soils with meat or melted cheese will be difficult or very difficult to remove

The action of heat can be difficult to predict. This is important because residual heat in the equipment or the effect of steam leaks can change the character and properties of the soil. With soups, for example, heat usually 'burns on' the soil, making it difficult or very difficult to remove. Exceptionally, it can cause a dried soup layer to peel and break away from the surface.

The viscosity of the soil is clearly important and can greatly increase the difficulty of removal both in pipelines and on surfaces. Dried gums or thickeners which have become wetted can take time to remove because they have to be 'dissolved' and washed away.

Where is the soil?

Obviously, soil trapped in inaccessible places is more difficult to remove and may take much longer. Sometimes unexpected difficulties occur, e.g. connective tissue in a meat slurry being handled through a worm and stator pump was found to be jammed behind the rotor 'worm'.

How much soil is there to be removed?

This question links with the previous one. Obviously, even if the amount is small but located in inaccessible places cleaning will be more difficult and take longer.

There is also the possibility of chemical reaction to be considered. If, for example, an acidic soil such as tomato puree is to be cleaned with an alkaline detergent, an inadequate pre-rinse will cause the alkali to be partly neutralized and so less effective. If sodium hypochlorite is used in this mix, toxic chlorine may be evolved resulting in increased corrosion as well as being a safety hazard to operators.

In what condition is the soil?

Is it solid, semi-solid, liquid, or a mixture of these? Are there parts of the equipment or environment in which the soil is in a condition which makes it more difficult to remove?

What are the constraints in the removal of soil?

These include such factors as:

- Design, construction, fabrication or installation of equipment or environment. A simple example is that strong alkalis should not be used on aluminum surfaces.
- Water hardness.
- Availability of detergent materials or cleaning aids, e.g. Clean-in-Place (CIP) circuits or High Pressure Low Volume (HPLV) hoses.

What standards of sanitation are required?

For each processing line, it must be clear what should be achieved by the cleaning operation, chemically, physically and microbiologically. Consideration of these factors leads to the design of the cleaning system and can be applied whether the job is large or small.

Dry cleaning

Although it is not usually considered in this context, the use of shovels, brooms, etc. does not require the use of water and is good practice when removing quantities of gross soil. Dry cleaning is appropriate when dry products or soils such as flours, cereals, dried milk powders, or chocolate are handled. Usually, surfaces are scraped or brushed or cleaned by vacuuming. 70% ethanol (ethyl alcohol) is often used as a sanitizer for surfaces and small parts but must be allowed to dry thoroughly before the equipment is re-used.

Equipment for handling dried foods, e.g. elevators for carrying nuts or cereal products, may well not be designed or adaptable for wet cleaning. Cleaning systems must therefore remove food residues and leave the equipment thoroughly dry before it is re-used.

Compressed air can be used to remove 'caked-on' residues, but it has the disadvantage that it moves the soil from one area of the equipment or environment to another. Unfortunately, there is no easy method of removing 'caked-on' residues. It should be remembered that steam jets are not 'dry'. They are most effective when used as a 'spitting' jet, i.e. a stream of steam with some droplets of water. The noise and clouds of steam may satisfy the operator but may not be as effective in the removal of soil as of lubricant. Steam also leaves a film of water on equipment which does not evaporate as quickly as is sometimes believed. For more than safety reasons, the use of steam jets should be strictly controlled.

Vacuuming has advantages over other methods of dry cleaning. It can remove imbedded dust and dirt and it picks it up so the soil is removed from the area. Imholte[1] has an excellent discussion of vacuum systems and discusses the merits of portable and central systems. He gives the advantages of the central system as:

- It has more power
- Several cleaners can use it at the same time
- It can be used to clean up product spills, assist in emergency dust control and safely clean equipment that otherwise might have to be shut down for cleaning. It must, however, be carefully designed and the inlet valves must be constructed so that they are not easily damaged and therefore leak as this would reduce the vacuum available.

Imholte[1] also gives an overview, which is well worth reading, on the design and use of vacuum systems (pp. 200–202). One key issue which he highlights is the number of inlet valves (hose points) which will be used at any one time. An undersized system will not be effective but an oversized system will be unnecessarily expensive.

He also advises limiting hose lengths to 7.5 m (25 ft) with a maximum of 15 m (50 ft) and states that 38 mm (1½ in) internal diameter hoses are usually satisfactory with 50 mm and 64 mm (2 and 2½ in) being used for large spills, large tanks, railroad cars and ships holds.

With regard to cleaning tools he makes the important points that 'the first and only pass should be at a speed which allows the tool to do an effective job' and that 'the effectiveness of a cleaning tool varies with the cube of the intake velocity'. Thus a tool entrance velocity of 2265 l/min (80 ft³/min) is two and a third times more effective than one of 1699 l/min (60 ft³/min).

Water

Water used in cleaning must be potable, i.e. fit for human consumption. It is therefore often considered 'pure'. Chemically, this is not correct since water falling as rain absorbs carbon dioxide from the air and becomes a weak solution of carbonic acid. This reacts with calciferous rocks such as chalk and limestone to form bicarbonates and dissolves other materials from the soil and rocks. These dissolved salts produce 'hardness' in water. This is apparent domestically by the formation from soap of a 'curd' with a hard water. Water hardness may be 'temporary' when it can be removed by boiling or 'permanent' if it cannot.

'Temporary' hardness, which can be removed by boiling, results in a precipitation of a white chalky material usually called 'fur' if soft and 'scale' if hard. It also reduces the efficiency of detergents. 'Permanent' hardness, which cannot be removed by boiling, reacts with carbonates, e.g. sodium carbonate or 'soda ash' often used in detergent mixes, to produce a precipitate. This is important in practice, as a detergent formulation which works excellently in a soft water area can produce a white chalky coating on contact surfaces in a hard water area. This needs acid treatment to remove it.

For cleaning purposes, the US Geological Survey Classification is usually used both in the USA and UK, although the German definitions are different. See Chapter 4 and Table 4.10. for further information on hardness of water. The effect of water hardness on detergent formulations is seen in Table 5.13., using data taken from Elliott in Katsuyama and Strachan[4].

Hardness in water for cleaning can be overcome either by using a detergent mix with sufficient sequestering (chelating) agent or by softening the water by the 'lime-soda' or ion exchange methods. The 'best' decision is primarily one of cost but in some locations where phosphates must be used in the detergent mix to condition (or soften) the water high phosphate levels are found in the plant effluent. These high levels can be unacceptable ecologically.

Significance of equipment design and materials of construction

These factors are very highly significant although the cost implications are not always properly appreciated by managers. This is because a careful estimate of cleaning costs over the expected lifetime of the equipment is rarely made. This estimate should include the cost of any modifications which may be needed for effective cleaning as well as taking account of alternative materials of construction which are offered, and equipment from other suppliers which could also do the job. Even if

Table 5.13 Effect of hard water on a typical CIP cleaner formulation

Ingredient	Function	Percentage change in weight of ingredient because of hard water
Sodium carbonate	Detergent alkalinity	−55
Sodium metasilicate	Detergent alkalinity and corrosion inhibition	−74
Sodium tripolyphosphate	Sequestering action (water softening)	+450
Wetting agent	Wetting, penetrating action	−66
Usage rates	Soft water 2.59 kg per 550 l (4.32 lb per 100 US gall) Hard water 3.85 kg per 550 l (6.43 lb per 100 US gall)	

Note. Soft water has the equivalent of 85.5 mg/kg or ppm (5 grains per US gall) calcium carbonate
Hard water has the equivalent of 342 mg/kg or ppm (20 grains per US gall) calcium carbonate

Table 5.14 Typical sanitation costs – from Marriott[5]

A typical manual cleaning operation has the following percentage breakdown of costs:

Labor	46.5
Water/Sewage	19.0
Energy	8.0
Cleaning Compounds and Sanitizers	6.0
Corrosion	1.5
Miscellaneous	19.0
Total	100.0

Marriott's[5] sanitation costs are very much an average or order of magnitude costing as Table 5.14. shows, labor is the largest percentage in manual cleaning operations. Labor in some instances may be well in excess of the percentage given. This is not so in a CIP system although many of these systems have an element of manual cleaning.

Statements have been made that labor can reach 80% and detergents 5% of costs. It is therefore prudent to examine design, construction and installation of equipment to determine what can be done to maintain standards and reduce costs by:

- Avoiding product (which becomes soil) being trapped and held, especially in inaccessible parts of the equipment.
- Insuring that product (which becomes soil) is accessible to the cleaning solution however it is applied. *Remember* Good design and installation can cut the time to reach equipment parts for manual cleaning and is essential if CIP is to deliver the required sanitary standards.

Warning In CIP systems take account of changes in the cleaning solution velocity caused by changes in the effective diameter (bore) of the pipeline. Romney[6] states that a change of up to 12 mm (½ in) can usually be accepted, but no greater change than this.

- Insuring that materials used do not restrict the use of fast acting, powerful cleaning materials. For example, hypochlorite must not be used in contact with nickel; strong acids or alkalis corrode aluminum or galvanized plate.

An overview is given by Thompson[7] of the effect of surface materials and detergents, given as Table 5.15. It should, however, be used with caution since, for example, there are quite variable properties among stainless steels. (See Campden Technical Manual No. 17[8].)

Choice of detergents

Introduction – points that experience highlights

When the variety of soil, equipment and environment is considered it is obvious that there is no universal detergent or method of application that can be used throughout the food industry. For this reason, it is important to understand the basic technology of plant cleaning so that optimal solutions to individual and specific problems can be found. Before considering detergents and their use, experience highlights a number of points:

- Keep the numbers used in a plant to the minimum and so reduce the chances of mistaken use or misuse. This also simplifies stock keeping and may, through bulk purchase, reduce costs.
- Clearly identify the different materials – one white powder is very like another. This may be done by using flake or powder forms to distinguish materials of similar color or by coloring

Table 5.15 Effect of detergent materials on surfaces (from Thompson [7])

Surface	Caustic Soda	Meta Silicate	Synthetic Detergents	Phosphoric Acid	Nitric Acid	Hydrochloric Acid	Sulphamic Acid	Solvents	Sodium Hypochlorite
Stainless Steel				TC	O	O	TC		
Mild Steel				TC	O	O	TC		T
Copper				TC	O	O	TC		TC
Zinc	O			O	O	O	O		O
Aluminum	O			TC	O	O	O		TC
PVC			C		TC	TC	TC	O	
Polypropylene			C		TC	TC	TC	O	
Oil Paint	O	C		–	–	–		O	
Emulsion Paint	O	C		–	–	–	–	O	
Floor Polishes	O	C		O	–	–	–	O	–
Concrete				O	O	O	O		
Terazzo				O	O	O	O		
Bitumen	TC	TC			TC	TC	TC	O	TC
Epoxy Resin	TC	TC		TC	TC	TC	TC	O	TC

Notes apply unless detergent is specifically inhibited against attack on surface.

Key O – Not suitable
   T – May be used at recommended temperatures
   C – May be used at recommended concentrations

materials or containers. Remember that a proportion of people are color-blind and that colors may seem different in daylight and artificial light. Remember also that it is easier to distinguish short product names than numbers; symbols such as a circle, square or triangle may also be used to assist identification.

- Weigh or measure amounts of detergents. Measures such as 'a large handful' or 'half a bucket' are unsatisfactory. Kilos (pounds) or liters (gallons) invite guessing. Ideally, use a small number of containers for a given unit volume such as the make-up tank or kettle.
- When installing liquid detergent lines assume all pipelines will leak at some time. Be sure that the leaks do not affect personnel, ingredients, products or packaging.
- Theory is a good guide to detergent selection and likely performance but confirmation on specific equipment or environment with specific soils is essential. It may be worth remembering that cannery experience (which may not be universally applicable) shows that about a 10–15% increase in the *use* concentration which works for an 'average to bad' soil will usually work with the 'worst case'.

Remember, 'costs are facts, costings are a matter of opinion'. Although, in theory, the minimum detergent cost would be obtained by each job having its own detergent mix, it would be foolish to attempt this. In practice, unless there are just a few materials with a few, simple, well-understood rules for make-up and use, mistakes will occur more than 'just occasionally'.

Costs of inadequate cleaning are more than that of wasted labor and materials. Either the cleaning is done a second time which may lose production time or there is an increased risk to safety and/or spoilage. This may lead to food poisoning, product recall or losses because sub-standard production has been made.

It is sound common-sense practice to use a 'reasonable margin of safety' in detergent concentration. To make savings in detergent costs, avoid the use of unnecessary ingredients.

Warning: detergents with perfume or odor have no place anywhere in a food plant.

Safety in use

Background

Detergent materials, especially when in concentrated form (e.g. the 'ingredients' for a mix), are powerful chemicals which can be dangerous to personnel and product if they are mishandled or misused. Obvious examples of this are:

- Mixing of sodium hypochlorite and an acid produces chlorine – a poisonous gas.
- Mixing solid alkalis, e.g. sodium hydroxide (caustic soda), with water produces heat. If mixed with 'boiling' or hot water it may 'boil over' and erupt, throwing out caustic spray. It is known that a man lost an eye through this occurrence in a food factory.
- Mistaken use of any detergent instead of the proper food ingredient in a product has obvious consequences.

Safety in cleaning operations results from applying knowledge gained from training in the use of well-thought-out systems. These systems will include the correct choice and use of:

- Cleaning chemicals
- Protective clothing
- Safe working

Cleaning chemicals

Every detergent supplier should be willing, and in some countries is required by law, to provide technical information relating to the safety of his product. When a detergent is selected for use in a plant this information should be incorporated into the appropriate Safety Manual which should be readily available on a 'need to know' basis. The Medical Department or First Aid Room *must* have an up-to-date copy. With regard to detergents the headings on the pages should include:

- Approved trade name(s)
- Description: e.g. white color, oil-bound granules to protect user from dust. Active ingredients 45% sodium hydroxide, 50% sodium orthosilicate, pH of 1% solution 12.9. Supplied in 25 kg bags or 200 kg drums marked . . . .

- Applications: e.g. 'cleaning in place'
- Preparation for use:
- Hazard category/categories: e.g. corrosive alkali
- Safety precautions:
  - Storage – e.g. store in original containers in cool dry place, separate from acidic materials.
  - Handling – e.g. wear goggles, rubber or plastic gloves, rubber boots and apron.
  - In case of fire –
  - If spillage or leakage occurs – e.g. hose to drain with much water.
- First Aid:
  - If swallowed –
  - If breathed –
  - If it contacts eye or skin –
  - Disposal and treatment as waste –
- Authorized users:

Remember, this is *not* a bureaucratic matter; human pain and suffering can be reduced by prompt, correct, remedial action.

Protective clothing

This will need to be specified (in a written document) and provided as appropriate. In addition to the usual goggles, rubber or plastic gloves, rubber boots and apron, special equipment will be needed for some jobs, e.g. protective hoods, gauntlet gloves and respirators. Where respirators are used, the proper size and type must be provided and full provision be made for their proper use and maintenance. Be aware that dry alkali powders can get inside a glove or shoe and cause burns. If dermatitis develops among individuals who have previously worked for some time without skin irritation, expert medical advice is needed to determine whether the affected worker is developing sensitivity to a substance or substances being handled.

Safe working

This requires both that a training system (or module or package) be developed for handling detergents and cleaning equipment and also that supervision is seen to insist on safe working practices. The needs and complexity of the training and retraining (on-going) programs will vary according to the needs of the individual plant, but must meet the test of 'can it be shown to be fully adequate?' This test should be used in addition to any local legislative requirement.

Marriott[9] has a useful discussion in his Chapter 5 on 'Alkali and Acid Hazards' together with First Aid for chemical burns which should be supplemented by your own local material.

Properties of detergent materials

Since no one detergent or combination of detergents possesses *all* of the properties required for the removal of *all* types of soil at minimum cost, a selection must be made of the most appropriate for the category of job that is being considered. Remember, however, that materials used must be acceptable to regulatory agencies especially in regard to biodegradability and the effect on the environment when discharged into surface waters. The effects on fishing interests should also be considered.

While local circumstances vary, many processors will have access to expert advice from detergent suppliers. Even if the processor does not mix his own detergents, it is important for an understanding of the cleaning system that the technologist knows the significance of detergent properties and the way in which they act. New developments are to be welcomed but their advantages need to be proved in factory trials and balanced against a management need to have the minimum number of materials on-site.

The properties include:

Dissolving action on the soil

This results mainly from the active alkalinity of the detergent mixture, and is usually expressed as percentage of Na_2O (sodium oxide). Because other properties are required, the required alkalinity is usually derived from a mixture of materials. Table

Table 5.16 Properties of alkalis found in cleaning chemicals

| Name | Chemical formula | Alkalinity as % Na₂0 | | pH of 1% solution |
		Total	Active	
Sodium hydroxide: caustic soda	$NaOH$	76.0	75.5	13.1
Sodium orthosilicate	$2Na_2OSiO_25H_2O$	62.1	60.5	12.8
Sodium sesquisilicate	$2Na_202SiO_211H_2O$	37.9	36.5	12.6
Sodium carbonate: soda ash	Na_2CO_3	58.0	29.0	11.2
Sodium metasilicate	$Na_2SiO_35H_2O$	29.2	28.0	12.4
Sodium sesquicarbonate	$Na_2CO_3NaHCO_32H_2O$	40.6	13.7	9.8
Trisodium phosphate	$Na_3PO_412H_2O$	18.0	10.0	11.9
Sodium tetraborate: borax	$Na_2B_4O_71OH_2O$	16.3	8.4	9.1
Tetrasodium pyrophosphate	$Na_4P_2O_7$	23.3	8.1	10.2
Sodium bicarbonate: baking soda	$NaHCO_3$	37.0	0.0	8.4

5.16. gives some of these properties for alkalis in common use.

As a general guide to active alkalinity, expressed as % Na₂0, the NCA[9] suggested for canneries:

300 ppm (0.03%) for hand cleaning operations where exposure is prolonged

500 to 900 ppm (0.05 to 0.09%) for hand cleaning operations

1000 to 1500 ppm (0.1 to 0.15%) for cleaning in place (CIP)

2000 to 3000 ppm (0.2 to 0.3%) for heavy-duty cleaning

Be careful with these figures, they are *not* the same as percentage ingredients in a mix. For each ingredient, the amount of Na₂0 must be known by calculation.

Warning Remember that caustic alkalis react with carbon dioxide. If used in a tank with a CO_2 blanket they will react, become degraded and may cause a vacuum to develop.

Of these alkalis, it is worth noting that:

Sodium hydroxide is the strongest and cheapest alkali. However, it is corrosive to a number of surfaces including human tissue, it precipitates mineral hardness from water, and has no buffering power. For these reasons, it is usually used as part of a mixture.

Sodium orthosilicate is caustic with no buffering power but is a better detergent than sodium hydroxide.

Sodium sesquisilicate is intermediate in properties between meta and ortho silicates. It is a good detergent, has fairly good buffering power and is useful in heavy duty cleaning of closed systems where grease and protein soils must be removed.

Sodium carbonate is similar to sodium hydroxide but weaker, less corrosive and has some buffering power. It is sometimes used to 'bulk out' or 'fill' a mix as it is relatively cheap.

Sodium metasilicate is widely used as a moderate alkali. Compared with the other silicates it is less likely to form precipitates in hard water, it has better rinsing properties and is less corrosive as it tends to inhibit corrosion. It has good penetrating, deflocculating and buffering power.

Sodium sesquicarbonate is a weak alkali with high buffering power and relatively low pH.

Trisodium phosphate is similar to sodium metasilicate in detergent properties but does not precipitate water hardness on surfaces. It has been widely used in dairies and washing machines.

Sodium tetraborate (borax) is used primarily not as an alkali but for its buffering power in hand-cleaning detergents to prevent skin injury.

Tetrasodium pyrophosphate is used primarily for its water conditioning (sequestering) properties.

Sodium bicarbonate may be used as a mild alkali 'filler'.

Wetting or penetrating action

This depends mainly on the ability to lower the surface tension of the solution thus penetrating the soil and loosening it. The most efficient compounds have a hydrophobic chain of 13 to 15 carbon atoms. This part of the molecule has an affinity for fats and oils. The other part of the molecule has an ionic group which is common to all wetting agents and is hydrophilic, i.e. has an affinity for water. These groups emulsify and disperse oils, fats, waxes and pigments. However, the effect of pH should be known, as some will precipitate at a pH of 12 and are unsuitable for use in strong alkali mixtures. Cleaning mixtures often contain about 0.15% wetting agent; a cost-effective level.

Wetting agents are classified as:

• Anionic, because they ionize in solution to produce negatively charged ions with wetting activity. They are not compatible with Quaternary Ammonium Compounds (QUATS or QATS) which are bactericidal agents. Elliott[4] states that they may be synergized with sodium sulphate, phosphates, sodium carboxymethyl-cellulose and some natural gums. He uses the Jennings classification of five groups.

- Cationic, because they ionize in solution to produce positively charged ions with wetting activity. In practice their use is as Quaternary Ammonium Compounds (QUATS). Their primary use is for their bactericidal and bacteriostatic properties as they are less effective as wetting agents than anionic and non-ionic agents and they can be adversely affected by mineral salts and soils.
- Non-ionic, because they do not produce ions. They are effective with oils, not much affected by water hardness and vary considerably in foam production. Elliott[4] uses the Jennings (1965) classification in his tables. Marriott[5] presents a useful Table 5.6 showing properties of synthetic detergents, which include wetting agents.

Water conditioning

In hard water, the presence of calcium and magnesium salts has two important consequences. It can reduce the effectiveness of the detergent mix and it can produce a white deposit of 'fur' or scale. Scale is not only unsightly but is objectionable because:

- It harbors and protects microorganisms.
- On heat exchanger surfaces, the deposit reduces the rate of heat exchange. This could lead to underprocessing, underpasteurization or under-sterilization.
- The presence of scale tends to increase corrosion.

Water conditioning is usually achieved by the use of phosphates as sequestering agents in a detergent mix. As a group of compounds, they also emulsify and disperse fats; suspend soils; deflocculate; peptize proteins and assist rinsability. However, if kept as a hot solution, polyphosphates gradually revert to orthophosphates and lose much of their sequestering power. Note also that phosphates may have adverse effects on effluent quality.

Another group of sequestering (or chelating) agents is more expensive but very effective so tend to be used for special applications. The agents are either ethylenediaminetetra-acetic acid (EDTA) and its salts, or organic acids (such as gluconic acid) and their salts. They are heat stable and since EDTA and its salts are compatible with QUATS, they may be used in the formulation of detergent-sanitizer mixtures. EDTA can also enhance the bactericidal effects of QUATS and amphoteric sanitizers.

Elliott[4] summarizes the characteristics of these materials in his Table 3 in Chapter 4.

Rinsability

This property is generally associated with wetting or penetrating powers. Silicates and phosphates may be chosen for this property.

Emulsifying power

This is the ability of a detergent to remove fats and oils from a surface and hold them in suspension as small droplets, thus preventing deposition on another surface of the equipment. Alkalis, including silicates, which combine with fats to form 'soaps', together with wetting agents may be chosen for this property.

Corrosion

This is an unwanted reaction between the surfaces of the equipment and the materials used for cleaning and sanitizing. The materials used for equipment construction (see Table 5.15) are important. For example, if strong alkalis are used on aluminum, pitting and excessive darkening of the surface can be expected. Although the possibility is not commonly recognized, a detergent solution may set up an electrical cell between dissimilar metals. This will accelerate corrosion. Corrosion can therefore be regarded as more or less inherent for any equipment using any detergents. Its extent is dependent on the physical-chemical reaction between detergent and surface materials and is also related to the temperature and time of the reaction.

In practice, its significance can be minimized by:

- Good design and installation so that the equipment can be quickly and effectively cleaned, rinsed and drained.
- Good construction and maintenance so that the materials used are compatible with the proposed range of products and the proposed cleaning system. Remember, the cleaning system comprises both the materials used for sanitation and the method(s) of application.
- Removal by abrasion of pits and rust stains which hold the cleaning and sanitizing materials and act as foci for corrosion.
- Removal of scale by chemical treatment, which can also act as a focus for corrosion.
- Electrically grounding (earthing) equipment.
- Using corrosion inhibitors when formulating highly acidic or alkaline detergents.
- Minimizing the contact time between detergents, sanitizers and equipment surfaces to that needed to achieve the objective.

• Being sure to give a thorough final rinse followed by draining and, where appropriate, drying by exposure to air. This is simple, practical and effective.

Irritability (irritation)

This is the effect primarily on human skin. Any material that splashes in the eyes must be dealt with as an 'accident at work'. However, especially in washing small machine parts, utensils or components, the operator will have his or her hands in contact with the detergent mix for a significant time. Because of the alkalinity, this contact will remove the natural oil from the skin. These detergents should therefore be buffered to around pH 10–11. Other materials such as chlorine and mineral acids also cause irritation. Marriott[5] makes the point that when dermatitis suddenly develops among individuals on a job in which they had been previously unaffected they should be sent immediately to an experienced physician for examination and tests.

Buffering power

The purpose of a buffer is to maintain the pH (acid/alkaline) level of the detergent mix during use, particularly due to loss by chemical reaction or on dilution. This reduces precipitation due to changes of pH and provides a 'reserve' of cleaning power. Sodium sesquicarbonate, polyphosphate and metasilicate are important ingredients with buffering power. Sodium carbonate and trisodium phosphate have some buffering power. In detergents used for hand washing of equipment, borax (sodium tetraborate) may be added as it reduces the pH to around 10–11.

Versatility

This is a highly desirable property because of the practical need to minimize the number of detergent mixes used in a plant. Elliott[4] states that a large, and by implication diverse, plant may require as many as five types. These are listed as (1) general purpose, (2) CIP, (3) manual washing, (4) acid, (5) heavy-duty alkaline.

The requirements of a specific plant will vary but the emphasis should be placed on having the minimum practical number of cleaners.

There may be pressure from detergent salesmen to buy their materials. This is usually either because it will 'do everything' or because it is very effective for a particular (usually small) job which is proving troublesome. Unless technologically sound reasons are forthcoming, such arguments should be treated with considerable scepticism. Even if there are good

reasons, they should be balanced against the management required for yet another material.

A further consideration is that since labor is the main constituent of manual cleaning costs, what is the true cost advantage of the 'new' material? Remember, the only cost that the salesman can be certain about is the cost of the material.

Foaming

This is a desirable property in detergents used for hand washing utensils and small parts because foam or suds indicate to the user that there is still an effective amount of detergent in the mix. Stable foams are the basis for foam application systems (see the following sub-section). For most other purposes in a plant, especially where the detergent is pumped, sprayed or where there is turbulence, foam is objectionable because it interferes with the proper functioning of the detergent.

Foams and gels

These materials consist of a carrier, the foam or gel, and the active alkali or acid cleaning material (detergent). The principle is that the detergent is held in contact with the surface for longer than it would be if it were sprayed or wiped onto the surface. It is useful especially for surfaces which are inaccessible or near-vertical. Foams are usually applied and then left for 10 to 20 minutes before being rinsed away with a hose or lance.

Gels, being more adhesive, have a longer residence time and are then washed away. The usual tendency is for the operator to overdose, so good control of the dispensing unit is important for cost effectiveness.

Cannery experience where soils are heavy and tenacious suggests that foam or the newer 'foam-gel' techniques work well with moderate soiling and 'cosmetic' cleaning of walls, ducting and inaccessible places.

'Cold' cleaners

These are new developments, which as their name implies can be used at lower than usual temperatures, with an obvious cost advantage. As with any cleaning system, the experimental period should be carefully monitored for a realistic time. Be especially careful when using with products containing fat or oil because of the build-up of film which could harbor microorganisms (bacteria and yeasts).

Costs

It is important to estimate costs carefully and to use a system which takes into account the cost of the whole cleaning operation over the working life of the equipment and not just the detergent cost. Remember, the choice of a detergent because it is 'the cheapest' may prove to be expensive if it is ineffective in performance, has excessive labor or utility requirements, must be used in large amounts, or causes corrosion.

Classification of detergents

There are a number of ways in which detergents can be classified, but in practice the following is often used:

- *Highly alkaline*. These include sodium hydroxide (caustic soda) and silicates having high Na_2O: SiO_2 ratios. They are for very heavy-duty use such as removal of dried-on or burned-on materials. Unless modified, they will be corrosive to many materials and invariably to skin at use concentrations and temperatures.
- *Moderately alkaline*. These are usually a mixture of alkaline materials such as sodium metasilicates, sodium phosphates of various kinds with added wetting agents, conditioners and corrosion inhibitors. With both highly and moderately alkaline detergents, the alkalinity is usually expressed as percentage or ppm Na_2O.
- *Mildly alkaline*. These are usually for hand cleaning of lightly soiled surfaces. They usually consist of wetting agents, water conditioners and mild alkalis, e.g. carbonates and phosphates. Borax may be added as a buffering agent.
- *Strong acids*. These are often used with a corrosion inhibitor and include nitric acid (used on stainless steel), sulphamic acid (which can be combined with a non-ionic wetting agent) and phosphoric acid.
- *Sodium hypochlorite*. This may be used as a cleaner when combined with a moderate alkali mixture. At an initial concentration of 150–300 mg per liter (ppm) chlorine, it is effective in canneries for heavy-duty cleaning of stainless steel equipment. Short contact times, high temperatures and very thorough final rinsing are important factors in its successful use.
- *Solvent cleaners*. These are usually used on petroleum-based soils and greases in and around workshop, pallet or fork truck areas. Their use should be strictly controlled.

Summary of functions and limitations of detergent types

Lewis[10] in his Chapter 14 in the ICMSF *Microbial Ecology of Foods*, Volume 1, gives a useful summary table, given here as Table 5.17. *Note*: A *Warning* has been added as a footnote to the table stating that *steel wool and metal 'chore balls' should not be used in a food plant*. Experience has shown

that they cause unacceptable damage to stainless steel surfaces by scratching. Subsequent cleaning is then made increasingly difficult.

Note that the figures given under the heading 'approximate concentrations for use' may not be representative of those needed for *your* application, as soils, equipment, and time available for cleaning vary widely between plants and product lines. For further information see NCA's Chapter 8 in reference 9 and Elliott's Chapter 4 in reference 4

Remember that technological considerations give a clear idea as to what is likely to work. The choice of material and system should be made as a result of carefully monitored tests in the plant using 'difficult', actual soils. It is important to have sound, positive evidence that the material and system are fully adequate under 'bad' running conditions.

Identification of detergents

In this context, identification does not mean chemical analysis but the means by which the correct detergent is known so that it can be used for a specific job. Remember that if mistakes *can* be made, they *will* be made – usually at the most inconvenient time.

Identification of stock material is helped, for example, by:

- Differences in physical form, e.g. if there are two similar-looking powdered materials, can one be obtained in flake form?
- Differences in color. Inactive coloring materials can sometimes be used, e.g. permanganate can be used to color sodium hypochlorite.
- Differences in packaging, e.g. size, shape, markings. Remember, however, that color-blindness is not uncommon among people and that artificial lighting can distort colors.
- Use of short distinctive names reduces the chances of mistakes, compared with either a number, alpha-numeric code or chemical name.

Remember, the person issuing or using a stock material must be *certain* of its identity.

In distribution of detergents, when issuing from store, it is prudent for technical and cost reasons to weigh or measure amounts of materials before issuing to the cleaning crew. Identification is helped by issuing the same material in the same container. With only a small number of different materials effective control is much easier. Materials stacked in containers on trolleys for distribution should, of course, be arranged so that they are easily identified and in a convenient order for use when off-loaded.

Delivery to point of use

When transporting and storing cleaning materials before use, they must be kept completely separate

Table 5.17 Types, functions, and limitations of cleaning agents used in the food industries[a] (from ICMSF[10])

Categories of aqueous cleaners	Approximate concentrations for use (%)[b]	Examples of chemicals used[c]	Functions	Limitations
Clean water	100	Usually contains dissolved air and soluble minerals in small amounts.	Solvent and carrier for soils, as well as chemical cleaners.	Hard water leaves deposit on surfaces. Residual moisture may allow microbial growth on washed surfaces. Promotes rusting of iron.
Strong alkalis	1–5	Sodium hydroxide Sodium orthosilicate Sodium sesquisilicate	Detergents for fat and protein. Precipitate water hardness. Produce alkaline pH.	Highly corrosive. Difficult to remove by rinsing. Irritating to skin and mucous membranes.
Mild alkalis	1–10	Sodium carbonate Sodium sesquicarbonate Trisodium phosphate Sodium tetraborate	Detergents. Buffers at pH 8.4 or above. Water softeners	Mildly corrosive. High concentrations are irritating to skin.
Inorganic acids	0.5	Hydrochloric Sulphuric Nitric Phosphoric Sulphamic	Produce pH 2.5 or below. Remove precipitates from surfaces.	Very corrosive to metals, but can be partially inhibited by amines. Irritating to skin and mucous membranes.
Organic acids	0.1–2	Acetic Hydroxyacetic Lactic	Remove inorganic precipitates and other acid-soluble substances from surfaces.	Moderately corrosive, but can be inhibited by various organic nitrogen compounds.
	0.1–2	Gluconic Citric Tartaric Levulinic Saccharic	Remove inorganic precipitates and other acid-soluble substances from surfaces.	Moderately corrosive, but can be inhibited by various organic nitrogen compounds.
Anionic wetting agents	0.15 or less.	Soaps Sulphated alcohols Sulphated hydrocarbons Aryl-alkyl polyether sulphates Sulphonated amides Alkyl-aryl sulphonates	Wet surfaces. Penetrate crevices and woven fabrics. Effective detergents. Emulsifiers for oils, fats, waxes, and pigments. Compatible with acid or alkaline cleaners and may be synergistic.	Some foam excessively. Not compatible with cationic wetting agents.
Non-ionic wetting agents	0.15 or less	Polyethenoxyethers Ethylene oxide–fatty acid condensates. Amine–fatty acid condensates.	Excellent detergents for oil. Used in mixtures of wetting agents to control foam.	May be sensitive to acids.
Cationic wetting agents	0.15 or less	Quaternary ammonium compounds.	Some wetting effect Antibacterial action.	Not compatible with anionic wetting agents. Inactivated by many minerals and some other soils.

Table 5.17 Types, functions, and limitations of cleaning agents used in the food industries[a] (from ICMSF [10])

Categories of aqueous cleaners	Approximate concentrations for use (%)[b]	Examples of chemicals used[c]	Functions	Limitations
Sequestering agents	Variable (depending on hardness of water)	Tetrasodium pyrophosphate Sodium tripolyphosphate Sodium hexametaphosphate Sodium tetraphosphate Sodium acid pyrophosphate Ethylenediaminetetra-acetic acid (sodium salt) Sodium gluconate with or without 3% sodium hydroxide	Form soluble complexes with metal ions such as calcium, magnesium, and iron to prevent film formation on equipment and utensils. See also strong and mild alkalis above.	Phosphates are inactivated by protracted exposure to heat Phosphates are unstable in acid solutions.
Abrasives	Variable	Volcanic ash Seismotite Pumice Feldspar Silica flour Steel wool[d] Metal or plastic 'chore balls'[d] Scrub brushes	Removal of dirt from surfaces with scrubbing. Can be used with detergents for difficult cleaning jobs.	Scratch surfaces. Particles may become imbedded in equipment and later appear in food. Damage skin of workers.
Chlorinated compounds	1	Dichlorocyanuric acid Trichlorocyanuric acid Dichlorohydantoin	Used with alkaline cleaners to increase peptizing of proteins and minimize milk deposits.	Not germicidal because of high pH. Concentrations vary depending on the alkaline cleaner and conditions of use.
Amphoterics	1–2	Mixtures of a cationic amine salt or a quaternary ammonium compound with an anionic carboxy compound, a sulfate ester, or a sulfonic acid	Loosen and soften charred food residues on ovens or other metal and ceramic surfaces.	Not suitable for use on food contact surfaces.[e]
Enzymes	0.3–1	Proteolytic enzymes produced in cultures of aerobic, spore-forming bacteria.	Digest proteins, and other complex organic soils.	Inactivated by heat. Some people become hyper-sensitive to the commercial preparations. Some have contained *Salmonella*.

[a] Based largely on information assembled by Elliott.
[b] Concentration of cleaning agent in solution as applied to equipment.
[c] Some regulatory agencies require prior approval.

[d] Steel wool and metal 'chore balls' should *not* be used on food plant.
[e] Some amphoteric *disinfectants* are used on food contact surfaces.

from food ingredients and products, in both space and time. If liquid detergents are transported by pipeline, it should be assumed that at some time leaks will occur. Care is therefore needed in routing to minimize danger to personnel, ingredients, products or buildings.

Clear identification of pipelines is important, as is correct sizing, so that the required quantities are delivered at the correct rates.

Application of detergents – theory

The cleaning process may be considered as the application of energy to a soil. The energy may be thought of as being available in three distinct forms:

- Mechanical (kinetic), which may be manual, e.g. scrubbing or scraping; mechanical, e.g. machine scrubbing; or hydraulic, e.g. hosing or pumping along pipelines. This is usually very effective.

• Thermal, which may be from hot water or liquids.
• Chemical, which may be from detergents.

Usually a combination of these forms is used, and getting the balance right is important if cleaning is to be successful and cost effective.

For the technologist who needs to understand more of the principles of detergency there are a number of useful references, including a review by Jennings[11], and his Theory and Practice of Hard Surface Cleaning[12] together with a very clear review by Koopal[13] of soil-removal mechanisms.

Application of detergents – practical

The practical application of detergents will be considered under the following headings:

• Personnel Safety
• General Principles
• Cleaning out of place (COP)
• Cleaning in place (CIP)
• Evaluation of cleaning systems

Personnel safety

Assuming that the design and installation of equipment are inherently safe then safe ways of working are primarily a matter of good management. Resources must therefore be allocated to:

• Develop safe working practices and codify them as appropriate procedures. This includes the specification and issue of protective equipment such as boots, aprons, gloves and bump caps. Remember, if protective equipment is not perceived to be effective and comfortable, operators will not use it.
• Train, and re-train as required, management, supervisors and operators in safe ways of working and the reasons for safety procedures.
• Supervise the cleaning operation and insist on safe ways of working.

Much of what is required is simple good practice:
– With hoses
  (a) When operating a steam/water mixer at a hose point, turn *water on before steam* and *steam off before water*.
  (b) After use, put a hose back on its reel. (Do not leave it lying on the floor where it may trip someone and cause injury.)
– Make sure that no one is working on those parts of the line where the cleaning cycle is about to start. (This is necessary but not always easy when the line is on more than one floor.)
– Close off or post warning notices around areas where hot detergent solution (whether acid or alkali) may splash other personnel.

Remember that cleaning chemicals are used because they work quickly and effectively. They are powerful and may cause severe injury to exposed skin or eyes. They may penetrate clothing to harm the skin. Hot liquids or steam may also injure. Good supervision insures that safety equipment is used and that the proper working practices are followed.

What needs to be done, therefore, is to identify pertinent hazards in the individual plant and then develop and enforce appropriate safety procedures.

General principles – outline of cleaning and sanitation processes

The main stages in cleaning have been described in Section 5.D.2 and may be summarized as: removal of gross soil, application of detergent(s), then rinse followed (if applicable) by sanitizer and the final rinse.

It is convenient to consider cleaning out of place (COP) and cleaning in place (CIP). These are not mutually exclusive, however, since most larger systems include both elements.

General principles – basis of choice of method

The choice of the way to apply detergents depends on:

• Design, construction, and installation of the equipment
• Nature and location of the soil
• Availability of cleaning equipment
• Other constraints, e.g. the time available for cleaning.

Some of the more important factors are indicated in Table 5.18.

Clean out of place (COP) applications

Equipment cleaned in this way is disassembled and/or moved from its normal location for cleaning. The term is also used for equipment that is opened up (i.e. made accessible) for cleaning manually or by using hoses. As a matter of convenience, COP application will be considered under the headings:

• Hoses
• Mechanical aids
• Cleaning baths
• Use of cloths and sponges
• Use of abrasives

Hoses

Hosing is one of the most usual ways of applying water which is used as a cleaning agent and, where appropriate, may be used to apply a detergent solution. The usual classification of hose systems is:

Table 5.18 Some factors affecting the choice of detergent application (modified from Table 6.1 in ICMSF *Microorganisms in Food* 4[14])

Factor	HPLV	LPHV	CIP	Manual/ mechanical	Foam/gel
Soil type:					
(Very tenacious	+ +	–	+	+ +	–)
Tenacious	+ +	+	+ +	+ +	–
Water soluble	+ +	+ +	+ +	+ +	+ +
Soil level:					
High	+ +	+	+ +	+ +	–
Low	+ +	+ +	+ +	+ +	+ +
Open equipment:					
Access – close	+ +	+ +	–	+ +	+ +
Access – distant (e.g. over 2 m)	+ (or –)	–	–	–	+ +
Horizontal surface	+ +	+	–	+ +	+ +
Vertical surface	+ +	– (or +)	–	+	+ +
Void spaces	+ +	+	–	+ +	–
Closed equipment (including lidded tanks):					
Absence of void space(s)	+ +*	–	+ +	+	–
Presence of void space(s)	+ +*	–	+	–	–

HPLV = High pressure – low volume
LPHV = Low pressure – high volume
CIP   = Clean in Place
*   = Vessel (or tanks) only; not pipelines
+ + = Ideally suited (likely to be suitable)
+   = May be suitable if managed (applied) effectively
–   = Inappropriate

Note: material shown in parentheses has been added.

- Low pressure–high volume (LPHV)
- High pressure–low volume (HPLV)

Low-pressure systems

These systems deserve more attention than they sometimes receive. The hose material must be strong enough to resist:

- Pressures of water and steam (if used)
- Temperatures both during operation and from steam leaks
- Detergents if used through the hose or which may contact it, e.g. when the hose is used to supplement the hot water supply to a kettle or tank
- Fats and oils present or on the floor in the area to be cleaned.

Additionally, the hose must be light enough in weight to be manageable by the operator and robust enough for heavy duty. These requirements make the choice of material a compromise, but reinforced neoprene is widely used in the UK.

Hoses are usually 15 to 21 m (50 to 70 ft) long and should always have a rack or automatic reel provided at the hose station. For really heavy duty, the rack may be preferable. Water is usually supplied at about 3½ to 5 bar (50 to 70 psig) and steam at up to 6 to 7 bar (100 psig). Be aware that there may be local regulations governing the use of steam. If automatic shut-off valves are used to save water they *must* be able to withstand the full water line pressure and be robust enough to be repeatedly dropped.

Warning – If shut-off valves were installed at the outlet of a steam/water hose this would make the hose a pressure vessel. This may be illegal and is very dangerous.

There are advantages in having different types of outlet nozzle – a flat 'fish-tail' or 'fin' will 'lift' soil from flat surfaces. The NCA[9] recommends that:

the outlets should be constructed so that nozzles of various types can be connected rapidly. The following interchangeable nozzles are suggested: a small jet type for cleaning deep cracks, a fin type for cleaning flat surfaces, a bent type for cleaning around and under equipment, and a spray head-brush combination type for cleaning surfaces where combined brushing and washing is needed.

Where a steam/water mixer is used to produce hot water for the hose Imholte[1] recommends installation of a direct-reading temperature gauge. As an alternative source of hot water, an accumulator tank may be installed with an outlet pump to provide the appropriate hose pressure. The tank, if of sufficient size, may be heated over a period thus reducing peak steam demand. Either water or detergent may be provided.

Steam/water mixers, sometimes called 'steam guns', can be obtained with a detergent pick-up which produces a jet at a slightly lower pressure than the steam in the line. Elliott[4] states that maximum pressure at the nozzle by steam injection of cold water is about double that of the steam.

For safety reasons, it is best to select mixers designed so that 'line steam only' cannot be drawn. The assembly should also be provided with a back-flow prevention device or non-return valve on the water line to prevent cross-contamination of the water supply system.

Note on steam hoses

These are used to remove certain types of debris, e.g. labeller glues in a cannery, or to blow off water from equipment after it has been cleaned. Marriott[5] states: 'Generally, this is not an effective method of cleaning because of fogging and condensation, and it does not sanitize the cleaned area.'

Management, supervisors and operators should be aware of safety hazards of this equipment as well as mechanical damage that can result from removal of lubricants such as greases and oil. Cleaners, however, seem to enjoy using this method, probably because the noise and fogging are impressive.

Central supply for hoses

These may be used to supply low or high pressure hoses from a central pumping station with hot or cold water and also detergents if required. These systems are discussed in some detail by Imholte[1] in his Chapter 8 and also by Marriott[5] in his Chapter 7. The pressures, temperatures, and volumes required, of course, are much the same whether supplied individually or from a central station.

High pressure systems

These are generally referred to as High Pressure–Low Volume (HPLV) systems. Pressures in the range of 40 to 85 bar or kg/cm^2 (600–1200 psig) are usual for food plant cleaning applications. This compares with about 5 bar (73 psig) for the low-pressure hose. Volumes are much lower with HPLV, usually between 4 to 10 liters per minute (1 to 2½ US gpm; 1 to 2 UK gpm) and 45 to 90 liters per minute (12 to 24 US gpm; 10 to 20 UK gpm). This compares with 450 to 500 liters per minute (119 to 132 US gpm; 100 to 110 UK gpm) from the conventional low-pressure hose.

For portable (mobile) units, Marriott[5] suggests delivery at 20 to 85 bar (290 to 1200 psig) at 4 to 10 liters per minute (1 to 2½ US gpm; 1 to 2 UK gpm) at 55°C (131°F).

Since the ability to remove soils depends primarily on the energy of impact, really stubborn soils will need higher pressures and/or volumes and/or time.

There seems to be a tendency in newer installations to use pressures around 70 Bar (1000 psig) at the pump. When considering HPLV systems it is essential for the user to appreciate the rapid pressure drop from the nozzle. To quote Imholte[1]: 'Many high-pressure systems are designed to produce the required pressure six inches (152 mm) from a nozzle with a 15° spray pattern. At a distance of 12 in (305 mm) the pressure drops to approximately one half; and at 18 in (457 mm), to approximately one quarter of the design pressure.' This relationship is shown in Figure 5.12 with the 'Low Pressure' hose at 5 bar (73 psig) marked in.

This pressure drop relationship is important when considering the design of the applicator guns (or wands). For safety reasons, a 'fail-safe' pistol grip which automatically shuts off the liquid stream is recommended. The materials of construction must resist corrosion from any cleaning material likely to be used. Wands of appropriate lengths are essential if adequate cleaning is to be achieved in reasonable time. Nozzles are usually available to give a range of spray patterns between 0° and 40°. Imholte[1] states:

> High-pressure low-volume systems need adjustable nozzles that can be easily adjusted to suit the cleaning requirement. A multiple spray pattern selection nozzle that offers several angles of spray (such as 0°, 15°, and 40°) and a selection of different orifices may be worthwhile if the cleaning tasks are somewhat different. Orifice size selection features allow chemicals to be applied at low pressures, which is desirable for eliminating splash back and chemical waste.

Remember, HPLV is a precision tool compared with a low-pressure hose but it is *not* effective in 'floating' debris, e.g. vegetable dice, along a floor to a drain.

Because of its power, HPLV must be used with care even when cold water is used. The NCA[9] states:

> The operator should wear waterproof clothing, insulated gloves, and a face shield if hot water or strong detergents are used. Training of operators is essential, not only for efficient use but also for safety.

Slow-acting, automatic shut-off valves are recommended to avoid jumping of the hose. Automatic

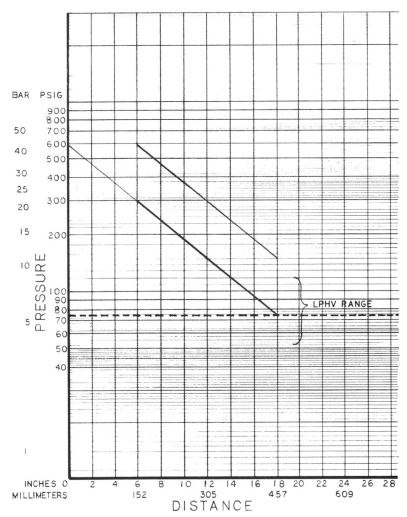

Figure 5.12 Pressure drop with distance from HPLV lance

shut-off valves prevent water loss and permit hoses to be moved without indiscriminate spraying of water. High-pressure hoses not over 15 m (50 ft) in length are suggested. Pumps should have corrosion-resistant surfaces when strongly alkaline or acid detergents are used. A high-pressure water clean-up should be preceded by a low-pressure wash to avoid scattering food particles.

One point worth noting is that when plant cleaners first use HPLV, because it is a powerful tool which can give excellent results, the time taken for a particular task may increase because cleaners can achieve a higher visual standard than was reached previously. More experience will reduce the time.

Mechanical aids – brushes and brooms

These are likely to be the first type of aid thought about when 'mechanical' aids are considered. In common usage, a broom has a long handle and is used for such tasks as floor cleaning while a brush is smaller and used on equipment.

Natural fibers in use include Bassine, which is very stiff, Palmyra (Palmetto) which is used for medium soils and Tampico, which may be used for light soils but nylon or other plastic, e.g. poly-propylene, is often preferred. Nylon fibers are strong, flexible, uniform in diameter, durable and non-absorbent. A major concern is that bristles when broken or detached become potential or actual contaminants. For this reason also the back or stock of the brush or broom should be of composition or solid plastic and not wood, which will splinter or chip. For similar reasons, the bristles should be imbedded or sealed into the stock and not glued or stapled.

A variety of types and sizes of brushes and brooms will be needed in a plant. Brushes for components or small equipment are best if they fit the contour of the surface being cleaned. The handles should be just long enough to enable the cleanup crew to reach the surface easily. Remember, the right brushes reduce time and labor needed for the job.

Warning – Wire brushes are unacceptable for food equipment cleaning because the wires may break, becoming potential or actual contaminants. Additionally, many equipment surfaces including stainless steel are scratched by wire brushes. Particles of wire embedded or trapped on stainless steel become foci of corrosion. In a food plant the proper use for wire brushes is by engineering staff removing paint etc. from a surface for maintenance or repair. Their use must be well separated from food preparation by space or by ample time to allow very thorough decontamination.

Solution-fed (or water-fed) brushes of various types are available for power and hand operation. The solution may be fed through a light hose from a thermostatically controlled tank.

Power brushes have a specialist application, e.g. in cleaning pipelines, but power brooms or floor scrubbers are commonly used.

Mechanical aids – scrapers and squeegees

These are useful aids. Scrapers made of plastic are a help when removing thick soils. The use of metal wallpaper scrapers, even on stainless steel, is not recommended as it can scratch or score the surface. Squeegees are useful when 'sweeping up' water or wet, light debris.

Steel wool or copper chore balls

Warning These scratch equipment and leave troublesome metal particles behind. *Do not use on food contact surfaces*

As an alternative, nylon pads will usually remove difficult soils. Woven stainless steel scouring cloths will remove very stubborn soils but *must* be discarded before wear results in loose particles which are potential or actual contaminants.

Cleaning baths (tank-type washers)

These are used for sanitary fittings, pipes and small equipment parts which may conveniently be cleaned in this way. As Imholte[1] states: 'Food manufacturing operations requiring a lot of disassembly of smaller equipment parts for out-of-place cleaning would benefit from equipping the cleaning room with a specially designed recirculating tank-type washer'.

Baths may be of a double-compartment stainless steel sink type. The first compartment is equipped with motor-driven brushes in which the cleaning solution is pumped through a sparge pipe to feed onto the brushes. Cleaning solutions are usually kept at 45 to 55°C (113°F to 131°F) by a thermostatically controlled heater. The second compartment is equipped with a spray of clean water to rinse the parts which are then stacked on racks to dry.

Baths may also be designed as tanks where turbulence provides the cleaning force (see Marriott[5]) to apply the detergent. The detergent solution may come from a CIP system. Imholte[1] gives a description of these tanks on page 186 and states:

> Important tank construction details are rounded corner junctions, automatic temperature regulating equipment, direct-reading thermometers, strainers on the water inlet, and valves to drain the tank. Optional equipment may include a chemical feed system and a heat exchanger for heating the water.

Baths using ultrasonic energy to generate pressure pulses to clean small parts are sometimes used. These may have an ultrasonic generator producing frequencies of 30 000 to 40 000 cycles per second in a tank of detergent solution at around 66°C (150°F). These baths tend to be used for small, often delicate, parts because the energy of the pulses is focused to give effective cleaning.

Cloths and sponges

These can be very real contamination hazards although they can be useful when properly used. Cloths made from old fabric or mill waste are likely to leave threads which are potential or actual contaminants. Those found and returned by consumers as a complaint may prove costly. Clean stockinet is satisfactory if it is discarded after use.

The use of paper cloths or towels, if permitted anywhere in the plant, must be carefully controlled. Experience shows that paper towels have a tendency to spread from washroom areas. Careless use results in potential or actual contamination.

Marriott[5] states that sponges are most effectively used for cleaning product storage tanks when they have insufficient volume to justify mechanized cleaning.

COP – hidden problems

These exist because brushes, cloths, sponges and other cleaning aids are not regarded as themselves in need of cleaning and sanitation (disinfection). It is instructive to run microbiological checks but be sure to use a quenching agent if appropriate and take the dilutions well above 1 in 1 million. The point is made by Troller[14]: 'Sponges and other porous cleaning devices should be avoided since they tend to stay moist for extended periods of time and may become a significant source of bacteria.' ICMSF in *Microbial Ecology of Foods Volume 1* [10] state:

All cloths, brushes, sponges, or other cleaning aids used for manual cleaning must be washed frequently and disinfected either by heat or by soaking for at least 10 minutes in a solution containing 200 ppm of available chlorine or 400 ppm of quaternary ammonium compound.

Manual cleaning is adaptable to all types and sizes of equipment, but its effectiveness depends heavily on the integrity of the worker. Thorough scrubbing, frequent changes of solutions, and special care to avoid cross contamination are essential.

Abrasives

These should be used sparingly, if at all, on food handling equipment. Typically, abrasives such as silica flour or pumice are mixed with a detergent and sometimes a chlorine-producing material, e.g. hypochlorite then applied as a paste. A nylon scouring pad may be used to increase the speed and effectiveness of cleaning. Abrasives will scratch surfaces, including stainless steel, and unless the abrasive together with any surface particles are thoroughly washed away they may contaminate foods.

Clean in place (CIP) applications

Historically, the dairy industry was aware at an early date of the importance of sanitary design. Since CIP requires equipment of a high standard of sanitary design, installation, and maintenance, as well as good engineering of the cleaning system, early and successful applications were made in dairy plants. The use of glass pipelines in dairies was, no doubt, a factor in CIP development because no one would wish to clean a glass line by dismantling and brushing it through with detergent.

CIP today is usually regarded as the method of choice for cleaning tanks, pipelines, pumps and, where applicable, valves. The aim is to minimize down-time on the line and reduce labor cost inherent in the dismantling, manual cleaning and reassembly of line equipment. To be successful, CIP application should be designed in from the start; attempts to 'retro-fit' a CIP system to equipment which already has been designed and installed are likely to prove expensive in time and money and are not fully effective.

It must be remembered that CIP is neither a magic formula nor universally applicable. It is the application of sanitary engineering principles and practices to provide a cost-effective cleaning system. CIP and COP are not mutually exclusive and in practice both are used in most 'CIP' applications. Tamplin[16] gives the practical benefits of CIP (for Soft Drinks) as follows:

- Cost savings. CIP insures optimum use of water, detergent, sterilant and steam to produce economy in operation.
- Increased plant utilization. Tanks and pipelines can be cleaned as soon as they are empty and refilled immediately after cleaning. Thus unproductive down time is minimized.
- Minimum manual effort. Manual operations can be reduced or eliminated entirely depending upon the degree of automation designed into the plant.
- Greater safety. Personnel have no need to enter vessels. The risk of falls on slippery internal surfaces is eliminated and personnel do not have to enter hazardous atmospheres.
- Improved hygiene. Cleaning schedules are adhered to exactly and consistently satisfactory results are obtained. Thus product quality is improved and shelf life extended.

Additionally, it has been claimed that CIP reduces disassembly damage and wear and reduces maintenance and repair costs.

CIP systems are regarded as standard practice in the dairy industry and are increasingly used in other food industries. Marriott[5] states:

The use of CIP systems in the meat and poultry industry is limited. This equipment is expensive and lacks effectiveness in heavily soiled areas. CIP cleaning has some application in vacuum thawing chambers, pumping and brine circulation lines, pre-blend/batch silos, and edible and inedible fat-rendering systems.

General principles

There is an overall pattern to the typical CIP cleaning cycle whichever system is used. It may be summarized as: removal of gross soil, application of detergents, then rinse followed (if applicable) by sanitizer and final rinse.

The close control of the CIP sequencing is critical. Often, both process and cleaning operations are controlled by microprocessor systems. While these are very reliable, they need effective inspection/ monitoring, regular routine checks and maintenance to prevent or detect malfunction.

Table 5.19 Comparison of typical features of CIP systems (from Tamplin[16])

Feature	Single-use	Re-use	Multi-use
Design concept for water and detergent	Use once then throw away	Re-use where possible	Some single-use, some re-use
Design type	Simple, often modular	Complex, 'one-off'	Complex, but modular
Extension to new equipment	Easy	More difficult	More difficult
Flexibility of detergent types	Flexible	Inflexible without modification	Flexible
Changes in detergent strength	Easy	Difficult without modification	Easy
Detergent make-up	As used	Stored hot	As used
Peak thermal load	High	Low	Moderate to high
Ability to deal with soiling classed as:	Heavy	Light to moderate	Moderate to heavy
Costs – capital	Low	High	Lower than Re-use
Costs – detergent	High	Low	Lower than Single-use
Costs – heat energy	High	Low	Lower than Single-use

Manual operation of control and sequencing is not satisfactory and accidental mixing of product and detergent has been known.

There are, of course, a great variety of CIP systems. Tamplin[3] summarized the position thus in 1980:

CIP systems have many forms. Originally it was customary to use detergents only once and then throw them away, but with the growth and complexity of plants and the number of circuits to be cleaned, the re-use of detergents was adopted.

It is not possible to describe and specify a system which will be optimum for all occasions, because there is no universal CIP system which can be specified for all applications.

There are two different basic techniques, single-use and re-use systems which have evolved with time. The latest development is to use combined systems which incorporate the best characteristics of the single-use and re-use units. These are generally known as multi-use systems.

Table 5.19 compares typical features of these three systems.

In order to understand the significance of Table 5.19 it is necessary to consider the systems in more detail. Reference should also be made to Tamplin[3] and [16] (which includes typical utilities for a multi-use module on pages 173 and 174), the new Society of Dairy Technology (SDT) publication[17] and to Timperley[18].

Single-use systems

The system shown in Figure 5.13 is from Tamplin[16] and page 162 onwards of his chapter gives further details. Detergents with a short active life, e.g. alkali-hypochlorite mixtures, are used and the modular unit can serve a number of circuits in sequence with different mixtures or strengths. If some circuits are heavily soiled, the single-use system is desirable because of difficulties in restoring the detergent mixture to its original strength and properties.

Note that Figure 5.13 shows heating by direct injection of steam; other heating arrangements are possible but would have higher capital cost. In order to avoid the cost of detergent and heat (thermal energy) losses, a limited recovery facility may be provided. This is an extra tank for detergent solutions and some rinses and it is used for the pre-rinse of the next cycle. This is shown in Figure 5.14.

Re-use systems

Providing that the detergent(s) are not degraded by use, then they may be re-used. This may well be so with dairy or soft drink applications. Tamplin[3] states that:

a typical dairy cleaning system may be required to work on the individual circuit 15 to 20 times a day. Therefore it is sometimes advantageous to add

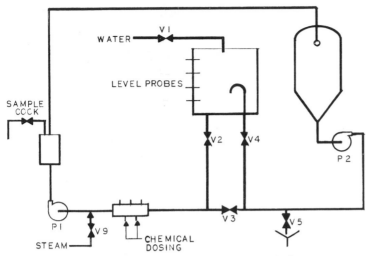

Figure 5.13 A typical single-use CIP system (from Tamplin[16])

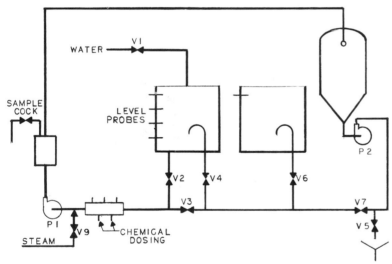

Figure 5.14 A single-use CIP system incorporating limited re-use (from Tamplin[16])

extra CIP feed pumps and thus enable two CIP circuits to be cleaned simultaneously. With this type of system, the tank capacity is defined by circuit volumes, temperature requirements and desired cleaning programs . . . With the recirculation facility of the hot-water option and the use of a return water tank, the water consumption of a re-use system can be optimized.

Such a system is shown in Figure 5.15.

These systems have a wide range of possible programs and with modern control systems can be made flexible. However, the multiple-tank units are large and adding on circuits can take the storage

tanks beyond their original design capabilities with consequent increase in capital cost.

Steam coils also have disadvantages. They must not be exposed during the cycle (to avoid damage to the heating coils) and temperature–time profiles are not easily altered. Other means of heating could be used but would have a higher peak steam demand.

The type and concentration of detergent and sanitizer are fixed; the heaviest duty determines these requirements, with 'overkill' on lesser duties.

The lack of flexibility also means that cleaning the occasional, exceptionally heavily soiled ('rogue') circuit involves preparing a fresh batch of appropriate CIP detergents. Other disadvantages are

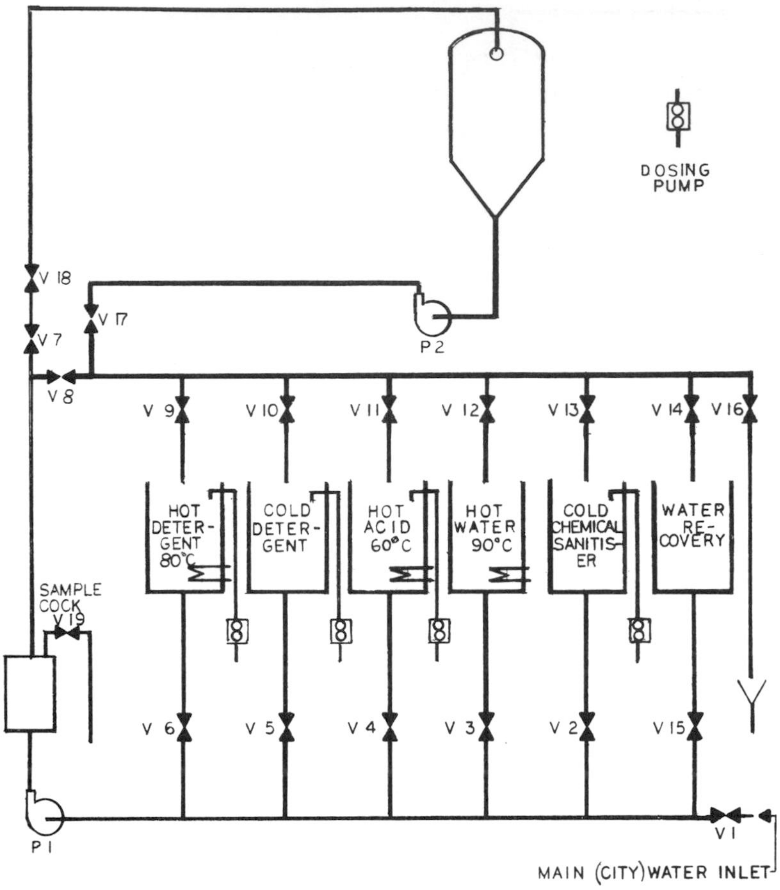

Figure 5.15 A typical re-use CIP system (from Tamplin[16])

discussed by Tamplin[16]. These led to the development of multi-use systems.

Decentralized re-use systems

Although many CIP systems have a single, centralized CIP unit there are advantages in using a decentralized system. The centralized system means that there is just one unit which, together with detergent supplies, can be located away from the process area. The long pipe runs for detergent and rinse supply and return mean that large volumes of liquid are involved with consequent heat losses and dilution of product and detergent. Timperley[18] in his review of CIP systems gives an excellent description of how decentralization improves the system design.

Essentially this means putting a pump, heat exchanger and small batch tank near to each group of equipment to be cleaned. This enables the centralized detergent supply to be retained but reduces the length of CIP pipework. This reduction

in pipework length means less water and detergent are used, heating requirements and heat losses are reduced and detergent dilution is minimized.

Multi-use systems

These modular systems are versatile for CIP sequence, detergent and sanitizer strength and temperatures. Tamplin[16] discusses the advantages of multi-use systems and gives a diagram of a typical system, shown here as Figure 5.16. He adds:

Reputable suppliers are usually able to offer:
1. 'No flow' inspection of the CIP fluid return, and associated audible/visual alarms.
2. Circular chart, 24h temperature recorder to visually display time, sequence and duration of all cleaning cycles.
3. Detergent neutralization facility – prior to discharge to drain.
4. Powder dosing equipment.
5. Automatic restricting or modulating valves to

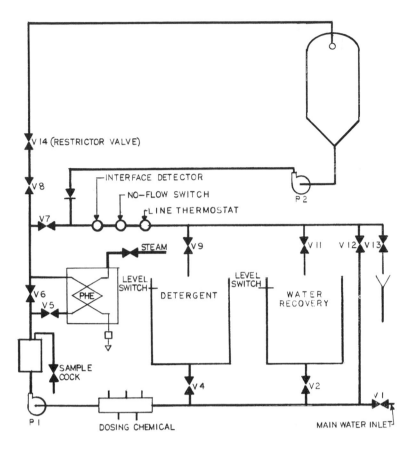

P H E   PLATE HEAT EXCHANGER

Figure 5.16 A typical modular multi-use CIP system (from Tamplin[16])

control initial flows of solutions at critical points in the CIP cycle.

6. Automatic detergent strength checking and control to monitor/display concentrations and indicate faults to the operator.

7. Air or water purge facilities.

8. Chemical test kits to enable the client to check and monitor cleaning efficiency for himself.

9. A range of spray balls and rotating devices to suit most vessels and tanks.

Since this was written, Programmable Logic Controller (PLC) developments have increased control flexibility and improved recording options.

The principle of a decentralized system can also be applied to multi-use systems. Optimization of design needs careful consideration of many factors and to get good results the design and installation of CIP systems to give good results needs expertise and experience.

'Tank' cleaning by spray balls and rotating jets

It is essential that the CIP liquids are distributed within the tank or other vessel. The cleaning effect is partly by direct impingement and partly by irrigation. The choice between spray ball and jet is one in which experience is needed both of the type and size of vessel and of the soil.

Tamplin[16] discusses the choice on pages 194 and 195. It is worth remembering that apart from standard sprayballs, it is possible to obtain specifically designed spray patterns to deal with difficult soil removal.

Other CIP engineering features

Tamplin[16] has an excellent discussion which should be read by those seriously interested in this subject. There are, however, two critically important design criteria which may be explicitly stated as:

- Absolutely positive and 'Fail Safe' separation of product and cleaning fluids. This has led, for example, to the rule that two valve faces *and* an air gap must separate product and cleaning fluid in adjacent lines. Reliance on pressure differences across valve faces which may leak under surge pulses is *not acceptable*.
- Positive assurance must be built into the design to show in each cleaning cycle that the sequence is operating correctly. This should be done in whatever way is appropriate to the system, e.g. by checking pump running time against a tank level probe, or in a more direct way, e.g. measuring flow, suction or output pressures. Additionally, it is important in complex PLC controlled systems to have the facility for easy confirmation of the sequence and timing (status) of operating commands.

Tamplin[16] gives highlights under the headings product safeguards, vessels, CIP feed and extraction pumps and pipelines. Considering each in turn:

- Product safeguards – this includes Key Pieces, Swing bends with microswitches (or proximity switches), Flow plates both manual and automatic, Safetybreaks (two valve faces and an air space) and layout of valve systems. Remember his comment: 'Considerable skill is required to achieve a satisfactory arrangement of valve combinations and orientations to insure that all valve seats are fully cleaned at some stage across the seat and at the same time permitting full drainage facilities within the system.'
- Vessels – This includes venting, which if not done correctly will cause collapse in a tank. He gives an example of what happens in a 22 725 liter tank (6000 US gallons, 5000 Imperial gallons) with 22.65 m^3 of air at 57°C after cleaning. If this is cooled by water sprayed at 10°C at 380 liters per minute, within 1 second the hot air shrinks approximately 1.45 m^3 – equivalent in occupied space to 1450 liters (383 US gallons; 319 Imperial gallons). This volume of air needs to enter the tank at approximately the same rate to prevent implosion of the tank.

Imholte[1] advises that spray assemblies should provide a minimum of 10 US gallons (38 liters) per hour per one square foot (9.29 dm^2) of surface.

He also discusses manways (manholes), agitators and baffles and the vessel slope and outlet. On slope, Tamplin[16] recommends a minimum, continuous floor gradient of 1 in 50 which for high solid loads may be increased to more than 1 in 33. He also states that 'The outlet size should be sufficient to provide optimum extraction. Typically, for a CIP extraction flow rate of l4 000 liters (3700 US gallons; 3080 Imperial gallons) per hour the minimum vessel outlet to be considered would be 50 mm (2 in) diameter.'

- CIP Feed and Extraction Pumps – These are usually centrifugal and for extraction 'abide by common sense engineering principles'. Tamplin quotes four:
  1. Pumps should be mounted as near to the extraction vessel as possible. A positive fall of 1 in 120 is considered essential on the pump suction.
  2. There should be a positive head between the tank outlet and pump suction.
  3. Starting the pump should be delayed to obtain a flooded suction.
  4. The net positive suction head (NPSH) should be calculated for each circuit.
- Pipelines – It is important to provide for visual and microbiological inspection at key points. Remember, if you can't see a surface how do you know that it is clean? This can be done by designing in a short length (or section) of pipe or a bend with convenient access in a hard-to-clean part of the circuit, e.g. a long horizontal section.

Warning: If a horizontal line is not filled during production and cleaning, heavy soiling is likely to be left on the top portion of the pipe.

Be sure that the size of pipe(s) and pump(s) is adequate for the proposed duty. Imholte[1] states that the minimum CIP solution circulation rate should be 38 liters (10 US gallons; 8.4 Imperial gallons) per hour for each 9.29 dm^2 (1 square foot) of tank surface to be cleaned. He also advises that large storage tanks of 11 355 liters (3000 US gallons; 2500 Imperial gallons) require a minimum circulation rate of 15 140 liters (4000 US gallons; 3300 Imperial gallons) per hour. It is important that during spray cleaning, solutions are extracted from the tanks fast enough to insure that the walls and base are completely washed. This may affect the pipe-sizing of part of a system. Romney[6] warns: *do not mix pipe sizes in the same circuit* and adds that in practice, a 12.7 mm (½ in) change, e.g. between 63.5 mm and 76.2 mm (2½ and 3 in) is the maximum acceptable when dealing with a heavy tenacious (dairy) soil. This statement gives an idea of the prudent margin of safety for design purposes but be aware of differences between soils.

Allowance must also be made for the expansion of pipework. Tamplin[16] states that typically a 30 m (98.5 ft) length of pipeline will expand about 25 mm (1 in) if the temperature rises from 20°C (68°F) to 80°C (176°F).

Be sure pipework slopes correctly to drain; aim for a pitch of 1 in 120. Pipework must be adequately supported to prevent sagging which allows solutions to lie static in the line. Imholte[1] suggests supports on 3 to 4½ meter (10 to 15 ft) centers.

Design out or minimize the number of dead-ends or tees since CIP cleaning will probably be ineffective in cleaning them. Flow directed into a tee is more effective than that away from a tee. Where a pocket or 'stub' tee has to be used, the rule-of-thumb is that it is likely to be acceptable if no deeper than one pipe diameter. (See Romney[6].)

With regard to CIP detergent flow velocity, it is the mean velocity of flow irrespective of pipe size which is important. Many years ago, work published by Unilever established the figure now commonly used in the food industry of 1.5 m (5 ft) per second as a minimum for hot CIP solutions. Flow dynamics in thin films, e.g. down a tank wall, are complex and very different from those in a pipeline. Payne[19] in the SDT Manual has an excellent account for the technologist of both of these in Chapter 6. It is important to remember that the nature of the soil, detergent formulation, temperature and contact time are also important factors.

Tamplin[16] states that: 'It has also been shown that there is little benefit in exceeding a flow velocity of 1.5 m per second (5 ft per second) under hot CIP conditions. At ambient conditions the CIP flow velocity can be increased to 2 m per second (6½ft) per second) to provide greater cleaning efficiency.' He illustrates the relationship between flow vel-

ocity, time and temperature in the diagram (Figure 5.17) taken from dairy industry experience.

Warning – This information is illustrative only. Do not use it to design your system without positive confirmation that it is fully applicable.

From this it may be concluded that successful CIP requires:

- Good equipment design
- CIP engineering based on experience of this equipment and soil types
- Good understanding of detergent action
- Good installation, commissioning, operation and maintenance

For it to be cost-effective, the true capital and running costs of alternative systems must be known.

For further reading on CIP systems in dairy plant, reference may be made to the SDT Manual[17].

Note on commissioning cleaning systems

Tamplin[16], who has considerable experience of CIP systems. states that:

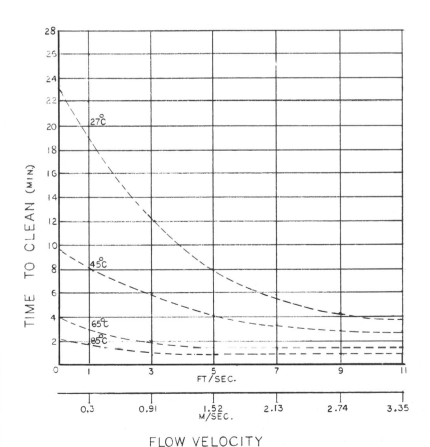

Figure 5.17 Flow velocity relationship versus temperature and time (from Tamplin[16])

In addition to any design, erection and project engineering responsibility, the client should insure that his contract includes terms for adequate commissioning.

As a basic requirement he should stipulate that the supplier undertakes to:

1. Insure that all equipment supplied under the CIP unit package is in good order.
2. Insure that the control panel is electrically and mechanically sound.
3. Check that all equipment designed to be operated or controlled from the control panel is functioning correctly.
4. Prove by means of a test program, that the programmer operates remote items correctly.
5. Produce a typical CIP program for a pipeline clean.
6. Produce a typical CIP program for *each* type of storage tank.
7. Train the customer to design suitable CIP programs.
8. Insure that the unit delivers the CIP fluids at the correct pressure to meet specification.
9. Insure that the unit delivers the CIP fluids at the correct rates to meet specification.
10. Insure that the return pump(s) meet specification regarding the rates.
11. Prove that the 'servicing' function of the CIP unit is operating satisfactorily, i.e., the filling, emptying, heating and dosing of chemical tanks is performed safely and automatically.
12. Supply the necessary manuals and documentation in the required language.

When commissioning complex systems which are PLC ('computer') controlled remember that:

- It is essential to check that the designed sequence occurs. With complex valve assemblies remote from the operating console, this needs careful pre-planning. This task is critically important – assumptions that 'it must be so because it is on the computer' can be very costly to later production.
- Experience shows that 'there is never enough time allowed for commissioning'. This means that for each phase the consequences of inadequate commissioning must be highlighted to management before commissioning starts. Done realistically, this should help planning and minimize both loss of time (which is money) and the frustration of wasted time.
- Expect that things will go wrong. Commissioning is a process of verification and making changes. Allowances need to be made for the consequences of this in both the cleaning and production programs.

Evaluation of cleaning systems

This is far more than a casual inspection of easy to see parts. When discussing assessment, ICMSF[14]

make the important distinction between 'monitoring' which is checking or inspection and 'verification' which is determining that the trend is satisfactory.

Monitoring usually means that all those parts of the equipment judged to be 'sensitive' or 'worst case' look, feel, and smell clean. Its value depend on the skill and experience of the inspector being able to interpret the findings in relation to the safety and quality of the food to be processed on the equipment. Unfortunately, apparent cleanliness can be misleading because 'clean' equipment may harbor microorganisms which affect the safety or quality of the food. Verification by microbiological testing shows whether the cleaning system is delivering what it is designed to do. It may therefore be regarded as an 'audit' which, if good records are kept, will show whether trends are developing which require remedial action. These data may also be used to establish sanitation criteria.

When considering the evaluation of a cleaning system, the HACCP approach is valid. Although not described as such, the HACCP strategy given in the American Frozen Food Institute Sanitation Manual[20] is to make a survey.

From this survey specific areas should be designated as sub-critical, critical, or highly critical for physical and bacterial contamination. These areas can thus be divided into those:

1. requiring continuous attention,
2. those requiring attention every two to three hours (at each break period),
3. every four to five hours (lunch break and end of shift),
4. every eight hours (end of shift),
5. daily, or
6. weekly.

When this strategy is used it is advisable to regard any doubtful areas as being in a 'more critical' category and then, when more results are available, adjust the category if necessary.

Whether used for verification or survey, the microbiological methods should be appropriate to the task, e.g. swabs, rinses, contact plates, or first product through the plant. ICMSF[14] make the valid and wise point that 'Microbiological samples must not be limited to flat surfaces that are easily accessible and permit measurements of a precise area. Verification programs limited to such locations can be misleading.' The rule must be to make tests where the information will be the most useful. This requires an understanding of the underlying microbial ecology. There is no substitute for this understanding which is fundamental to safe food processing. If necessary, equipment should be modified to enable checks to be made.

Another technique which has been used on liquid lines is to run chlorinated water, e.g. at 50 mg per liter (ppm), through cleaned lines to see if any loss of chlorine occurs. To supplement this information it may be appropriate to determine the microbial status in or on the food after all handling and preparation is completed.

Flow sheet sampling

Flow sheet sampling is an extension of end product sampling and is described by ICMSF[14] as:

measuring the microbial levels in or on food samples collected after each step in the sequence of preparation. When samples are collected of the first food coming into contact with 'cleaned' equipment, it is possible to measure the contribution of microorganisms from each piece of equipment that the food contacts. Flow sheet sampling is particularly useful for measuring hygiene in complex equipment and enclosed systems. The method of analyzing flow sheet samples depends upon the type and source of contamination that will most likely occur. For example, surface contamination would be expected for solid foods moving along a conveyor.

Environmental samples

Samples taken from the food processing environment are seen as being increasingly important in the control of pathogens such as *Salmonella* spp. and *Listeria* spp. ICMSF[14] states with regard to dry foods:

Results from the analysis of samples of fines that collect in the drying equipment and swabs from drains or dust from air filters and dust collectors are a means of verifying hygiene in an environment devoted to preparing dry foods (e.g. dried eggs, non-fat dry milk, chocolate). This is particularly useful for assessing the control of *Salmonellae* in a food processing environment.

Air intakes, buildings, e.g. overheads, walls, floors, drains, equipment and utilities e.g. air, water are all part of the environment.

Be aware of the importance and usefulness of environmental samples for any 'non-terminally sterilized food'.

Notes on assessment – general

It is prudent to remember that while theory, experience and perhaps small-scale trials have led to the choice of detergents and operating conditions, it is commissioning which must show that the system works under 'worst case' practical conditions. On the full-size plant, it will be more difficult to control detergent concentrations and temperatures at the point of application than under laboratory conditions.

Sometimes in a commissioning program it is difficult to arrange 'worst case' conditions on demand. Although these conditions must be used for an assessment the rule-of-thumb for cannery soils is that a 10 to 15% increase in detergent strength over that needed for 'average' conditions will usually deal with the 'worst case'.

Measurement of temperature is usually either from instruments already installed on the equipment or by hand-held electrical thermometer with sufficiently long leads. *Do not use mercury-in-glass thermometers* because of the contamination hazard. For the same reason, avoid the use of laboratory glassware or plastic which can shatter or break.

Tracking hypochlorite and alkali concentrations

Where alkali and hypochlorite are used, it is instructive during commissioning (or development) to track the usage of these materials throughout the cleaning cycle. In a large or complex system it is important to know where usage is greatest. This is easily done using standard chemical methods as follows:

- Hypochlorite (chlorine) Put 10 ml sample into a beaker, acidify (preferably with acetic acid) and add excess potassium iodide – a 200 mg or 1 g tablet is convenient. Titrate against 0.01N sodium thiosulphate until the color is yellow. Then add 4 or 5 drops of starch indicator solution and titrate slowly until the blue color is lost. The titre × 35 = ppm chlorine.
- Alkalinity. Put a 10 ml sample into a beaker and add 4 to 5 drops of phenolphthalein solution. Titrate against 0.02N hydrochloric acid until the violet-pink color is lost. The titre × 0.008 = % NaOH. This is convenient for low concentrations, i.e. 25 ml titre = 0.2% NaOH. For higher concentrations, O.IN hydrochloric acid and a factor of 0.04 may be used, i.e. 25 ml titre = 1% NaOH.

When problems arise

Assessment of cleaning is important first to develop and prove the cleaning system chosen for specific equipment and/or areas. Afterwards the monitoring and verification programs provide the necessary positive assurance that the system is in control. From time to time problems will arise. It will then be useful to remember that the words of J. G. Davis[21] on sanitation in dairies have wide application.

Whenever a problem in hygiene (sanitation) arises, one should always look for the simplest explanation, which is usually that someone has not done his appointed job, or the 'fallibility of the human element'. Common examples in the dairy industry are:

Failure of a supervisor to be present.

Failure of an operative to be there in time to perform a given operation.

Turn-over of labor resulting in misunderstanding, failure to instruct or train the new man – broadly speaking, lack of communication.

Change in method, e.g. sequence of operations, introduction of new type of detergent or sterilant, or even size of scoop etc. used for measurement.

More subtle causes are:

Assumption that a given cleaning/sterilizing method is effective because 'it must be' or 'always has been'.

Complete reliance on the advice of a salesman, often given in good faith, because 'it works in other dairies'.

Breakdown of a component and replacement by another which, because of different design, introduces a new problem.

Faulty repair of a broken down component.

References

1. Imholte, T. J. (1984) *Engineering for Food Safety and Sanitation*. Crystal, MINN: Technical Institute of Food Safety
2. Watrous, G. H. Jr (1975) Food soils, water hardness and alkaline cleaner formulations. *Journal of Milk and Food Technology*, **38**, 163–165
3. Tamplin, T. C. (1980) CIP technology, detergents, and sanitizers. In *Hygienic Design and Operation of Food Plant*, edited by R. Jowitt, pp. 183–225. Chichester: Ellis Horwood
4. Elliott, R. P. (1980) Equipment. In *Principles of Food Processing Sanitation*, edited by A.M. Katsuyama and J. P. Strachan, pp. 105–154. Washington, DC: The Food Processors Institute
5. Marriott, N. G. (1989) *Principles of Food Sanitation*, 2nd edn. An AVI book published by Van Nostrand Reinhold, New York
6. Romney, A. J. D. (1988) Designing for hygiene. *Food Science & Technology Today*, **2**, (4), 268–271
7. Thompson, R. (1971) Table 7: Effect of detergent materials on surfaces. In *Hygiene and Food Production*, edited by A. Fox. Edinburgh/London: Churchill Livingstone
8. Thorpe, R. H., and Barker, P. M. (1987 – publication continuing) Technical Manual No. 17. *Hygienic Design of Liquid Handling Equipment for the Food Industry*, The Campden Food Preservation Research Association, Chipping Campden, Gloucestershire GL55 6LD
9. National Canners Association (1973) Sanitation. In *Quality Control for the Food Industry*, Vol. 2, edited by A. Kramer and B. A. Twigg. An AVI book published by AVI Publishing Company, Westport, CONN, acquired by Van Nostrand Reinhold, New York
10. Lewis, K. H. (1980) Cleaning, disinfection & hygiene. In *Microbial Ecology of Foods. Vol.1 – Factors Affecting Life and Death of Microorganisms*, Inter-national Commission on Microbiological Specifications for Foods, New York: Academic Press
11. Jennings, W. G. (1963) An interpretive review of detergency for the food technologist. *Food Technology* (Champaign), **17**, (7), 873–881
12. Jennings, W. G. (1965) Theory and practice of hard surface cleaning. *Advances in Food Research*, **14**, 325-459
13. Koopal, L. K. (1985) Physico-chemical aspects of hard – surface cleaning 1. Soil removal mechanisms. *Netherlands Milk and Dairy Journal*, **39**, 127–154 (In English)
14. ICMSF (1988) *Application of the Hazard Analysis Critical Control Point (HACCP) System to Ensure Microbiological Safety and Quality, Microorganisms in Food 4*. International Commission on Microbiological Specifications for Foods, Oxford: Blackwell Scientific
15. Troller, J. A. (1983) *Sanitation in Food Processing*, New York: Academic Press
16. Tamplin, T. C. (1981) Cleaning In Place (CIP) systems and associated technology. In *Developments in Soft Drinks Technology* 2 edited by H. W. Houghton, pp. 139–199. Barking: Applied Science
17. Romney, A. J. D. (ed) (1990) *CIP: Cleaning in Place*, 2nd edn. Society of Dairy Technology, Huntingdon, Cambridgeshire PE18 6E2
18. Timperley, D. A., (1981) Modern cleaning and recovery systems and techniques. *Journal of the Society of Dairy Technology*, **34**, (1), 6–14
19. Payne, J. (1990) Fluid flow dynamics. In *CIP: Cleaning in Place*, 2nd edn, edited by A. J. D. Romney, pp. 104–121. Society of Dairy Technology, Huntingdon, Cambridgeshire PE18 6E2
20. AFFI (1970) *Sanitation Manual for Frozen Food*. McLean, VA formerly at Washington, DC: American Frozen Food Institute
21. Davis, J. G. (1971) Hygiene in the dairy industry. In *Hygiene and Food Production*, edited by A. Fox, p. 87. Edinburgh/London: Churchill Livingstone

Further reading

Holah, J. T., Timperley, A. W. and Holder, J. S. (1990) Technical Memorandum No. 590. *The Spread of Listeria by Cleaning Systems*. The Campden Food and Drink Research Association, U.K.

Seiberling, D. A. (1978) Process/CIP engineering for product safety. In *Food Protection Technology*, edited by C. W. Felix, (pp. 181–199). Chelsea, MICH: Lewis Publishers. Although mainly referring to liquid milk plants operation, the information is of value in all plants which process liquids.

5.D.4 Sanitizers

Introduction

This section deals with practical aspects. Theoretical background is found in Chapter 8.C.

Sanitizers (disinfectants) are usually considered to be chemicals, but for some equipment steam, either under pressure or at atmospheric (free) pressure, or hot water may be used. Remember that effective use of heat means that the required temperature must be maintained for at least the minimum specified time. Large pieces of metal, e.g. valves, will take longer to reach the required temperature than, for example, pipework. The relationship between instrument readings and microbiologically 'worst case' conditions must therefore be known. As ICMSF[1] state:

> The proper use of steam and hot water depends upon maintaining adequate temperatures for a sufficient length of time at the surface of the equipment being treated. In enclosed systems (e.g. milk processing equipment), covered tanks (e.g. starter culture vats, butter churns), tanks for immersing items to be cleaned, mechanical dishwashers and steam chambers, the time and temperature can be adequately controlled to provide disinfection. By contrast, attempting to disinfect equipment (e.g. conveyor belts) in a refrigerated work room by spraying steam or hot water from a hand-held hose is usually ineffective because it is not possible to control the time and temperature at the surface of the equipment to assure disinfection.

Of the chemicals, the five most commonly used are:

- Chlorine and chlorine compounds
- Iodophors
- Quaternary Ammonium Compounds (QUATS, QATs or QACs)
- Acid anionic surfactants
- Amphoteric surfactants

Others such as peracetic acid ($CH_3COO-OH$) and hydrogen peroxide are used for special applications mostly with aseptic processing or packaging equipment.

Comparison of sanitizers

Factors influencing the effectiveness of sanitizers include:

- Concentration and contact time
- Temperature at the point of use
- pH of the sanitizer solution
- Water hardness
- Presence of residual detergents
- The protection given to target organisms by residual soil(s)
- Numbers and types of microorganisms to be killed and the presence or absence of spores.

Apart from effectiveness of the sanitizer, other factors affecting selection include:

- Is there regulatory approval for the proposed use?
- Toxicity
- Corrosiveness
- Effect of residual sanitizer on the food being produced.
- Whether residual activity is desired or whether a (potable) water rinse is needed to remove it.
- Staining of equipment by the sanitizer
- Effect on environment and on effluent treatment plant.
- Cost

In making a comparison between commonly used sanitizers, ICMSF[1] omits reference to amphoteric compounds (See Table 5.20). Hugo and Russell[2] describe them as of mixed anionic–cationic character and state that their bactericidal activity remains virtually constant over a wide pH range and they are less readily inactivated by proteins than QACs. For practical purposes, they are considered to be wide spectrum, i.e. effective against Gram-positive and Gram-negative bacteria. They are also high foaming so are unsuitable for CIP applications. Tamplin[3] gives a list of sanitizing agents for CIP systems which includes a detailed discussion on sodium hypochlorite which 'is very popular because of its effective bactericidal spectrum and cheapness'.

There is a detailed discussion by Elliott of chlorine and its compounds[4]. Readers wanting further information will find useful material in Marriott[5] Chapter 6, Harper[6], Tamplin[7] (who refers to sanitization as 'sterilization'), and Davis[8].

Which sanitizer?

In practice, the choice is usually determined by the nature of the job to be done – what is to be killed? bacterial spores? yeasts? molds? – and the nature of the equipment to be sanitized, e.g. are there aluminum parts? or plastic that cannot be heated? It clearly makes good sense to use wherever possible a wide spectrum sanitizer, which will not select for particular kinds of organisms. This is because the arguments for using the minimum number of detergents apply also to sanitizers. However, it will usually be found that there are a number of 'special' applications which need a different material. This requires careful management, training and supervision because mistakes resulting from ineffective sanitization can be costly. As with detergents, theoretical considerations can be a good guide to a choice, but experimental evidence *must* be obtained to show that the choice is effective.

Useful guides have been published by, among others, Forewalter[9], and Marriott[5] who quotes Lentsch in a sanitation notebook written for the seafood industry. Harper[6] gives a list of manufacturers' recommendations for concentrations and time of exposure.

The information given by the these authors is intended for general guidance. There are other materials used for specific purposes, e.g. a polymeric biguanide with average molecular weight of

Table 5.20 Comparison of the more commonly used disinfectants (sanitizers) (from ICMSF[1])

	Steam	*Chlorine*	*Iodophors*	*QACs/QUATS*	*Acid anionic Surfactants*
Effective against					
Gram-positive bacteria (lactics, clostridia, *Bacillus*, *Staphylococcus*)	Best	Good	Good	Good	Good
Gram-negative bacteria (*E.coli, Salmonella*, psychrotrophs)	Best	Good	Good	Poor (can select for *Pseudomonas*)	Good
Spores	Good	Good	Poor	Fair	Fair
Bacteriophage	Best	Good	Good	Poor	Poor
Properties					
Corrosive	No	Yes	Slightly	No	Slightly
Affected by hard water	No	No	Slightly	Type A – No Type B – Yes	Slightly
Irritative to skin	Yes	Yes	Yes, some people	No	Yes
Affected by organic matter	No	Most	Somewhat	Least	Somewhat
Incompatible with:	Materials sensitive high temperatures	Phenols, amines soft metals	Starch, silver	Anionic wetting agents soaps, wood, cloth, cellulose, nylon	Cationic surfactants and alkaline detergents
Stability of use solution	N/A	Dissipates rapidly	Dissipates slowly	Stable	Stable
Stability in hot solution (greater than 66°C)	N/A	Unstable, some compounds stable	Highly unstable (Best used below 45°C)	Stable	Stable
Leaves active residue	No	No	Yes	Yes	Yes
Tests for active residue chemical	Unnecessary	Simple	Simple	Difficult	Difficult
Maximum level permitted by USDA and FDA w/o rinse	No limit	200 ppm	25 ppm	200 ppm	a
Effective at neutral pH	Yes	Yes	No	Yes	No

a 400 ppm dodecyl benzene sulphonate, 200 ppm sodium salt of oleic acid.

Notes 1. Comments in parentheses are additional.
2. Unilever research concludes that EDTA added to QACs and amphoterics synergizes the disinfectant and prevents development of resistance.
3. Comparative costs have been omitted from this table as they vary locally.

3000 which Hugo and Russell[2] state 'because it is not a surface active agent, it can be used in the brewing industry as it does not effect head retention on ales and beers'.

It is worth remembering that new materials and formulations are being developed and that these necessarily will not appear in the older literature.

Balanced against this, a material is not necessarily 'better' because it is 'new'. It should be kept in mind that, as with detergents, effective management is easier with few different kinds.

Manufacturers' recommendations carry an implied assurance that if followed unwanted effects, e.g. corrosion, will not occur. For this reason they will be conservative recommendations and the user, on his own responsibility, may well use other conditions in his plant. Tamplin[3] gives an example of this on pages 180–181 for (alkaline) hypochlorite-based sanitizers on stainless steel. Provided that thorough rinsing is given, *some* easing of concentration, temperature and time conditions may be made. This is in line with cannery plant cleaning experience where, with the correct grade of stainless steel, alkaline hypochlorite-detergent mixtures can be used with an initial concentration as high as 150 to 300 mg per l (ppm) chlorine.

The optimal choice of sterilizer (or detergent) is made by taking the most experienced advice available for the product range and equipment concerned and keeping the choice under microbiological and engineering review.

A note on detergent sanitizers. These are usually for lighter duties, e.g. for hand or dish washing or for use with small utensils or machine parts. They are classified into three types by Elliott[4]:

- Alkaline detergent, e.g. sodium carbonate, trisodium phosphate or tetrasodium pyrophosphate with a Quat and non-ionic wetting agent.
- Acid detergent, e.g. sulfamic or phosphoric acids.
- Ampholytic (amphoteric) compounds.

Although sometimes called detergent sterilizers this is not a true description and detergent sanitizers is the better description.

References

1. ICMSF (1988) *Application of the Hazard Analysis Critical Control Point (HACCP) System to Ensure Microbiological Safety and Quality, Microorganisms in Foods 4*, International Commission on Microbiological Specifications for Foods, Oxford: Blackwell Scientific
2. Hugo, W. B. and Russell, A. D. (1982) Chapter 2. Types of antimicrobial agent. In *Principles and Practice of Disinfection, Preservation and Sterilization*, edited by A. D. Russell, W. B. Hugo, and G. A. Ayliffe. Oxford: Blackwell Scientific
3. Tamplin, T. C. (1981) Cleaning in Place (CIP) systems and associated technology. In *Developments in Soft Drinks Technology – 2*, edited by H. W. Houghton, pp. 139–199. Barking: Applied Science
4. Elliott, R. P. (1980) Cleaning and sanitizing. In *Principles of Food Processing Sanitation*, edited by A. M. Katsuyama and J. P. Strachan, pp. 74–77. Washington, DC: The Food Processors Institute
5. Marriott, N. G. (1989) *Principles of Food Sanitation*, 2nd edn. An AVI book published by Van Nostrand Reinhold, New York
6. Harper, W. J. (1980) Sanitation in dairy food plants. In *Food Sanitation*, 2nd edn, edited by R. K. Guthrie. Westport, CONN: AVI Publishing
7. Tamplin, T. C. (1980) CIP Technology, detergents and sanitizers. In *Hygienic Design and Operation of Food Plant*, edited by R. Jowitt, pp. 183–225. Chichester: Ellis Horwood
8. Davis, J. G. (1968) Chemical sterilization. In *Progress in Industrial Microbiology, Vol. 8*, edited by D. J. D. Hockenhull. London: J. and A. Churchill
9. Forwalter, J. (1980) 1980 Selection guide cleaning and sanitizing compounds. *Food Processing*, **41**, (2), 40–44, 46

5.D.5 Notes on exceptional cleaning requirements

There are occasions when good sense indicates that the 'normal' cleaning system is not sufficient. Examples are:

- After major alterations or refurbishing of equipment or environment.
- After an incident in which the existing system may have been inadequate.

Clearly, the first thing to be checked is that the cleaning system was properly applied. However, problems may remain. While each instance is unique there are some general principles worth remembering.

After major alterations

Accept as inevitable that there will be, in both equipment and environment, contaminants which will harm equipment and product. These include swarf, builders' material, rubble, dust, small pieces of wire, unused nuts, bolts, washers, etc., small and large pieces of plastic, cigarette ends, cloth, fibers, and debris brought in on footwear. Although these may not in fact be present, it is prudent to plan as if they were. Remember that it has been known for:

- Drains to be blocked with cement
- Motors to have been cross-wired so that they operate in the reverse direction
- Pipe and valve connections to be made incorrectly, and
- Many other 'impossible' things to occur.

The major concern is to protect pumps and other equipment from damage caused by debris which may be swept along the line, and to remove contamination from contact with the product.

The usual strategy is:

- Clean environment and 'overheads', e.g. beams, light fittings, ledges, where nuts, bolts, etc. may have been left.
- Isolate pumps, tanks and other major items and manually clean using hoses, then inspect thoroughly. Further cleaning, e.g. of paint spots, may be needed.

Run water 'full bore' through pipelines dropping to waste on the floor. Remember that debris may be in the lines, so do not operate pumps unless there is protection, e.g. by a sieve immediately upstream.

- If pumps are not used, and pipework does not drain naturally to the floor, disconnect lines at lowest point and flush with hose to remove debris.
- Clean the equipment at least twice using the normal cleaning system.

Carefully inspect or discard the first product through the line on start-up. Remember, it is not unknown for a viscous or 'thick' product to dislodge debris that thin liquids such as water or detergents failed to move.

If the cleaning system seems inadequate

It is always hard when under pressure to go back to 'first principles' but in reality this is likely to provide the quickest long-term solution. Remember the old saying that time spent in reconnaissance is rarely wasted. This means careful inspection of the whole environment and line including the 'hard-to-get-at' parts *before* taking decisions on remedial action. Always *remember – the first thing found to be wrong may not be the only thing wrong*.

Check out CIP velocities as well as the functioning of equipment such as spray-balls. This is additional to checking the sequences.

There are a large number of possibilities in a 'real life' situation but it is worth remembering that if the existing cleaning system is inadequate then simply repeating it more frequently may *not* resolve the problem.

Strategies worth considering are:

- Increasing the detergent contact time.
- Increasing the detergent strength, e.g. to double 'normal' usage, and consider whether an alkaline hypochlorite mix (using 300 ppm chlorine) could be used at least as a 'one off'.
- Increasing the detergent contact temperatures. Some chemical reactions double in speed for a 10°C rise in temperature.

These strategies can be used singly or in any combination according to the problem.

Always check thoroughly the physical and micro-biological status of potential problem areas after each 'emergency' clean. This information is needed before making permanent changes to the cleaning system.

Note that where there is a problem of inadequate disinfection (sanitization) the use of heat, as a 'one off' temporary strategy or a permanent change, may be the only satisfactory solution.

5.E Foreign material control

5.E.1 Introduction to the problems

Foreign material may be defined as objectionable material which should not be present. It may be derived from the ingredients, e.g. stones in legume seeds (pulses), or from processing or packaging, e.g. oil or grease, cardboard fibers.

In reality, foreign material from the ingredients will inevitably be present at some low level. The tasks of the processor are first, to reduce this low level to that minimum imposed by the 'state of the art' of control measures applicable to his products and second, to minimize the addition of foreign material during processing, packaging and, where applicable, distribution. To do this, it is helpful to understand the extent of the problems and the means of solving them. What follows can only be an overview of a complex situation which each processor has to resolve in his own individual circumstances.

Published figures from a retail chain in the UK

Heinz Corporate Policy requires the production of clean, wholesome food in a sanitary manner – which means that foreign material (foreign bodies, UK) must be controlled. Published information by Robson and Barnes[1] for complaints received in 1977 by Marks and Spencer (a retail chain) showed that of the 13 477 received 6289 were due to defects in equipment (machinery, UK) either because of the way in which it was operated or from its design. 1166 were due to insects, some of which resulted from poor factory or machinery design. An analysis of the 7455 complaints is given for:

- Bread
- Biscuits (cookies and crackers)
- Cake
- Confectionery
- Dairy products
- Delicatessen
- Meat pies
- Savory snacks

which are summarized as percentages of the total in Table 5.21.

Other published UK data

It is interesting to compare these figures to those given by Arthey and Reynolds[2] (see Table 5.22) which were recorded by the regulatory authorities

Table 5.21 Foreign body complaints due to all causes (Marks and Spencer annual figures for 1977)

	%		%
Dirt	9.7	Fibers	3.9
Insects	8.7	Foreign dough	3.8
Metal	8.3	Oven blacks	3.1
Plastic/rubber	7.1	Stalks, thorns	2.5
Human hair	6.3	Glass	1.9
Oil/grease	5.6	Dirty dough	1.9
Stones	5.6	Paper	1.6
Bones	4.9	Bristles	1.5
String	4.4	Cigarettes/matches	0.3
Wood	4.2	Others	14.7
		Total	100.0

Note: Robson and Barnes do not quote complaint rates, but simply numbers and percentages of complaints received in different categories

Table 5.22 Incidence of contaminants (numbers)

	1978	%	1979	1980	1981	1982	1983	1984
Metal	108	20.7	101	105	145	105	115	109
Insects[a]	121	23.2	115	99	132	101	110	117
Glass	52	9.7	43	31	52	48	36	48
Animal	33	6.3	36	25	37	24	24	28
Dirt/debris	32	6.1	23	37	26	28	21	42
Fabrics	20	3.8	25	21	22	22	16	18
Rubber°	7	1.3	4	6	–	16	5	–
Tobacco	37	7.1	15	14	18	12	20	21
Plastics	7	1.3	10	6	11	6	8	12
Wood (matchsticks)	10	1.9	10	9	7	8	9	15
Paper	10	1.9	5	6	10	6	19	16
Oil (rust)	12	2.3	6	–	5	4	5	3
Assorted	73	13.8	70	58	55	39	42	115
TOTAL	522	99.4 due to rounding off						

[a] Includes larvae, slugs and snails

°1984 – category dropped and 'stones' included for the first time.

for the period 1978 to 1984. For comparison with Table 5.21. the complaints for 1978 are also shown as a percentage of the total for that year. Together these give a picture – although incomplete – of the size of the problem and its complexity.

It must be remembered that the assessment of foreign material complaint rates depends both on the accuracy of recording and classification, and also upon whether or not the consumer finds the material and then complains when it is found. For this reason, the foreign material (body) rates quoted by Arthey and Reynolds[2] of:

- 1 per million for soft drinks in glass
- 2 per million for manufactured dry goods
- 17 per million for imported canned fruits
- 10 to 30 per million for recipe products

should be regarded as an approximate consensus or 'order of magnitude' figure of complaints known in the UK in the early 1980s and *not* used as a target or an excuse.

Rationally, a business should determine the appropriate categories of foreign material likely to be found and then the rates at which it is willing to trade. When this has been established, the cost of equipment, personnel and systems to deliver these rates is justified. However, the desire simply to maintain existing rates, or to improve them by an arbitrary amount; or the emotional response to a specific contamination problem, inhibit or prevent a logical approach. Be aware that in a recall or near-recall situation the response to publicity may tend to overwhelm logic.

Anderson and Dennis in their paper 'The Systematic Approach to the Reduction of Foreign Bodies (materials) in Foods'[3] define a 'foreign' material as 'anything in a food not of the nature or substance of the food *expected* by the consumer'. Using this definition, control of foreign materials requires detection and extraction or removal from the food. To be effective, as well as using appropriate equipment which is correctly installed, maintained, operated and calibrated, it also requires organization (systems), and personnel who must be trained, vigilant, motivated and supervized. Systems must also be developed to use calibration and performance data to modify procedures and practices as necessary in order to achieve and maintain effective control.

Principles of control

The equipment employed for foreign material control will detect a difference in physical properties between the foreign material and the product material in order to separate them. This may be shape, size, color, density, magnetic properties or opacity to X-rays. Additionally, barriers are used to prevent the foreign material contaminating the product. These may be physical, e.g. the use of lids on tanks, or spatial, e.g. the separation of 'clean' and 'dirty' materials in a meat operation.

Control of foreign materials requires a systematic hazard analysis of product and process, including cleaning and operating environment. It is therefore a specific study. What follows here is an overview together with some comments from experience which it is hoped will be helpful. This is given under the headings:

- Means of control
- Note on establishing standards
- Notes on application of control methods

5.E.2 Means of control

For any specific material there may often be alternative methods for detection and removal of foreign materials. The optimal choice and the equipment used depend on specific situations. Some commonly used methods are given in Table 5.23 for some usual contaminants. One item of equipment may, of course, use more than one method, e.g. vibration plus magnets.

It is not always appreciated that the best strategy with ingredients is to remove the foreign material at an early stage of processing. This is because processing often includes reducing the size of both wanted and contaminant materials, making the task of foreign material removal more difficult.

Effective use of foreign material control equipment requires:

- An understanding of the basic scientific principles which make it work.
- Expert knowledge of the calibration, use and cleaning of the individual item of equipment. (This is clearly outside the scope of this chapter.)
- An effective system for recording calibration, use and performance of the equipment.

This may be illustrated by considering devices in common use – magnets and metal detectors.

Magnets

Imholte[4] has an excellent Chapter 11, 'Foreign Material Control', which should be consulted by those interested in equipment application of magnets, metal detectors, scalping and sifting or aspirators. His concern is primarily with dry materials but his information is widely applicable. He points out that 'Magnets can also be helpful in protecting facilities that contain explosive food ingredients'. Note that any *very finely divided* organic material, e.g. sugar or flours, may be explosive. He continues 'By catching metal contaminants before they reach the explosive, dusty

Table 5.23 Control of foreign materials

Material	Visual [1] sorting belt or table	Vibration	Sifter (dry) or Strainer (wet)	Air separation	Water separation	Color sorting	Magnets	Metal detector	X-ray [2]	Fluidized bed
Dirt/earth				+		+				+
Insects			+	+						
Tramp metal	+						+	+	+	+
Ferrous metal [3]							+	+	?	+
Non-ferrous metal								+	?	+
Plastic/rubber	+			+					?	
Stones	+	+	+	+	+	+			?	+
Bones	+				+				?	
Paper/string	+		+	+						
Wood	+		+	+		+				
Stalks/thorns	+		+	+						
Glass [4]	+					?			+	+
Bristles	+		+							
Foreign seeds	+					+				

+ = usual, ? = possible use

Notes:
1. Light must be good, color of belt/table may be used to enhance discrimination or minimize eye strain. Operators must have good vision with spectacles if needed. Rotation of inspection duties is usual to avoid fatigue.
2. Image analysis by electronic means seems likely to increase X-ray application.
3. Not all stainless steels are magnetic.
4. Glass fragments in jars may be detected by on-line, high-speed equipment.

atmosphere found in some equipment, magnets can prevent ignition by sparking'.

As an example of the working out of basic principles, consider two which apply to magnets. The first is that the strength of a magnet falls off very rapidly with distance. It lessens according to the 'inverse square law', i.e. it varies with the inverse square of the distance from the magnet. Thus in the design of a 'pipeline magnet' the food, e.g. sauce or soup, is spread out in a thin layer over the magnetic surface to maximize efficiency.

The second principle is that magnetic materials are attracted with a force proportional to their size, thus large pieces of metal are more easily removed than small ones, e.g. staples or filings. Pipeline magnets are therefore usually installed in a series or 'cascade' so that the turbulence of the liquid flow brings particles near the surface of one of the magnets thus increasing the chances of trapping even small metal pieces.

When magnets are installed in either solids or liquids remember that they have to be inspected regularly, perhaps several times during the day or shift. Access to the magnet and ease of opening are therefore important. It should not be necessary to have to crawl under or over equipment in order to gain access.

Imholte[4] has useful advice for removing fine (small fragments of) metal contamination:

Fine metal contamination is difficult to remove from most magnets, particularly from grate magnets. A good sticky masking tape is one of the best ways to remove fine metal contamination from any type of magnet.

Making a record of contamination found

When metal contaminants are found, a record should be made in a log book. Remember that pieces of loose paper on which records were made tend to be easily lost or mislaid.

Do not attempt to use the log book for more than one specific compact area or department. Remember, log books are cheap and that logged information becomes valuable when it is used. It is usually best to use the log book to record by exception, i.e. only when something is found. This way it is easier to detect a developing problem or malfunctioning of upstream equipment. Logging systems which record results every time the magnet is checked tend to obscure a trend because most entries will be 'OK', meaning that nothing was found.

To record that the scheduled inspection has been done a check off sheet may be used. This should include the magnet location, the date and time of inspection and the identity, e.g. initials or clock number of the inspector. For ease of operation, the location may be pre-printed but it is most unwise to pre-print times so that the inspector only has to mark or tick a 'box' thus showing that the inspection has been done. Circumstances may well mean that the actual time of inspection is vitally important information.

Trapped material should be placed in a small envelope or bag for further examination. This may be useful when trying, for example, to reconstruct a dicer blade which has broken. It may also be useful to weigh the material when making an assessment of the amount of contamination.

Imholte[4] writes this on inspection and cleaning of magnets.

- Frequency. The frequency with which magnets are inspected and cleaned depends on two important factors; the food product and the location of the magnet.
- Food Product. If the food product is a raw commodity such as grain, the contamination level may be more suspect. Raw commodities have greater exposure to contamination in the fields and in the transporting process. Magnets used on raw commodities should be checked and cleaned often.

As a rule, magnets used on processed ingredients do not need to be inspected as frequently. Experience should establish inspection and cleaning frequency.

- Location. Magnets located on the head end of a system where incoming ingredients or raw commodities are brought into the plant for the first time should be inspected after every lot is unloaded. In the case of raw commodities, it might be necessary to make very frequent checks.

He adds the reminder that:

Magnets protecting equipment such as knives, dies, and grinders should also be inspected frequently. Trapped contaminants should be removed so they are not washed from the magnet back into the equipment.

Magnets need to be marked with an identification number so that they may be calibrated or checked to confirm that they have at least the specified minimum holding and pulling powers. This should be done at appropriate regular periods and the results logged against each magnet.

Both magnets and metal detectors are needed for an anti-contamination program. Magnets will not trap non-magnetic material, e.g. non-ferrous metals, and metal detectors have size and sensitivity limits.

Imholte[4] gives an excellent description of how metal detectors work in his Chapter 11, (page 254 onwards) which should be consulted by those interested in more information.

It is important, however, to understand that there are a number of factors which influence sensitivity. Among these are:

- Properties of the product in which the metal is suspended. If it is a good conductor, e.g. has a high moisture and salt content, then the product itself will generate a signal. This can be reduced by use of a lower frequency.
- Conductivity of the object, its dimensions and orientation in the product stream.
- Size of the aperture in the detector. Imholte[4] gives as an approximate guide – a 1 mm sphere is sensed in a 102 mm (4 in) aperture and 2 mm in one of 152 mm (6 in).
- Background interference or 'noise' from circuitry, product, or mechanical vibration will limit sensitivity.

Be aware that vibration may come from a belt conveyor going through the detector and that electro-magnetic induction can come from metal shafts, pulleys or bearings being too close to the search head.

Liquid products, e.g. soups, are taken through the search head in a suitable non-metallic pipe.

Metal detectors

Performance

Since a metal detector removes contamination by rejecting contaminated material, it is essential that both the search head and reject mechanism are proved to be working correctly. This means that a suitable test routine must be applied at appropriate intervals, e.g. before start, at middle and end of shift.

With regard to performance, Imholte[4] states that 'consistent rejection of a 3/64 in (1 mm) diameter steel sphere is not too much to expect of an efficiently operating unit'. The USDA Meat and Poultry Equipment MP1–2[5] offers a 'classification standard' on a voluntary basis. This is based on detection of at least 9 out of 10 passes of a '316' stainless steel sphere of specified size. Grade 'A' is when a sphere of 0.8 mm (1/32 in) is detected.

These criteria give an idea of what may be expected, but the processor will need to adopt a detailed test methodology, which should include specifying how the sphere is mounted in a plastic wand and the way in which it is passed through the search head with or without product being present. Remember to take the nature of the product into account, e.g. high moisture and salt products affect the performance.

When this is done, the operation of the reject device is observed. This can be, for example, a flow diversion valve for liquids in a pipeline, a plough on a conveyor belt or a package marker or displacement device.

As with magnets, it is essential to keep a log book of operation (or 'triggering') which may be combined, with records of calibration during the shift. Any reject material should be examined.

Note on sieves or strainers

These are valuable means of contamination control. However, careful and appropriately frequent inspection must be made to check that they are intact. If they are not, they should be judged to have been ineffective since the previous inspection and the missing piece(s) *must* be regarded as serious foreign material contamination. Where line strainers of different diameter perforations but of rather similar appearance are used, mistakes will be made unless the different types are clearly identified.

5.E.3 Notes on establishing criteria

At some time criteria of 'freedom' from foreign material will be needed. It needs to be recognized that assessing low or very low levels of contamination is difficult. Statistical advice will be needed and it is very important to get the best available. Unless the statistician is familiar with the problem the greatest difficulty is not mathematical, but to achieve mutual understanding of the situation under discussion.

Assessing the amount of contamination

The first problem is to assess the amount of foreign material. It is important to know the accuracy and repeatability of the method. The use of 'marked' contamination, e.g. stones colored to make them easier to distinguish from blanched beans, needs to be done with care. As an experimental tool it is useful and valid. It is unwise to use it to test the performance of an operation in the removal of contamination when producing product for sale. For this purpose a representative selection of 'current contamination' is preferred.

Assessing the distribution of contamination

The second problem is to understand how the foreign material is distributed in the ingredient or product. For example, in a specific lot it may always be there in greater or lesser amounts and be distributed at random. An example of this is stones in legume seeds (pulses). Note, however, that some seasons and some regions of production may be better than others at any one time. As another example, the foreign materials may occur sporadically and in a particular way, e.g. metallic contamination due to machine wear. This information on distribution is needed to develop the Sampling Inspection Scheme which comprises the rules determining the sampling procedure.

It is also important to be certain that the criteria developed are based on sound experimental work and realistic sampling. To take an absurd example,

it is foolish to specify '*n* per tonne' if it is impractical to take and examine more than 50 kg of material and where 50 kg in fact is the quantity used for test purposes. A good criterion will therefore have:

- A clearly stated Sampling Inspection Scheme. (Rules for taking the sample.)
- Clearly defined test method(s).
- A numerical standard which is sound statistically.

It will *not* have statements such as:

- 'in accordance with best commercial practice'
- 'be free from all *x*'
- 'as far as practicable be free from *x*'

5.E.4 Notes on the application of control methods

The words of Anderson and Dennis[3] should be remembered whenever the control of foreign material is being considered.

> However good detection methods may be, they are never perfect, nor for that matter do they exist for all materials in all circumstances. Similarly preventative methods are not completely effective, but every foreign body prevented from entering the production chain represents a 100% success and that cannot be bettered by any system of detection.

These words put detection and removal into their content. What follows is not a treatise on application but a few brief examples of things which may be overlooked when particular applications are being considered.

- Personnel
- Ingredients
- Preparation and Manufacture
- Packaging

Personnel

People may be a source of foreign material as well as being an important means of its removal. As Anderson and Dennis[3] state:

> It is amazing what can drop off people, anything apparently from diamond earrings to chips of nail varnish. Writing instruments from supervisory staff can slip out of a pocket unnoticed. First Aid dressings are not always immediately missed by the shedders of them and the fact of the matter is, that cigarette ends and a whole variety of other items are not unknown in manufactured foods.

Problems may be caused unintentionally by thoughtless actions. For example, laboratory glassware is perfectly satisfactory in a laboratory but should *never* be used in ingredient storage or production areas because of the contamination hazard. This is also true for plastic materials which may shatter or break.

What is needed to minimize the chances of personnel being the source of foreign materials is:

- Provision of 'the right equipment for the job'
- Training in 'the right way to do the job'
- Motivation to 'do the job the right way first time' – which includes explaining the consequences of doing it in the wrong way.
- Self-discipline and supervision to insure that 'the rules are obeyed'. This means, for example, that when protective clothing such as hats and coats should be worn, top management and visitors obey the same rules as personnel in the department.

Ingredients

By definition, if the wrong ingredient was used, then the wrong ingredient is 'foreign material'. This may have serious consequences as Troller[6] observes:

> Many white crystalline materials or powders may appear identical yet be vastly different chemically and toxicologically. A supplier of a perfectly wholesome ingredient for food use may also manufacture toxic chemicals of identical appearance for other purposes, for example, pesticides. The chance that a mixup could occur is very real if labeling were to be confused, ignored, or incorrect. As a result, many companies now utilize white powder tests, which consist of simple chemical tests that provide some assurance that a raw material is what it is supposed to be.

The chances of mistakes being made are also clearly higher if human food and animal food or non-food grades of the same material are made and packaged in the same plant. An important control measure in such instances would be to ask the producer to demonstrate the systems that are used to prevent mistakes and then to make a reasoned judgment on their adequacy.

Not all foreign materials are the expected ones. Troller[6] gives as an example a partial list of objects found in food-plant raw materials:

Umbrella handle
Tennis and golf balls
Hat
Rubber boot
Half of one Bible
Glass beads
Miscellaneous pieces of farm machinery
Pair of eyeglasses
Athletic supporter
Typewriter

To this may be added a passport, dentures, riding spur (nearly complete), etc. However, these unusual objects should not divert attention from the more likely ones.

Plastic wrapping of frozen blocks of meat can present foreign material problems. This is caused by entrapment of the plastic when the meat is unwrapped for manufacturing use. Ideally, the plastic would be colored for easy recognition and be of a sufficiently thick gauge to resist tearing.

Given that a level for a specific foreign material in an ingredient has been established, e.g. stones in legume seeds (pulses), there are two ways in which this may be achieved. Either the supplier applies controls or the buyer must. *Note*: *These are not mutually exclusive.* It is usually considered that the lower the level of foreign material which is acceptable, the more likely it is that controls will be applied 'in-house'. This leads to greater confidence and assurance that the control measures have been applied in a way that is fully known. In a situation where the statistics of sampling mean that a very large amount of material has to be checked to gain meaningful assurance, such a policy is understandable.

When dealing with a supplier in order to achieve a specific level one may be told 'other people are happy with it as it is'; the implication being that one is unreasonable to ask for something different. One answer is that 'it's what we need to buy'.

When auditing an operation where visual sorting is done, especially if it is done on piece-rates, it should be remembered that the inspector or 'checker' involved may well be under pressure by the group. These pressures may be not obvious. A system of measuring *individual* performance, e.g. weighing bone from diced chicken meat, can do much to reduce pressures. Measuring the performances of the *group* as a whole is not as useful as it averages out what may be importantly variable individual efficiencies.

Preparation and manufacture

Both buildings and equipment may be the sources of foreign material. Robson and Barnes[1] discuss this in some detail in their Chapter 7 which is well worth careful study. Obvious examples of foreign materials are flaking paint from ceilings and excessive wear on maladjusted machine parts. Less obvious examples are:

- Material picked up on the bases of buckets or bins which are then tipped into process vessels. Standing utensils on wet or damp floors increases this risk.
- Use of unprotected mercury-in-glass thermometers in ingredients or products

- Nuts, bolts, springs, clips, etc. which work loose from equipment
- Pipeline gaskets cut or damaged because of poor alignment prior to tightening the coupling
- Damaged or broken sieves or strainers
- Oil or grease from bearings and hydraulic fluid losses from overhead lines getting into ingredients or products
- Threads of cleaning cloths or pieces of (misused) paper towels

This list could be continued, but these examples should be sufficient to make the point.

The processor will, of course, take positive steps to avoid contamination. These include covering bins, vats and open food or container (can or jar) conveyors. This will avoid both metallic and non-metallic contamination.

Additional and stringent precautions have to be taken when handling glass containers. These precautions take account of 'worst case' effects of glass breakage and include the provision of special 'goose neck' dump bins.

Sometimes the inherent design of equipment will cause a problem unless it is operated in a specified way. Consider a pump with a metal 'worm' working inside a rubber material stator. When wet, minimal wear occurs but if run dry the stator may become broken, shedding pieces into the product. The most effective control to apply is to use an 'anti-dry running' system which prevents operation when there is no liquid feed to the pump.

Packaging

As Troller[6] in his interesting chapter on packaging sanitation states, 'packaging is the forgotten ingredient' in food. It may be the source of foreign material in the product if, for example, insect or other contamination is present, such as glass fragments from or in a glass jar. If the packaging is defective then leakage outward or inward may occur. This could result in chemical changes, e.g. in color, as well as microbial contamination. Oxygen, in this context, is to be regarded as being a foreign material.

Even with closed containers there is the possibility of deliberate and malicious interference. This criminal activity is difficult to prevent but increasingly 'tamper-evident' features are designed into the packaging system to make such activity obvious. This is an example of a specialized and unusual form of foreign matter control.

References

1. Robson, J. N. and Barnes, G. (1980) Plant design to minimize non-microbial contamination, In *Hygienic Design and Operation of Food Plant*, edited by R. Jowitt, pp. 121–135. Chichester: Ellis Horwood
2. Arthey, D. and Reynolds, A. (1985) Foreign bodies in foods – a review of the data. In *Proceedings of a Symposium held on Wednesday, June 19, 1985*, Campden Food Preservation Research Association, Chipping Campden, Gloucestershire GL55 6LD
3. Anderson, K. and Dennis, P. (1985) The systematic approach to the reduction of foreign bodies in foods. In *Proceedings of a Symposium held on Wednesday, June 19, 1985*, Campden Food Preservation Association, Chipping Campden, Gloucestershire GL55 6LD
4. Imholte, T. J. (1984) *Engineering for Food Safety and Sanitation*. Crystal, MINN: Technical Institute of Food Safety
5. USDA (1990) *Accepted Meat and Poultry Equipment MP1–2*, FSIS Directive 11220.1. Washington, DC: US Government Printing Office
6. Troller, J. A. (1983) *Sanitation in Food Processing*, New York: Academic Press

Chapter 6
Personnel

6.A Hygienic practices

Introduction

'Hygienic practices', 'hygiene', and 'sanitary' are words used in legislation, regulations and guidelines concerned with production of safe food but are seldom fully defined. The implication is that 'everyone knows what is intended'. The definition of hygiene, however, has two aspects:

1. The presentation or promotion of health.
2. Sanitary science.

The *Oxford Dictionary* definition of Sanitary is 'pertaining to conditions affecting health *with reference to cleanliness and precautions against infection*'.

The italicized words form the basis of the hygienic procedures which must be applied to food manufacture, including the personnel working in the factory or plant. However, there are two factors contained in the word 'health'. The obvious one when considering food safety is that the food which leaves the plant must not cause illness to the consumer. When applied to plant personnel, however, it must be remembered that certain raw foods, e.g. meats, can carry pathogens, although in low numbers, and so will present a risk of illness to the factory personnel handling raw foods. The risk is, of course, increased if pathogens are allowed to multiply. It is the responsibility of management to inform personnel of this risk and to provide hygienic procedures which will protect them against illness, as well as insuring that the personnel themselves do not contaminate equipment or product.

For instance, where a 'high risk' food, e.g. a baby food, is being made in a location where the environment presents clear and serious problems, extreme measures may have to be taken. These may include the provision of showers, a complete change of clothing together with high standards of hand and foot sanitation. They may also include color coding of clothing, trash bins, etc. to prevent cross-contamination.

Regulations and guidelines

Most countries have Food Acts or Regulations dealing with hygiene, and the food manufacturer must comply with these. Regulatory jurisdiction covers both the health and cleanliness aspects of hygiene. For instance, employees must notify their employer when suffering from a 'notifiable' or specified disease. Food poisoning is included in this category. In turn, the employer may be required to notify the local medical authority and the affected person is then requested to discontinue work (UK Food Act, 1984[1]). Similarly, an employee with cuts or sores is either not permitted to work (USA[2, 3]) or can only continue to work if open cuts and sores are covered with a waterproof dressing (UK[4]).

Cleanliness, as it affects employees, is dealt with by requirements on handwashing; prohibitions on spitting, smoking and eating in food handling areas; and the wearing of protective clothing.

Published legislation often gives little detail as to how a processor may best comply with the hygiene requirements within his plant. However, governments may either prepare their own codes of practice, e.g. in the USA the Federal Register Current Good Manufacturing Practice in manufacturing, packing or holding human food[3], or encourage industry to draw up codes or guidelines. These describe in some detail how a reputable processor may achieve the goal of good hygiene in a plant (or factory). Although such guidelines are not in themselves regulatory, governments expect that processors will follow the recommendations given in these documents in order to insure hygienic manufacture. Well-drafted regulations allow for alternative practices providing they can be demonstrated to meet the required criteria.

Several guidelines have been published since the 1985 *Salmonella* outbreaks in the USA (pasteurized milk) and the UK (infant dried milk), and following the realization that *Listeria monocytogenes* has been responsible for food-borne outbreaks of illness. The guidelines include those of the Dairy Trade Federation[5] and the Chilled Foods Association[6] in the UK and those of the American Meat Institute[7] and the National Food Processors Association[8] in the USA. Personal hygiene, of course, forms only part of these documents, and is not always detailed enough to enable management to set up effective procedures in a plant simply by following what is written.

Management program

The challenge to management presented by setting up effective personal hygiene procedures is precisely because many of the requirements are, in fact, simple, straightforward and obvious. They appear self-evident and so do not require management time or technical skills to enforce, in contrast to those needed by a complex processing operation. However, the increase in food poisoning outbreaks over the last few years indicates that simple, self-evident procedures must be managed as thoroughly as any manufacturing process. The HACCP system is valid in determining a good hygiene program, and is necessary to establish the true costs of failure of sanitary status.

Perhaps the most effective way to carry this out is to present all new employees with a comprehensive induction program then reinforce it through means of posters, clear instructions in toilet blocks, changing rooms and hand washing facilities in the plant. Regular group sessions, which can include videos, are also helpful[9]. Additionally, there must be sufficient on-going supervision of personal hygiene procedures in production departments to insure that everyone complies with these procedures. Employees are quick to judge how seriously their management regards hygienic practices. For this reason, management themselves must set a good example if they intend to achieve their proclaimed standards.

Selection of employees

Personal hygiene really starts with the selection of employees. Their attitude towards cleanliness and tidiness forms part of selection, as hygienic working practices depend upon their cleanliness and the way in which they work. A medical examination is routinely given to a new employee but this may not give management useful information regarding hygiene. People may, for example, intermittently excrete food poisoning organisms, e.g. *Salmonella*, yet have no symptoms of illness. A survey of 60 people who had Salmonellosis showed that 24 were still carrying the infective *Salmonella* strain after 12 months[10].

Another pathogen, *Staphylococcus aureus*, is present in the nose or on the skin of between 10% and 40% of the population. In one study[11], the numbers of *Staph. aureus* present in the nose have been 10^4 or more per swab.

The main emphasis to prevent contamination of food by the handler must be to enforce sound hygienic procedures. Hand washing before handling food, as well as after using the toilet, is the principal means of preventing symptomless excreters contaminating food. However, it is good practice that when an employee has had food poisoning, stool testing should be carried out before they work again in a food handling department.

Induction training

Every company should have induction training in personal hygiene for all employees. Everyone should be aware of the required standards, even if they do not work directly in a manufacturing department. At one time or another, most people in a plant have to enter a food handling department, and they should know the correct procedures before they enter – no one should just walk in without proper preparation.

An induction program should include:

● Personal responsibility
● Protective clothing requirements and use (see Section 6.B)
● Hand washing requirements
● Prevention of cross-contamination from raw materials to finished product areas

Personal responsibility

Each employee should know and follow the correct hygienic procedures. These include the following universal *Do not's*.

Do not:

Pick your nose

Scratch your head or face

Cough or sneeze onto food

Wipe your nose with hands, fingers or overalls

Spit

Smoke or take snuff except in designated areas

Wipe dirty hands on your overalls

Eat any ingredient or product in a food handling department

Appropriate *Do's* of hand washing, protective clothing etc. need to be prepared for each department in a plant. It is important to get acceptance of the following basic truth:

That if the correct procedures are not followed at all times, then an employee may be the direct cause of a food poisoning outbreak. This outbreak will make people fall ill, and is likely to result in loss of production from the factory which in turn may threaten people's jobs. It is therefore in everyone's interest to work in a safe and hygienic way.

Protective clothing

Requirements and use are given in Section 6.B.

Hand washing

These requirements apply to both operatives and management. Hands must be washed before leaving the toilet block and each time a food handling area is entered. Within an area, hands should be washed when they get dirty. Management must provide accessible, easily operated hand wash stations or basins at the entrance to each work area. There must be a plentiful supply of hot water, soap, clean towels or hand driers, and a dispenser of a hand sanitizing solution, together with a barrier cream as appropriate. Clear instructions should be placed close to each basin, or group of basins.

It is also management's responsibility, through supervision, to insure that everyone washes and sanitizes their hands on entry or re-entry to their workplace. For 'high care' areas it is worth considering a fail-safe installation, whereby a simple barrier at the entrance will only open when a person uses the sanitizer dispenser. In this way, entrance to the 'high care' area is not easy unless hands are washed and sanitized. Such a barrier has been installed, and seems to work well, in a French factory producing mold-ripened soft cheese – a product susceptible to *Listeria monocytogenes* contamination.

Prevention of cross contamination

Although many factors will contribute to cross contamination within a plant, the prevention of cross contamination from raw to finished product areas must include controlling the movement of people so that they do not go directly from a raw material department or area into a finished product area. This applies to all people – management, operatives, contractors and any visitors. Entry to the finished product area should certainly not be possible without either a change of footwear or going through a sanitizing footbath, although it must be recognized that floor mats and foot baths have been found to contain *L. monocytogenes* (US Dairy Plant data, 1989). In view of this, disposable overshoes may be the most efficient solution to cross contamination, but it is difficult to insure that the overshoes are non-slip.

It is critically important to avoid cross-contamination where 'high risk' products are being manufactured. These products comprise the Hazard Categories A to F as defined in the HACCP system document prepared by the NAC[12]. They include foods prepared for infants or old people, foods which have no processing step in manufacture, foods open to contamination between processing and packaging and foods with a substantial potential for abuse in handling during distribution. It should be remembered that cross contamination prevention includes more than people's movements. It applies to traffic of ingredients, containers and packaging into the area, the microbiological quality of the air supply as well as structural requirements. For a fuller description of procedures to be applied in a 'high care' area (which may also be known as a 'high risk' or 'clean' area) see Sections 4.2 and 4.3 of the DTF Guidelines[5].

References

1. HMSO (1984) The Food Act, (see Chapter 30). London: HMSO
2. ADMI (1970) *Recommended Sanitary/Quality Standards Code for the Dry Milk Industry*, Bulletin 915, American Dry Milk Institute Inc. Chicago, Ill.
3. CFR (1987) *Current Good Manufacturing Practice in Manufacturing, Packing, or Holding Human Food*, Title 21 Code of Federal Regulations Chapter 1 part 110.10, Washington, DC: US Government Printing Office
4. HMSO (1970) *The Food Hygiene (General) Regulations 1970, Statutory Instruments*, S1 1970 No. 1172 London: HMSO
5. DTF (1989) In *Hygiene & Health Guidelines for Good Hygienic Practice in the Manufacture of Dairy-based Products*, pp. 50–52, The Dairy Trade Federation 19, Cornwall Terrace London NW1 4QP
6. CFA (1989) *Guidelines for Good Hygienic Practice in the Manufacture Distribution and Sale of Foods to be Eaten Cold or after Re-heating Relying upon Chilled Storage to Maintain their Safety*, Chilled Foods Association, 6 Catherine Street, London WC2B 5JJ
7. AMI (1988) Interim Guidelines for Microbial Control During Production of Ready-to-Eat Meat Products, 2nd edn. Washington, DC: American Meat Institute

8. NFPA Bulletin 42–L (1989) *Guidelines for the Development, Production, Distribution and Handling of Refrigerated Foods.* Washington D.C.: National Food Processors Association

9. Sprenger, R. A. (1983) Training and education of food handlers, and personal hygiene. In *Hygiene for Management*, pp. 57–59, 60–66. Rotherham: Highfield Publications

10. Kotova, A. L., Kondratskaya, S. A. and Yasuitis, I. M. (1988) *Salmonella* carrier state and biological characteristics of the infectious agent. *Journal of Hygiene, Epidemiology, Microbiology and Immunology*, **32**, no. 1, 71–78

11. Polledo, J. J. F., Garcia, M. L., Moreno, B. and Menes, I. (1985) Importance of food handlers as a source of enterotoxigenic *Staphylococci. Zentralblatt fur Baktenologie und Hygiene, Abteilung 1, Originale Reihe B*, **181**, 364–373

12. NAC (1990) *Hazard Analysis and Critical Control Point System*, HACCP Working Group of the National Advisory Committee on Microbiological Criteria for Foods, USDA/FSIS, Room 3175, South Ag. Building, Washington, DC 20250

Further reading

Hobbs, B. C. and Roberts, D. 1987 *Food Poisoning and Food Hygiene*, 5th edn. London: Edward Arnold

6.B Facilities and protective clothing

Facilities

The provision of hygienic facilities is normally governed by legislation or regulations, as are the hygienic practices demanded of employees and management (see Section 6.A). However in both the USA and the UK the legal requirements refer to the minimum provision of facilities. Manufacturers committed to Total Quality Management should not be satisfied with providing the minimum; they should recognize, as Hayes[1] states, 'that morale suffers if employee service and welfare facilities are substandard'. If morale is low, then management cannot expect personal responsibility to be accepted by all employees, and personal responsibility is the key to hygienic work patterns throughout production.

Facilities which must be provided by management are:

• Space to change clothes and to hang outdoor clothes
• Washrooms
• Handwash stations (washbasins) in process areas
• Eating and smoking areas
• First aid
• Litter receptacles

Changing/locker room area

The changing/locker room area should be designed so that it cannot serve as a general storage area, and the decoration should be pleasant and easy to maintain. It should be large enough to insure that conditions are not cramped when it is being used at changeover of shifts.

Washroom facilities

The toilet facilities, again, need to be well decorated. Special attention must be given to the floor and wall finish: it must be durable, impervious to water, have coved junctions between wall and floor and the wall angles rounded (see Hayes[1]). Local regulations may forbid the use of hand-operated taps, and govern the supply of water to the washbasins. Toilet areas need to be ventilated, and thought must be given to the siting of the ventilation pipes with reference to the food production areas of the plant so that extraction is away from food areas.

Handwash stations in process areas

Handwash stations in process areas are covered in Section 6.A.

Eating areas and smoking facilities

Eating areas and smoking facilities must be provided since eating and smoking are not permitted in food or packaging handling areas. As these areas are used by employees in their rest periods they should be attractive in design and decoration and, if at all possible, in their outlook. As with all other areas, they should be well maintained and kept litter free, so choice of finish of room and furniture should be made with this requirement in mind.

First-aid facilities

First-aid facilities comprise an area set aside for this purpose, people on-site who have recognized first-aid training and qualifications, and first-aid kits in each process area. These kits must be clearly displayed and labeled, and must be kept fully stocked with adequate first-aid equipment and instructions. Each process area should have a list of qualified 'first-aiders' on site, with their telephone numbers. This list must be kept up to date, so needs regular reviewing, e.g. each month, by a designated management nominee.

Litter receptacles

Throughout the whole factory there should be receptacles for litter and garbage which are adequate for their anticipated loading. There should be a scheduled procedure for these receptacles to be emptied. Do not forget that in 'high care' (or 'clean') areas this may mean that it is workers in the 'high care' areas who must place them outside the area so that they may be emptied. The people who empty litter and garbage bins throughout the factory should not be allowed into 'high care' areas because of the risk of cross contamination.

Protective clothing

Although wearing protective clothing is a regulatory requirement in both the UK and the USA the detail included in the legislation or regulations is insufficient to help management decide how best to achieve product safety. Food hygiene books and guidelines include sections on protective clothing, and both the UK Institute of Food Science & Technology[2] and an article in *Food Manufacture*[3] give useful detail. However it is often the individual processor who has to decide on what is good protective clothing, leading to many interpretations of what is good design for such clothing and what is good practice in use. The following points give the requirements for a well-managed, safe operation.

Protective clothing comprises a coat, hair covering, and sometimes trousers and gloves. Footwear is also included, and must be capable of being sanitized. The coat must be comfortable, it must cover all the worker's own clothes, it must have no outside pockets and fastenings must be non-detachable. Hair covering includes snoods or nets as well as a hat. It is important that all the hair should be restrained; this means use of beard covering as well as hair nets.

Coats may be long sleeved or short sleeved. Short-sleeved coats enable wrists and arms to be washed more readily, but obviously long-sleeved shirts cannot be worn underneath. Cuffed long-sleeved coats also enable wrists to be easily washed. Protective clothing must be changed regularly. Daily change is preferable, and if required it should be changed within the day. However, the use of protective disposable aprons for particularly dirty jobs within a clean area may be a more cost-effective way of tackling this situation. The laundering of protective clothing needs to be a high-temperature wash, preferably with use of disinfectant in the rinse. Such laundering should be a specialist operation – employees must not take their protective clothing home to be included in a domestic household wash.

Gloves can present contamination problems when used for product handling, and their use needs careful thought. However, when the worker needs protection against heat, cold, or injury then heavy-duty gloves must be worn, and suitable sanitary treatment given to the gloves to avoid contamination risks.

For use of gloves when food safety is the main purpose, plastic disposable gloves are probably best. However, it must be remembered that they can be punctured in use, and so will lose their ability to protect the food. If plastic gloves are used, they should be discarded when the wearer leaves the work area, and a new pair used when work is recommenced. Specific warnings against the use of cotton gloves are given by the AMI in its guidelines[4]. In addition to these warnings, cotton gloves have been known to carry low levels of *Escherichia coli* as delivered to the user factory.

Provided that hand washing is enforced as it should be, the need for gloves when handling product is limited, except to protect the worker against temperature extremes or injury.

Remember that protective clothing is a requirement for every person who enters the process areas – this includes visitors and maintenance workers. Visitors may be given disposable overalls or coats, but these must be strong enough not to tear in use. Maintenance workers, e.g. engineers, need to have clean overalls which must be changed regularly.

In conclusion, the provision of protective clothing, its use, and its cleaning should form part of the hygiene schedule of a factory, and should not be overlooked as a trivial detail, and it should have its own HACCP system applied.

References

1. Hayes, P. R. (1985) *Food Microbiology and Hygiene.* London/New York: Elsevier Applied Science
2. IFST (1989) *Food and Drink Manufacture – Good Manufacturing Practice: a Guide to its Responsible Management.* 2nd edn. London: Institute for Food Science and Technology (UK)
3. Rowan, C. (1989) Workwear: more than just fashion. *Food Manufacture,* **64**, 8, London: Morgan-Grampian (Process Press)
4. AMI (1988) *Interim Guidelines for Microbial Control During Production of Ready-to-Eat Meat Products,* 2nd edn. Washington, DC: American Meat Institute

6.C Training for product safety – management, supervision and operators

Introduction

A single factor underlies many of the topics dealt with in this book – it is people who put into practice all the many aspects of hygienic food processing which together give product safety. Hand-in-hand with this factor is that if people are to carry out their tasks properly, and particularly 'to do it right first time', they must be educated and trained so that they can do their jobs well. Education and training apply to everyone in a food processing factory, whether they are in management, supervision, operators on the process floor or clean-up crew. There will obviously be differences in approach and in content of the education and training for each group of people, and for each group of operators, depending upon what department they are in or what equipment they use within the factory.

It should be recognized that training begins when someone starts work in a company – induction training in fact can be regarded as both education and training, since most companies include an industry overview as well as hygiene and health care as it applies to the worker. A good induction course will also include how equipment sanitation relates to food safety and the effects of cross contamination; these topics will be covered in general terms before particular training is given in the department where a person will be working. It must be emphasized that induction training is not enough; training and retraining must be part of the training program of a company. Retraining ideally should not be a repetition of a particular training program but may well take the form of an audit of a person, or a group of people, using the original training program as the reference for the audit. This will establish how well the people have understood the skills which they learned while training. What must be determined is how well each person is applying those skills to their particular task or tasks within the factory.

Training is generally carried out as group exercise for most purposes. The reasons for this are partly economic, as no company can afford to train each worker individually, and partly because group training can encourage team-building within a factory and also make each member of the group realize how their job relates to the overall running of the factory. They become aware that the way they do their task can affect another person's work – if they do their job correctly this means that the next person in the production line is able to do the same.

Education and training

Both these words are used when an overall learning program is being applied, and it is important to appreciate what is meant by each of these terms. Useful definitions have been given by Richardson[1] in the Contemporary Quality Assurance Seminar of the International Dairy Federation and USNAC held in Chicago in 1987:

> Training is the term used to describe the imparting and acquiring of skills to perform specific tasks. Education on the other hand is the term applied to the gaining of a broad understanding of principles and objectives, of appreciating factors involved in situations and processes and in developing judgmental capabilities and is usually of a longer term nature.

Table 6.1 Specimen technical training for personnel operating/monitoring the safe processing of HTST pasteurized milk

	Recommended proficiency level for			
	Laboratory technician	*Manager of Quality Assurance*	*Line operative*	*Manager of production*
Operation and control of HTST heat exchange plant; validation of operating procedures; record-keeping	O	P	T	T/P
Heat exchange principles; instrumentation; correct diversion of under-heated milk	O	P	T	P
Hygienic plant design	NA	P	O	P
Plant maintenance	NA	P	O	P
Sanitation checks, including CIP solutions	T	P	T	O
Laboratory procedures (plating, media/ reagent preparation, phosphatase test, etc)	T	P	O	O
Knowledge of pathogen (and spoilage) microorganisms	T/O	P	O	O
Thermometer calibration	T	P	O	O

O = orientation (familiarization)
T = technical
P = professional
NA = not applicable

Table 6.2 Recommended administrative training

I. COMMUNICATION

   Proper record-keeping
   Writing skills
   Listening skills
   Manuals

II. FINANCE

   Basic economics
   Inventory control principles
   Various operational cost reports
   Quality costs

III. MANAGEMENT

   Concepts of quality
   Basics of quality control
   Characteristics of quality
   Measuring of standards
   Basic and advanced management skills
   Project management
   Creative problem solving
   Supplier visitation programs

He elaborates on training to say that 'In the context of quality control and assurance, training relates to tasks described in regulations, standards, techniques and procedures, and tends to be presented in programs of relatively short duration, dependent upon the complexity of the subject'.

It is also important to realize that there are different aspects of technical training as well as differences between technical and administrative training. Lyons[2] makes these points clear in a paper on 'Required personnel training in plant quality control'. Table 6.1 (after Table 1 in his paper) shows the technical training aspects for the safe processing of HTST pasteurized milk as applied to the personnel involved. Requirements for administrative training are given in Table 6.2 (Lyons, Table 2). In fact what Lyons calls administrative training includes topics which may be regarded as education, e.g. basic economics, concepts of quality, etc., illustrating that on occasions the distinction between education and training is not always clear.

Education can be described as 'finding out the *why* of a topic' and training as 'finding out and putting into practice the *how* of a topic or task'. To take an example – maintaining the cool chain in product distribution. The manager needs to know *why* the chain must be kept cool. The reason for this is that unless it is kept cool, microbial growth in the food product will occur more rapidly, leading to increased spoilage and possibly becoming unsafe for the consumer. It will be helpful to the manager, however, to know *how* microorganisms multiply in numbers and some training in basic microbiology will be an advantage to him. The truck driver, even though the main content of his training will be the *how* of correct refrigeration and temperature monitoring in his truck, will benefit from knowing *why* these tasks must be done.

Unless people understand their tasks they will not realize the true significance of what they are doing, and therefore the consequences of *not* doing their job correctly. The education of a manager should have sufficient technical content so that he appreciates the importance of each processing stage, and the consequences which may occur when the process deviates from normal. For example, he should know that a pasteurization or sterilization stage consists of holding product at a specified temperature for a specified time if pathogens or their spores are to be destroyed. He needs to know how the heating is achieved, e.g. by hot water in a plate heat exchanger or by steam heating under pressure for cans in a retort. Unlike the operative, he does not need to know exactly which button is pressed in order to achieve heating. However, the manager does need to know how to read a process chart so that he can evaluate the seriousness of a deviation from the safe scheduled process.

Training programs

Setting up training programs is not new. In 1962 Reynolds[3], writing on quality control training programs, said 'modern manufacture requires that people throughout the company be equipped with the necessary attitudes, knowledge and skills to deal with the quality problems'. Safe food production is perhaps the most important aspect of quality.

More recently Dedhia[4] details what most training programs in industry comprise. The four headings are:

- Indoctrination of new employees, i.e. induction course
- Core program of job-related training
- Career/personal development, such as presentation techniques, writing skills
- Self-development options

Indoctrination of new employees has been covered earlier in this chapter. Of the other three headings, it is the core program of job-related training which forms the bulk of a company's training program, since this should cover every employee whatever their job, or their aspirations, might be.

Development of job-related training is not easy, as has been acknowledged by Duke[5] in a paper delivered to the IDF/USNAC Seminar in Chicago in 1987. As she says:

Having designed and engineered the factory and its processing areas to allow for the application of good manufacturing practices according to product requirements, having established the routines in written schemes with the what, why and how, the third most crucial and, sometimes, most difficult step must be realized in order to achieve that goal of consumer satisfaction. This is *training personnel*. In fact, in the introduction it was called participative training because personnel have to feel and be totally involved for a successful result.

Experience has shown that there are some very special aspects of training which are quite difficult to achieve:

- Development of appropriate training material
- Identification of an adept trainer

It may be helpful to include here what she says about the choice of a trainer, as well as how a successful training group should operate, before considering the development of appropriate training material.

The trainer can be at manager or supervisor level but it is essential that this person has easy contact with people and a facility to explain. Use of the supervisor to train the group for whom he or she is responsible can be very motivating and the closeness of their levels eases dialogue. Often it is not the expert who is the best trainer. For example, when training in good manufacturing practices for hygienic processing, it is not always the microbiologist who is best suited to the task, sometimes the talks become too scientific.

The scientific approach will destroy the aim of making training a participative affair. This leads to the topic of quality control circles which are formed so that personnel can take their part in assuring good quality processing. During recent years everyone has been talking of Q/C circles but the mode is fading. In fact, it has often been realized that Q/C circles are not something unique but actually, a meeting of personnel with a chance for dialogue regarding improvements or practices to assure quality. Participative training, stimulated and coordinated by top management, is also aimed at allowing dialogue. Apart from training, many other types of personnel participative groups have been formed with various titles, e.g. *Salmonella* committees, hygiene committees, good housekeeping study groups. It all boils down to one result – that is the feeling of involvement of the personnel with, hopefully, the incentive to carry out the required practices.

If personnel feel this incentive to follow the practices and have been trained to understand how and why, it is hoped that a system of auto-control will develop. As stated in the introduction, this means that the factory is run by many dedicated observers following good manufacturing practices. The management task then becomes one of surveillance, continuous innovation and improvement of systems. [5]

Development of training material

Training material may be developed within a company by a department dedicated solely to the training needs of the company, or by an outside agency which must, however, work closely with the company in developing the training program. Outside agencies will include specialist consultancies, training/technology departments of colleges or universities, and training/specialist technologists of research associations. Whichever route is chosen for developing training material, the company management must have a clear objective or objectives as to what it requires from training. For safe processing of food, the requirement is that everyone in the company understands that it is *their* responsibility to perform their task in a manner which assures food safety. To do this, they need to know what comprises food safety (Chapter 1) and they need to know in broad outline both how food safety is achieved and what will compromise it: the do's and don'ts of manufacturing practice. This general information should form part of the training/education of everybody – managers, supervisors and operators, not forgetting the maintenance staff and contractors, who may be regarded as temporary employees of the company, and so in need of training in 'safe working'.

The best way to present training material is to small groups, and successful training will develop the active participation of group members. This may be more difficult with general background material than it is with the more detailed job-related training, where group members can contribute from their own experience in a way which they may consider that they are unable to when discussing more general aspects of food safety. However, the use of training videos or slide/tape presentations as well as booklets available to each group member all help active involvement. Remember, it is easier to pause and allow discussion with a slide/tape than it is with a video. It is important to keep each training session short. This is both so that interest in the learning topic is maintained without the boredom level being reached and also so that training is not regarded as an unwelcome interruption to production. A session of between 25 and 35 minutes is usually sufficient to enable the training objectives to be achieved for the topic in hand.

A booklet for each training topic will help operatives feel 'ownership' of the material being presented. This should include some questions relating to the session which are answered by the owner of the booklet. These questions should highlight the key points and standards required, so that the commitment of the person answering the questions is reinforced. Reading the answers to the whole group encourages participation of the group members and helps to emphasize the key points of the training topic.

Technical training modules

For certain jobs within a plant, a training module may be more suitable than a booklet. Laboratory technicians will serve as a good example, as the analyses they do are well defined and follow approved standard methods, whether National or International. A Technical Training Module has been developed by Grand Metropolitan Foods Europe to provide a record of competence for each laboratory technician. It lists the analyses carried out within a laboratory and the Standard Methods used. The Laboratory supervisor or manager will train the technicians then assess each technician regarding knowledge of a method and competence in performing the analysis, including recording the results. Providing that the analysis has been correctly carried out, the technician's module is signed, which then serves as a permanent training record.

Such a laboratory module can also be linked to National Accreditation schemes which are being implemented to a greater or lesser extent in the USA, France, Italy, Australia, New Zealand and Germany as well as in the UK. The aim of these schemes is to insure that laboratories are competent to carry out the testing schedules appropriate to the factory or process concerned. As well as receiving accreditation upon joining the scheme, there needs to be renewal of accreditation – in other words, re-auditing procedure to check that the laboratory maintains the correct functioning role needed to verify processing control. Accreditation is being increasingly demanded by major retail customers in the UK and other countries.

Heat processing training has been recognized as vital to both canning and pasteurizing technologies. In the USA, retort operators must operate under a supervisor who has attended a Better Process Control School and received certification of competence before the plant can run a product. The same applies to HTST pasteurization plant operators in some states in the USA.

In the UK, the National Dairymen's Association (NDA) in its February 1989 *Processor's Bulletin* [6] has the following section on Operator Training under the heading 'The Pasteurization Process':

(a) It is essential that all pasteurizers are fully trained in the operation of the plant and have a basic understanding of the principles of pasteurization.
(b) Retraining of operators needs to take place every two years.

Although there is no regulatory requirement in the UK for licensing HTST pasteurizing plant operators, the Food Safety Act of 1990 currently includes an enabling power for training of workers in the food industry.

Conclusions

Management must have the right training, retraining and education programs for its technical experts so that the company stays ahead of inevitable changes in technology. Training programs must include the whole workforce so that necessary changes in work patterns to insure continuing safe production of food are accomplished smoothly. With technically sound training programs production of safe food should be accompanied by profitabilty.

References

1. Richardson, P. A. (1988) Training and development of quality assurance professionals. In *Contemporary Quality Assurance Bulletin of the International Dairy Federation no. 229*, pp. 69–76, IDF, 41 Square Vergote, B – 1040 Brussels, Belgium
2. Lyons, R. L. (1984) Required personnel training in plant quality control. *Food Technology*, **38**, (4) 105–110
3. Reynolds, E. A. (1962) Training programs in Quality Control. In Quality Control Handbook, 2nd edn, edited by J. M. Juran, pp. 7–13 to 7–16. New York: McGraw Hill
4. Dedhia, N. S. (1985) New demands for quality, changing job patterns and advances in technology have created new needs for quality education and training. *Quality Progress*, January, 14–15
5. Duke, M. (1988) Good manufacturing practices – an essential ingredient of quality and safety. In *Contemporary Quality Assurance Bulletin of the International Dairy Federation No. 229*, pp. 27–34.IDF, 41 Square Vergote, B–1040 Brussels, Belgium
6. N.D.A. (1989) *Processor's Bulletin S 21/89*. London: National Dairymen's Association

6.D Training aids and materials – some commercially available examples

David Bates (1987) *Food Safety – an international source list of audiovisual material*, WHO/EHE/FOS 87.1.

This catalog has been prepared for the WHO Food Safety Program by the British Life Assurance Trust Center, BMA House, Tavistock Square, London WC1H 9JP

Training package

A 'Food Safety' package has been prepared by Grand Metropolitan Foods Europe. It consists of a video, a supporting booklet and additional material for use by trainers. The package is available from Technical Department, 430 Victoria Road, South Ruislip, Middlesex HA4 OHF.

Individual audio visual aids

Yours Disgustedly   Video
Waldegrave Films, UK

Hygiene in the Food and Beverage   Slide tape
Industry   or Video
Diversey, UK

Hygiene – the professional touch   Video (in
J. Lyons and Co. Ltd, UK   three parts)

The Handling of Chilled Dairy   Film
Products
(Maintain the Chain)
Dairy Trade Federation, UK

Milk Pasteurization Operations   Video
Dairy Trade Federation, UK (Part of
a training package 'Getting it Right'
which includes Milk Tanker
Reception Operations and a Training
Diary and Trainers Guide)

Hygiene and the Engineer   Slide tape
Voss Training Services Ltd, UK   or video

Good Sanitation Practices   Slide tape
and
handbooks

On the Line   Video/film
Food Processors Institute (part of the
NFPA) USA

Sanitation Part 1. It All Depends on   Slide tape
You
Salad Manufacturers Association,
USA

Retort Operations   Slide tape
Campden Food and Drink Research
Association, UK

For the Retort Operator   Slide tape
Food Processors Institute, USA

Foreign Matter in Food   Slide tape
Voss Training Services Ltd, UK   or video

The Food Processors Institute also offers audio-visual material on Workers Safety: titles detailed in Schools, Publications and Audio-visuals (1989) FPI 1401 New York Avenue, N.W., Suite 400, Washington, DC 20005.

Chapter 7
Microorganisms – an outline of their structure

Introduction

This brief part of a short chapter is intended to give some background material to answer the question 'what are microorganisms like?' It does not attempt a systematic treatment nor is it any substitute for a textbook.

The striking characteristic about microorganisms is that they are small to exceedingly small on a human scale. As a generalization, molds are the largest although a magnifying glass or low-power (total magnification × 100) microscope is needed to see the hair-like structures (hyphae) that make up most of the mass visible to the human eye. Yeasts are smaller in size and a magnification of about × 400 is used for their study. Bacteria are even smaller and a high-power microscope (× 1000) is needed to see them. Viruses are much smaller and an electron microscope must be used to make them visible.

Other groups of microorganisms, e.g. protozoa in a water supply, may sometimes be of concern to the food microbiologist but these too are small microscope creatures.

The small size of microorganisms is of significance. This is partly because of the tiny space needed for microorganisms to grow to very large numbers. It is also because very small differences of location, by the human scale, in the solid or liquid environment where the organism lives can have a marked effect on growth.

In spite of their small size, microorganisms are complex living things both structurally and physiologically (see Chapter 8). They are also inherently variable to an extent that is hard to visualize if higher plants and animals are regarded as providing a 'normal' criterion. Although microorganisms have been studied for over 100 years much remains to be learned. It is also worth remembering that it is the relatively few kinds of microorganisms which make an obvious impact on the perceived interests of mankind which have received detailed attention. The only certain prediction for the future is that microorganisms will continue to provide surprises and problems for management and microbiologists.

Further information on what microorganisms do and how they behave is given in Chapters 8, 10 and 11.

7.A Viruses

Size

Viruses are very small infective particles. Their dimensions range from 20 to 300 nm, that is, from about one hundredth to one tenth the size of a bacterium. Most viruses can only be seen by using an electron microscope. The size of those viruses commonly associated with causing gastroenteritis in humans is in the range of 22 to 38 nm; Rotavirus, a less common cause of food poisoning, measures 70 nm. (See Table 7.1.)

Table 7.1 Food-borne viruses

Type	Size (nm)	Notes
Gastroenteritis viruses 'Small Round Structured Viruses'	30–35	This group includes Norwalk and morphologically similar but antigenically distinct viruses.
Astrovirus	28–30	So called because they resemble a five- or six-pointed star. Cause gastroenteritis mainly in children under 1 year.
Candidate human parvovirus	18–25	A small round virus associated with gastroenteritis from cockles. It is known as the 'cockle agent'.
Hepatitis A (HAV = Hepatitis A virus)	20–30	Outbreaks associated with shellfish and infected food handlers.

Structure

This is simple compared with that of bacteria, and a virus particle consists of a single molecule of deoxyribonucleic acid (DNA) or of ribonucleic acid (RNA) surrounded by a coat made from protein. The protein coat protects the nucleic acid, which is the means by which the virus multiplies, and also serves to attach the virus to the surface of the host cell. True viruses will only multiply within living cells.

The viruses known to cause food poisoning contain RNA, not DNA. Coxsackieviruses, Echoviruses, Small Round Structured Viruses (including Norwalk virus) and Hepatitis A virus, contain a single strand of RNA, but Rotaviruses contain a double strand of RNA. All these viruses are 'icosahedral' in shape, that is, they are a polyhedron of 20 sides, 12 vertices and 30 edges.

Multiplication/replication

Viruses are unable to reproduce themselves outside another living organism. They are obligate parasites of all living organisms; bacteria, fungi, algae, protozoa, higher plants, invertebrate and vertebrate animals. Once the protein coat becomes attached to the surface of the appropriate host cell, the host cell either engulfs the virus particle or the nucleic acid is injected from the virus particle into the host cell, as is the case with bacteriophages active against bacteria. The viral nucleic acid then uses the nuclear apparatus of the host to induce both the production of viral nucleic acid and of the coat protein. These are produced separately, but assemble within the host cell as new viral particles, and are released gradually from the host cell. The duration of the replication cycle in a single cell may take from 8 hours to more than a day, although bacteriophage can replicate in as little as 30 minutes.

In animals some infected host cells may die, but others survive infection with the virus, and resume their normal function. It is not necessary for the host cells to die for the host organism, in the case of man, to become ill. The fact that the host cells are not performing their normal function is what causes the illness. Once the normal function is re-established, then recovery from illness occurs.

Detection

The inability of viruses to reproduce themselves outside the host cell, together with their small size, leads to problems in the isolation of viruses from foods suspected of being the cause of illness in humans. Because of this, the connection between the disease in humans and the causative food is most often epidemiological, i.e. tracing the implicated food through what was consumed in common by those people who became ill. Many outbreaks have been so investigated, and the food implicated as the vehicle carrying the virus because it was prepared by someone who was ill, or recently ill, and who was found to be excreting the virus. In these cases, the food prepared received no processing between the time it was handled by the infected person and the time it was consumed by the people who became ill.

Note: There is no evidence of the HIV (AIDS) virus being transmitted by foods.

In contrast to the difficulties in detecting pathogenic food-borne viruses, detection of bacteriophage in lactic acid bacteria cultures used in dairies is fairly easy, and is part of factory quality control procedures.

Further reading

Cliver, D. O., Ellender, R. D. and Sobsey, M.D. (1984) Food-borne viruses. In *Compendium of Methods for the Microbiological Examination of Foods*, 2nd edn, edited by M. L. Speck, pp. 508–541. Washington DC: American Public Health Association

Cliver, D. O. (1983) *Manual on Food Virology* VPH/83.46, Geneva: World Health Organization

Cliver, D. O. (1987) Addendum to the *WHO Manual on Food Virology* VPH/83.46, Geneva: World Health Organization

Halligan, A. C., and Auty, M. A. E. (1988) *Literature Survey No. 15. Foodborne Viruses – a survey of the recent literature*, Leatherhead Food Research Association, Leatherhead, Surrey

7.B Bacteria

Shape and size

Characteristically, bacteria are unicellular organisms which multiply by transverse binary fission and are either spherical (coccus, plural cocci), cylindrical (rod(s)) or helical. Cocci may be isolated or aggregated two by two (diplococci), in chains (streptococci), in more or less regular clusters (micrococci, staphylococci) or in regular cubic packs (sarcina).

A coccus has a diameter of about one micron or micro-meter. Rod-shaped cells are cylindrical or ellipsoidal, about 1 micron in diameter and 2 to 5 microns in length. They can occur singly, in chains, or in palisades. They may be curved once (vibrio) or several times (spirillum).

The bacterial shape is usually a stable character of the species, although dimensions may change in response to environmental conditions or because of the age of the cell. Appearance may be atypical, for example, in old cultures or on first isolation from a hostile environment such as spoiled food.

Bacteria are divided into families and genera, depending upon common characteristics of shape and the overall type of metabolism of each grouping, and the members of each genus are divided into species. Each species differs from another by possessing one or more different characteristics. Certain species are further subdivided into strains. Some of these characteristics are of great importance in food microbiology, since one strain may be more heat resistant than other strains of the same species, as happens with certain *Salmonella* spp., or in cheese starter cultures one strain may produce acid more rapidly in cheese making than another and so be more useful commercially.

Structure

The fundamental structure of the bacterial cell comprises an external wall (with appendages and slimy material), a cytoplasmic membrane, and cytoplasm, which includes the genetic material. There are appendages associated with the cell wall and membrane which enable the bacterial cell to be motile. They are called flagella and pili.

Cell wall

The cell wall is a rigid structure 10 to 80 nm thick. Usually, bacteria are divided into Gram-positives and Gram-negatives, according to the color taken when subjected to the Gram-staining treatment. This is related to the structure and composition of the cell wall.

In Gram-positive bacteria the wall is essential for maintaining the shape and structural integrity of the cell. Fundamentally, it is a sack-like structure made up of peptidoglycan molecules often known as murein. Details of this are to be found in Appendix 1. Other molecules with uncertain properties may be present in the peptidoglycan network of Gram-positive bacteria, for example, teichoic acids, teichuronic acids, different polysaccharides and proteins. These are likely to be responsible for the antigenic properties of the cell wall.

In Gram-negative bacteria the cell wall is more complex. It consists of two main layers. The inner layer is of peptidoglycan (PG) and the outer one is made up of phospholipids, proteins and lipopolysaccharides. The outermost layer is made up of some proteins but mostly of a lipopolysaccharide (LPS). (For further details see Appendix 2.)

The LPS molecules of several Gram-negative bacteria are toxic to man and animals. Since it is a structural constituent of the cell wall it is continuously released in small amounts into the environment surrounding the bacteria. *Salmonella* spp. are good examples of such organisms.

Cytoplasmic membrane

This membrane is between the inner peptidoglycan layer and the cytoplasm of the bacterium. It has a relatively simple structure which can be described as a phospholipid bi-layer associated with a large number of protein molecules. (For further details see Appendix 3.)

Proteins associated with the bi-layer are of particular importance to the living organism. They are involved in respiration, synthesis and transport of the cell component molecules, excretion of extracellular enzymes and toxins together with transport of nutrients and catabolites across the membrane.

Flagellae

Most rod-shaped and just a few spherical bacteria are motile in fluid media. Motility occurs in organisms provided with thin filamentous structures called flagella. The cell may have a single flagellum (monotrichus) or a tuft of flagella (lophotrichus) at one or both poles, or many flagella (peritrichus) over the entire surface. Flagella are unbranched helical filaments of 10 to 20 nm diameter and up to 15 000 nm length, anchored to the cell wall and membrane. Movements of flagella, supported by energy expenditure, allow the movement of the cell through the liquid in which it is suspended. The liquid need only be present as a very thin layer to allow movement, or motility, to take place.

Pili

Freshly isolated Gram-negative bacteria may have one or more (up to 300) straight appendages of less than 1000 nm length and 10 nm diameter, called pili. They have several functions: (i) as structures enabling the bacteria to adhere to one another, to red blood cells or to the intestinal epithelium of higher animals, or (ii) as tubes for transfer of nuclear DNA between cells during conjugation (sexual pili).

Capsule and slime layers

Some bacteria secrete a slimy polymeric material of differing composition (omo- or etero-poly-saccharides, polypeptides, polynucleotides) as a function of the species and of the environmental conditions. According to the density and degree of association with the outer surface of the cell wall, the polymeric material can be seen microscopically as an envelope around the cell (true capsule), or as material loosely associated with the cell (slime). *Acetobacter aceti* the 'mother of vinegar' is unusual in that it produces a cellulose material around the cell.

Cytoplasm

This consists of 70% water. Dissolved or suspended in this intracellular water there are inorganic salts, low molecular weight organic compounds, high molecular weight polymers such as proteins, nucleic acids, glycogen and other polysaccharides, and ribosomes.

The cytoplasmic proteins are mainly enzymes which are primarily involved in the metabolic activity of the cell. There are about 2500 types of protein per cell.

Other components of the cytoplasm are vacuoles, metachromatic granules, lipid droplets, and inclusions such as mitochondria. Also present are plasmids which are extrachromosomal genetic elements of autonomously replicating cyclic DNA molecules. These carry factors such as the F factor, the first fertility factor which generates bacterial recombination, and the R factors involved with resistance to antibiotics and bacteriocinogenic factors.

The bacterial spore

Bacteria belonging to genera *Bacillus, Clostridium, Desulphotomaculum, Sporolactobacillus* and *Sporosarcina* may form a single characteristic structure within the cytoplasm, called a spore or endospore. The spore is the bacterial particle most resistant to all anti-microbial physicochemical agents, with few exceptions. One main exception is that some vegetative bacteria, for example, some staphylococci, show a resistance to ionizing radiation (irradiation) which equals, or sometimes exceeds, that of spores.

The morphological changes which occur in the vegetative cell during spore formation take place in the following manner over about an 8-hour period. DNA migrates to one end of the cell. Invagination of the cytoplasmic membrane forms a pre-spore septum (wall) followed by a second invagination process to form a second membrane layer. A thick peptidoglycan layer, or cortex, is then synthesized by the inner membrane and accumulates between the two membranes, while the outer membrane synthesizes a rigid cheratin-like bilayer with inner and outer coats. The spore becomes more refractive with loss of water (dehydration) and eventually the wall of the vegatative cell lyses and the free spore is released. The spore can remain dormant in adverse environmental conditions for long periods, up to years in some cases. However, when environmental conditions suitable for growth occur, the spore will germinate. An activating stimulus, which may be physical, chemical or mechanical, causes the spore to absorb water, lose both its refractivity and its heat resistance, and it develops into a vegetative cell, with accompanying cell multiplication.

Reproduction

Details of this process, which is complex, are outside the scope of this chapter. For growth kinetics see Chapter 8.

Identification

It is not always necessary to confirm the identity of an organism which has been isolated from food to the species or strain level. Presumptive identification, e.g. of a *Salmonella*, is usually sufficient for appropriate remedial action to be taken. Identification is usually by a combination of:

- Structure and Gram-staining
- Biochemical tests
- Immunological reactions

Taxonomic niceties are of little help to the practical food microbiologist; it is the action which must be taken that is important.

Further reading

Bergey Trust (1986) *Bergey's Manual of Systematic Bacteriology*, 1st edn of 4 sub-volumes, Volumes 1 and 2, editor-in-chief J. G. Holt. Baltimore, MD: Williams and Wilkins

Davis, B., Dulbecco, R., Eisen, H. and Ginsberg, H. (1980) *Microbiology*, 3rd edn. New York: Harper and Row

Hawker, L. E. and Linton, A. H. (1979) *Microorganisms – Function, Form and Environment*, 2nd edn. London: Edward Arnold

Appendix 1 The structure of Murein

Murein is a long linear polymer of N–acetyl–gluco-samine and N–acetyl–muramic acid, cross-linked with a tetrapeptide to form a rigid two- or three-dimensional net-like structure.

```
M       M==--M-  - - = M    M=----M

:       :     :      :      :      :

G       G     G      G      G      G

:       :     :      :      :      :

M--==  M=---- M==--  M      M      M

:       :     :      :      :      :

G       G     G      G      G      G

:       :     :      :      :      :

M      M--==  M       M----- M----=M

:       :     :      :      :      :
```

G = N-Acetyl-glucosamine

M = N-Acetyl-muramic acid

Appendix 2 Outline of structure of the Gram-negative bacteria cell wall

```
O       O        O        O        O        O
S       S        S        S        S        S
S       S        S        S        S        S
S       S        S        S        S        S
LLLppppLLLpppppLLLpppLLLppppppLLLppppLLL
111ppppl11ppppp111ppp111ppppp111pppp111
     pp       ppp pp         pppp      ppp
11LLlpp11111LLlpp1111LLlpp111111p111LL11
ffPPfppfffffPPfppffffPPfppffffffpfffPPff
   pp pp      PP pp     PP              p  PP
GMGMGMGMGMGMGMGMGMGMGMGMGMGMGMGMGMGMGMGM
```

• •

Lipopolysaccharide:

```
O  } somatic antigen
S
S  } polysaccharidic core
S
LLL } Lipid
111
```

Proteins:
```
pp    ppp   p
pp    ppp   p,  etc.
pp    pp    ppp
```

Lipoproteins:
```
LL    LL
PP    PP
PP    PP
```

Phospholipids:
```
11111 11 1 111
fffff ff f fff
```

Peptidoglycan layer: GMGMGMGMGMGMGMGMGMGMGMGMGMGMGM

• •

Appendix 3 Outline of cytoplasmic membrane structure

```
 PP        PP       PPPPP          P
ffPPffffffffPPffPPffPPPPPffffffffPPfff
11PP111PPP1111PPPP11PPP1111111PPP111
        PPP       PPPP    P            PPP
1111111PPP1111PPPP1111111PPP11PPP111
fPPffffPPPfffffPPffffffPPPPPffPPffff
 PP                        PPP     P
```

```
                ]
Phospholipid:   f, f
                ]
```

```
                            ffffffffff
                            1111111111

Phospholipids bilayer:

                            1111111111
                            ffffffffff
```

```
                        PP
            PP    PP     PPP
Proteins:   PPPP  PP     PPPPP,  etc.
            PP    PP     PPPP
             P           PPP
                          P
```

7.C Fungi – yeasts and molds

Cell structure

The fungal cell is an eukaryotic cell (as that of higher animals and plants) and it differs from the bacterial cell (a prokaryotic one) mainly by the possession of one or more nuclei enclosed in a perforated nuclear membrane; by containing typical mitochondria; by having more complex membrane structures, larger dimensions and other typical features which are discussed below. Some fungi may have variable numbers of nuclei in a cell.

Cell wall

The cell wall of fungi consists of a network of microfibrils embedded in a matrix containing carbohydrates, lipids and proteins. The microfibrils in molds are made of chitin (and/or cellulose) and glucans, or only of non-cellulosic glucans sometimes containing chitin in yeasts.

Cytoplasmic membrane

The cytoplasmic membrane (plasmalemma) of fungi, although having the general structure of a typical biological membrane (bimolecular leaflets of lipids interspersed with proteins and glycoproteins), differs from the bacterial cytoplasmic membrane by

the presence of complex invaginations into the cytoplasm. The cytoplasm, characteristically for eukaryotic cells, contains a number of additional elements. These are the endoplasmic reticulum, Golgi apparatus, vacuoles and secretory vesicles which represent a functional continuum between them and the plasma membrane. They are metabolic sites for different materials and, through various degrees of temporary association between them, the way by which materials are transferred both into and out of the cell.

Nucleus

This contains the chromosomes and is surrounded by an envelope formed by two membranes separated by a (perinuclear) space. This envelope has a number of randomly distributed pores with both membranes united at the edges of the pores. Some pores may have a diaphragm composed mainly of proteinaceous material.

Vacuoles

These are very easily seen in the fungal cell. They are vesicles of varying shapes and sizes delineated by a single membrane called a tonoplast. Apart from water, vacuoles contain a variety of hydrolytic enzymes, e.g. proteases, phosphastases, sulfatase, invertase, glucanase, cellulase, RNA substances that have been degraded, newly synthesized materials such as extracellular polysaccharides or storage products mostly containing nitrogen and phosphate.

Mitochondria

The size, shape, number and composition of mitochondria vary widely in different fungi. They are intracytoplasmic bodies characterized by the presence of inner and outer membranes, the inner membrane forming flattened sacs (cristae) containing the enzymes of the respiratory system.

Yeasts

These are unicellular fungi. The cells are usually globose, ellipsoid or almost cylindrical and about 3 to 5 microns (μm) in size. Under favorable environmental conditions, the yeast cells divide rapidly but the generation time is slower than that of bacteria, with a minimum of 2 to 3 hours. Most yeast species of interest to food microbiology multiply by budding: a small protuberance or bud grows out from the mother cell, enlarges to reach about the same size of the mother cell and eventually separates from it becoming a new individual. At the detachment point is left a scar. Each mother cell usually undergoes several budding processes. Some yeasts multiply by a binary fission mechanism analogous to that of the bacterial cell. Some intermediate reproduction modes are also known.

In defined environmental conditions, certain species produce chains (branched or not) of cells failing to separate and resembling a mycelium (the thallous of filamentous fungi). Some yeasts produce spores (ascospores or basidiospores) through a sexual cycle, or asexual spores (conidia, ballistospores). However, the spores are formed in a manner different from those of bacteria and do not possess the high resistance typical of bacterial spores, although ascospores in particular are more resistant than vegetative cells.

Molds

These consist of tubular cells, ranging from 30 to 100 μm in diameter called hyphae, which may or may not be branched and which form a macroscopic mass called a mycelium. The hypha may be a single multinucleate cell called coenocytic or non-septate hypha, or a filament containing crosswalls or septa, which form the individual cells, called septate hypha. However, with septate hyphae there is some protoplasmic continuity from cell to cell since each septum has a more or less complex pore in its structure through which cytoplasm and nuclei can pass.

In coenocytic hyphae cytoplasm is free streaming along the tubule and is readily seen with a light microscope. In senescent septate hyphal cells the pore is usually plugged with oblong bodies called Woronin bodies.

The hypha grows in length by apical extension. Branches develop at some distance from the tip of the main hypha and grow equally by apical extension. Single hyphae of the same mycelium may differ in size, in color, in the mode of branching, etc., according to the physiological, nutritional or reproductive condition. So, hyphal elements which penetrate the substrate (rhizoidal hyphae) are quite different from simple aerial hyphae and from those bearing fruiting bodies.

Furthermore, under particular environmental conditions fungal cells may more or less completely lose the typical hyphal appearance and look like the budding cells of yeasts.

Reproduction

Many molds reproduce both by asexual spores and sexually formed spores. Identification of molds is primarily by the structure of these fruiting bodies. This term includes both spores and the associated specialized hyphae, whether produced asexually or sexually. The following are the main types of asexual spores formed:

(i) Sporangiospores, which develop in multispored sporangia at the top of specialized hyphae called sporangiophores surrounded by a thin wall.

(ii) Conidia, which develop as single spores at the tip of a fertile hypha called a conidiophore. They develop from cells called sterigmata and are usually produced in enormous numbers. Because they weigh so little and are therefore carried by slight air movement they are very efficient in spreading the mold to new environments.

(iii) Clamydospores, which develop within a hypha by thickening of the cell wall.

(iv) Blastospores, which develop by budding.

(v) Oidia or arthrospores, which are formed by fragmentation of the mold mycelium.

Sexual spores are produced less frequently and may, in fact, be a response to adverse environmental conditions. The following are the different types of sexual spores:

(i) Zygospores, which are formed by the fusion of two similar hyphae, called gametangia; produced as terminal segments of specialized branches. After fusion of one or more pairs of nuclei the fusing cells develop into a thickwalled spore. Meiosis, i.e. the reduction in the number of chromosomes in the nucleus, usually occurs when these spores germinate.

(ii) Ascospores, which form in specialized hyphal segments, called asci. This is the site of nuclear fusion and meiosis. Asci are often enveloped in a relatively dense mass of sterile hyphae, called a cleistothecium, of characteristic shape and size, depending on the species.

(iii) Basidiospores, which differ from ascospores in that the basidiospores are formed externally to sporogenous cells, rather than internally.

(iv) Oospores are formed by fusion of two differentiated gametangial cells called the antheridium (male cell) and the oogonium (female cell). The antheridial nuclei migrate through tubes into the oogonium, and after fertilization form oospheres. Following this, the cell wall thickens and the zygote matures into a number of oospores.

When environmental conditions are suitable, the germination of these sexual spores occurs, initiating the formation of a new mycelium.

Note that classical identification methods rely heavily on the reproductive structures. Recent developments of interest to the food microbiologist make more use of biochemical tests.

Further reading

Yeasts

Rose, A. H. and Harrison, J. S. (1968) *The Yeasts*. London/New York: Academic Press

Skinner, F. A., Passmore, S. M. and Davenport, R. R. (eds) (1980) *Biology and Activities of Yeasts*, Society for Applied Bacteriology Symposium Series No. 9, London: Academic Press

Molds

Burnet, J. H. (1968) *Fundamentals of Mycology*. London: Edward Arnold

Pitt, J. I. (1988) *A Laboratory Guide to Common Penicillium Species*, 2nd edn, Commonwealth Scientific and Industrial Research Organization, North Ryde, Australia

Raper, K. B. and Fennel, D. I. (1965) *The Genus Aspergillus*. Baltimore, MD: Williams and Wilkins

7.D A general glossary of terms of form, function and attribute

Introduction

The glossary is not intended to be either an academic or extensive one. The terms have been chosen to help both for general and training purposes. Only a selection of words can be given and the reader who does not find a wanted word here is advised to consult a specialist dictionary such as that of Singleton and Sainsbury[1] or a microbiologist.

The explanations are not rigorous but are intended to illustrate the meaning or uses of the terms within the food industry.

Glossary

Absent Microorganisms are said to be 'absent' if they have not been found using a particular technique and a known size of sample. It does not necessarily mean that they were not there – they just were not found. It is good practice to state 'absent in' and then state the quantity tested, e.g. in ml or g.

Adaption or Adaptation Changes in a population by which the organisms become more suited to a particular environment. As an example, *Zygosaccharomyces (Saccharomyces) bailii* in a sauce or pickle processing plant may take one, two or more years to adapt to adverse conditions of acidity, salt or preservatives. It may then cause commercially significant spoilage in conditions in which it was not thought possible for it to grow.

Aerobe An organism which normally grows in the presence of air (20% oxygen) or oxygen. Some aerobes can only grow if oxygen is present and are called 'strict' aerobes. If they also grow in anaerobic conditions (i.e. in the absence of oxygen) they are called 'facultative' anaerobes.

Agar A solidifying agent widely used in microbiology. It is a carbohydrate material obtained from species of marine algae and the gel has the useful property of 'melting' at about 100°C and solidifying at about 40 to 45°C. The gel can be incubated at temperatures at which the organism gives optimal growth, including thermophilic temperatures of 55 to 65°C. Because of this wide difference of 'melting' and 'solidifying' temperatures it is more widely used than gelatin, which will only remain solid at temperatures below 25°C. Other materials, e.g. silica gel, are used for special purposes. *Note* An unsolidified culture medium is usually called a 'broth'.

Algae These plants are a heterogeneous group which include uni- and multi-cellular forms. They have chromosomes in a nucleus and contain chlorophyll. As an example, typical seaweeds are algae. They differ from higher plants in significant ways. They can become established in effluent plants and cooling towers.

Anabolic Reactions by which more complex cell components are built up from precursors which may be inorganic or organic. This process requires energy, e.g. light or chemical energy.

Anaerobe An organism which normally grows in the absence of air (20% oxygen) or oxygen. Anaerobes which are poisoned by oxygen are called 'strict' or 'obligate' anaerobes. Anaerobes which can grow under aerobic conditions are called 'facultative' aerobes. This group includes organisms of interest in food microbiology, e.g. *Bacillus macerans*, *B. polymyxa* and some members of the *Enterobacteriaceae*. 'Micro-aerophilic' organisms grow best in an environment where only a small amount of oxygen is present compared with that of air.

Anaerobic jar A container in which cultures, e.g. inoculated agar plates, are grown in a gaseous atmosphere other than air. Hydrogen or hydrogen + carbon dioxide mixtures are commonly used. A catalyst is used to remove the last traces of oxygen using the reaction hydrogen + oxygen forms water.

Background flora The mixture of microorganisms in a sample or product from which a particular type of organism, e.g. *Salmonella* spp., is being isolated or studied. In foods, the background flora can both be variable and change rapidly.

Bacteriophage or phage A virus which attacks a bacterial cell. Bacteriophages may be highly specific for their hosts and strains of a particular species can therefore be 'phage typed'. This can be important during the investigation of a food-poisoning outbreak. Bacteriophages can also attack lactic acid bacteria used in cheese making, causing a 'slow make'.

Broken heating curve When the temperature of canned food is plotted against time in the sterilization process, some products, e.g. starch-containing soups, show a marked change in slope called the 'break point'. Such a product is said to have a 'broken heating curve'.

Candle (or candled) jar A container, usually an anaerobic jar or dessicator, in which a lighted candle (or night-light) is placed to reduce the percentage of oxygen and increase that of carbon dioxide. The jar must be sealed once the lighted candle is in place. This technique is useful when culturing some lactobacilli.

Carbohydrate An organic compound containing carbon, hydrogen and oxygen in which the hydrogen and oxygen are in the proportions found in water. Typical examples are sugars, starches and celluloses.

Catabolic The opposite to anabolic, i.e. a reaction by which complex cell components, e.g. sugars, are broken down into simpler compounds with the release of energy.

Catalyst A substance that increases the rate of a chemical reaction and which may be recovered almost unchanged when the reaction is complete. Enzymes are considered to be complex organic catalysts. A negative catalyst slows the reaction.

Cellulose An important structural cell wall polysaccharide in plants including algae and some fungi. Extracellular cellulose is produced to form a pellicle or 'skin' by the 'mother of vinegar', *Acetobacter acetii*. Chemically, it is a polymer of B (beta)-D-glucose units linked in chains of 1000 to 4000 units.

Chitin A structural polysaccharide found in animals such as marine invertebrates and arthropods where it is the principal part of the exoskeleton, e.g. the 'shell' of prawns. It is also found in the cell walls of most fungi and yeasts. Chemically it is a polymer of mostly unbranched chains of B (beta)-(1-4)-2-acetamide-2-deoxy-D-glucose (also called N-acetyl-D-glucosamine).

Class In classification, e.g. of microorganisms, it is a taxonomic rank above 'order' and below 'division'. With fungi the suffix 'mycetes', e.g. ascomycetes, is used.

Coccus A more or less spherical bacterial cell. The plural, which is usually used, is 'cocci'. They may occur according to species; singly, in pairs (Diplococci), in chains (*Streptococci*), in clusters with a resemblance to a bunch of grapes (*Staphylococci*), in groups of four (Tetrads) or in irregular clusters.

Coenocyte A multinucleate cell or mass of protoplasm, formed by repeated nuclear division without division of the cell(s).

Conduction heating pack In canning, where the heating of the contents of the container is by conduction. This is a slower rate of heating than 'convection' heating or a pack with a 'broken heating curve'.

Convection heating In canning, where the heating of the contents is by convection. This is the quickest rate of heat transfer compared with 'conduction' or 'broken heating curve' packs. Note that cooling of a container at the end of a process often begins with internal 'boiling', which gives a very rapid rate of heat transfer.

Co-valent Interatomic bond, hence 'co-valent bond', in which each of two atoms contributes one electron to form a stable pair.

Criterion A yardstick for making a judgment. A rule or test which is used for assessment. The plural is criteria. Microbiological criteria are regarded as 'standards', 'guidelines' and 'specifications'. See also Chapter 13.

D value This is the decimal reduction time, i.e. the time required at a quoted temperature to reduce the numbers of vegetative organisms or spores of a specific microorganism in a particular substrate or suspending medium to 10% of the initial number. With spores the temperature is usually quoted in minutes but it is essential to specify the unit of time just as it is to give the temperature and suspending medium.

Degradation The process of transforming organic compounds to others with a lower number of carbon and hydrogen atoms. This is usually an oxidative process.

DNA The short form of 'Deoxyribonucleic acid'. This is an organic molecule which carries genetic information. It is a long linear polymer of deoxyribonucleotide repeating units (purine + 2-deoxyribose + phosphate or pyrimidine + 2-deoxyribose + phosphate) linked together through the phosphate group joining the 3' of a sugar moiety (or equal part) to the 5' of the next one. Most DNA molecules are double stranded with the two strands kept together by hydrogen bonds between facing pairs of purine/pyrimidine bases. Double-stranded DNA has a right-handed helix structure.

Enzyme An organic catalyst usually consisting of a protein with a non-protein group, e.g. a co-enzyme, prosthetic group or metal atom(s).

Eukaryotic Organisms in which the chromosomes are in the nucleus of the cells, e.g. fungi. This is in contrast with prokaryotic organisms where they are not, e.g. bacteria.

F value An arbitrary unit of lethal heat equivalent to 1 minute at 121°C or 250°F assuming instantaneous heating and cooling. F_c is the center F value, i.e. at the center of a container which is assumed to be the slowest point of heating although this is not necessarily so. F_o is the F value calculated using a z value of 10°C or 18°F. Remember, an F value always has a reference temperature and a z value. Note that the letter 'F' may be used in other ways, which can be confusing.

Family In classification, e.g. of microorganisms, it is a rank below a 'Class' and above 'Genus' and 'Species'. Family names for bacteria and fungi end in -'aceae'.

Fat protection The presence of fat or oil droplets in a food may be associated with increased heat resistance of microorganisms. This is referred to as 'fat protection'.

Gamete A cell or nucleus involved in sexual reproduction, e.g. in fungi. It has half the number of chromosomes usual for the species.

Gametangium A structure which forms gametes.

Germ Popular name for a bacterial cell, usually used as the plural 'germs'.

Germ tube A short tube-like structure in fungi or sometimes yeasts which develops from some types of spore.

Glucans A polysaccharide made up of D-glucose units (or 'residues'), e.g. starch, cellulose.

Glycoprotein The name given to conjugated proteins which contain both carbohydrate and protein molecules.

Gram stain A stain developed originally by Christian Gram which should strictly be called Gram's stain. Bacteria can be classified, as an initial step in identification, into 'Gram-positive' which retain the stain and 'Gram-negative' which do not. Young, clinical isolates usually stain well and so give clear results. Old cultures and those taken from foods where the organism is 'stressed' may be 'Gram-variable'. Constant practice is required to get good results.

Helix A spiral, often used as helical, i.e. in the form or shape of a spiral.

Ionizing radiation A beam of particles or photons are used so that in collision with matter, the electrically neutral atoms gain or lose electrons becoming ions. For example, beta rays (electrons), gamma rays and X-rays can be used for sterilization of medical and biological equipment. Irradiation of food in 'pasteurizing' or 'sterilizing' doses is subject to regulatory and organoleptic (sensory) limits.

Lethal heat Heat which kills microorganisms. It is defined by exposure times and temperatures. Note that for a given time and temperature moist heat is much more effective than dry heat. This is illustrated if the laboratory 'sterilization' of glassware and culture media is compared.

Lipids These are organic compounds containing aliphatic hydrocarbon and derivatives like fatty acids, alcohols, etc. Fats and waxes are lipids.

Lipopolysaccharides These are conjugated polysaccharides formed from a lipid linked to a carbohydrate.

Lipoprotein These are conjugated proteins formed from a lipid linked to a protein.

Log reduction This term is usually used in connection with reducing microbial numbers by, for example, heat or chemical agents. The microbial numbers are expressed as numbers to the power of 10, e.g. 2000 is 2×10^3. A tenfold reduction is described as a log reduction of one, or one log cycle; it is also known as reducing the numbers by an order of magnitude.

Medium Used in the sense of 'culture medium' it is a nutrient mixture which allows growth of microorganisms. The plural is media. Diagnostic media show, e.g. by color change or colony appearance, that a particular kind of organism is present. 'Selective media' are designed to select for a particular group of organisms, i.e. they encourage the wanted organisms and discourage the unwanted ones.

Meiosis A cell-division process which results in the number of chromosomes being halved, e.g. in the formation of gametes.

Microaerophilic An organism which grows best when oxygen is present at a level significantly lower than that of air. The term is often used in place of 'microaerobic'.

Mesophilic Usually regarded as an organism which grows best at moderate temperatures, e.g. 20 to 45°C, but the limits may be considered as 15 to 45°C.

Micron Usually represented by the Greek letter μ it is one millionth of a meter (micrometer) or one thousandth of millimeter. Note that the European spelling of meter is metre.

Mitosis Cell division into two cells each of which has the same number of chromosomes as the original. This is in contrast to meiosis.

Mitochondrion The plural 'mitochondria' is usually used for these granules, rods or filaments which occur in most eukaryotic cells. They have a complex structure and composition and are important in the cells' chemical reactions (metabolism).

Morphology A study of form and structure.

Nucleoside A compound of a purine or pyrimidine base with a pentose (five carbon atom) sugar.

Nucleotide An ester of phosphoric acid with a nucleoside.

Order In classification, e.g. of microorganisms, it is a rank below a 'Class' and above a 'Family'. Order names for bacteria and fungi end in -'ales'.

Order of magnitude Indication of approximate microbial numbers usually expressed as a number to the power of 10, e.g. $5 \times 10^4 = 50\,000$.

P value An arbitrary unit of lethal heat used in assessing a pasteurization process. Like an '*F* value', it has a reference temperature and a *z* value. The term PU for Pasteurization Unit is also used.

Parasites Animals, plants, or microorganisms deriving nutrients from living organisms of another species upon which or in which they live. Saprophites live on dead organisms.

Phage See Bacteriophage – a virus infecting a bacterial cell.

Phospholipid A molecule of phosphoric acid combined with one or two molecules of fatty acids, an alcohol and a nitrogenous base.

Physiology A study of biological processes especially physico-chemical processes, activities or phenomena. The adjective is 'physiological'.

Plates The usual colloquial word for Petri dishes. In another form the word 'plate' is used in the phrase 'plating out' or inoculating Petri dishes.

Polymer A large or 'giant' molecule made of simple molecules (the monomers) joined to form a chain or web. Important biological materials, e.g. starches, cellulose, proteins, are polymers.

Polynucleotide A large molecule made from a linear sequence of nucleotides.

Polypeptide A large molecule, usually having a lower molecular weight than a protein, made of a chain of amino-acids linked by 'peptide bonds'.

Presumptive This is the first stage of identifying an organism in a processor's laboratory. It is usual to take action on this information, e.g. a presumptive *Salmonella* spp. diagnosis initiates corrective action. Further confirmatory and typing, e.g. phage typing, tests may be run if appropriate.

Prokaryotic Describes an organism in which the chromosomes are not separated from the cell contents (or cytoplasm) in a nucleus. Typically, bacteria are prokaryotic.

Protein This is a large, high molecular weight organic compound made up of α-amino acids linked together by peptide linkages.

Protozoa Microscopic, typically unicellular, eukaryotic animals. As an example, *Giardia lamblia* is a water-borne intestinal pathogen.

Psychrophile Literally, this means 'cold loving'. An organism which grows best at fairly low temperatures, e.g. optimum 10 to 15°C with a range of 0 to 20°C. Sometimes used to include psychrotrophs.

Psychrotroph An organism which can grow at low temperatures, e.g. 0 to 5°C but will grow better at moderate temperatures, i.e. 20 to 25°C. They can be regarded as a sub-group of mesophiles, since they

will grow at 30°C or, e.g. *Listeria monocytogenes*, at 37°C.

Quenching agent Because some disinfectants (sanitizers) show a bacteriostatic effect (i.e. multiplication is prevented) at high (great) dilutions, a neutralizing or 'quenching' agent must be used when taking swabs or rinses. If this is not done the results will be erroneously 'good'.

Refraction The change in direction of a propagating wave, usually a beam of light, produced by passing through a transparent medium. Refractivity is a measure of refraction. This can be used when examining spores microscopically.

Respiration This is an energy-producing oxidative process which may take place aerobically or anaerobically since it involves electron transfer. The yield of energy is greater than would be obtained by fermentation of the given substrate.

Rhizoid A root-like structure found in some fungi and algae.

Ribosome A particle in the cytoplasm which may be free or membrane-bound and is usually found in large numbers in a cell. It is made up of proteins and ribonucleic acid (RNA) and is involved in protein synthesis.

RNA A long chain of molecules of ribonucleotides, usually single stranded, which is involved in protein synthesis.

Rods A rod-shaped bacterium. Usually used in a brief description of an organism with its Gram straining reaction and whether it forms spores, e.g. 'It's a Gram-positive spore forming rod'.

Saprophyte Animals, plants or microorganisms deriving nutrients from dead or decaying organisms. Microorganisms do this by absorbing soluble compounds.

Somatic Refers to a non-reproductive cell, structure or function. A somatic antigen is located in the body of the cell, usually at the surface, in distinction from one which occurs in a particular part of the cell, e.g. in the flagella or the capsule.

Species A taxonomic group of similar organisms which in higher plants and animals is sufficiently alike to be able to interbreed. With microorganisms, the species concept is a difficult one which relies for appropriate criteria on a consensus of expert judgment. For the processor, the strain, serotype or phage type which is part of a species can be important information especially when dealing with an environmental problem. As an example, many slightly different organisms belong to the *Salmonella* species but to locate the source of a particular contaminant a closely defined 'fingerprinting' system is needed.

Spore Some, but not all, microorganisms produce spores. Bacterial spores are characteristically resistant to heat, acid, salt and other adverse conditions. A bacterium which can produce spores is called a spore former, and one bacterium will produce a single spore. Mold spores can be produced in enormous numbers which are carried by slight currents of air and are an efficient means of distributing the organism. Molds are identified using the spores and spore-producing structures as important criteria.

Spore suspension A high concentration, usually over 10^6 per ml, of spores of a particular strain used for experimental purposes, e.g. establishing 'sterilizing' conditions for product or packaging.

Strain A microorganism distinguishable in some way(s) from other similar organisms, usually associated with a particular use or concern. Thus a strain may be particularly heat resistant, or make acid quickly, or produce a desirable flavor. Because microorganisms may readily change their properties when sub-cultured, techniques such as freeze-drying are used to 'preserve' wanted strains.

Synthesis A biological process of building complex and/or different molecules usually using simpler molecules.

Taxonomy This is a system or the science of biological classification which aims to reflect natural affinities and to serve as a system of naming organisms. Different systems may be used for different purposes and in recent years numerical taxonomy and molecular biology methods have been used to develop objective criteria and reflect fundamental relationships.

Thermoduric Describes an organism which tolerates or survives time/temperature conditions which would be expected to kill organisms of a similar type. An alternative description is thermotolerant. In dairy usage, it refers to non-sporing organisms surviving the pasteurization process.

Thermophile A term which has been rather imprecisely used. It may be preferable to apply it to organisms whose optimum growth temperature is over 45°C (113°F). Thermophilic spore formers of interest to canners may have optimal growth temperatures in the 55° to 65°C range.

Variability (of organisms) The extent of this is easily underestimated as a culture, whether in the laboratory or in a food, will adapt to the environment. In the laboratory, cultures can be 'trained' to have 'atypical' or unusual characteristics. These characteristics, however, may not be stable and this can present problems in biotechnology applications as well as in the maintenance of cultures with desirable properties or in identifying microorganisms isolated from foods.

Virion A virus particle.

z value The number of degrees (C or F) required for a tenfold change in the *D* value of a microorganism in a particular substrate. Be aware that both °C and °F are used in the published literature.

Reference

1. Singleton, P. and Sainsbury, D. (1987) *Dictionary of Microbiology and Molecular Biology*, 2nd edn. Chichester: John Wiley

Chapter 8
Biological factors underlying food safety, preservation and stability

8.A Numbers of microorganisms in foods and their growth kinetics

8.A.1 Contamination

Living microorganisms exist everywhere in the environment, e.g. in plant equipment, ingredients, and packaging, unless action is taken to kill them. It is important for management to understand that plant and equipment that are cleaned, so that they can take a justifiable pride in its appearance, are not necessarily sterile. The residual microorganism are defined as contaminants and their presence explains why contamination of food and the growth kinetics of the contaminating microorganisms are important issues.

Contamination of foods

Testing shows that microorganisms are so widespread in the environment that all substrates, including foods, can be regarded as contaminated. The significance of contamination depends on the numbers and types of microorganisms, and the opportunity for growth.

Upper limits

Levels of contamination will range from calculated upper limits to extremely low ones. Using the dimensions of two guide microorganisms (*Escherichia coli* representing bacteria and *Saccharomyces cerevisiae* yeasts) the maximum level of contamination when the organism occupies *all* the available space may be calculated to be of the order of:

1 or 2×10^{12} bacterial cells per ml or g

and

1×10^{10} yeast cells per ml or g

However, in practice the counts found in contaminated foods or in broth cultures are rarely more than 10^9 per ml or g of bacteria and 10^6 per ml or g of yeasts.

The question may well be asked, at what level of contamination does a food become spoiled or fermented? When numbers reach 10^9 per ml or g, food will certainly be spoiled. Spoilage can occur at much lower levels, for example at 10^5 or 10^6 per ml or g, depending on the type of organism present in a particular food. At these lower numbers the taste, texture or appearance of the food may be affected by certain types of bacteria, in which case spoilage has occurred.

Lower limits

The minimum level of contamination found in food may be less than 1 cell per ml or g, and can be so low

that contamination cannot be detected by usual microbiological analytical methods. This is the case with levels of one microbial cell per 1000 kg of substrate, or 1 in 10^{12} containers of substrate, as is aimed for when applying a botulinum cook to foods. Since this level of contamination is too low to verify, the food will be commonly regarded as 'sterile'.

There is, however, the possibility of misunderstanding the word 'sterile' particularly when phrases like 'almost sterile' or 'semi-sterile' are heard. To promote the concept of safety it is helpful to use the word 'sterile' to mean the absence of *all* living organisms. Microbiologists will know that absence means 'not found' in a particular weight, volume or area when using specified methods for examination.

At intermediate levels of contamination a food may be regarded as safe or not, depending on the types and numbers of microorganisms present. Unwashed fresh foods may contain about 10^5 cells per g of saprophytic, i.e. non-pathogenic, microorganisms. When carefully washed, such foods may carry only 10^2 to 10^3 microorganisms per gram, and until the numbers of microorganism increase, such foods will not suffer microbiological spoilage. Providing that pathogen numbers are low, i.e. below an 'infective dose' or toxin has not been produced in amounts which cause harm, such foods are regarded as safe.

The infective dose of some bacterial pathogens may be as low as one single cell, and has been shown to be 50 to 100 *Salmonella* from contaminated chocolate and <10 *Salmonella* from cheddar cheese. For most pathogens, however, higher numbers are required to cause illness. This is particularly the case with bacteria which produce toxins, when a count of at least 10^5 per ml or g must usually be present to produce sufficient toxin in the food to cause illness. This order of magnitude of bacteria is also necessary in food for *Clostridium perfringens*, when the pathogen produces toxin rapidly in the intestine following the contaminated food being eaten.

8.A.2 Growth kinetics

Inevitably in suitable environmental conditions of nutrition and temperature the number of contaminant microorganisms increases. This means that:

1. Pathogens may reach a dangerous concentration even though starting from low contamination levels.
2. The saprophytic flora may reach very high levels and so spoil the food.

In order to give a picture of the way in which microbial populations change, a simple idealized model will be described. However, it is a good model and applies to pure culture or food situations. It is usually referred to as the microbial growth curve.

Phases of growth

The growth curve is characterized by four basic phases (see Figure 8.1).

I.   The Lag Phase
II. The Exponential Phase
III. The Stationary Phase
IV. The Death Phase

I. Lag Phase – During this phase the cell numbers do not increase. The microorganisms adapt to the environment, and intra-cellular metabolic pathways are established which are important during the growth cycle.

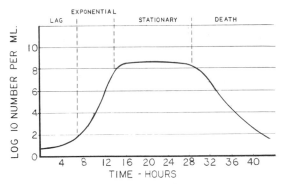

Figure 8.1 Growth curve of a population of microorganisms showing phases

In principle, the food manufacturer would wish to process his products during the Lag Phase. Unfortunately, the variety of microorganisms found in a typical ingredient or product range makes this an ideal rather than a fully achievable goal.

II. Exponential Phase – This is the true growth phase during which the microbial numbers increase at the maximum rate allowed by environmental conditions. The length of this phase varies; usually it lasts at least two hours, but may extend over several hours.

Bacteria multiply by binary fission, characterized by the duplication of cell components within each cell, followed by prompt separation to form two daughter cells. The new cells then undergo the same process. Each duplication takes what is known as a generation time to complete. Theoretically, under favorable environmental conditions, a single cell may produce a progeny of about 10^9 cells within 7.5 hours when the generation time is 15 minutes.

Most yeasts multiply by budding. A small outgrowth takes place from a cell, which enlarges in size until it equals the parent cell, and then divides off to form a new individual cell. Generation time of yeasts is slower than that of bacteria and a typical time is 2 to 3 hours in foods leading from an original contamination of 1 yeast per ml or g of food to spoilage in about 40 to 60 hours.

The overall process of multiplication involves both the absorption of nutrients from the substrate into the cell (for synthetic – anabolic – purposes) and the removal of waste or catabolic products from the cells into the substrate.

Thus microbial growth unavoidably results in physico-chemical changes of the environment. These become great enough to make the environment unsuitable for further growth. In this condition, the microbial population has reached senescence or the stationary phase.

III. Stationary Phase – This is characterized by no net increase in numbers. The maximum cell concentration in this phase equals about 10^9 cells per ml or g of food. The length of the stationary phase (usually ranging from 2 to 4 hours to more than 30 days) depends upon both the availability of energy sources for the maintenance of cell viability and the degree of pollution in (hostility of) the environment.

In fact, the physico-chemical characteristics of the environment are altered so profoundly by the growing process that the survival of the populations is affected. These conditions result in the death phase.

IV. Death Phase – This is characterized by a decrease in the number of living organisms very often according to an exponential decay. The length of this phase may range between about 24 h and more than 30 days, depending on the environmental conditions and type of organisms. For example, the 'die off' period of *Salmonella* in cheddar cheese is considered to be at least 60 days and can be as long as 9 months.

The shape of the growth curve is, of course, affected by factors such as temperatures higher or lower than the optimum, limitation of nutrients for the organism and the presence of natural antimicrobial compounds or disinfectants.

8.B. Factors affecting the growth curve

Background – the most important factors

Physico-chemical factors, including temperature, affect microbial growth and produce changes in the shape of the growth curve. Sub-optimal conditions are usually characterized by an increase in length of the lag phase, a decrease of the growth rate in the exponential phase and/or by a lowering of the maximum level of cells reached in the stationary phase. These changes are dependent on the environmental factors found in the product concerned. In extreme situations, growth may be prevented completely. The most relevant physico-chemical factors affecting growth in foods are:

1. Water
2. Temperature
3. pH
4. Oxygen tension
5. Nutrients (substrate composition)

In practical food-preservation, a combination of factors is usually used. This is the basis of the 'Hurdle' concept first proposed by Leistner[1] and may include the use of 'chemical' or 'artificial' preservatives.

There are, of course, many other factors which can affect the shape of the growth curve; among these are anti-microbial compounds which occur as part of a product formulation, such as some spices and 'essential oils'. However, for this section attention will be given to the five factors listed above.

Reference

1. Leistner, L. (1978) Microbiology of ready-to-serve foods. *Fleischwirtschaft*, **58**, 2008–2111

8.B.1 Water

Introduction

Microbial growth occurs only in the presence of water. This does not necessarily need to be an obvious, visible, aqueous layer. For example, microbial growth can cause problems in aircraft fuel systems and yeasts and molds will grow on dried fruit. Individual types of microorganisms have different requirements for inorganic and simple or complex organic compounds and thus, because microorganisms absorb their food through the cell wall, they will grow in aqueous solutions which provide for their needs.

However, the mere presence of water in a solution does not insure water availability, since this depends on the solute concentration. Since the process of solution involves the setting up of linkages between the water and solute molecules, as the quantity of solute increases, the quantity of water involved in these linkages also increases. It follows, therefore, that increasing the solute concentration decreases the availability of water and thus the probability of growth to any microorganisms in this environment.

A criterion is therefore needed which measures the availability of water to the organism other than that of percentage water in the substrate. The one which is used is 'water activity' with the symbol a_w.

Water activity a_w – definition

Water activity, a_w, is equal to the ratio (p/po) of the vapor pressure of the solution (p) to that of pure water (po) at the same temperature and pressure, and equals the fractional Equilibrium Relative Humidity ERH $\div$ 100 in the atmosphere above the solution.

Theoretically, dilute solutions obey Raoult's Law:

$$a_w = N_w/(N_w + N_s) \qquad \text{(equation A.1)}$$

where N_w and N_s are the numbers of moles (or molecules) of water and solute, respectively. It follows that in pure water the above relationship becomes:

$$a_w = N_w/(N_w + 0) = 1$$

and in absence of water:

$$a_w = 0/(0 + N_s) = 0$$

Therefore the a_w values necessarily range between 0 and 1. The theoretical a_w of a solution could be computed using equation (A.1). However, some deviation from the predicated a_w of solutions does occur in practice, mostly at high solute concentrations. The a_w of complex substrates cannot be calculated using equation (A.1) since in solutions containing many solutes there are interacting effects among molecules that equation (A.1) does not take into account. Several relationships have been derived which enable the calculation of the a_w of solutions containing different solutes to be made. The equations of Norrish[1] and that of Ross[2] have wide acceptance. However, water activity of complex solutions and of foods is better measured by means of appropriate instruments such as electronic hygrometers. Accurate measurement is not easy, and no better than ±0.02 is to be expected in routine determinations.

Most natural products have high water activity, usually near $a_w = 0.99$. Water activity is usually lowered by adding solutes such as salt or sugar or by removing some water from the food product by concentration or drying.

The probability of microbial development is very high at a_w values near 1 and it decreases as the a_w decreases.

Water activity and growth of organisms

Microorganisms may grow only in substrates having an a_w higher than 0.61. However, they could be classified as steno-osmotic (steno- means obligate) and euri-osmotic (euri- means facultative), in that some microorganisms grow only in a well-defined a_w range, while others grow in the entire range from 0.999 to 0.61. A few organisms grow only in, or prefer substrates having, low a_w that is between 0.80 to 0.61.

If the main a_w lowering solute is a salt (usually NaCl) microorganisms able to grow are called halophiles, while if it is sugar or a mixture of sugars they are called osmophiles. The so-called xerophiles are molds growing only at low a_w.

As shown in Table 8.1 each microorganism has a limiting a_w below which it is unable to grow.

As a general rule, most bacteria are unable to grow at an a_w less than 0.90 and none at an a_w less than 0.75. Yeast and molds can grow over a wider range of a_w values and some grow at an a_w as low as 0.61 or 0.62.

Effect of a_w on growth curve

Apart from the complete inhibition of growth in substrates with a low enough a_w value, the main effects produced by lowering a_w can be regarded as:

- prolongation of the lag phase, e.g. from 1 to 2 h to more than 4 to 20 h.
- lengthening the generation time in the exponential phase, e.g. lengthening the generation from 0.5 to 2 h to more than 10 to 20 h.
- lowering the maximum level of microorganism numbers reached in the stationary phase.
- increased speed of death during the death phase.

These effects are more marked as the a_w approaches the limiting value (Figure 8.1).

In food technology the most important a_w lowering solutes are sucrose, invert sugar, salt (NaCl) and dextrose syrups. The relationship between sugars or salt concentration and water activity of the substrate is shown in Tables 8.2 to 8.4.

Generally, fresh foods contain less than 5 to 10% of most important a_w lowering solutes; that is, sugars and salts. With such foods the a_w reached by adding sugar or salt is not significantly lower than the value expected in an aqueous solution containing only the added sugar or salt. With concentrated or dried foods, instrumental measurement is the method of choice for determining a_w of these

Table 8.1 Lower limiting water activity of microorganisms

GENUS	Limiting a_w	
	Fungi	Bacteria
Schizosaccharomyces	0.984	
Campylobacter		0.98
Leuconostoc		0.96
Rhizoctonia	0.96	
Agrobacterium		0.958
Acinetobacter		0.95
Aeromonas		0.95
Arthrobacter		0.95
Alcaligenes		0.95
Corynebacterium		0.95
Citrobacter		0.95
Flavobacterium		0.95
Klebsiella		0.95
Yersinia		0.95
Enterobacter		0.943
Serratia		0.943
Listeria		0.94
Proteus		0.94
Pseudomonas		0.94
Torula	0.94	
Stachybotrys	0.94	
Microbacterium		0.935
Candida	0.931	
Botrytis	0.93	
Clostridium		0.93
Escherichia		0.93
Mucor	0.93	
Rhizopus	0.93	
Salmonella		0.93
Lactobacillus		0.928
Streptococcus		0.924
Sarcina		0.92
Bacillus		0.90
Saccharomyces (not osmophilic)	0.90	
Trichothecium	0.90	
Hansenula	0.882	
Endomyces	0.88	
Willia	0.88	
Vibrio		0.86
Staphylococcus		0.85
Byssochlamis	0.84	
Paecilomyces	0.84	
Debaryomyces	0.83	
Micrococcus		0.83
Aspergillus (not osmophilic)	0.77	
Halobacterium		0.75
Wallemia	0.75	
Aspergillus (osmophilic)	0.70	
Chrysosporium	0.69	
Penicillium	0.62	
Saccharomyces (osmophilic)	0.62	
Eremascus	0.61	
Xeromyces	0.61	

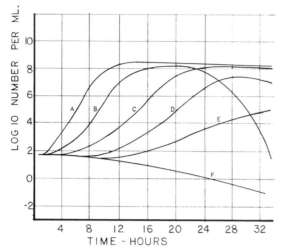

Figure 8.2 Changes in the growth curve produced by changing the water activity. *Note*. This is a generalized series of curves with *A* being high water activity and *F* being low

Table 8.3. Isotherm of invert sugar at 25°C

g/100 g Solution	a_w	g/100 g Water	a_w
10	0.989	20	0.980
15	0.982	40	0.958
20	0.975	60	0.937
25	0.966	80	0.915
30	0.956	100	0.892
35	0.945	120	0.870
40	0.932	140	0.848
45	0.916	160	0.826
50	0.897	180	0.804
55	0.874	200	0.783
60	0.846	220	0.762
65	0.811	–	–
70	0.765	–	–

Table 8.2 Isotherm of sucrose at 25°C

g/100 g Solution	a_w	g/100 g Water	a_w
10	0.994	20	0.989
15	0.990	40	0.977
20	0.986	60	0.964
25	0.981	80	0.950
30	0.975	100	0.935
35	0.968	120	0.919
40	0.959	140	0.904
45	0.949	160	0.887
50	0.935	180	0.870
55	0.919	200	0.853
60	0.897	220	0.836
65	0.868	–	–
70	0.828	–	–

Table 8.4 Isotherm of NaCl at 25°C

g/100 g Solution	a_w	g/100 g Water	a_w
2	0.993	3	0.989
4	0.984	6	0.975
6	0.973	9	0.959
8	0.961	12	0.940
10	0.946	15	0.919
12	0.928	18	0.896
14	0.909	21	0.871
16	0.887	24	0.845
18	0.863	27	0.819
20	0.836	30	0.791
22	0.808	33	0.763
24	0.776	–	–
26	0.743	–	–

complex substrates as they contain significant concentrations of several a_w-lowering solutes.

Effect of a_w of foods on microbial growth

Table 8.5 shows the a_w usually found in several foods. By combining Tables 8.1 and 8.5, we obtain Table 8.6, showing which groups of microorganisms are able or unable to grow in different foods.

For further information, see the chapter by Christian in the ICMSF book *Microbial Ecology of Foods*[3].

References

1. Norrish, R. S. (1966) An equation for the activity coefficients and equilibrium relative humidities of water in confectionary syrups. *Journal of Food Technology*, **1**, 25–39
2. Ross, K. D. (1975) Estimation of water activity in intermediate moisture foods. *Food Technology*, **29**, (3), 26–34
3. Christian, J. H. B. (1980) Reduced water activity. In *Microbial Ecology of Foods*, Vol. 1, Factors Affecting the Life and Death of Microorganisms, International Commission on Microbiological Specifications for Foods, pp. 70–91. New York: Academic Press

Table 8.5 Examples of water activity of foods (decreasing order)

Product	a_w
Truffles – ground	0.99
Fresh poultry	0.985
Fresh fish	0.985
Fresh vegetables	0.98
Gnocchi	0.98
Fresh fruit	0.98
Fresh meat	0.98
Pudding	0.98
Wurstel	0.976
Cheese – spreadable	0.975
Mortadella	0.97
Eggs	0.97
Truffles – (white) paste	0.96
Salami – Hungarian style (upper value)	0.942
Truffles – (black) paste	0.94
Bread – white	0.94
Olives – ground (upper value)	0.939
Coconut ground and sugared (upper value)	0.935
Salami – Hungarian style (lower value)	0.931
Mayonnaise – 80% oil	0.925
Olives – ground (lower value)	0.921
Salami – fermented	0.91
Mint-flavored syrup (upper value)	0.905
Maple syrup	0.9
Chestnut jam	0.895
Mint-flavored syrup (lower value)	0.875
Mint-flavored syrup (spoiled)	0.87
Coconut – ground and sugared (lower value)	0.864
Chestnut jam – Invert sugar 59%	0.848
Peach jam	0.84
Jelly	0.84
Milk – condensed	0.83
Chestnut jam	0.825
Bilberry jam	0.81
Almond paste – sugared	0.81
Orange-flavored syrup (upper value)	0.805
Mandarin jam	0.801
Tamarind pulp	0.8
Black cherry-flavored syrup	0.8
Chestnut heavy syrup	0.8
Apricot jam	0.8
Rice	0.8
Cake	0.8
Quince jam	0.79
Cherry jam	0.79
Orange juice conc. (upper value)	0.79
Cheese – parmesan	0.785
Mint-flavored syrup (lowest value)	0.783

Product	a_w
Orange-flavored syrup (lower value)	0.78
Prunes – dried	0.77
Benzoin paste	0.76
Molasses	0.76
Orange juice conc. (lower value)	0.74
Figs – dried	0.72
Glucose syrup	0.7
Dates – dried	0.7
Malt extract – Invert Sugar 38%	0.69
Cream – chocolate filled	0.69
Chocolate candy	0.69
Sultanas	0.66
Flour (spoiled)	0.651
Apricots – (dried)	0.65
Toffee	0.61
Caramels	0.6
Honey – crystallized	0.589
Flour	0.575
Honey	0.56
Pasta	0.44
Soup – dried	0.42
Bread crust	0.3
Biscuits	0.3
Milk – dried whole	0.2
Vegetables – dried	0.2
Sugar	0.19
Cereals	0.15
Crackers	0.1

Note: Several products of the same type were found to have different water activities. This table is intended only as a guide. Remember that the accurate determination of a_w is not an easy technique and that the range of the determinations is important in assessing product stability.

Further reading

1. Caurie, M. (1985) A corrected Ross equation. *Journal of Food Science*, **50**, 1445–1447
2. Hocking, A. D. (1988) Molds and yeasts associated with foods of reduced water activity: ecological interactions. In *Food Preservation by Moisture Control*, conference proceedings, edited by C.C. Seow, pp. 57–72. Barking: Elsevier Applied Science
3. Labuza, T. P. (1980) The effect of water activity on reaction kinetics of food deterioration. *Food Technology*, **34**, (4) 36–41+59
4. Resnik, S. L. and Chirife, J. (1988) Proposed theoretical water activity values at various temperatures for selected solutions to be used as reference sources in the range of microbial growth. *Journal of Food Protection*, **51**, 419–423

Table 8.6 Water activity, foods and spoilage organisms

GENUS	Limiting a_w	Affected foods
Schizosaccharomyces	0.984	Fresh meat, poultry, fish, milk; fresh vegetables, fruits, wurstel, mortadella, pudding; gnocchi, cheese (spreadable), eggs, truffles (white) paste
Campylobacter	0.98	
Leuconostoc	0.96	
Rhizoctonia	0.96	
Agrobacterium	0.958	
Acinetobacter	0.95	
Aeromonas	0.95	
Arthrobacter	0.95	
Alcaligenes	0.95	
Corynebacterium	0.95	
Citrobacter	0.95	
Flavobacterium	0.95	Processed cheese, gouda cheese, canned cured meats[c]
Klebsiella	0.95	
Yersinia	0.95	
Enterobacter	0.943	
Serratia	0.943	
Listeria	0.94	
Proteus	0.94	Truffles (black), bread (white), salami Hungarian style (upper value), olives (ground), coconut ground – sugared (upper value)[c]
Pseudomonas	0.94	
Torula	0.94	
Stachybotrys	0.94	
Microbacterium	0.935	
Candida	0.931	
Botrytis	0.93	
Clostridium	0.93	
Escherichia	0.93	
Mucor	0.93	
Rhizopus	0.93	Mayonnaise, salami Hungarian style (lower value), olives (ground), salami Italian style, mint-flavored syrup (upper value)[c]
Salmonella	0.93	
Lactobacillus	0.928	
Streptococcus	0.924	
Sarcina	0.92	
Bacillus	0.90	
Saccharomyces[a]	0.90	
Trichothecium	0.90	Maple syrup, chestnut jam[c]
Hansenula	0.882	
Endomyces	0.88	
Willia	0.88	
Vibrio	0.86	Coconut ground – sugared; mint-flavored syrup, chestnut jam (invert sugar 59%), peach jam, fruit jelly, condensed milk[c]
Staphylococcus	0.85	
Byssochlamis	0.84	
Paecilomyces	0.84	
Debaryomyces	0.83	
Micrococcus	0.83	
Aspergillus[a]	0.77	Jams – chestnut, bilberry, mandarin, cherry, quince, apricot, orange-flavored syrup, black cherry-flavored syrup, chestnut in heavy syrup, mint-flavored syrup (lower value), rice cake, molasses, cheese (parmesan), prunes (dried), benzoin paste, almond paste, tamarind pulp[c]

Table 8.6 Continued

GENUS	Limiting a_w	Affected foods
Halobacterium	0.75	Orange juice concentrate, dates (dried), figs (dried), dextrose
Wallemia	0.75	syrup[c]
Aspergillus[b]	0.70	
Chrysosporium	0.69	Malt extract, chocolate candy, chocolate (cream filled)[c]
Penicillium	0.62	Sultanas, caramels, flour (spoiled), apricot (dried), toffees[c]
Saccharomyces[b]	0.62	
Eremascus	0.61	
Xeromyces	0.61	

[a] non-osmophile species
[b] osmophile species
[c] 'All the above products and in addition:'

Products having a_w <0.61:

Honey	0.59	Bread crust	0.30	Sugar	0.19
Pasta	0.44	Biscuits	0.30	Milk (dried)	0.15
Soup (cube)	0.42	Vegetables (dried)	0.20	Cereals	0.15
				Crackers	0.10

8.B.2 Temperature

Introduction

Microorganisms of interest in food microbiology may grow at temperatures in a range from about −12 to 42°C or above, e.g. in hot can vending machines. No single microorganism is able to grow over this entire range.

They are therefore classified in three arbitrary major categories or groups according to the range of temperature within which they may grow. These groups are psychrophilic, mesophilic and thermophilic organisms. Each group is characterized by three cardinal values: the minimum, optimum and maximum temperatures of growth. The minimum and maximum temperatures are those beyond which microorganisms are unable to grow, although at temperatures lower than the minimum microorganisms may survive for very long periods. At temperatures higher than maximum they die at a rate proportional to the temperature. The optimum temperature is determined mostly on an empirical basis, e.g. according to the rate of growth or of metabolism, the amount of recovered metabolic by-products, etc.

As so often happens in microbiology, there are exceptions to the rule; in this instance that temperatures higher than the optimum result in death, i.e. a diminution of count. With some spore suspensions, initially heating above the optimum (known as a 'heat' shock) increases the count. This gives rise to the 'shoulder' on the thermal death time curve. Further heating shows a logarithmic order of death.

Data on cardinal temperatures

The data presented in Table 8.7 are for 'typical' or 'usual' strains. It is only intended as a guide and the experienced microbiologist may well have encountered strains which grow outside the given limits. With the development of new product/package/processing combinations it is important to know exactly how limits were determined before building them into a predictive model.

Warning Be sure the data you propose using are fully valid for your purposes.

As can be seen from the data in Table 8.7., microorganisms growing at low temperature can be ascribed to the Psychrophilic (sometimes called the Psychrotrophic) group, those able to grow at high temperature to the Thermophilic group, and those able to grow only at intermediate temperature (the majority of organisms) to the Mesophilic group. Psychrophilic and Thermophilic organisms may be further sub-divided into obligate (or steno-) or facultative (or euri-) groups. The obligate group has a narrower growth temperature range than the faculative group.

Table 8.7 shows cardinal temperatures of some common microorganisms. As can be seen, most microorganisms are unable to grow at temperature lower than about 5°C or higher than 46° to 48°C; those able to grow at the lowest temperatures are in general, the molds and yeasts, although among pathogenic bacteria *Listeria monocytogenes*, *Yersinia enterocolitica* and *Clostridium botulinum* Type E and non-proteolytic B and F strains can grow at 3.3°C.

Table 8.7 Growth range temperatures of some microorganisms(°C)

Organism	Minimum	Optimum	Maximum
Debaryomyces hansenii	−12.5	–	37 F
Cryptococcus albidus	−12	–	35 F
Trichosporon aculeatum	−10	–	41 F
Molds	−10	25	58 F
Trch. pullulans	−8.5	–	34 F
Cladosporium	−8	–	– F
Candida frigida	−7	15	20 F
Pseudomonas fragi	−6.5	24	– b
Arthrobacter glacialis	−5	15	20 b
Moraxella	−5	–	42 b
Ps. fluorescens	−4	26	40 b
Ps. rubescens	−3	–	– b
Ps. putrefaciens	−2	25	– b
Rhodotorula glutinis	−2	–	41 F
Botrytis cinerea	−1	20	30 F
Leucosporidium scottii	−1	–	35 F
Vibrio anguillarum	−1	28	40 b
Aeromonas	0	28	42 b
Brochothrix	0	25	33 b
Candida humicola	0	25	38 F
Clostridium putrefaciens	0	20	35 b
Mucor mucedo	0	–	28 F
Nadsonia elongata	0	–	25 F
Saccharomyces cerevisiae	0	28	44 F
Xanthomonas	0	28	40 b
Alternaria solani	2	27	45 F
Chromobacterium lividum	2	–	35 b
Propionibacterium	2	28	45 b
Rhizoctonia solani	2	28	35 F
Pichia membranefaciens	3	20	30 F
Acetobacter	5	28	42 b
Acinetobacter	5	28	48 b
Arthrobacter	5	28	37 b
Brevibacterium	5	28	42 b
Candida lipolytica	5	25	40 F
Kurthia	5	28	45 b
Lactobacillus	5	28	55 b
Lypomyces starkey	5	–	40 F
Micrococcus	5	28	45 b
Rhizopus	5	25	28 F
Streptococcus cremoris[a]	5	28	40 b
Torulopsis candida	5	22	32 F
Verticillium atrum	5	25	35 F
Proteus	6	28	43 b
Gluconobacter oxydans	7	28	41 b
Trichophyton mentagrophytes	8	30	40 F
Bacillus subtilis	10	28	51 b
Chromobacterium violaceum	10	–	44 b

Table 8.7 Continued

Organism	Minimum	Optimum	Maximum
Leuconostoc	10	28	42 b
Microbacterium	10	28	40 b
Streptococcus lactis[b]	10	28	42 b
Humicola thermoides	24	45	56 F
B. stearothermophilus	28	55	72 b
B. coagulans	30	45	60 b
Cl. thermosaccharolyticum	30	55	71 b
Desulfotomaculum nigrificans	30	55	71 b
Mucor pusillus	–	43	– F
Penicillium rubrum	–	26	– F
Hansenula	–	40	50 F
Kloeckera	–	28	– F
A. fumigatus	–	35	50 F

F = Fungi (molds or yeasts)
b = bacteria
[a] *Lactococcus lactis* subsp. *cremoris*
[b] *Lactococcus lactis* subsp. *lactis*

Table 8.8 shows cardinal temperatures of some pathogenic microorganisms. As can be seen, pathogenic microorganisms belong to the mesophilic range of growth, with a few exceptions.

Growth rate models

The food microbiologist is usually more concerned commercially with the rate of growth than with the extreme limits of growth. This is so that predictions can be made as to when an organism or group of organisms are likely to cause a problem or to provide an explanation for a problem. There are two models which are often used, the 'Arrhenius' and the 'Square Root'. Using such a model demands both microbiological data and mathematical skills. It does, however, enable predictions to be made with more precision than could otherwise be made. In the current state of the art it is probably better to regard them as 'research tools' which hopefully will soon become a 'development' method, much as 'Sterilization Theory' is used in canning.

Arrhenius model

Within a defined range of temperature, microbial activity is affected in a number of ways, e.g. in respect of nutritional requirements, anabolic and

Table 8.8 Growth range temperatures of some pathogenic microorganisms (°C)

Organism	Minimum	Optimum	Maximum
Listeria monocytogenes	0	28	45
Yersinia enterocolitica	3	32 to 34	44
Aspergillus flavus	3	25	44
A. parasiticus	3	25	44
Bacillus cereus	3 to 4	30	48
Cl. botulinum, non-proteolytic	3.3	25 to 37	40 to 45
Vibrio parahaemolyticus	5	37	43
Pseudomonas aeruginosa	5	25 to 30	42
Salmonella	5 (8 to 12 in product)	37	46
Staphylococcus aureus	7 (8 to 15 in product)	37	48
Staph. aureus toxin production	10	40 to 45	48
Cl. botulinum, proteolytic	10 to 12	30 to 40	48
E. coli, EEC	<10	30 to 42	44
Cl. perfringens	15	43 to 45	52
Campylobacter	30	42	45

Note: Some *B. cereus* strains can grow at 3° to 4°C. At 4°C, two log cycle increases occur in 7 days and 3 log cycles in 10 days.

catabolic pathways, morphology, rate of reproduction, etc. An important effect of temperature upon microbial growth is on the generation time.

As a general rule, the effect of temperature upon the growth curve is analogous to the effect produced by water activity. As the temperature increases the lag and stationary phases shorten, the slope of the exponential and death phases is steeper, and the maximum cell level attained during the stationary phase increases. At temperatures a few degrees above the optimum cardinal value the growth rate reaches the maximum value and decreases very rapidly at temperatures only a little higher. Plotting the generation time (g_t) in the exponential phase of growth against temperature (T) yields curves of the type shown in Figure 8.3. As may be seen, growth rate changes suddenly at temperatures higher than optimum, but more slowly with decreasing temperature. It is well known from chemical kinetics that

reaction rates usually increase with temperature. See Section 8.C.2 for a discussion on thermal death rates.

Accordingly, the microbial growth rate can be described by plotting the logarithm of the growth rate (k) against $1/T$. As shown in Figure 8.4, the shape of the curve is analogous to that shown in Figure 8.3, although inverted. The linear portion of the curve can be described by a relationship of the type described below:

Explanation of symbols:

Ln = Log_e and is followed by a symbol
k = Reaction rate
E_a = Arrhenius activation energy
μ = the temperature characteristic of the organism (See E_a above)
R = the gas constant
T = the absolute temperature in degrees Kelvin
N_t = Concentration of cells at time t
N_0 = Concentration of cells at time zero
t = time
A = the 'frequency factor' (see below)
K = absolute reaction rate
exp = exponential
* = multiply

$$Ln\,k = Ln\,A - \mu/RT$$

where the value of k is obtained from the general equation of the exponential growth:

$$N_t = N_0 * \exp(k * t)$$

and then:

$$Ln\,N_t = Ln\,N_0 + k * t$$

so that:

$$k = (Ln\,N_t - Ln\,N_0)/t$$

The μ value is a constant called the temperature characteristic of the microorganism. The parameter μ is equivalent to the Arrhenius (apparent) activation energy (E_a) of chemical kinetics, obtained from the classical relationship:

$$K = A^* \exp(-E_a/RT)$$

after which:

$$Ln\,k = Ln\,A - E_a/RT$$

where T is the absolute temperature (degrees Kelvin), R is the gas constant and A is the 'frequency factor' linked to the entropy of the system. Higher μ values are expected in thermophilic microorganisms than in psychrophiles.

For this equation to fit the data closely, μ should have a constant value over the temperature range studied. Experimental data show that this is not so, and the value decreases with increasing temperature. Schoolfield et al.[1] have taken account of this and their model is used in a discussion of predictive

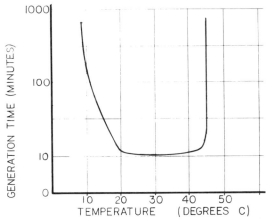

Figure 8.3 Relationship between generation time and temperature

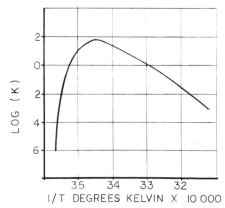

Figure 8.4 Relationship between growth rate and absolute temperature

food microbiology by Baird-Parker and Kilsby[2]. They refer to work on growth kinetics taking the factors of water activity, pH and temperature separately and in combination. They conclude that 'it is in situations such as this, where a number of variables interact to affect the lag time and the specific growth curve that this modeling technique becomes of real use to the food microbiologist'.

A table which they give shows the growth of *Cl. botulinum* in vacuum packed minced meat between 15°C and 30° C. Predictions are given using the Ratkowsky (square root) and Schoolfield models for lag time and generation time. They concluded that compared with the observed data, 'the predictions agree fairly well, but there are differences at the lower growth temperatures, and the Schoolfield predictions are closer to the actual data'.

This view has been strengthened with later work, e.g. by Adair *et al.*[3] using both pathogens and spoilage bacteria. They concluded that significant errors may result at low (chill) temperatures if the square root model is used.

Analysis by Davey[4] used a modified and 'additive' Arrhenius model applied to temperature and water activity effects on microbial growth and applied it to published data spanning 50 years. This included low (chill) temperature data used for Schoolfield and square-root modeling. He found this his prediction agreed well with the original experimental data.

Modeling is therefore an active field of work world wide with considerable and continuing efforts to refine and apply predictive modeling to microbial growth rates.

'Square root rule'

The 'square root rule' or 'square root plot' was developed by Ratkowsky *et al.*[5,6] and may be considered easier to use than the Schoolfield model. Examples of the range of products which have been studied are:

- Dairy products – UHT and pasteurized skimmed milk, full-fat milk, double cream (48% fat) and UHT single cream (18% fat) by Phillips and Griffiths[7].
- Chilled products including delicatessen salads, pizzas, cheese and minced beef by Leatherhead Food RA[8].
- Chilled fish by Storey[9], Jorgensen *et al.*[10].
- Yeast growth in fruit drink by Cole *et al.*[11].

For chilled product studies, the most important temperature range is approximately 0°C to 15°C or 20°C. The model chosen must therefore fit well over this range.

The Leatherhead RA workers concluded that for spoilage predictions in a variety of chilled products the Ratkowsky (square root) model is useful.

The square root rule in its simple form is:

$$\text{square root of } r = b\,(T - T_0)$$

where r = the reciprocal of the time taken to achieve a specific increase in numbers (e.g. time for a 1, 2, 3 or 4 log cycle increase) or the growth rate constant.

  b = the slope of the regression line between square root of r and T.

  T = the temperature in °K (Kelvin).

  T_0 = the theoretical or notional temperature in °K below which the organism cannot grow (assuming no lowering of a_w because of ice formation) or conceptual temperature of no metabolic significance.

Note that absolute temperatures in °K are used to avoid the inconvenience of negative temperatures.

To get the best fit for the growth curve when plotting colony count data against time Stannard *et al.*[12] and others including Phillips and Griffiths[7] used the Nelder–Mead Simplex Minimization Procedure Computer Program.

The conclusion from earlier work on the square root rule seemed to be that although an empirical rule it seemed to work in a variety of situations and to be accurate enough to be useful.

At temperatures higher than 15°C the mesophilic spoilage flora seem increasingly important which requires a change in the equations used to describe behavior. However, McMeekin and Olley[13] believe that the next generation of electronic time–temperature integrators may well incorporate relative spoilage rate curves based on Ratkowsky's *et al.*'s[6] expanded equation:

$$\text{square root of } k = b\,(T - T_{min})$$
$$(1 - \exp\,[C\,(T - T_{max})])$$

where k = rate of deterioration
  b = a constant
  c = a constant
  T = the temperature in °K

  T_{min} = the temperature (°K) at which $k = 0$ crosses the temperature intercept. It is obtained by extrapolation of the regression line linking rate and temperature.

  T_{max} = the theoretical maximum temperature (°K) for growth. McMeekin and Olley[13] also indicate that the square root rule is valid for temperatures at different a_w values.

Before using this or any other model, it is essential to be sure that the chosen model is appropriate for its intended use. This means that the mathematics must be thoroughly understood and the limitations of the model appreciated. Particular care must be taken when modeling pathogens as the safety of the consumer has priority over all other considerations.

Other models have been proposed. For example, Gibson *et al.*[14] experimentally determined the response of a mixed inoculum of *Salmonellae* to five concentrations of salt (NaCl) between 0.5 to 4.5%; 5 pH levels between 5.6 to 6.8, and five storage temperatures between 10°C and 30°C. Sigmoid curves of the Gompertz form were fitted to the data and the curve parameters used to produce a polynomial model from which predicted growth curves could be generated for any combination of NaCl, pH, and storage temperature *within* the limits studied. From those growth curves, values for growth rate, generation time, lag time and other values such as time to a thousandfold increase in numbers were derived. They also discussed some problems in fitting curves to microbial growth data and of modeling such data.

A recent paper by Baker and Genigeorgis[15] is an excellent illustration of the value and utility of a good model. The title is self-explanatory – 'Predicting the safe storage of fresh fish under modified atmospheres with respect to *Cl. botulinum* toxigenesis by modeling length of the lag phase of growth'. They used data from over 900 experiments using nearly 19 000 samples. The data showed that temperature was the most important limiting factor (or hurdle) and that their model, when applied to published data from different countries, could be used to predict the time before toxigenesis in inoculated fish stored under different modified atmospheres.

Models of growth for pathogens (safety) and spoilage (quality) are currently being very actively developed in a number of countries. This is both because of their power in prediction and, in certain instances, as an audit tool, under various product/process/packaging/distribution conditions. Their cost-effectiveness in microbiological applications insures that resources will be made available for further developments.

As a generalization, current growth models are empirical or pragmatic because they aim to give a good fit to observed data. This means that the user must be sure that the choice of model is appropriate. Future developments will increase the choice available. However, the following questions highlight the most important criteria:

- Is microbial growth the critical factor? Instances are known where shelf life is not microbiologically determined.
- For each factor, can the model predict with sufficient accuracy over the whole of the desired range of values? *Note*. This is critically important where pathogens are modeled .
- Are the limitations of the model thoroughly known and are they acceptable without compromise of safety?
- How easy is the model to use?

Warning: Whatever model is used it is unsafe to base a commercial process on extrapolated, insufficient or unconfirmed data.

References

1. Schoolfield, R. M., Sharpe, P. J. and Magnuson, C. E. (1981) Non-linear regression of biological temperature-dependent rate models based on absolute reaction-rate theory. *Journal of Theoretical Biology*, **88**, 719–731.
2. Baird-Parker, A. C. and Kilsby, D. C. (1987) Principles of predictive food microbiology. In *Changing Perspectives in Applied Microbiology*, edited by C. S. Gutteridge and J. R. Norris, pp. 42S–49S. Society of Applied Bacteriology Supplement Series No. 16, Oxford: Blackwell Scientific
3. Adair C., Kilsby, D. C. and Whittall, P. T. (1989) Comparison of the Schoolfield (non-linear Arrhenius) model and the square root model for predicting bacterial growth in foods. *Food Microbiology*, **6**, 7–18
4. Davey, K. R. (1989) A predictive model for combined temperature and water activity on microbial growth during the growth phase. *Journal of Applied Bacteriology*, **67**, 483–488
5. Ratkowsky, D. A., Olley, J., McMeekin, T. A. and Ball, A. (1982) Relationship between temperature and growth rate of bacterial cultures. *Journal of Bacteriology*, **149**, 1–5
6. Ratkowsky, D. A., Lowry, R. K., McMeekin, T. A., Stokes, A. N. and Chandler, R. E. (1983) Model for bacterial culture growth rate throughout the entire biokinetic temperature range. *Journal of Bacteriology*, **154**, 1222–1226
7. Phillips, J. D. and Griffiths, M. W. (1987) The relationship between temperature and growth of bacteria in dairy products. *Food Microbiology*, **4**, 173–185
8. Gibbs, P. A. and Williams, A. P. (1990) Using mathematics for shelf life prediction. In: *Food Technology International Europe*, edited by A. Turner, pp. 287–290. London: Sterling Publications International
9. Storey M., (1985) Time–temperature function integration, its realization and application to chilled fish. 11R Commissions C2 – D3, Aberdeen UK, 239–243.
10. Jorgensen, B. R., Gibson, D.M. and Huss, H. H. (1988) Microbiology and shelf life prediction of chilled fish. *International Journal of Food Microbiology*, **6**, 295–307.
11. Cole, M. B., Franklin, J. G. and Keenan, M. H. J. (1987) Probability of growth of the spoilage yeast *Zygosaccharomyces bailii* in a model fruit drink system. *Food Microbiology*, **4**, 115–119
12. Stannard, C. J., Williams, A. P. and Gibbs, P. A. (1985) Temperature/growth relationships for psychrotrophic food-spoilage bacteria. *Food Microbiology*, **2**, 115–122
13. McMeekin, T. A. and Olley, J. (1986) Predictive microbiology. *Food Technology in Australia*, **38**, 331–334.
14. Gibson, A. M., Bratchell, N. and Roberts, T. A. (1988) Predicting microbial growth: growth responses of *Salmonellae* in a laboratory medium as influenced by pH, sodium chloride and storage temperature. *International Journal of Food Microbiology*, **6**, 155–78.
15. Baker, D. A., and Genigeorgis, C. (1990) Predicting the safe storage of fresh fish under modified atmospheres with respect to *Clostridium botulinum* toxigenesis by modeling length of the lag phase of growth. *Journal of Food Protection*, **53**, 131–140 + 153

8.B.3 pH and acidity

Introduction

In routine factory quality control work, acidity is often measured by titration against a standard alkali solution using an indicator or a pH meter to determine when neutrality is achieved. The results are usually expressed as a percentage assuming that all the acid is present as that which is in the largest quantity, e.g. citric, lactic or acetic acids.

However, when considering the behavior of microorganisms it is more convenient to express acidity on the pH scale. This runs from 0 which is acid, through 7 which is neutral to 14 which is alkali. It is a logarithmic scale and conveniently covers the microbial growth range.

pH can be defined as equal to $\log_{10}$ of the reciprocal of the hydrogen ion concentration (in gram molecules per liter). This can also be written as pH $= -\log_{10}$ [H+].

This is quite different from the term pK which is defined as equal to $\log_{10}$ of the reciprocal of dissociation constant of a weak electrolyte. This can be written as pK $= -\log_{10}$ of the dissociation constant.

Effect of pH

Microorganisms have a characteristic range of pH values over which they can grow. Most bacteria have an optimum pH near 6.8 and may grow at pH values ranging from 4 to 8; a small number of bacterial species can multiply at either pH less than 4 or pH greater than 8. Yeasts and molds can sometimes grow at pH less than 2. Table 8.9 shows typical pH values for fruit juices, which explains why mold rather than bacterial spoilage is the more usual.

Usually, the growth rate decreases as the pH drops below the optimum value. Approaching the lower limiting pH for growth, microorganisms are first inhibited and eventually killed. Figure 8.5 shows growth/inhibition/death curves typically obtained in media of different pH values.

As can be seen, the curve describing microbial growth at optimum pH becomes increasingly flattened as the pH decreases from this value. This is because the lag phase lengthens, the slope of the exponential growth phase decreases, the maximum cell count drops, the length of the stationary phase shortens, and the slope of the death phase increases. At very low pH values a true death curve occurs.

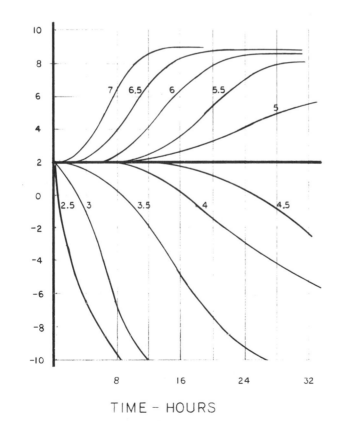

Figure 8.5 The response of microorganisms to environmental pH. *Note.* This is a generalized series of curves with the pH shown in 0.5 steps

Effects of acids

In initially suitable environmental conditions, the end of the exponential growth phase, the stationary and the death phases are regarded as being caused by the increased concentration of acids (mainly lactic, acetic, formic) produced during metabolism, rather than by the exhaustion of nutrients.

The exact mechanisms of microbial inhibition by acids are unknown. However, it is well known that most macromolecules are active in a narrow range of pH and that the intra-cellular pH is maintained in a condition of relative independence of the environmental one (Conway and Downey[1]; Neal *et al.*[2]; Harold and Baarda[3]), and through a regulated flow of protons (H^+) across the microbial membrane (Hamilton[4]). Several membrane-bound enzymes regulate this flow (Garland[5]), hydrogen ions or protons being unable to flow freely across membranes. Nevertheless, as environmental pH decreases it can be expected that the pH of the cell interior will decrease by the action of at least two mechanisms:

1. Directly, through the lessening of the proton-impermeability of the membrane due to the high external proton pressure, and
2. Indirectly, through the penetration of the cell by molecules undissociated in the acidic environment but which dissociate at the higher interior pH of the cell. In both cases the intra-cellular pH will be higher than the environmental one. As the interior pH decreases, an increasing fraction of macromolecules becomes inactive and the growth rate likewise decreases. When the exterior pH is low enough, the pH of the cell interior drops to levels incompatible with physiological activity. On the basis of available experimental data it is not possible to distinguish conclusively between the effect of pH and that of the undissociated molecule of acids.

Despite this the following conclusions may be drawn:

1. The degree of inhibition increases as pH decreases.
2. The relationship between the degree of inhibition (Relative Biological Effectiveness or RBE) and pH (or the logarithm of the molarity, Log M) is linear.
3. The inhibition curves can be regarded as parallel, irrespective of the acid used.
4. The degree of inhibition for any given pH value increases linearly with the pK of the acid used.
5. The rate of change of the RBE as a function of pK or of Log M is equal and about 0.8 times the rate of change of RBE against pH.

The difference in inhibiting and/or lethal effects of acids having different pK values is well known.

Citric (pK 3.08) is the principal acid in citrus fruits, strawberries, cranberries, blueberries, tomatoes; malic acid (pK 3.4) predominates in apples, pears, apricots, and peaches; malic and tartaric acids (pK 2.98) are found in grapes. Oxalic, succinic, fumaric and isocitric are some of the other organic acids that can be found in fruits. Acetic acid (pK 4.75) is usually present only as a trace. It is, however, a very effective preservative and vinegar has traditionally been used in pickles, sauces, etc.

Table 8.9 – Typical pH values for various fruit juices

Fruit	pH
Lemon	2.2–2.4
Lime	2.2–2.6
Raspberry	2.7–3.3
Blueberry	2.8–2.9
Grapefruit	2.9–3.4
Grape	3.0–4.0
Strawberry	3.1–3.9
Blackberry	3.2–3.4
Cherry	3.2–3.9
Apple	3.3–3.5
Orange	3.3–4.0
Apricot	3.4–3.6
Pineapple	3.4–3.7
Peach	3.4–4.2
Cocktail	3.6–4.0
Plum	3.7–4.3
Pear	3.7–4.7
Mango	3.8–4.7
Tomato	3.9–4.6

Table 8.9 shows the range of pH usually found in fruit juices. As can be seen, even the higher pH values are mostly lower than 4.5. In canned foods a pH of 4.5 (4.6 in US Federal Regulations) is used as borderline between acid and low-acid foods, that is, foods not requiring and those requiring, respectively, the minimum botulinum cook (12*D*). This is because of the view that only a limited spectrum of microorganisms can grow at pH <4.5 and these are mostly non-pathogenic. However, this assumption does not take into account the ability of *Cl. botulinum* to grow at pH levels near 4 in particular and very specialized environmental conditions, nor the ability of *Staph. aureus* and several *Salmonella* strains to grow at pH <4 which would also be of significance if container leakage occurred when these organisms were present.

References

1. Conway, E. J. and Downey, M. (1950) An outer metabolic region of the yeast cell. *Biochemical Journal*, **47**, 347–355

2. Neal, A. L., Weinstock, J. O. and Lampen, J. O., (1965) Mechanisms of fatty acid toxicity for yeasts. *Journal of Bacteriology*, **90**, 126–131
3. Harold, F. M. and Baarda, J. R. (1968) Inhibition of membrane transport in *Streptococcus faecalis* by uncouplers of oxidative phosphorylation and its relation to proton conduction. *Journal of Bacteriology*, **96**, 2025–2034
4. Hamilton, W. A. (1975) Energy coupling in microbial transport. *Advances in Microbial Physiology*, **12**, 1–53
5. Garland, P. B. (1977) Energy transduction and transmission in microbial systems. *Symposium of the Society of General Microbiology*, **27**, 1–21

Further reading

Corlett, D. A., Jr and Brown, M. H. (1980) pH and Acidity. In *Microbial Ecology of Foods*, Volume 1, pp. 92–111. International Commission on Microbiological Specifications for Foods, New York: Academic Press

8.B.4 Oxygen and oxidation-reduction potential (Redox)

Introduction

Oxygen is required by some microorganisms, but not all. It is toxic to some. The amount of oxygen available to the organism is therefore an important factor in microbial ecology.

Oxygen

As previously stated, the growth of microorganisms is characterized by the increase in the numbers of cells or biomass. The energy required for the synthesis (anabolism) of new cellular materials is obtained through the degradation (catabolism) of nutrients present in the environment.

The most relevant catabolic process is respiration. This is an oxidative process which is best understood as hydrogen transfer from natural reduced (highly hydrogenated) compounds, e.g. sugars, to oxygen or other oxidized inorganic compounds. All saprophytic microorganisms that are able to transfer hydrogen as H^+ and e^- (i.e. electrons) to molecular oxygen (O_2) are called aerobes, and they can therefore live in the presence of air. All microorganisms that are able to grow *only* in the presence of oxygen are obligate aerobes.

The process of molecular oxygen reduction leads to the formation of superoxide free radicals. All aerobes that have been tested contain the enzyme Superoxide dismutase (SOD). This is able to catalyze the reaction:

$$2O_2^- + 2H^+ = H_2O_2 + O_2$$

thus destroying superoxide free radicals that are very toxic. The degradation of the hydrogen peroxide (H_2O_2), also a toxic compound, is carried out by the enzyme catalase:

$$H_2O_2 = H_2O + \tfrac{1}{2}O_2$$

Microorganisms which do not possess SOD and catalase are unable to grow in the presence of oxygen. These are called anaerobes, and oxygen is more or less toxic to these organisms according to the species and strains. Oxygen is so toxic to obligate anaerobes that they cannot be recovered from foods when usual routine laboratory methods and diluents are used.

Many microorganisms are facultative anaerobes in that they can grow both in the presence and absence of oxygen, being provided with SOD and catalase and being able to transfer hydrogen to the molecular oxygen and other oxidized compounds. Microaerophilic microorganisms prefer a reduced oxygen tension in the environment.

Oxidation/reduction potential

A substance can be oxidized, i.e. it can lose electrons (become a donor) if there is in the system a substance that can be reduced, i.e. can accept electrons. The oxidation/reduction (OR) potential, or Redox potential, is a measure of the tendency of a system to give or receive electrons, that is, of the prevalence of either reduced or oxidized compounds. It is proportional to the logarithm of the ratio of the two types of compound according to the Nernst equation:

$$Eh = E_o + \frac{RT}{nF} \ln \frac{\text{Oxy}}{\text{Red}}$$

where *Eh* is the OR potential and is usually expressed in millivolts (mV)

E_o = the standard OR potential at pH O, but with the other solute components present at 1 M concentration
R = the gas constant
T = the absolute temperature in °K
n = the number of electrons involved
F = the Faraday quantity of electricity
$\ln$ = the natural logarithm
Oxy = the concentration of oxidized compounds
Red = the concentration of reduced compounds

In the presence of air, i.e. with oxygen present, *Eh* is a positive value. Aerobic microorganisms which are non-photosynthetic therefore reduce the oxygen during respiration, thus lowering the *Eh* value of the substrate. Anaerobic microorganisms transfer electrons to substances other than oxygen which has the same effect on the *Eh* value. In different microbial cultures the *Eh* ranges from about $-420\,\text{mV}$ with anaerobes to $+300\,\text{mV}$ with aerobes.

Eh is generally measured using an inert metal electrode, usually platinum, in a circuit with a reference electrode. However, the nature of most food products makes it difficult to obtain an accurate steady reading, due to drift. *Eh*, however, can be measured using microelectronic methods and is the subject of current research.

The tendency to have a low Redox potential is a general condition of all living cells. So food products usually have a low Redox potential due to the natural content of reduced compounds such as – SH containing proteins in meats, reducing sugars and ascorbic acid in vegetables, etc. However, the Redox potential of a food is not uniform. Usually, the surface has a higher Redox potential than deeper layers. It should be remembered that the world of the microbe is very small.

The access of atmospheric oxygen (and then the Redox potential changing capacity) depends on (1) the oxygen tension in the environment, (2) the compactness of the food structure and (3) the concentration and type of reduced compounds naturally present in a food which are able to exert a buffering action over the Redox potential drift. So, the *Eh* of meat may range from $+250$ to $-200\,\text{mV}$ and that of vegetables from $+400$ to $-500\,\text{mV}$.

Obviously, small particles of food are more exposed to oxygen effects than large ones, due to the high surface/volume ratio. So, squeezing, trimming, stirring, mixing, etc. all increase oxygenation of foods. Applying vacuum and packing with materials impervious to oxygen may substantially contribute to maintaining a low Redox potential when this is required.

Further reading

1. Banwort, G. J. (1981) *Basic Food Microbiology*. Westport, CONN: AVI Publishing
2. Brown, M. H. and Emberge, O. (1980) Oxidation-reduction potential. In *Microbial Ecology of Foods, Vol. 1, Factors Affecting Life and Death of Microorganisms*. pp. 112–125, International Commission on Microbiological Specifications for Foods, New York: Academic Press
3. Lehninger, A. L. (1975) *Biochemistry: The Molecular Basis of Cell Structure and Function*, 2nd edn. New York: Worth Publishers
4. Speck, M. L. (ed) (1984) Section 58.14 Redox potential (Eh). In *Compendium of Methods for the Microbiological Examination of Foods*, 2nd edn, pp. 791–792. Washington, DC: American Public Health Association

8.B.5 Nutrients

Background

Growing organisms all require chemical elements and compounds necessary for the synthesis of cellular constituents. A typical bacterial cell, e.g. *E. coli*, is made of about 28% organic compounds, 70% water and only 1% inorganic ions. Carbon, hydrogen, oxygen, nitrogen, sulfur and phosphorus are the main elements required by the actively dividing cell; other elements (Mg, Mn, Ca, Na, K, Cl, Cu, Fe, Zn, Co, Mo, etc.) are required at very low levels, e.g. a few mg per kg (ppm).

The required elements may be available in the environment although in a state not directly utilizable by all microorganisms. As a typical example, nitrogen, the fundamental constituent of proteins, is very abundant in normal atmosphere but is utilizable as such, to build nitrogenous organic compounds, only by a very limited number of species. Moreover, microorganisms may be unable to synthesize molecules, e.g. one or more amino-acid, purine, pyrimidine, vitamin or lipid, etc., and these must be absorbed from the environment as pre-formed entities.

This may be a factor in determining the microbial succession found in food spoilage, when certain bacteria will not develop until metabolites released by the growth of bacteria or other microorganisms on the food are available.

Microorganisms are usually classified nutritionally according to their ability to use inorganic or organic compounds as an energy source. Autotrophs are able to use carbon dioxide as their carbon source and oxidizable inorganic compounds as their energy source, whereas heterotrophs can only use organic compounds for both carbon and energy requirements. It is this latter group of heterotrophs that is important in food microbiology.

8.C Physico-chemical factors affecting survival

Background

Microorganisms exist either as vegetative forms which with a few exceptions are not usually resistant to physical or chemical agents or as spores. Bacterial spores especially are much more resistant than the corresponding vegetative forms. Both forms are affected by adverse factors and the difference is largely one of degree.

The most important factors can be regarded as heat, radiation and lethal chemical compounds (disinfectants). Ultrasonic waves, antibiotics, hydro-static pressure, etc. are less relevant in food processing and preservation technology.

Examples of the use of these factors to destroy organisms are:

- Heat – pasteurization and commercial steriliza-tion of canned foods
- Ionizing radiation – sterilization of packaging materials. It may also be used to inhibit sprouting of onions, potatoes and garlic, delay ripening and mold growth and destroy insects.
- Ultra-violet radiation – air and water sterilization
- Disinfectants – sanitization of surfaces

It should be noted that some level of microbial destruction always occurs in environmental con-ditions which do not allow growth although it may be at very low rates, e.g. frozen or dried storage.

Several physico-chemical factors acting together may dramatically increase the inactivation rate. This is illustrated by the 'Hurdle Concept' of food preservation in which a multifactorial system is effective but in which the individual factors have insufficient or comparatively little preservative effect.

8.C.1 The rate of death

Factors affecting death rate – introduction and background

One of the important tasks of the food microbiolog-ist is to control microbial numbers by the use of treatments, e.g. heat. As a general rule, when the number of surviving organisms N_s is plotted against treatment values (t) the graph shows the typical concave curve of logarithmic decay. By plotting the logarithm of the number of surviving organisms (Log N_s) against treatment values (t) a straight line is obtained.

A straight line is also obtained by plotting the logarithm of the fraction N_s/N_0 of the original population (N_0) surviving lethal treatment. The curve describing survival behavior of the micro-organisms can be treated both mathematically and graphically. By taking into account that the equation of a decaying straight line is represented by:

$$Y = a - b * X \qquad (1)$$

(*Note.* * means multiply throughout this section)

the microbial survival curve can be transformed into the following equation:

$$\text{Log } N_s = \text{Log } N_0 - k * t \qquad (2)$$

where N_s = number of survivors k = a constant
N_0 = original numbers t = the treatment value

As examples of t, in heat inactivation curves t is the time (generally minutes) at the constant treatment temperature T; in radiation inactivation curves, t is the dose (in rad or gray); in the disinfection process, t is the contact time of the organisms with a lethal concentration of the chemical.

The law of logarithmic decay followed by the microbial death process implies that:

- equal fractions of the cell population are killed by treatments of equal value
- absolute sterility can theoretically never be reached no matter what value of lethal treatment is used

Both of these conclusions can be visualized numerically as in Table 8.10 below.

As can be seen, assuming that treatment value D equals 1 is the lethality to destroy 90% of a microbial population, each time the treatment value increases by D, 90% of the surviving population is killed.

Surviving numbers lower than 1 (e.g. 0.1, 0.001, 0.0000001, etc.) do not mean that only a fraction of a cell is surviving. These figures are to be interpreted on a probability basis. A survival of 0.1 cells/sample means there is likely to be one cell surviving in each 10 samples; 0.01 cells/sample means one cell each 100 samples; 0.000004 cells/sample means one surviving cell each 250000 samples; and so on. Accordingly, one can lower the survival probability at will by increasing the treatment or time as necessary, without ever reaching the absolute certainty of sterility. Never-theless, if the survival probability is as low as 0.00001 cells/sample, for instance, and the number of treated samples (e.g. cans, bottles or packs) is 100, then treated samples can be expected with a certain statistical confidence (while not with abso-lute certainty) to contain no living organisms. This is because the quoted survival probability means that only one sample out of 1000 lots of 100 samples is expected to contain one surviving organism.

Table 8.10 Example of death rate

Value of treatment (*D*)	Number of cells Surviving (*N*ₛ)	Killed	Log *N*ₛ	% of cells Surviving[a]	Killed[b]
0	1000	0.0	3	100.00	0.00
1	100	900	2	10.00	90.00
2	10	90	1	1.00	99.00
3	1	9	0	0.10	99.90
4	0.1	0.9	−1	0.01	99.99
..	. . . . .	. . . . .	. . . . .	. . . . .	. . . . .
7	0.0001	0.0009	−4	0.00001	99.99999

[a] $100 (N_s/N_0)$
[b] $100 (N_0 − N_s)/N_0$

8.C.2 Heat death – logarithmic death rates

Heat resistance

High temperature is the most often used and is a very effective physical lethal agent. The microbial heat inactivation curve is a straight line in semi-log plot, as previously seen, and the equation describing the curve is usually represented by:

$$\text{Log } N_s = \text{Log } N_0 − (1/D) * t \qquad (3)$$

where $1/D$ equals k of the general equation (2) and t is the time (minutes) of treatment at the temperature T.

D is the decimal reduction time at temperature T. It is usually written D with the temperature T as a subscript, e.g. D_{121} is the time required at 121°C to kill 90% of organisms. The value of D can be obtained by calculating the parameters of the inactivation equation (3) or graphically, which may be done, although with lower accuracy.

D is one of the two most important parameters characterizing heat resistance of microorganisms. By knowing the D value at the temperature T one can define the time necessary at that temperature to obtain a required sterilization probability by a very simple computation. Instantaneous heating and cooling of the sample is assumed.

As an illustration of the consequences of logarithmic order of death, consider the following example

Let the initial contamination of a liter of milk be 100 000 bacteria. As we wish to pasteurize 1000 one-liter bottles, then the total initial number, N_0 equals 100 000 multiplied by $1000 = 10^8$ bacteria, so the Log No. = 8. To reduce the surviving probability to one cell each 10 000 treated bottles, i.e.: $N_t = 1/10 000 = 0.0001$ survival and therefore, Log $N_t = −4$, the kill required equals $(8 + 4) *D$ which is a 12D kill. Using equation (3):

$$t = (\text{Log } N_0 − \text{Log } N_t) * D \qquad (4)$$

so, $t = (8 − (−4)) * D = 12 * D$

Thus if the $D(T)$ of the microorganism under consideration equals 1, 5 or 8 minutes, for instance, the milk will need to be heated for 12, 60 or 96 minutes, respectively.

Conversely, knowing the applied treatment time, one can infer the survival probability level for any microorganism of known D value.

Let, for instance, 15 seconds at 71.7°C be the applied treatment to a milk sample (a pasteurization process approved by FDA and one of the legally specified pasteurization processes in the UK). Assume that it is contaminated with *L. monocytogenes* at the level of one cell in 50 ml. Without knowing the heat-resistance parameter (D value) of the actual strain, we must make the 'worst case' assumption and use the highest heat resistance data reported in literature:

(a) Freely-suspended cells: $D(71.7) = 3.3$ seconds

(b) Intra-cellular *Listeria*: $D(71.7) = 4.1$ seconds

Accordingly, in the first case (a) the pasteurization treatment yields $15/3.3 = 4.6$ decimal reductions, and in the second one (b) $15/4.1 = 3.7$. If the contamination level is 1 in 50 ml or 20 per liter (N_0) then the expected rate of survival should be 1 per 2000 cartons and 1 in 250 cartons respectively when the milk is packed in one-liter cartons. This can be shown by using equation (4):

$$t/D = \text{Log } N_0 − \text{Log } N_t$$

or:

$$\text{Log } N_0 − t/D = \text{Log } N_t$$

so for freely suspended *L. monocytogenes* cells,

$$1.3 − 4.6 = \text{Log } N_t$$

$$− 3.3 = \text{Log } N_t$$

that is: $N_t = 10^{−3.3} = 0.0005 = 1$ in 2000

Similarly, using the same equation (4), intracellular *L. monocytogenes* would result in 1 in 250 packs of one liter being contaminated with surviving *Listeria*.

Heat resistant organisms have higher *D* values than heat labile ones. Nearly all vegetative bacteria, yeasts and molds have a *D* value of a few seconds at 70°C. Some heat resistant bacterial spores have similar *D* values at around 130°C to 140°C.

Factors affecting heat resistance

Introduction

The most important factors affecting microbial heat resistance are usually regarded as (i) temperature, (ii) relative humidity (a_w) (iii) substrate composition including pH.

Temperature

Microbial heat resistance decreases with increasing temperature. Plotting Log *D* against temperature we usually obtain a straight line of slope 1/z:

$$\text{Log } D\,(T_r) = \text{Log } D\,(T) - 1/z^* \,(\text{Tr} - T) \qquad (5)$$

after which:

$$z = (T_r - T)/(\text{Log } D\,(T) - \text{Log } D\,(T_r)) \qquad (6)$$

where T_r is the reference temperature and T is the temperature under consideration.

'z' is therefore the second important parameter defining microbial heat resistance.

'z' is the number of degrees change in temperature required to achieve a tenfold change in $D(T)$, so that: at $T - z°$ the *D* value will be 10 times that at T, and similarly at $T + z°$ the *D* value will be 1/10 that at T. This is expressed in

$$0.1 * D\,(T - z) = D(T) = 10 * D\,(T + z) \qquad (7)$$

The z value can be, and in the USA usually is, expressed in degrees Fahrenheit. However, with wider use of the metric system it is becoming more usual to work in degrees celsius. As a rule-of-thumb, yeasts and vegetative bacteria have z values between 5 and 8°C whereas many bacterial spores have z values between 8 and 11°C when measured in liquid around neutral pH. Remember that for any organism the z value is dependent on the substrate. A particular instance of this is the resistance under 'dry' conditions, i.e. where the a_w is low. Although it is difficult to determine heat resistance under these conditions, some bacterial spores have a z value of 40°C.

The z value is linked to the familiar temperature coefficient Q_{10} often used in chemistry by the relationship:

$$z = 10/\text{Log } Q_{10} \qquad (8)$$

or

$$Q_{10} = 10^{10/z} \qquad (9)$$

In addition to a knowledge of *D* and z values for microorganisms of interest it is necessary to establish the killing efficiency of a heat sterilization process.

A knowledge of *D* and z values is needed to establish the lethal effect of a heat sterilization process. For instance, knowing z, one can state that the previously cited heat (pasteurization) treatments of 12, 60 or 96 minutes at the temperature T are expected to yield exactly the same result (i.e. the same percent destruction) if the treatments are carried out at the temperature $T + z$ for 1.2, 6 or 9.6 minutes, or at the temperature $T - z$ for 120, 600 or 960 minutes, according to equation (7).

Since equation (5) is analogous to the following:

$$t(T) = t(T_r) * 10^{((\text{Tr} - T)/z))} \qquad (10)$$

if the treatment time *t* needed at one temperature T_r, i.e. $t(T_r)$, is known, then the equivalent treatment time required at another temperature, $t(T)$ can be computed.

As an example, given that for milk the pasteurization conditions are:

Temperature 71.7°C; time, $t(T_r)$ 15 seconds; equivalent pasteurization times in seconds at temperatures of, say, 65°, 70.5°, 78° and 82.4° may be calculated using equation (10). To simplify the example, z values of 10°C and 5°C are used although other numbers could be chosen.

Using z = 10°C:

$t(65) = 15* 10^{((71.7-65)/10)} = 70.2 \text{ seconds}$
(1 minute 10.2 seconds)

$t(70.5) = 15* 10^{((71.7-70.5)/10)} = 19.8 \text{ seconds}$

$t(78) = 15* 10^{((71.7-78)/10)} = 3.5 \text{ seconds}$

$t(82.4) = 15* 10^{((71.7-82.4)/10)} = 1.3 \text{ seconds}$

Using z = 5°C:

$t(65) = 15* 10^{((71.7-65)/5)} = 328.2 \text{ seconds}$
(5 minutes. 28.2 seconds)

$t(70.5) = 15* 10^{((71.7-70.5)/5)} = 26.1 \text{ seconds}$

$t(78) = 15* 10^{((71.7-78)/5)} = 0.8 \text{ seconds}$

$t(82.4) = 15* 10^{((71.7-82.4)/5)} = 0.1 \text{ seconds}$

If the correct z value is not known, it is considered a safe practice to use a lower z value for treatments at temperatures exceeding the reference one (T_r), and a higher z value for treatments at lower temperatures.

Stumbo's textbook[1], which was written for canners, should be consulted for further aspects of sterilization theory, e.g.integrated F values, and allowing for different initial temperatures.

Water activity

As a general rule, heat resistance of microorganisms increases as the water content and/or relative humidity decreases. The increase in heat resistance may be considerable and involves both D and z values. For example, *E. coli*, which has a $D_{60} = 1.5$ min in a fully hydrated substrate, may have a $D_{120} = 2$ to 3 min in butter fat having less than 1% water.

For a few organisms there is a linear relationship between water activity in the environment and $\text{Log } D$. In general, higher heat resistances are associated with lower water activities, but the substrate composition is very important. As an example, the z value may rise, at low water activity, from 5° to 10° to 30° to 50°C. There is not, however, an accepted relationship between D and/or z values and the water content in the environment.

pH

Usually, microbial heat resistance reaches the highest value in media having a pH close to 6.8. Yeasts and molds have their highest resistance at pH values near 5.0. The heat resistance (as shown by the value of $D(T)$) of all organisms decreases as pH decreases.

It was found that the decimal reduction time of many microorganisms is lowered 10 times by a decrease of 2 pH units. In alkaline substrates heat resistance also decreases.

Application of pasteurization theory

Introduction

Where the pasteurization process is legally defined, e.g. with liquid milk for retail purposes, there is no need for further discussion, since time and temperature must conform to legal requirements. Where mild heat treatment is used to achieve product safety or stability, usually in combination with one or more factors as 'hurdles', it becomes important to understand the technological basis of the process.

Temperatures above the optimum for the growth of an organism have an inhibitory or lethal effect which is very substantially affected by the aqueous environment in which the organism is heated. As an example, the lethal effect of heat can be enhanced or synergized by the presence of acids and diminished or antagonized by sugar which lowers the water activity. There may also be a protective effect of fat or oil globules which can be considered as effectively exposing the organisms to 'dry' heat

which is much less effective at a given temperature than 'moist' heat.

In heterogeneous foods, it should be remembered that there may well be a number of different environments in any of which organisms should not be assumed to be randomly distributed.

Approaches

With acid packs, i.e. those with foods of pH below 4.5 or 4.6 (USA), there are two main approaches to handling pasteurization processes:

(a) Using temperature as the key parameter. The instructions might be 'heat to at least 70°C'. In order to insure that this minimum temperature is reached the instructions may well be 'heat to at least $x°$ and hold for at least y minutes or seconds'. The figures used may be traditional or based on thermal death time data for the main known spoilage organism, e.g. *Zygosaccharomyces bailii* (also known as *Saccharomyces bailii*). Some fruit juice processes are of this kind and have been used for some time. The rationale, is therefore, that they work in the market place and this is a powerful argument. The limitation is that if conditions change, e.g. composition, pH, equipment used to administer the process, no predictions are logically possible.

(b) Using the classical sterilization theories which were developed for bacterial spores. These are described by Stumbo[1] and have been widely used for pickles and sauces. An outline of the rationale is given below.

Significance of process administration

The usual textbook treatment is to deal in turn with the organisms, process evaluation and then process establishment. The reason for considering process administration first is because the real practical difficulty in dealing with the subject is to get sufficient reliable data about the process as it is administered. It is not unusual for the technician beginning work on pasteurization problems to think that the mathematics is the main difficulty. This is not so and the apparent precision given, e.g. by the use of tables where figures are quoted to four decimal places, can give an illusion of achievable accuracy. The concept of building up the error budget is therefore useful in this work. Consider the likely 'errors' in administration; if the product is packaged and pasteurized underwater in 'tanks' or 'containers' their design may well mean that temperature distribution through the load shows a much greater variation than the canner would expect from well-vented steam retorts. In tanks there may be a 3° to 4°C spread because of inherent design features so it is important to know and work with the 'worst case'. Different stacking patterns resulting from pack size changes may give a different temperature distribution pattern, and the location and extent of 'cold spots' may not always be consistent for any one size. Use of different perforation patterns or margins for divider plates may also affect results.

If the product is packaged and pasteurized in a continuous pasteurizer the temperature distribution of the heating medium would be expected to be much better. However, inadequate steam supply or poor maintenance and calibration of equipment can result in conditions being administered different to those which were intended.

If a 'Hot fill and hold' system is used, the hot product is used to pasteurize the inner surfaces of the containers and the air in the headspace. It might be reasonable to allow for a 3° to 5°C drop in temperature between the filler pan thermometer and the temperature in the container. However, this figure should be reliably determined by experiment over a range of operating conditions.

If an aseptic packaging system is used the expectation would be that the pasteurization process would be very well controlled. The instrumentation should be designed to give accurate measurement, control and recording and provide a positive assurance of process and aseptic safety. This needs to be confirmed during commissioning and on an on-going basis during normal production.

If pasteurization is one of the 'hurdle factors', e.g. in baked goods, cooked hams or vacuum packed smoked fish, the likely temperature distribution variation in the heating oven or 'cooker' needs to be known accurately in order to use the lethal effect of heat in a controlled manner.

Process evaluation

This is described by Shapton *et al.*[2] as assigning to the process a numerical value in arbitrary units of lethal heat which enables the lethal effect of different time–temperature combinations to be compared in the same units. There are two stages in this evaluation. The first is to measure the time–temperature history or profile of the food and the second is to calculate its worth in the chosen units termed P (for Pasteurization) values. These P values are analagous to F or F_o units used in canning.

Assuming that the first step is taken and the errors of the whole measuring system are known, the units chosen should give the result in convenient numbers (i.e. not all to the right of the decimal point nor in the high tens or hundreds).

The choice of unit depends on the reference temperature which may conveniently be at or near the 'holding' temperature. One unit or P value is

often one minute at this temperature assuming instantaneous heating and cooling. Other units of time may be used. To make use of these units, the z value of the organism needs to be known over the range of temperatures found in process administration.

Perhaps the simplest way of thinking about the z value is to realize that it is a property of a specific organism heated in a specific substrate. Because the determination of actual values needs a great deal of work, an assumed value which is believed to be appropriate for the work in hand is often used for routine purposes.

In dealing with unknown organisms or mixtures of organisms an assumption is often made about the appropriate z value to use. Dakin[3] quotes figures which suggest that a z of 10°C (or 18°F) is a suitable value for some pickles where yeasts are known spoilage organisms. Put and de Jong[4] working with *Z. bailii* found that the z value was approximately 5°C. Gaze *et al.*[5] found a value of approximately 6° to 7°C for *L. monocytogenes* in some non-dairy foods.

Since z values are often assumed, what happens if the wrong assumption is made? To get convenient numbers of P values, the reference temperature for the P values is likely to be that of the 'holding' temperature at which the process is administered. In this case, the use of a higher assumed z value than is actually true for the organism and substrate is a 'safe' assumption.

Since the P value units are arbitrary it may be quite satisfactory to use an assumed value for Quality Assurance purposes, but for process establishment development work use accurate values of D and z. This is to enable accurate estimates to be made of the probability of microbial survivors.

When a suitable reference temperature has been chosen and the z value assigned, the next step is to construct and use a table showing the worth in units of lethal heat of temperatures other than the reference temperature. This can be done by calculation from the equation (10) or the formula given by Stumbo[1]. *Note: Use either degrees celsius or fahrenheit consistently throughout all your work.*

An example of such a table is given in Table 8.11. Because the numbers in the body of the table depend on the z values it is convenient to have tables based on the z values likely to be used and fill in the appropriate temperatures for use on a photocopy of the appropriate table. Note the numerical difference made by a change in z value in Table 8.11.

Alternatively, for the determination of lethal rates, a simple graphical method may be used which is often accurate enough. Cut and stick together three sheets of three cycle semi-log paper to make one wide sheet. P values (lethal rates) are on the log scale and temperatures on the linear one. The

Table 8.11 Table of P values per minute

Temperature[a]	$R \pm °C$	z = 10°C		z = 5°C	
		T = 0.0	T = 0.5	T = 0.0	T = 0.5
60	$R - 10$	0.100	0.112	–	–
61	$R - 9$	0.126	0.141	–	–
62	$R - 8$	0.158	0.178	–	–
63	$R - 7$	0.200	0.224	–	–
64	$R - 6$	0.251	0.282	–	–
65	$R - 5$	0.316	0.355	0.100	0.126
66	$R - 4$	0.398	0.447	0.158	0.200
67	$R - 3$	0.501	0.562	0.251	0.316
68	$R - 2$	0.631	0.708	0.398	0.501
69	$R - 1$	0.794	0.891	0.631	0.794
70	R	1.00	1.12	1.00	1.26
71	$R + 1$	1.26	1.41	1.59	2.00
72	$R + 2$	1.58	1.78	2.51	3.16
73	$R + 3$	2.00	2.24	3.98	5.01
74	$R + 4$	2.51	2.82	6.31	7.94
75	$R + 5$	3.16	3.55	10.00	12.59
76	$R + 6$	3.98	4.47	15.85	19.93
77	$R + 7$	5.01	5.62	25.12	31.62
78	$R + 8$	6.31	7.08	39.81	50.12
79	$R + 9$	7.94	8.91	63.10	79.4
80	$R + 10$	10.00	–	100.00	–

Reference temperature = R°C. Temperature under consideration = T, values rounded at whole and 0.5°

[a] Any appropriate temperature range spanning 20°C may be entered in this column.

Notes.

1. Using Patashnik's method[6], temperatures are measured at fixed time intervals, e.g. 5 minutes. Temperatures are converted to P values per minute. The first and last figures are halved. The sum of the P values is found and multiplied by the time interval to give the P value of the process[2].
2. Always specify reference temperature and z value of P values, e.g. P_{65}^5 and P_{77}^9.
3. As an example, if a lethality of 1 minute at 70°C is required it will be seen from the table that where z = 10°C, 10 minutes' heating is required at 60°C or 0.1 minute at 80°C for equivalent lethality.

reference temperature has a *P* value of 1 and since there is a tenfold increase or decrease in *P* value for each *z* degrees, the 'slope' – which is a straight line – may be easily drawn. It is usual for a graph of 2*z* degrees below reference and *z* degrees above to be fully adequate.

Having listed the lethal rates at the time intervals of measurement throughout the process, the next step is to estimate its 'worth'. Any standard method of intregration under the curve can be used, e.g. Lethal rate paper (see Stumbo[1]) but the method of Patashnik[6] is fully adequate for nearly all Quality Assurance purposes.

Process evaluation can be used not only for Quality Assurance but also to calculate equivalent processes using different times and temperatures. This can be done using standard 'sterilization' calculations (Stumbo[1]). It is a safe assumption to use the process at the slowest heating portion of the food rather than the integrated values. Remember, for accurate work it is essential to take full account of the variation in *P* values found in the original process.

Process establishment

This can be done using equation (4) in this section or alternatively the equivalent Stumbo equation[1]:

$$F = D \log 10^n$$

where F = the required lethality of the process (in *P* values)

D = *D* value at the reference temperature (used for *P* values)

n = the number of log cycles which specify the reduction in microbial numbers which is required, e.g. 6, 8, 10.

To use either equation properly, a considerable amount of thermal death time data is needed. Both *D* and *z* values of the strains of organism(s) in the specific substrate(s) under consideration need to be known. In addition, a decision must be taken on the reduction of microbial numbers that is required. This will depend on whether pathogens or spoilage organisms are being considered, and have due regard to the requirements of the market.

If microbial growth occurs either before the heat treatment or afterwards in the pack because of container defects or abuse, this increases the chance of the customer finding an unsatisfactory pack. It is better to make efforts to prevent this happening than spend technical and management time estimating the resulting increase in failure rates.

Once the lethality has been determined in P values at one (reference) temperature and z value, it is easy to calculate equivalent processes for other temperatures and times using standard 'sterilization' methods. (See equation (10) in this section or Stumbo[1].)

8.C.3 Radiation

Electromagnetic radiation is characterized by a frequency, a wavelength and an energy range, according to the band name (long-wave radio, micro-waves, radar, infra-red, ultra-violet, soft X-rays, gamma rays, etc.).

In mathematical terms, the relationship between frequency (f) and wavelength (w) is:

$$c = f * w \qquad (11)$$

where $c = 2.998 \times 10^{10}$ cm/s is the speed of radiation.

Energy (E^+) of a quantum of radiation is:

$$E^+ = h * f = h * c/w \qquad (12)$$

(* means multiply in this section)

where h is the Plank's constant (6.624×10^{-27} ergs/s), frequency is given in cycles/second, and wavelength in cm.

Ionizing radiation has:

w less than 30 Angstrom
f greater than 10^{17} cycles/s
E^+ greater than 1 keV
Note: 1 Angstrom = 10^{-8} cm
       1 eV = 1.59×10^{-12} ergs
       1 keV = 1×10^3 eV.

Ultra-violet and ionizing radiation are the sterilizing sources of radiation employed in food technology.

Ultra-violet radiation

Ultra-violet radiation is a low energy radiation (a photon of $w = 2500$ Angstrom UV light has 5 eV energy) but is recognized as one of the most potent agents inducing mutation of microorganisms. UV radiation is usually obtained from mercury vapor lamps, made of quartz or special formula glass. They produce a discontinuous spectrum with most energy released at 253.7 nm. A commercial 30 watts lamp yields about 7 watts of 253.7 nm radiation. The maximum radiation intensity is at the tube surface.

For a distance equivalent to about half the length of the tube the intensity can be assumed to fall off as the inverse of distance, and beyond this, as the inverse square of the distance.

Smooth surfaces including those likely to be found in a food plant all show some reflectivity as shown in the Table 8.12.

Table 8.12 Reflectivity to UV light

Material	UV fraction reflected by the surface (approximate %)
Aluminum	50 to 80
Chromium plating	50
Stainless steel	25
Ceiling paint	4
Glass	5

Microbicidal lamps may cause burning to the skin but not tanning; 250 and 280 nm wavelengths being most dangerous.

Remember that operators must be shielded from both direct and indirect radiation, because UV wavebands are able to cause severe eye irritation (conjunctivitis) and skin inflammation (erythema). Lower wavelengths usually produce ozone (which is toxic) in irradiated air. The penetration ability of UV rays is very limited, as shown in Table 8.13.

Table 8.13 Penetration of UV light

Substrate	Distance which UV will penetrate (cm)
Air	300 to 500
Water	30
Window glass	<0.1
Transparent plastic foil	<0.01
Fruit juice	0.1
Fluid milk	0.01

High-intensity sources of UV radiation are very effective in killing unprotected microorganisms suspended in air, or in thin films of transparent liquids, or on the surface of materials and products. Covered, shaded or shielded organisms will not be affected. It follows that UV sterilization technology is employed for plant surfaces, air and water sanitization.

There are, however, two issues to remember. The first is that UV is only effective when it strikes the microorganisms, i.e. there is no residual effect, so growth may occur in 'dead ends'. The second is that, as a rule, UV lamps show a 'fall-off' or reduction in power with time which should therefore be monitored with sufficient frequency to enable remedial action to be taken.

The major absorption band of nucleic acids is in the region of 260 nm, so nucleic acids are the primary target of UV radiation. Microorganisms are not killed instantaneously. The killing efficiency increases with the dose, the power of which is usually expressed as milliwatt (mW) or microwatt (µW) UV energy per second per square cm. Both vegetative cells and spores are sensitive to UV radiation.

The inactivation kinetics are usually exponential until about 99.99% of organisms are destroyed. The dose required to destroy 90% of the population (D) is sometimes expressed as the time (seconds or minutes) of treatment at a defined distance from a source of known geometry and potency. This is not really a better way of expressing the dose than simply milliwatts or microwatts *per se* per square cm. Be aware that often the straight line portion of the semi-log plot is followed after 3 to 5D reductions in numbers by a tail. This has implications for any proposed use.

Microbial resistance to UV radiation increases in the order: vegetative bacterial cells < yeasts < bacterial spores < mold spores. D values range between about 10 and 20 mW s/cm^2 for vegetative bacterial cells, and close to 25 mW s/cm^2 for yeasts. D values of spores range between about 25 and 35 mW s/cm^2 for bacterial spores and from about 100 to more than 2000 mW s/cm^2 for spores of *Aspergillus* and *Penicillium*.

Ionizing radiation

Although its use may be technically justified, legal and beneficial, prospective users should be aware both of public distrust of anything 'nuclear' and of flavor changes induced in a number of products even at low doses. These flavor changes are caused by free radicals produced by irradiation especially if fats are present. Unacceptable 'off flavors' may be developed at surprisingly low doses. Legislation which permits irradiation of food may limit the dose to 1 M rad or 10 k Gray. This dose is considerably less than the equivalent of the 12D thermal process applied against spores of *Cl. botulinum*, but is effective against other organisms, e.g. *Salmonella* spp.

Electromagnetic waves and subatomic particles having enough energy to cause ejection of electrons from impacted atoms are collectively named ionizing radiation. The most common source of ionizing radiation employed in sterilization technology are X-rays and gamma-rays. X-rays represent secondary emission of a target bombarded with high velocity electrons; gamma-rays result from nuclear transformations in radioactive elements.

X-rays are produced when a stream of high-velocity electrons impinge on a metallic plate, (called the anti-cathode and usually made of tungsten) in a suitable vacuum tube operating at a suitable voltage: about 5% of the energy is transformed into non-corpuscular electromagnetic radiation of continuous broad spectrum wavelengths (from soft, 1 keV, to hard, 10 MeV, rays). Monochromatic radiation can be obtained using a double refractometer.

Cesium-137 and Cobalt-60 are the active isotopes most commonly employed for sterilization purposes. These may come either from waste products of nuclear plants or from materials deliberately processed in such plants. They produce gamma rays of up to 1.33 MeV. Note that to induce radioactivity the radiation energy must be higher than about 10 MeV.

The absorbed energy is measured in rads; 1 rad being equal to 100 ergs energy absorbed/g material (0.01 J/kg), and multiples commonly used are kilorad (krad, 10^3 rads) and Megarad (Mrad, 10^6 rads). The new unit of absorbed dose for food irradiation is the gray (1 Gy = 1 J/kg = 100 rads) and 10 kilograys equal 1 Megarad.

Inactivation kinetics of radiation sensitive organisms are usually exponential, so radiation resistance can be measured by the decimal inactivation dose D.

Inactivation curves of radiation resistant organisms are often exponential after a first phase of no or slow inactivation at lower doses (giving a 'shoulder' to the inactivation curve); the exponential phase is often followed by a phase of slow or no inactivation, giving a 'tail' to the inactivation curve. These curves are usually called 'sigmoid'. Though found to some extent in inactivation curves of all physical or chemical lethal agents, sigmoid curves are to be regarded as typical of microbial inactivation by radiation. For microorganisms having a sigmoid radiation survival curve the radiation resistance cannot be represented safely by the decimal reduction dose D. A better way of expressing the dose required is that it should destroy 10^n organisms where n is usually between 6 and 12.

Vegetative bacterial cells are more sensitive, as a rule, than bacterial spores. Relevant exceptions are some species belonging to the genera *Micrococcus*, *Moraxella*, *Acinetobacter* and also Viruses.

Cl. botulinum is the most radiation resistant pathogenic bacterium. The spores of *Cl. botulinum* type A may have a D of 0.37 Mrad (3.7 k Gy). Accepting the 12D concept, the minimum radiation sterilization dose required for foods would be about 4.5 Mrads or 45 k Gy. Be aware that this is considerably more than the 'pasteurization' or 'radicidation' dose of 1 Mrad or 10 k Gy allowed in certain countries.

Although expert opinion is that irradiation of food within the proposed limit of 10 k Gy is safe for the consumer, its wide use requires public acceptance as well as acceptable sensory properties and favorable economics for the processing of any particular food.

A note on the effect of microwaves

Microwaves are frequently used domestically and in catering operations to 'defrost' frozen foods or to heat food to serving temperatures. It should be understood that the temperature produced in these applications is most unlikely to be sufficiently uniform for this to be regarded as an adequate pasteurization process.

Processors using microwave heating for pasteurization of, e.g. ingredients, should regard the lethal effect as due only to the temperatures achieved in the food and not to any lethal properties of microwaves.

8.C.4 Lethal chemical agents

Introduction

Lethal chemical agents comprise a wide range of compounds employed in the microbial inactivation process called disinfection. Among the most important of these are halogens, quaternary ammonium compounds, nascent oxygen, hydrogen peroxide, acids and alkalis. Phenol is an effective disinfectant but should not be used anywhere in a food plant because of the risk of taints especially if chlorine or hypochlorite are used. Although not further discussed here, peracetic acid, formaldehyde and glutaraldehyde are also used as disinfectants in some countries. See also Chapter 5.D for additional material on some applications of disinfectants.

The lethal activity of chemical compounds is closely linked to their concentration. At low enough concentrations of these compounds, microorganisms are not affected at all (the inactivation rate is very low and approaches zero); at higher doses the death rate increases as the dose itself increases; at the highest concentrations, the rate of death does not increase further since it is sorption-limited.

The concept of 'effective concentration' means that the compound must reach and accumulate at the site(s) of action, whether on the surface of, or within, the microbial cell. Dose–effect curves are thus expected to be characteristically sigmoid. However, following the first attempts to describe disinfection phenomena in 1907–1908, the first-order inactivation kinetics was the basis of subsequent investigations. Accordingly, there are two series of equations.

The first series describes lethal activity rate as a function of the contact time with a fixed concentration (C) of a lethal compound:

$$N_t/N_0 = \exp(-k(t) * t) \tag{13}$$

where N_0 is the initial number of organisms
  N_t is the number of organisms after the contact time t
  $k(t)$ is the inactivation rate constant,
(* means 'multiplied by' in this section)

Equation (13) is more usually written as:

$$N_t/N_0 = 10^{(-k'(t) * t)} \tag{14}$$

The symbol ' means to base 10 instead of to base e, thus $1/k'(t)$ is the time required to obtain 90% destruction of microbial particles using the concentration C. The value of $k(t)$ is expected to change with the concentration of the lethal compound according to the following equation:

$$k(t) = R * \exp(K(c) * C) \tag{15}$$

This equation is more usually written as:

$$k'(t) = R' * 10^{(K'(c) * C)} \tag{16}$$

where $K'(c)$ is the rate of change of inactivation rate $k'(t)$ per unit change in the concentration C of the lethal compound, and R is a constant.

The second series of equations describes the lethal rate obtained by treating, for a fixed contact time, microorganisms subjected to increasing concentrations of chemical compounds. Accordingly:

$$N_d/N_0 = \exp(-k(d) * d) \tag{17}$$

where N_0 is the initial number of organisms, N_d is the concentration of microorganisms surviving the dose d, and $k(d)$ is the inactivation rate constant. Equation (17) is usually employed in the form:

$$N_d/N_0 = 10^{(-k'(d) * d)} \tag{18}$$

where $1/k'(d)$ is the concentration of the lethal chemical agent required to obtain 90% destruction of tested microorganisms after a fixed treatment time.

All inactivation rate constants, $k(t)$, $k(c)$, $k(d)$, are expected to increase with increasing treatment temperature, according to the Arrhenius law, in the form:

$$k(t, c, d) = F * \exp(-E_a/R * T) \tag{19}$$

where F is a frequency factor changing according to the rate constant considered, i.e. ($k(t)$, $k(c)$ or $k(d)$.

The lethal activity of disinfectants is affected by a number of factors, e.g. the type of organism(s), their physiological state including the effects of age, the pH of the solution, and the type and amount and kinds of organic material present.

Halogens

Of the halogens, the elements chlorine and iodine together with their compounds are the most widely used in food plants. Bromine is not used as such but chlorine/bromine disinfectants are in commercial use.

Chlorine – mode of action

When chlorine (Cl_2) or hypochlorites ($NaOCl$, $Ca(OCl)_2$) are added to water they hydrolyze to form hypochlorous acid ($HOCl$):

$$Cl_2 + H_2O = HOCl + H^+ + Cl^-$$

$$NaOCl + H_2O = HOCl + NaOH$$

$$Ca(OCl)_2 + H_2O = 2\,HOCl + Ca(OH)_2$$

Hypochlorous acid will dissociate to form hydrogen ion (H^+) and hypochlorite ion (OCl^-) according to pH, temperature and ionic strength:

$$HOCl = H^+ + OCl^-$$

At low pH, HOCl predominates; at high pH, OCl^- does. Chlorine is regarded as more effective at low pH when hypochlorous acid is responsible for the rapid bactericidal action on vegetative forms. Spores are more resistant.

The decimal reduction time (D) of vegetative bacterial cells suspended in a 1 mg/l (ppm) free-available-chlorine (FAC) solution usually ranges between some seconds and one minute. The following relationship can be regarded as a useful approximation:

$$\text{Log}\,D = -0.93 - \text{Log}\,(\text{mg/l FAC}) \qquad (26)$$

For bacterial spores suspended in a 100 mg/l free-available-chlorine, D is usually close to one minute. The following relationship may represent resistance of most *Bacillus* spp. spores (except for spores of *B. stearothermophilus*):

$$\text{Log}\,D = 2 - \text{Log}\,(\text{mg/l FAC}) \qquad (21)$$

For Clostridium spores, the following equation applies:

$$\text{Log}\,D = 0.97 - \text{Log}\,(\text{mg/l FAC}) \qquad (22)$$

A recommended treatment for sanitizing clean equipment is treatment for 2 to 5 minutes with water having a concentration of 200 mg/l chlorine. Using equations (20), (21) and (22) this would be expected to give a reduction of more than $4D$ for *Bacillus* spp. spores and over $40D$ for vegetative pathogenic organisms. However, a treatment for 10 to 20 minutes with a concentration of 50 to 100 mg/l is more often used to minimize corrosion of equipment.

Hypochlorous acid reacting with organic matter (amines, amides, imines or imides) forms chlor-amines which have a slower disinfectant action than free chlorine. Different commercial chloramines have higher or lower bactericidal activity than chlorine at equivalent concentrations.

Chlorine dioxide (ClO_2), unlike chlorine, does not hydrolyze in aqueous solution. Chlorine dioxide is more effective than chlorine, mostly at high pH values. It is less affected by organic matter than chlorine and because of this it has been used in vegetable flumes to minimize the microbial load of yeasts, molds and bacteria. However, problems have been experienced in maintaining effective residual levels in some applications and there may be regulatory restrictions on its use.

Iodine

Microorganisms are very sensitive to iodine. Hypoiodous acid is regarded as the active compound of iodine solutions, while the free elemental iodine is also considered to be active. Usually, iodine-based disinfectants are an association of the halogen with polyvinylpyrrolidone, or different surfactants, and phosphoric acid. These mixtures are called iodophors. Iodophors show greater bactericidal activity at acidic pH where hypoiodous acid is produced. Its activity is less affected by organic matter than chlorine. Activity of the iodine against vegetative microbial cells is equal to that of chlorine, but is lower against bacterial spores. At increasing temperatures, iodine becomes more sporicidal than chlorine, probably because it is a specific – SH reagent. This contrasts with chlorine, which is a more general oxidizing agent.

Bromine

Bromine is a more effective sporicide than either iodine and chlorine, is less affected by pH, but is also more toxic. Usually, organic bromine or chlorine–bromine compounds are equally effective as organic chlorine ones. Used in combination, bromine increases the effectiveness of chlorine solutions.

Quaternary ammonium compounds

The general formula of quaternary ammonium salts (also known as QUATS, or QACS) is the following:

$$R1 - \underset{\underset{R4}{|}}{\overset{\overset{R2}{|}}{N}} - R3 - X$$

Where R stands for an organic molecule ranging from CH_3 to $C_{18}H_{37}$ or phenyl groups; X is an anion such as Cl, Br, acetate, etc.

Bactericidal properties of QUATS are less pronounced than those of halogens, while QUATS are less affected by pH and organic matter, and are less corrosive and toxic. They also show a bacteriostatic effect at high dilutions, and are therefore often used as an equipment rinse which is left overnight. Gram-positive bacteria are usually more sensitive than Gram-negative bacteria, bacterial spores, phages, etc. It should also be remembered that bacteria may acquire a resistance to the action of QUATS, and so the germicidal action is no longer effective.

Hydrogen peroxide (H₂O₂)

At low concentrations and temperatures, hydrogen peroxide solutions are bactericidal but not highly sporicidal. To obtain rapid sporicidal activity, relatively high concentrations (35%) of hot peroxide must be used.

The following relationship holds for destruction of spores of *Bacillus stearothermophilus*:

$$\text{Log } D = 1.7 - 1.3 * \text{Log} (\% \text{ } H_2O_2) \qquad (23)$$

where D is the decimal reduction time in minutes.

Temperature significantly affects the sporicidal activity. A $Q_{10} = 1.5$ to 2.5 (average $z = 40°C$) is often reported. In some aseptic systems, the sterilization of the packaging material is achieved by using a high H_2O_2 concentration coupled with a subsequent heating by hot air at 125°C.

Under acidic conditions the sporicidal activity is enhanced. It is worth noting that ultrasonic waves and ultra-violet light have a synergistic effect with peroxide on the destruction of bacterial spores.

Acids

It is well known that bacteria, phages and viruses are susceptible to strong acidic conditions in the environment. However, detailed information about this topic is lacking. As a general rule the inactivation rate in acid medium increases with decreasing pH. In substrates having pH values lower than 3.5, the inactivation rate changes ten times (tenfold) for approximately each 0.3 pH unit change. The rate of death is also affected by the pK of the prevailing acid in the solution. Death rate increases as the dissociation constant of the acid decreases, increasing the pK. The effect of pK is increasingly evident as the pH falls, though about five times (fivefold) less important than pH. Strong hydrochloric acid is rapidly lethal to most bacterial species (eight decimal reductions in a population of *Staphylococcus spp.* are expected to occur after 30 minutes in a HCl solution at pH 1.9). Citric acid (pK = 3.08) is nearly equally effective; lactic acid (pK = 3.87) is a little more bactericidal; acetic acid

is the most lethal of commonly used acids at any pH lower than 5.

Acids may kill bacterial spores, though at a pH lower than 4 the decimal reduction time may be a month or more at room temperature. Death rate, of course, increases as storage temperature rises.

Alkalis

It is well known that a pH higher than neutrality (over 7) affects the survival probability of micro-organisms. However, there are few reports on inactivation kinetics in the literature, so the phenomenon cannot be described with any useful accuracy. A 0.5% NaOH solution has about the same microbicidal activity as a 100 mg/l chlorine or iodine solution. The activity increases with the NaOH concentration and temperature.

8.C.5 Indirect action of environmental conditions

Introduction

The main purpose of some technologies is to maintain the desirable physico-chemical (sensory) characteristics of foods, without having any definite expectation about lethal activity against microbial population, e.g. freezing, chilling, drying, reducing water activity.

Alternatively, there may be some expectation of lethal activity, for example with salting or fermentation. However, most environmental conditions applied in food preservation technology cannot be regarded as lethal *per se*, although only at very low temperatures, e.g. −200°C, can prolonged survival of microbial populations be expected to occur. At higher temperatures, when environmental conditions are not suitable for growth, some destruction of the microbial population always occurs, although possibly at a very low rate.

These factors contribute to an indirect action of the environment. Usually the lethal effects produced by growth-inhibiting conditions, e.g. frozen or dried storage, become more pronounced as environmental conditions depart from these which allow growth. Examples of these environmental conditions are pH, temperature, and presence of inhibiting or lethal chemicals.

Inactivation kinetics of indirect action by environmental conditions cannot be regarded as exponential and cannot be reported unless descriptively, as there is no currently accepted mathematical interpretation. However, as a general rule, the first nD (n being 1 to 3) occurs within weeks, while any subsequent decimal reductions require months.

Some conditions commonly encountered in food technology are now very briefly considered.

Frozen storage

It is generally found that in suitable environmental conditions microorganisms may survive between one and two years in liquid nitrogen. However, frozen foods are usually held at $-18°C$, and at this temperature some death occurs although at a very low rate.

The lethal rate in the frozen condition increases with decreasing pH and as the storage temperature approaches $0°C$. The lethal rate is slower as the storage temperature decreases, and is much affected by the chemical composition of the environment, e.g. presence of 'protective agents', etc.

Dried storage

Lyophilization is a widely used means for preserving microorganisms, as is storage in liquid nitrogen. However, several factors affect survival, and among the most relevant are the water activity value of the lyophilized substrate (the lowest death rate is expected to occur at $a_w = 0.1$ to 0.2) and the amount of light (at about 360 nm wavelength) to which the culture is exposed. The death rate is lowered by storage in the dark.

The lethal rate in dried foods may increase by: decreasing the pH and/or increasing the storage temperature although this may adversely affect the sensory properties, and/or increasing the RH to 70%, and/or increasing the oxygen concentration in the environment (both during and after the drying process).

Use of salt

High salt concentrations have some microbicidal activity. Nearly all microorganisms survive in media containing less than 3% NaCl, but most organisms do not tolerate 5 to 10% salt. *Staphylococcus* spp. can survive and grow in substrates having more than 15% NaCl; other pathogenic microorganisms do not survive for a long time at salt concentrations higher than 10% (w/w). However, it must be noted that *L. monocytogenes* can survive in 20% to 30% salt for 10 days at $22°C$, and in brine solutions of between 10.5% and 30%, survival has been found for 100 days at $4°C$.

Lactobacilli tolerate 10% salt with difficulty. Salt-loving bacteria, e.g. *Halobacterium* and *Halococcus*, survive for some time following growth, at levels approaching saturation (about 26%).

Curing and fermentation

Several chemicals usually employed in meat seasoning, such as salt, nitrite, flavors and organic acids, increase the probability of inhibition and death of microorganisms naturally present in meat. Rates of inhibition are affected by the concentration of the inhibitors, temperature, water activity and metabolic activities of resistant organisms including competition for nutrients and the production of acetic, lactic and formic acids and ethanol.

References

1. Stumbo, C.R. (1973) *Thermobacteriology in Food Processing*, New York: 2nd edn. Academic Press
2. Shapton, D.A., Lovelock, D.W. and Laurita-Longo, R. (1971) The evaluation of sterilization and pasteurization processes from temperature measurements in degrees celsius (°C). *Journal of Applied Bacteriology*, **34**, (2), 491–500
3. Dakin, J.C. (1962) Pasteurization of acetic acid preserves. In *Recent Advances in Food Science*, Vol. 2 edited by J. Hawthorn & J. M. Leitch. London: Butterworths
4. Put, H. M. C. and de Jong, J. (1980) The heat resistance of selected yeasts causing spoilage of canned soft drinks and fruit products. In *Biology & Activities of Yeasts*, edited by F. A. Skinner, S. M. Passmore and R. R. Davenport, Society for Applied Bacteriology Symposium Series No. 9. London: Academic Press
5. Gaze, J. E., Brown G., Gaskell, D.E. and Banks, J .G. (1989) *Technical Memorandum No. 523. Heat Resistance of Listeria monocytogenes in Non-Dairy Food Menstrua*, The Campden Food and Drink Research Association, Chipping Campden, Gloucestershire GL55 6LD
6. Patashnik, M. (1953) A simplified procedure for thermal process evaluation. *Food Technology*, **7**, 1–6

Further reading

Pflug, I. J. (1987) *Textbook for an Introductory Course in the Microbiology and Engineering of Sterilization Processes*, 6th edn, Environmental Sterilization Laboratory, 100 Union Street, Minneapolis, MINN 55455

Campden R. A. (in press) *Technical Manual no. 27 Guidelines to Food Products Stabilized by Mild Heat (pasteurization) Treatments*, Campden Food and Drink Research Association, Chipping Campden, Gloucestershire GL55 6LD

For the effect of heating oily foods, references may be made to Campden Food Preservation Research Association *Technical Memorandum No. 345, Heat Resistance and Recovery of Cocci and Spores Heated in Oil and Oily Canned Foods*, J. E. Gaze, July (1983)

Gill, J. and Selman, J. (1989) Technical Bulletin No. 71. *Food Irradiation: an Overview*, Campden Food and Drink Research Association, Chipping Campden, Gloucestershire GL55 6LD

Walker, S. J., Bows, J., Richardson, P. and Banks, J. G. (1989) Technical Memorandum No. 548. *Effect of Recommended Microwave Cooking on the Survival of Listeria Monocytogenes in Chilled Retail Products*, Campden Food and Drink Research Association, Chipping Campden, Gloucestershire GL55 6LD

Bows, J. R., (1989) Technical Memorandum No. 527. *The Influence of the Thermal, Electrical and Physical Properties on the Quality of Food Heated by Microwaves*, Campden Food and Drink Research Association, Chipping Campden, Gloucestershire GL55 6LD

Bows, J. R. (1990) Technical Memorandum No. 573. *The Influence of the Thermal, Electrical and Physical Properties on the Quality of Food Heated by Micro-waves*, Campden Food and Drink Research Association, Chipping Campden, Gloucestershire GL55 6LD
(Note: these two Technical Memoranda report continuing work from this research project.)
Bernstein, M. (1990) The chemistry of disinfectants. In *CIP: Cleaning in Place,* edited by A. J. D. Romney, Society of Dairy Technology, 72 Ermine Street Hunting-don, Cambs PE18 6EZ
Casolari, A. (1981) A model describing microbial inactiva-tion and growth kinetics. *Journal of Theoretical Biology,* **88**, 1–34
Casolari, A. (1988) Microbial death. In *Physiological Models in Microbiology,* Vol. II, edited by M. J. Bazin and J. I. Prosser, pp. 1–44. Florida: CRC Press

8.C.6 Sulfites – a note on their mode of action and application

This is included as an example of the material which may be produced either in response to a manage-ment request for background briefing or for use in training. This note was written for readers in the USA but those in other countries should find both the approach and the data interesting and useful.

Sulfites have a long history as food ingredients. Until very recently they were a common preserva-tive both in foods and in some drugs. They have been listed as Generally Recognized as Safe (GRAS) in the USA since November 10 1959 (24 Federal Register 9368) for use as chemical preserva-tives.

'Sulfite' is a term applied to several sulfur-based substances including sulfur dioxide, sodium sulfite, sodium and potassium bisulfite and sodium and potassium metabisulfite. All six of these sulfiting agents were considered GRAS at one time but their status in the USA has now been changed[1].

Sulfites generally act as preservatives or anti-oxidants in foods but have other permitted uses. Sulfur dioxide can be applied as a bleaching agent for food starches. Sodium sulfite and sodium metabisulfite are used in the production of cel-lophane used to package food. Sodium sulfite and sodium metabisulfite are combined with other chemicals to prevent rust and scaling in boiler water used in making steam that would eventually make contact with food. Sodium and potassium metabisul-fite, sodium bisulfite, and sulfur dioxide are commonly used as sterilizing agents in wine making.

Sulfites show a wide range of useful technical effects as shown in Table 8.14.

As previously stated, sulfite addition to foods was always considered safe for consumers. For instance, the IFT Scientific Status Summary 1975[2] and in 1976 the Federation of American Societies for Experimental Biology (FASEB)[3] after a review of

Table 8.14 Functional properties of sulfite in food

Function	Example
Prevents enzymatic browning	Retards darkening of cut fruit, e.g. apples, and vegetables, e.g. potatoes
	Prevents black spot formation in shrimp
Prevents non-enzymatic browning	Controls the browning of dehydrated potatoes
Controls microbial growth	Prevents bacterial growth and acid formation during wine fermentations
	Controls bacterial growth during the wet milling of corn
	Limits mold growth during shipment of table grapes
Acts as a dough conditioner	Used specially in frozen doughs
Prevents oxidative rancidity	Possible but not used extensively for this purpose
Serves as a bleaching agent	Used in processing of maraschino cherries

the scientific literature up to 1975 concluded that sulfites posed no health hazard to the general population at the levels they were being used in food. However, in 1981 work by Baker *et al.*[4] and Stevenson and Simpson[5] implicated sulfites as precursors to asthmatic reactions in a very small subgroup of the asthmatic population.

Although in 1985 a more recent report of the scientific panel of the FASEB[6] again concluded that sulfites are safe for the majority of consumers and pose no hazard to the general population, it also demonstrated that there was sufficient evidence to show a severe hazard to consumers sensitive to sulfites. Because of these disclosures the Food and Drug Administration (FDA) was prompted to take additional steps to assure the safety of the public.

The FDA has for some time required that packaged foods containing sulfites present as preservatives be so labeled. These foods include lemon juice, maraschino cherries, grape juice, packaged fresh mushrooms, some canned soups, dried fruits and vegetables (but not dark raisins or prunes) and dehydrated vegetables (but not garlic or onions).

The FDA has prohibited the use of sulfites in foods that are an important source of thiamine (vitamin B_1) such as enriched flour and breads, because of the adverse effect of sulfites on this essential nutrient. Sulfites are also banned from addition to meats because of possible consumer deception. The loss of red color in meat can be restored by the use of sulfites, and by doing so can give a false impression of freshness.

The FDA proposed two regulations relating to sulfites – FR50 #64, p. 13306, April 3 1985 and FR50 #157, p. 32838, August 4 1985. One proposal required declaration of sulfite on the label when the residual sulfite content in the food exceeded 10 mg/kg or parts per million (ppm) as total SO_2. If used as a preservative it would have to be labeled no matter what the amount. The second regulation proposed rescinding the GRAS status of sulfites for use on fruits and vegetables intended for consumption in the raw state (except potatoes). The first regulation will help sulfite sensitive people avoid packaged foods such as cookies that have sulfites used as a dough conditioner, and although the SO_2 would be dissipated in the baking process it would still leave behind a discernible residue. The 10 mg/kg (ppm) level was arrived at in the proposed regulations because it was the lowest level possible at which sulfites could be measured accurately enough for law enforcement by the FDA. The second regulation will revoke the GRAS status of sulfites used on raw produce. It would ban the use of sulfites on fruits and vegetables used on salads and at salad bars and at other food-service outlets as well as produce sold in supermarkets.

The regulations were passed and appear in the Federal Register of July 9 1986 and took effect January 9 1987. The 1985 FASEB report[6] demonstrates that these two regulations will have a profound effect on the use of sulfites in the USA from what was previously permitted.

Sulfites when added to foods react readily with a variety of food components. Their action is dependent on the nature of the food, the type and extent of processing used, and the length and type of storage conditions, the permeability of the package and the level of addition (see Taylor *et al.*[7]). Sulfites will react with reducing sugars, aldehydes and ketones, and proteins to form various combined sulfites. Sulfites can be oxidized to sulfate, a quite harmless product. In foods with acidic pHs (pH less than 4.0), sulfur dioxide can be volatilized and dissipated from sulfited foods. In many foods, only a small fraction of the added sulfite remains as free, inorganic sulfite in the finished product. In hard sweet biscuit doughs, 63% of the added sulfite reacts with the flour constituents, 30% is oxidized to sulfate, and less than 0.2% remains as free, inorganic sulfite (see Thewlis and Wade[8]). In jams and jellies in the UK 98.5% of the absorbed sulfite was volatilized as SO_2 during cooking, or reacted with jam components. Less than 1.5% remained as

inorganic sulfite (McWeeney *et al.*[9]). There is no sulfiting of jams or jellies in the USA. The majority of sulfite added to lettuce (no longer permitted) remained as free, inorganic sulfite (Martin *et al.*[10]). This preponderance of free sulfite was the major cause of many asthmatic reactions. The fate of added sulfite has important implications and it is for that reason the labeling of the individual sulfiting agent is now required.

Not all products in the USA come under the jurisdiction of the FDA. For instance, wine and beer are administered by the Federal Bureau of Alcohol, Tobacco and Firearms Agency. In 1985 they also proposed legislation that would require sulfite labeling for wine, distilled spirits and malt beverages if the sulfite levels were 10 or more mg/kg (ppm).

One of the sulfites, sulfur dioxide, is used as a fungicide on grapes and as such comes under the authority of the US Environmental Protection Agency (EPA). Thus far, EPA has not taken any action against sulfites.

The killing action effect of sulfites on microorganisms was studied extensively at the University of Wisconsin by Strong *et al.* in 1971 to 1974[11]. The susceptibility of different microorganisms, e.g. *E. coli*, *Lactobacillus delbrueckii*, *Acetobacter acetii*, *Acetobacter suboxydans* and *Saccharomymes cervisiae*, is dependent on both time and sulfite concentration. None of the organisms tested was resistant to more than 190 mg of SO_2 per liter at a pH of 4.0. Experiments revealed that free SO_2 is the bactericidal agent. The resistance does not appear to depend upon the inherent physiological differences in bacterial organisms. The killing power of SO_2 with regard to *S. cervisiae* can be best described as 'multiple hit' kinetics. A relationship exists between cell concentration and the concentration of free SO_2 required to kill a population of organisms. The energy metabolism is inhibited immediately after the addition of SO_2. Cell death is never immediate. The experimental run seems to point to the cytoplasmic membrane as the primary site of action of sulfite. However, it was shown that under anaerobic conditions, yeasts treated with SO_2 were at least ten times more resistant than those under aerobic conditions.

Therefore there is reason for concern about the use of sulfites in foods. The use of Good Manufacturing Practices and informational labeling may be sufficient to control the existing hazards for sulfite-sensitive asthmatics. The selective banning of sulfites, however, does not appear to be justified except for the direct use of sulfites in restaurants. This opinion is based on the greater probability of abuse of sulfites by untrained or unskilled kitchen workers. Abusive use on lettuce can lead to enormous sulfite residue levels which would trigger asthmatic reactions in the uninformed sensitive individual[12].

References

1. Taylor, S. L., Bush, R. K. and Busse, W. W. (1985) The sulfite story. *Associated Food Drug Officials Quarterly Bulletin*, **49**, 185–193
2. IFT Scientific Status Summary (1975). Sulfites as food additives, Institute of Food Technologists. *Food Technology*, **29**, (10), 117–120
3. FASEB (1976) *The Examination of the GRAS Status of Sulfiting Agents*. Bethesda, MD: Federation of American Societies for Experimental Biology
4. Baker, G. J., Collett, P. and Allen, D.H. (1981) Bronchospasms induced by metabisulfite-containing foods and drugs. *Medical Journal of Australia*, **2**, 614–616
5. Stevenson, D. D. and Simon, R. A. (1981) Sensitivity to ingested metabisulfites in asthmatic subjects. *Journal of Allergy and Clinical Immunology*, **68**, 26
6. FASEB (1985) *The Re-examination of the GRAS Status of Sulfiting Agents*. Bethesda, MD: Federation of American Societies for Experimental Biology
7. Taylor, S. L., Higley, N. A. and Bush, R. K. (1986) Sulfites in foods; uses, analytical methods, residues, fate, exposure assessment, metabolism, toxicity, and hypersensitivity. *Advances in Food Research*, **30**, 1–76
8. Thewlis, B. H. and Wade, P. (1974) An investigation into the fate of sulfites added to hard sweet biscuit doughs. *Journal of the Science of Food and Agriculture*, **25**, 99–105
9. McWeeney, D. J., Shephard, M. J. and Bates, M. L. (1980) Physical loss and chemical reactions of SO_2 in strawberry jam production. *Journal of Food Technology*, **15**, 613–617
10. Martin, L. B., Nordlee, J. A. and Taylor, S. L. (1986) Sulfite residues in restaurant salads. *Journal of Food Protection*, **49**, 126–129
11. Strong, F. M. (1974) The effect of sulfites on microorganisms. In *Annual Report*, Food Research Institute, University of Wisconsin, Wisconsin
12. Taylor, S. L., Martin, L. B. and Nordlee, J. A. (1985) Detection of sulfite residues in restaurant salads. *Journal of Allergy and Clinical Immunology*, **75**, 198

Further reading on toxicological aspects

Strong, F. M. and Gibson, W. B. (1971) Metabolism and elimination of sulfite by rats, mice and monkeys. *Foods and Cosmetics Toxicology*, **10**, 185–198
Strong, F. M. and Gibson, W. B. (1973) Accumulation of ingested sulfite and sulfate – Sulfur and utilization of sulfited proteins by rats. *Foods and Cosmetics Toxicology*, **12**, 102–106

Chapter 9
Insights into microbiological control methods

9.A Microbiological examination

Scope

Common reasons for microbiological testing or examination include:

- Investigational testing:
  - To determine if a microbiological hazard is present or indicated
  - To ascertain whether GMP has been effectively applied (either by a supplier or 'in-house')
  - To determine or monitor sanitary status
  - To verify that critical limits are met in the HACCP system
  - To determine the effect of a modification or change to process, equipment or ingredient or product
  - To establish microbiological trends
  - To determine positive spoilage, e.g. in low acid foods
  - To determine if microbial metabolites including toxins or large numbers of dead cells suggest poor handling previous to a subsequent 'kill step'
  - To conduct challenge testing to predict safety in use or predicted abuse situations

- To conduct challenge testing to predict shelf life
- Acceptance testing:
  - To comply with legal or contractual requirements
  - To determine compliance with purchase specifications
  - To confirm that desired levels and kinds of microorganisms are present, e.g. starter cultures, yogurts
  - To confirm that starter cultures are 'pure', i.e. not contaminated, and are of the required activity
  - To demonstrate the exercise of 'due diligence' in the purchase, manufacture or selling of food
  - To determine appropriate use or disposal where quality varies
  - To assess microbiological status or keeping quality at any stage of manufacture or in the finished product
- Professional reasons:
  - Taking part in assurance testing/ring (round robin) trials to establish methodology
  - Comparison of alternative methodologies
  - As part of research or development
- To meet specific educational or training needs

255

Introduction

Given that quality assurance systems are being applied within a production department as part of Good Manufacturing Practice (GMP), there is still a real and justifiable need for microbiological examination to verify that correct practices and procedures are being followed, even though there is a delay in obtaining results.

It is, of course, true that any usable information on the microbiological status of a plant, process or product is useful for its management. There will, however, be times when there is no substitute for specific testing for organisms whether for kinds, e.g. pathogens; for numbers, e.g. total viable count; or for numbers of specific types, e.g. spoilage yeasts.

Microbiological examination includes the environment in which production occurs, ingredients from which products are made and the product itself. Whatever the type of examination, or the reason for it, there are safety aspects affecting the laboratory technician, as well as accuracy aspects concerning the methods applied during the examination.

Safety aspects

There may or may not be safety regulations in force which apply to laboratories. In the UK, compliance with the Control of Substances Hazardous to Health Regulations (COSHH)[1] is mandatory. These require that the employer must insure that all control measures to protect workers from hazardous substances are followed. Pathogens are included among the substances which create a hazard to the health of a person, and a Code of Practice relating to pathogens (and their containment) was published in 1984[2]. However, do not forget that in a microbiology laboratory hazardous substances also include powdered culture media; reagents used in diagnostic testing; alcohol and other solvents; undiluted stock sodium hypochlorite solution.

In Canada, the Workplace Hazardous Materials Information System (WHMIS) exists as a system for classification and identification of hazardous materials. Canadian legislation states, any worker who is or could be exposed to a controlled product in the workplace must have proper training. Other components of this mandatory system include specific labelling requirements for all hazardous materials and issuance of a Material Safety Data Sheet.

Whether or not there is a regulatory requirement, the approach taken in COSHH ought to be applied within each microbiology laboratory. The five main steps which management must take are:

1. For each substance which poses a health hazard, assess the risk of injury.
2. Introduce appropriate measures to control the risk.
3. Insure that these control measures are used, equipment is maintained and correct procedures are followed.
4. Monitor exposure of worker(s) to hazardous substances or radiation where this is necessary.
5. Inform, instruct, train and direct the laboratory workers about the hazards, risks and what precautions must be taken.

It is of particular importance that a safety cabinet is used when known or suspected pathogens are being handled. There are three classes of cabinets as defined by British Standard (BS) 5726[3], and the USA[4] and Australia[5] also have national standards. Cabinets should comply with local national standards. Campden Technical Manual 21[6] and Collins *et al.*[7] both give good practical summaries on their use.

Do not confuse a safety cabinet with a laminar flow clean air workstation. The clean air work station protects the work inside the station from contamination but does not protect the worker from contamination from the materials being examined. Again, see Collins *et al.*[7] and Campden Technical Manual 21[6].

For both kinds of cabinet the air flow pattern must be correct for the proper functioning of the cabinet. Monitoring of the air flow pattern should be carried out according to the manufacturers instructions. Leaks within the cabinet caused by faulty welds or gaskets must be prevented by regular maintenance.

Even if lawful, as a matter of prudent policy, in no circumstances should a plant laboratory test for *Clostridium botulinum*. Specialist research or medical laboratories should be used for any examination of suspect *Cl. botulinum* in foods.

Pathogen examination should not be carried out in plant premises where there is a risk, even if small, of finished product contamination from any laboratory operation, including the disposal of waste.

'Standard' methods and accreditation of laboratories

Because of the inherent variability of microbiological methods, over the years there has been a desire to standardize methods. This standardization includes media and equipment used as well as the procedures (or protocols). It is important to be clear about what the method is intended to deliver. There may well be a conflict between a desire for sensitivity, i.e. maximum recovery or count, and reproducibility within and between laboratories. Standard methods, either national or international, should be used when they exist and if they are appropriate. Remember the general rule that a lot of simple, quick tests give more useful information than that from one or a few complicated ones.

Even when standard methods are used in different laboratories, as in collaborative or ring (round robin) tests, the results between laboratories can be surprisingly different. These collaborative tests demand resources to design and perform, and also to analyze the results. They are therefore expensive to carry out and can be technically difficult, particularly when inoculated material must be prepared. In addition, they should not be carried out on a single batch of material, but should be repeated at appropriate intervals.

To provide assurance that laboratories can achieve consistent results within the 'state-of-the-art' there have been moves towards 'accreditation' of laboratories and, in the USA and Canada, of microbiologists. As yet, few microbiology laboratories have received accreditation, although several countries have set forth the requirements for such accreditation. One scheme is that of the UK, which functions within the National Measurement and Accreditation Service (NAMAS). However, there is a tendency for important customers to want results from accredited laboratories, and processors need to be aware of this. In the near future, accreditation of your own laboratories may be justifiable and cost-effective.

Sampling

To be meaningful, the information given by microbiological examination must be a fair and accurate assessment of the microbiological status of the material being examined. There are two main aspects to achieving this aim:

1. The sample must be as representative as possible of the environment or of the product.
2. The sample must be taken aseptically, that is, without any contamination occurring during sampling. Sample handling between the time that it is taken and the time that it is examined should prevent significant microbiological multiplication or death, as any change of numbers will give an incorrect result.

Remember, changes of microbiological populations in environments or products are complex. Growth and death can occur rapidly and at different rates with different kinds of organisms leading to major changes in the make-up of populations. Remember also that this means there is no opportunity to resample at leisure, so it is important to 'get it right first time'.

Representative sampling

This must not be confused with the unacceptable practice of taking samples from the easiest or most convenient points to make up the specified number. An illustration of this point is the taking of random samples from delivered pallets constituting a 'lot'. In practical terms this is expensive both in time and labor and is unlikely to be done properly. The solution is to operate a Supplier Quality Assurance scheme. Included in the specification for the purchased material is the requirement that sampling is done while pallets are being loaded. The alternative of sampling packs from the edges or outside of the delivered pallets should not be done if truly random samples are required. Information from such sampling will not necessarily be representative of the microbiological quality of the batch or lot of material.

To be successful, microbiological and, wherever appropriate, statistical principles must be applied to the sampling process. For a fuller discussion on sampling and sample taking see Speck[8] (Chapter 2) 'Sampling plans, sample collection, shipment and preparation for analysis', Garfield's paper 'Sampling in the Analytical Scheme'[9], Steiner[10] in his chapter on 'Statistical Methods in Quality Control' and the recent text by Jarvis[11] which is written for microbiologists.

Environmental samples

To obtain information about the factory environment, sampling is often carried out at places known to be critical for, or prone to, contamination by pathogens. Examples of these, with reference to contamination by *Listeria monocytogenes*, are the use of drain swabs, floor swabs of entrances to production areas, samples of brine in the manufacture of soft mold-ripened cheese, swabs of wet areas of floors and the undersides of pallets. In the case of *Salmonella* in powder or other dried material product, as well as floor or conveyor belt samples, powder samples from the vacuum cleaning system form the basis of samples to be taken.

As an example of production equipment sampling, representative samples are obtained by swabbing known areas of surfaces immediately before production begins. Swabs are taken at selected parts of the plant so that the efficiency of cleaning can be assessed. To illustrate the care needed in selection of swab points, it is necessary to swab valve seatings as well as storage tank surfaces. Remember, the information required relates to the whole plant, not just the surfaces which are easy to reach.

As well as examining swab samples, samples of the first product through the plant may be taken. If the microbial count on these samples is higher than usually obtained, then plant cleaning has not been properly carried out or alternatively, contaminated materials have been used. For further information on methods, see Chapter 3 in Speck[8].

Product samples

As well as taking representative samples of product it is also prudent to sample ingredients according to a plan which will enable the probability of accepting or rejecting an unsatisfactory lot to be quantified. Shapton[12] describes such a plan for thermophilic spores in flours, etc. It must be understood that when examining a food for microorganisms, homogenous distribution of microorganisms is seldom found, even in liquids, which makes quantifying the microorganisms in a production lot a difficult task. However, by using either a sampling plan which takes a given number of random samples from a lot or by using experience of the factory microbiologist or quality expert to select the sampling points in the production line, it is possible to assess the microbiological quality of product when the results of the microbiological examination are known.

When sampling product from a valve or 'sampling point', this must be sanitized before taking the sample. If this is done with a chemical sanitizer, be sure an appropriate quenching agent is used.

Aseptic handling of samples

As contamination may introduce serious errors which could lead to the wrong action being taken, aseptic handling must be applied from the time that the sample is taken until the microbiological examination is completed. All the apparatus used for sampling and laboratory examination must be sterile. The person taking the sample and the person examining it must know and use the correct technique to avoid contamination of the sample. This will involve training by a qualified (professional) microbiologist of both the sampler and the laboratory technician. It is also necessary to be sure that good practices, once learned, are maintained. As examples of the care needed, the sample container must be opened and closed without touching the inside of the container with fingers, and the container lid (if there is one) must not be placed on the floor or other contaminating surface while the sample is being taken. The sample container must not be over-filled, as this prevents correct closing of the container and thus contamination may occur.

Once the samples are taken they must be sent to the laboratory immediately, or stored away from the production area. To minimize microbial population changes, liquid samples must be held below 5°C although dry samples may be held at ambient [room] temperature provided this is less than 20°C.

In the laboratory, liquid samples should be held under refrigeration, and all samples examined as soon as possible after receipt. Examination of samples usually means preparing serial dilutions, so that accurate counts of microorganisms may be made. For each dilution used in making the dilution series a separate sterile pipette is used, or pipette tip when using Eppendorf pipettes. Care must be taken using Eppendorf pipettes, as, if the end of the pipette body gets wet and is used while it is wet, cross contamination of samples will occur, even though separate sterile pipette tips are used for each sample.

Laboratory procedures

As well as correct sampling and sample handling, the laboratory procedures must use good (sound) microbiological technique(s) and practice(s) to insure that the result is meaningful. Procedures will be considered under the following headings:

• Principles of safety and containment
• Design of the laboratory
• Safe working procedures
• Laboratory equipment and apparatus
• Staining
• Dilution of samples
• Counting of microorganisms
• Growth of microorganisms and culture media
• Isolation of microorganisms
• Identification of microorganisms

Principles of safety and containment

Both the design of the laboratory and the equipment and working practices must insure:

• That personnel are protected from infection.
• That contamination of work, e.g. samples, is avoided.
• That disposal of laboratory materials, e.g. samples, dilutions of samples, pipettes or pipette tips, plates, broth cultures, takes place after decontamination within the laboratory. This prevents contamination spreading outside of the laboratory and is particularly important when cleaning or garbage removal is done by non-technical personnel.

Warning The first rule for safety is to act as if all organisms are pathogens.

Design of the laboratory

As a basic principle, sanitary design should include separation of 'clean' areas, such as sample reception, storage of sterilized apparatus and culture media and preparation of culture media, and 'dirty' areas, such as working areas for bench examination of samples and incubators and disposal of incubated materials. Within the working area of the laboratory, it is preferable to have a separate area for working with pathogens, even when such examination is made using safety cabinets to protect the personnel.

The working surfaces and floor surfaces of the laboratory should be easily cleaned, and should be resistant to the disinfectant solution (which may be hypochlorite) used to wipe down benches and floors as a regular part of safe working procedures.

Space immediately outside the working area of the laboratory must be provided, so that outdoor coats may be properly hung up and laboratory coats kept there, to be put on before personnel enter the laboratory area. A hand washbasin should be provided at the entrance to the laboratory, and disposable paper towels should be used to dry hands. Taps on the basin should be operated by foot, knee or elbow.

However the laboratory space is arranged, it is worth remembering that the preference of most microbiologists for a 'laboratory at the end of a corridor' is based on bitter experience of problems that arise when the laboratory is used as a passage to other areas.

Safe working procedures

To prevent sample contamination, strict aseptic technique must be followed throughout the whole of the testing procedure of each sample. The safety aspects as far as the workers are concerned must include training in the correct procedure to follow if spillage occurs, and in the correct handling of incubated material – which is to *carefully* put incubated plates into autoclavable bags; broth cultures into baskets or clean empty cans which will prevent spillage of broth; and loading of the autoclave with the incubated material so that the whole of the load receives a full and proper sterilization treatment. Also, personnel must be made aware that during examination of samples, certain aspects of aseptic technique may in fact present hazards. Formation of aerosols, which can carry living microorganisms, can occur during pipetting, and also during transfer of incubated broths by means of sterile loops. Useful information on the amount of such contamination is given in Table 1 (p. 633) of Steere[13]. Correct techniques to overcome these hazards are described in Chapter 5 of Collins[14]. Further details of design and procedures to insure safety are given in Miller[15], with special reference to Chapters 16 to 24. These and other literature references should be consulted to have a good understanding of the requirements which need to be met so that safety is an integral and on-going part of laboratory practice.

As a reminder of items which may be overlooked:

- The most likely accidents in a food microbiology laboratory are minor cuts and burns. Alcohol and other solvents are flammable, and if spilled and ignited may cause serious burns. Glacial acetic acid is also flammable. All flammables should be kept in small containers on the bench. Containers should have a screw-top lid, not one which will fall off if the bottle is knocked over. Cooling a hot metal wire inoculating needle by plunging it into a beaker of alcohol is inherently dangerous! 'Sharps', e.g. scalpel blades, should be disposed of in containers designed for this purpose.

- The disinfectant solution used in pipette discard jars can only be effective if it reaches the organisms. However, it may not be effective even then. Salesmen's claims that the 'new wonder material is infallible' should be checked after actual use, as food materials interfere with the disinfection process. Hypochlorite is a good general-purpose disinfectant in a food microbiology laboratory but it must be strong enough to be effective at the end of its period of intended used (e.g. a day or half-day). The color added, e.g. permanganate, is intended for identification purposes and as a rough and ready guide to dilution. It is *not* an indicator of efficiency. (See Collins[14].)

  Remember that it is sound practice to:
  1. Make sure the solution is deep enough to cover the top graduations but not the cotton-wool plugs.
  2. Change the solution daily or more often if needed.
  3. *Not to overload discard jars.*

- Be sure that no material leaves the laboratory unless it is known to be safe. Remember, burning may not be effective unless a design of incinerator with an afterburner is used correctly.

It is good food microbiology practice to autoclave materials before disposal but be sure that the scheduled safe process temperatures are achieved *throughout* the load and that the full time is given. Air is an insulator, so proper venting to remove air from the autoclave is necessary before timing of the process is started.

Laboratory equipment and apparatus

The types of equipment which will be discussed here are thermometers, autoclaves, hot air ovens, incubators, waterbaths, refrigerators, pH meters and microscopes.

Thermometers

In these and other applications accurate temperature measurement is an essential requirement. It is worth remembering that a platinum resistance thermometer is the usual primary standard or reference thermometer in National Standard laboratories. It is also worth remembering that uniform temperature throughout a liquid (e.g. oil bath) is easier to achieve when cooling than when heating.

If you use mercury-in-glass thermometers which have been calibrated by your National Standard laboratories as your 'lab standard' be sure that:

- They are used correctly as intended by the National Standard Laboratory.
- The correct range is used – do interpolate not extrapolate over the range to be checked.
- The correct immersion depth is known and used.

If you use digital thermometers, be sure that the batteries are kept charged or changed when needed and that they are regularly checked against thermometers of known accuracy.

Each piece of equipment which is designed to operate at a given temperature needs to have measurement, control and indicating and/or recording instruments which operate within known and acceptable limits of accuracy. Remember that what is important is the accuracy of the system as a whole and not just the accuracy of the temperature sensing element.

For equipment which is to be maintained at a given temperature, e.g. incubators, water baths, refrigerators, each should have its own thermometer which is checked and recorded at least daily. As a principle, it is only the ability to demonstrate accuracy over the relevant period of time that justifies real confidence in temperature data.

Autoclaves

In food microbiology laboratories these are usually simple steam autoclaves, without vacuum as part of the cycle. They are used to sterilize some kinds of 'glassware', apparatus and culture media at a temperature around 121°C (250°F) equivalent to 1 bar (15 psig). For precise operation an accurate (bled) thermometer should be fitted. Since pressure governs the temperature reached in the autoclave, the vessel must have intact seals (gaskets). An accurate pressure gauge should be fitted. Before the vessel is sealed by closing the vent valve, the air inside the autoclave and load must have been completely displaced by steam, otherwise the correct temperature is not reached even when the correct pressure is reached.

As a safety feature, there must be a valve on the autoclave which releases the pressure when it reaches a predetermined pressure, e.g. 2 bar (30 psig). Unless local law is more stringent, the autoclave should be inspected once a year by safety engineers to check that the pressure gauge is accurate and that the safety valve is correctly set, in addition to the safety of the vessel itself. Periodic checks on thermometer accuracy are required – it should be within 0.5°C (1°F) of the master.

Although microbiological methods, e.g. spore strips, are available to test the adequacy of the autoclave process, the best method is to use thermocouples distributed throughout the load to determine the temperature–time profile. Lethality can then be assessed using standard cannery techniques. Results of such tests may well show 'cold spots' due to venting problems and the adverse effects of close packing a load.

Hot air ovens

These are used to sterilize apparatus which it would be impractical to autoclave, e.g. glass bottles with glass stoppers, test tubes and glass pipettes in canisters. The temperature should be between 170°C to 180°C. The holding time recommended by the UK Department of Health is 18 minutes at 170°C and 7½ minutes at 180°C, and the come up time to reach these temperatures is at least 1 hour. Whether heated by gas or electricity, the door must not be opened for at least one hour after the heating has been turned off, otherwise glassware may crack.

Other time/temperature combinations of equal lethality have been proposed. These may be needed if charring of cotton-wool plugs occurs at 170°C to 180°C. This charring may inhibit microbial growth of sensitive species.

Incubators

These are used to allow the microorganisms to grow on the nutrient agars (media) so that a count of their numbers can be made. They are also employed to grow microorganisms in broth, which is used for presence/absence tests in products and for some diagnostic testing. It is essential that the incubators work at the selected temperature, and that this temperature is maintained throughout incubation. Each incubator, therefore, must contain a thermometer. Daily thermometer checks must be made for each incubator, and the temperature adjusted if it is incorrect. Temperature recorders are usually used in research work.

An important design feature is air circulation. With this the temperature distribution within the load is greatly improved and the rate of heat transfer may be increased.

It is instructive to use thermocouples to determine temperature distributions throughout the incubator and load and to determine the rate of heating in plates, bottles and flasks. When this is done the benefits of air circulation become apparent.

It is also worth remembering that, as with autoclaves, temperatures are measured at one point and this value is assumed to be true for the whole working volume.

If a trail of colonies is seen across one or more plates then an infestation of insects or mites is likely to have occurred. Thorough cleaning and disinfection are needed and a mite infestation requires extreme measures.

Waterbaths

These are used in several ways in a microbiological laboratory, e.g. to hold growth media at the correct temperature before use; to carry out heat resistance determinations of microorganisms; or to culture microorganisms, but whatever the purpose for which they are used, there must be accurate temperature control. This control usually includes water agitation. Each bath should have its own thermometer, and the temperature recorded daily.

To insure the containers held in the waterbath are at the correct temperature, the water level of the bath must be at least 50 mm (2 in) above the level of the liquid held in the containers. The water level must not be too high, otherwise water may enter the container and so contaminate the sample or the growth medium. The water in the bath must be kept clean. This can be done by replacing the water weekly, and cleaning the bath. Do not use quaternary ammonium compounds or other sanitizers with a residual bacteriostatic or bactericidal action.

Water from waterbaths can be a source of contamination of agar plates. All containers of agar must be wiped dry as soon as they are removed from the bath and before any plates are poured. A disposable paper towel is best for this purpose.

Refrigerators

These must be kept clean, and the temperature of each refrigerator monitored daily. Any reagents, media or chemicals stored in the refrigerator should be labeled and date coded to allow stock rotation. Monthly 'clean-outs' show whether this is done.

pH meters

The correct use of pH meters together with centrifuges and balances is dealt with in Campden Technical Manual No. 21[6], in sections 3.6, 3.7, and 3.8, in the US regulations 21CFR 114.90[16] and in other standard texts. However, it should be noted that the membrane in the glass electrode of a pH meter can become 'poisoned' by precipitated protein in a food, for example yogurt, or by fat in high fat samples. This will have the effect of causing drift in the pH measurement when a large number of samples are examined. The drift may not be large, but when small changes in pH values are being looked for to indicate spoilage in a product, then there is a need to recalibrate the pH meter during the examination, as well as at the beginning and end.

In present-day instruments, the electronics are usually very reliable. To avoid 'poisoning' the electrode, it must be cleaned. After washing with water, either dropping acetone from a dropping bottle or gently wiping the electrode with paper tissue soaked in acetone will usually prevent poisoning or restore the electrode to its normal performance.

Microscopes

These are often less well used and understood than they should be in a food microbiology laboratory. They are most often used during the isolation of specific microorganisms, and during identification of these types. However, it may be necessary on occasions to examine a preparation of the laboratory sample, whether it be a solid or liquid.

When examining a food sample, great care must be taken not to confuse small particles of the food itself with either bacteria or yeasts which may be present in the sample. Also, remember that food material present on the microscope slide may lie on top of the microorganisms so that they cannot be seen.

It should also be remembered that if yeast has been incorporated as an ingredient in a food, the dead yeasts will be seen, since both living and dead cells are stained during sample preparation, and may be indistinguishable from any living yeast cells which are food contaminants. Note that large numbers of cocci present in a food may indicate the presence of *Staphylococcus aureus*. Even though these may be dead, having been killed by heat processing, toxin may have been produced in the food at levels which would cause food poisoning.

For most purposes a microscopic examination of the food sample or a dilution, using oil immersion, will be accompanied by culturing of the sample on agar or in broth. The advantage of a microscopic examination is that if the microorganisms are clearly visible, a rapid observation can be made to determine if a sample is contaminated. However, only grossly contaminated samples are detected by direct microscopic examination, as only a small quantity of food material is spread onto the slide (0.01 ml) and only a limited area of the slide is examined. For most direct examinations, 1 bacterium or yeast per field of view represents a count of around 500 000 microorganisms per ml. Using techniques such as DEFT, the procedure includes a concentration of the sample (incorporating treatment to enable fat and protein to be membrane filtered) and enables the sensitivity to be increased to detecting approximately 5000 microorganisms per ml or g. It would not be practicable to detect microbial contamination at levels lower than this.

Microscopes must be used and maintained correctly if they are to give useful results. They should be kept covered when not in use. The manufacturer's instructions must be followed when setting up the microscope for use, and the workers using it must become thoroughly familiar with it, so that they are able to use it easily and quickly. It is important that microscopes are kept clean and serviced regularly. If this is not done, then the lens system loses its clarity, and sharp focusing is not possible. When this happens, then identification of bacteria and yeasts cannot be easily carried out and is a considerable disincentive to the proper use of microscopy.

The Gram stain

As a comment on the widely used Gram's stain, there are a number of modifications to the method, e.g. Speck[8] gives two. This indicates that it is not a simple or easy matter to get clear, unambiguous results. The most difficult part of the technique is to decolorize sufficiently without over-decolorizing. Alcohol (ethanol) is slower than acetone and is sometimes preferred but continuing practice is essential for good, reliable results. The stain works best with young (18 to 24 h), actively growing cultures, preferably from clinical material. Food microbiologists, however, want to examine directly materials where the organisms may be quite old and highly stressed. This is why it is good practice to have a known Gram-positive and Gram-negative culture on either side of the 'unknown(s)' on the slide.

Dilution of samples

In the preparation for microbiological examination all solid and most liquid food or ingredient samples will be diluted in a liquid. The choice of diluent is important and depends on the particular type of examination and on the composition of the food. The method of mixing sample and diluent is also important. Mixing must be efficient, but it may also cause clumps of bacteria to come apart and mold hyphae to break. The count will therefore be affected. A standard mixing procedure must be adopted so that results may be comparable both between different workers and between different batches of the same product.

There are two main types of examination, both based on dilution, which will be considered:

• Enrichment
• Counting

Enrichment

Enrichment is a technique which is used when low numbers of a specific group of microorganisms are being looked for, and generally these microorganisms are pathogens. A known quantity of food or liquid is added to a given volume of broth (usually the ratio is 1:10 sample to broth), the inoculated broth is incubated, then examined for the pathogen under test. Remember that this makes the added food material a significant part of the culture medium, and may explain why some protocols or formulations are more successful than others for a particular application.

Since pathogens in certain foods, for example those which have a low a_w, may be 'stressed', that is, they may take a longer time to multiply in the broth than unstressed organisms, the method by which the food is added to the broth and the composition of the broth may be very important. If the correct procedure is not followed, the pathogens may not be detected even though they are present in the food.

This process of 'resuscitation' is well documented – Mossel and Corry[17], Beckers *et al.*[18] – but unfortunately a common method cannot be applied for one pathogen or for any one foodstuff. Chocolate products use milk as the enrichment medium for *Salmonella*, and dried milk responds better when rehydration in broth is carried out in two stages. The microbiologist must keep up to date with current research to be aware of improvements in procedures which may be applicable.

Counting and dilution

This method is used when it is anticipated that the food or liquid contains microorganisms at levels in excess of 10 per g or ml. The food or liquid is diluted, a 1 ml sample of each dilution pipetted onto a plate, an agar growth medium added at 42°C to 45°C, and the colonies (which develop after incubation from each microorganism or clump of microorganisms) are counted. The count of microorganisms in the food or liquid can be calculated as follows:

Colony count per plate × the dilution factor plated = the count per g or ml.

Remember that:

1. It is easy to make a mistake in the dilution factor especially when an 'unusual' or 'unfamiliar' one is used.
2. In counting where the food itself, e.g. some spices, is inhibitory, the organisms may only be able to grow at high dilutions, e.g. 10^{-3} or 10^{-4}. Where an unfamiliar or inhibitory material is under examination it is therefore prudent to take the dilution series further than would be necessary for routine counting.

Choice of diluent

Dilutions of the sample are made through a series of tubes or containers. Usually, the dilutions are decimal dilutions, and may be made as 1 in 10 dilutions or as 1 in 100 dilutions at each stage. Whatever the volume used for each dilution step, the choice of diluent is important and so is the total time taken to complete the procedure. During dilution, there should be no death of microorganisms, nor should multiplication occur. This means that the diluent should contain salt, and possibly peptone, at a level which prevents any osmotic shock damaging the microorganisms and possibly killing them. It also means that the time between making the original dilution of the sample, and adding the growth medium to the plate should be not more than 30 minutes.

If diluted samples are held longer than this, then multiplication could occur in the dilution tubes in which there is more food present (1 in 10 or 1 in 100 dilution tubes), whereas in the dilution tubes of 1:1000 or greater, even though the diluent is designed to prevent osmotic shock, some microorganisms may die. Others, such as *Pseudomonas* species, may in fact be able to multiply even in diluent. To overcome these problems, therefore, the dilution series procedure should be completed as soon as possible. When large numbers of samples are being examined, it is advisable that these should be divided into batches of no more than 10, and the total examination of each batch of 10 completed before the next batch is begun.

Certain specialist examinations, for example osmophilic yeasts, may require the use of a specific diluent which is formulated to prevent death of microorganisms occurring once dilution of the food takes place.

In the same way, strict anaerobes require 'de-gassed', i.e. anaerobic, diluent as well as special handling during examination. For some purposes microbiologists prefer to reduce 'thermal shock' during dilution by using diluent at ambient temperature rather than straight from chilled storage.

However, for most applications a simple peptone/saline diluent is used for a wide variety of foods and liquids, which may not be the best possible diluent for every foodstuff, but which, when used correctly, gives a satisfactory recovery rate of microorganisms from food samples. Always check to see that peptone diluent is non-turbid before use. Turbidity indicates either gross multiplication of microorganisms or poor preparation of diluent. Sterility of a diluent batch should always be checked (see under 'The colony count').

Counting of microorganisms

In the same way that a sample taken for microbiological analysis must be representative of the 'lot' or batch of food, and that it must not become contaminated during the sampling procedure, so must the counting of microorganisms be made as accurate as possible. The result of examination can then be relied upon when judging the quality of the food, or plant surface, examined, and so determining whether the food is acceptable for use or whether the plant/equipment is suitable for production.

Counting is usually made by the following methods:

- The colony count
- Most probable number technique
- Direct microscopic count (DMC), and its variants
- 'Rapid' methods

The colony count

The most usual method used to estimate the count of microorganisms in foods or liquids is the colony count, which is both simple and versatile. This has the advantage that by using a series of dilutions it is possible, by a single examination, to obtain a result even though the anticipated microbiological status of the sample is not known. It is also possible to obtain counts for specific microorganisms. A single dilution series may be plated onto several selective agars, for example, Violet Red Bile Agar (VRBA) for coliforms; Baird-Parker Agar for *Staph. aureus*; and Acidified Malt Extract Agar, Oxytetracycline-Glucose-Yeast Extract Agar or Chloramphenicol Rose Bengal Agar for yeasts and/or molds. By carrying out this procedure, often accompanied by plating on a non-selective agar and incubating aerobically and/or anaerobically at different temperatures to obtain the 'total' bacterial count, much information is available. This enables a judgment to be made of the microbiological status of the food, indicating both the efficiency of processing and the hygiene of food handling process.

However, for this information to be useful it must be accurate. To achieve accuracy, the colonies which arise from microbial growth must be carefully counted and the correct dilution factor applied so that no wrong conclusions are made regarding the microbiological status of the foodstuff under test. Also, the media and apparatus used during preparation of the sample must be sterile, and the technique used by the technician must be completely aseptic, otherwise colonies will develop from contaminating microorganisms, not just from those present in the foodstuff. To confirm sterility and asepsis, control plates should be incubated, that is, plates containing the agar only, or agar with diluent.

Accurate counting is easier when the following conditions apply

There are not too many colonies on the plate. It is difficult to count a plate when there are more than 250 to 300 bacterial colonies present, or where there are more than 150 to 200 yeast colonies. Although it is possible to count colonies when there are more present, the colony size is smaller because of the overcrowding, and colonies over-lie one another so that undercounting of colonies occurs. On occasions, this undercounting can result in as much as tenfold reduction of the correct count.

On selective agars, which often contain diagnostic agents in their formulation, too many colonies results in loss of the correct diagnostic change. A good example of this is VRBA with coliforms. When the number of colonies is less than 100 per plate, the coliform colonies are purple-red in color and surrounded by a zone of precipitated bile acids. When the colonies are overcrowded, no purple-red color develops in the colonies, neither is there any precipitation, and so the sample would be recorded as coliform free, when in fact it had an unacceptably high count.

There is no background of food material on the plate. This is not possible to achieve when low dilutions of food samples are plated, and when the particles are small. For example, certain cream samples have fine protein particles which resemble colonies, and yogurt samples contain acid precipitated protein, and these particles may be counted as colonies. One recommendation to overcome this problem is to incorporate 2,3,5 triphenyltetrazolium chloride (TTC) into the agar medium immediately before the plates are poured (see Speck[8]). Most bacterial colonies are colored red, and so can be easily counted. However, TTC can be inhibitory to colony development of certain Gram-positive bacteria, and not all the colonies which develop give a red color, so would be missed rather than counted. This lack of red color may also occur when the TTC is added to the plates following incubation.

Another method is to pour an extra plate, and to refrigerate this plate during the time which the other plates are incubating. When the plates are counted, a careful comparison of the incubated plates with the refrigerated plate for each sample should enable a colony count to be made. However, it must be recognized that counting of these foods which have small particles is not easy, and it would be wise for more than one experienced technician to check the counts where these problems occur.

The microorganisms do not form small colonies (pinpoints). Some microorganisms never form large colonies, and the technician *must* use magnification on plates to enable a count to be made. They must also thoroughly scrutinize the plate, as a superficial examination makes it likely that the colonies will not be seen, and so not counted.

No colonies spread across the agar surface. When 'spreading' occurs, colonies of other microorganisms may be prevented from developing, or they may not be seen. This happens occasionally in the thermophilic spore count, when spreaders producing

ammonia give the plate a 'clear' appearance. In either case, undercounting of the sample results. Although there are agar formulations which inhibit the formation of spreaders, they can also have the effect of reducing the colony count of the sample, so should not be adopted without investigation on any foodstuff which is causing problems.

Providing the spreader does not cover more than half of the area of the plate, a count may be made of the clear portion of the plate, equal to half the area of the plate, and the count obtained as follows:

Colony count × 2 × dilution factor = count per ml or g

Most probable number (MPN) count

This technique is not as precise as the colony count but does enable estimates to be made of concentrations of microorganisms below one per ml or g of the food material or of the 1 in 10 dilution of the food materials.

To improve the accuracy of the technique, replicate tubes are used. Usually a total of nine tubes of culture media are inoculated with the material under test; three tubes are inoculated with 10 ml of either the liquid food or the 1:10 dilution of a solid food; three tubes with 1 ml of the liquid or the 1:10 dilution of a solid food, and three tubes with the equivalent of 0.1 ml of the liquid or the 1:10 dilution of a solid food. The inoculated tubes are incubated, and after incubation the tubes showing growth, or characteristic growth if a selective culture medium has been used, are recorded as positive. Other numbers of tubes, volumes of inocula and ranges of dilutions may be used for particular ranges or precision of counts. MPN tables, e.g. those of deMan[19], are then used to determine the most probable number of bacteria per ml or g of food, depending on how many tubes at each inoculation level show a growth response.

Compared with the colony count more effort is needed at the bench to carry out an MPN on a food sample, but it does enable a reasonable estimate of the numbers of bacteria present. The most usual application of this technique is to determine low numbers of coliform bacteria, and to confirm whether any *Escherischia coli* are present among these coliform bacteria, but it is also used for detecting low numbers of *Staph. aureus*, *Salmonella* spp. and *Vibrio parahaemolyticus*. It can be applied to other groups of microorganisms provided there is a selective broth available for the organism under test, and provided that when growth occurs, a characteristic reaction is given in the selective broth. For example, coliforms in McConkey broth give an acid + gas reaction and in Lauryl Sulfate Tryptose broth they give a gas reaction. *Staph. aureus* in Tellurite Mannitol Glycine broth (Giolitti and

Cantoni broth) gives a characteristic blackening, although confirmation of coagulase production must be carried out on bacteria isolated from positive tubes. It is also very useful in counting gas-producing anaerobes that split the agar in plates or tubes.

Direct microscopic count (DMC)

One of the oldest and most widely known of these counting methods is the Breed smear, in which 0.01 ml of liquid, or a food suspension, is spread over a one cm square of a glass slide. After drying and fixing the suspension is stained and a number of fields are chosen at random. The stained bacteria in each field are counted, the average number per field is calculated and the number per g then calculated from the formula below.

The diameter of the field is measured by use of the stage micrometer, and the radius found by dividing the diameter by 2. The microscope factor (MF) is then calculated as follows:

$$MF = \frac{10\,000}{\pi r^2} = \frac{10\,000}{3.1416 \times r^2}$$

where r = the radius of the field (mm)
10 000 = the number applied when 0.01 ml is spread over 1 centimeter square of the slide, i.e. 100 mm^2 (the area of the smear) divided by 0.01 (the volume of material examined)

If the area of the slide covered by the material under examination is changed, or the volume of material is changed, then the MF must be recalculated. Failure to do this new calculation will give the wrong count of microorganisms.

Direct microscopic examination can be useful when high counts are anticipated. Typically, the sensitivity of the Breed Smear method is 5×10^5/ml on undiluted liquid, i.e. counts of less than 5×10^5 give an average count per field of less than 1. If a 1/10 dilution of the food material or liquid must be made before examinating, then the sensitivity will become 5×10^6/ml, i.e. counts of less than this will not be able to be assessed with any accuracy. The count can be made with a minimum of apparatus, and will be available within approximately twenty minutes. However, the count made will not differentiate between living and dead cells, so may be misleading in many practical situations.

DMC variants

To overcome this lack of differentiation between viable and non-viable cells, certain dyes may be applied to the test material which react differently depending on whether the cells are living or dead. The food application so devised has also been linked to increasing the sensitivity of the test and was developed using raw milk. Two ml aliquots of milk are treated so that they can be filtered, and the retained bacteria then stained with acridine orange. Using an ultra-violet light source, the membrane filter is examined under the microscope; the living bacteria fluoresce orange, whereas the dead bacteria are green. As can be seen from this brief description of the method, it is more complex than the Breed Smear examination, and a specialist UV pack needs to be attached to the microscope in order to count the bacteria. A full description of the method is given by Pettipher (who first developed it), in *The Direct Epifluorescent Technique* published in 1983, and its wider application, although with varying success, has also been investigated[20, 21].

Note on accuracy of counts

Accuracy has always been a matter of concern especially since the classic work of Wilson *et al.* in 1935[22]. There are a number of rules of thumb which seem to apply in particular instances, e.g. 'The true count is likely to be the plate count ±2 × the square root of the mean' or 'a significant difference is likely to be at least 0.5 of a log cycle'. The microbiologist needs to consider what accuracy is needed in a particular situation to make good decisions. The method(s) chosen and the amount of work necessary follow from this consideration. It is much better to do this than to use a 'standard method' without any idea of its accuracy.

Rapid methods

These are used for assessing microorganism levels in foods or on equipment surfaces. The disadvantage of most of the established (or conventional) methods for measuring levels of microorganisms is that an incubation must be given to the material under test, and this incubation period is usually at least 48 hours, and may extend to 96 hours, or even 1 week. The exception to this is the coliform count: using Violet Red Bile Agar a result is available in 20 to 24 hours. Thus, conventional methods may be used for quality control, since they give information on the status of product or equipment, but they are not particularly successful as part of a Quality Assurance program, as the necessary steps to improve processing or handling of material in the factory cannot be taken in time to minimize any period of 'out of specification' production. Direct Microscopic examination, the Breed Smear, is a rapid test, but, until recently, was unable to give the information as to whether the microorganisms were living or dead, and also would not detect accurately any counts below 10^5 per ml or 10^6 per g. Specifications of microbiological levels in the majority of foods are often for counts of less than

100 or 1000 per ml; and for pathogens, or equipment contaminants such as coliform, are generally 'absent from 1 ml or g,' or, for example *Salmonella* and *Listeria*, 'absent from 25 ml or g of the food product.'

Over the past decade there has been a search for methods which would shorten the time before a result was available, which would be able to differentiate between types of microorganisms and would be as reliable as the accepted counting methods in determining product quality.

Four main methods have emerged:

1. Instruments which measure the changes in electrical charges which occur in culture media as microorganisms multiply.
2. Instruments and reagents which measure the Adenosine Triphosphate (ATP) which all microorganisms possess, and which, therefore, is a measure of the number of microorganisms in a sample.
3. Direct Epifluorescent Technique (DEFT), which improves the sensitivity of the Breed Smear and by choice of stain, differentiates between living and dead cells.
4. Flow cytometry, which can enumerate particles the size of bacteria or of yeast in a fixed volume of liquid.

All of these methods require sample preparation which, on the whole, is more than that given when either plate counting or broth inoculation is being carried out. This means that the rapid methods are generally more labor intensive per sample than the conventional methods which are in use at present, and the timing of the stages of sample preparation is more critical than that in use for the stages of plate counting. An outline is given for these four methods below and some examples of their application are also included.

1. *Changes in electric charge*; or impedance/conductance/capacitance changes. As microorganisms multiply in culture media, the electric charge is changed, and the time which elapses before the change in the electrical charge becomes exponential is inversely related to the number of microorganisms in the media. That is, the more microorganisms there are in a sample under test, the shorter is the time before the detectable change in electrical charge occurs. This change is detected by regular monitoring of the inoculated incubation chamber throughout the incubation period, which means that electric probes are present in the incubating medium throughout the period of incubation. The monitoring is carried out in an instrument which is specifically designed for the purpose, and which is microprocessor controlled with suitable software, so that the time of change in the medium is recorded onto a screen display.

Before these instruments are used for routine microbiological monitoring, calibration of samples must be carried out and the time of change of electrical charge, or detection time, must be related to the microbial loading of each sample. At least one hundred samples should be calibrated to relate the detection time to the colony count, and the regression coefficient found for these samples. The range of counts should cover 3 log cycles for the calibration to be valid.

An application which does not need calibration is where sterility is being assessed, as in UHT products. The packs of product are incubated at 30°C for sufficient time to allow growth of contaminating organisms, usually for a minimum of 7 days, then sampled into broth and monitored in an instrument. Detection time occurs within 4 to 6 hours of inoculation and sterile samples show no detection time during the 18 hours of incubation.

Salmonella detection has also been carried out using those instruments, but a pre-enrichment incubation is necessary, and incubation in the instrument must be done in a selective broth. As with conventional methodology, presumptive positives, that is, those samples showing a detection time, must be confirmed, but the electrical method shows a saving in time of 24 hours compared with the traditional methodology.

2. *Adenosine Triphosphate (ATP) measurement* All living organisms contain ATP, and in the presence of a luciferin–luciferase enzyme system, a reaction takes place which produces light. The light can be measured, and the amount of light released is proportional to the number of microorganisms. However, certain liquids such as milk contain somatic cells of animal origin, as do meat and vegetables, which also contain ATP, and so before bacterial cells can be measured, the somatic cells must be removed from the sample. Thus, a sample treatment is necessary before the enzyme reaction can be allowed to occur. Although the light reaction is very rapid, the sample pre-treatment together with the care needed in handling all the apparatus used during sample examination to prevent ATP from the operator contaminating the sample means that, as yet, this is not an easy or reliable test to introduce into a routine laboratory.

3. *Direct Epifluorescent Technique – DEFT* This technique is an adaptation of the Direct Microscopic Count as developed by Breed. The changes have:

(a) Increased the sensitivity, so that counts of 5×10^3/ml can be detected instead of 5×10^5/ml.
(b) Allowed differentiation between living and dead cells by use of acridine orange staining of bacteria.

The increase in sensitivity has been achieved by treatment of the sample so that it can be filtered; the bacteria are retained on a membrane filter and are stained prior to microscopic examination. The technique was developed for examination of milk, so the pretreatment uses trypsin to digest the protein and a surfactant to break up the fat globules; other treatments can be applied to different foods or food suspensions to enable them to be filtered.

The microscope must have a Ploem pak fitted to it to enable UV light to detect the fluorescing yellow/orange living cells. When applied to raw milk (e.g. farm milk) there is a good correlation between the count of yellow/orange cells and the colony count at 30°C. However, when products are examined which contain heat stressed cells, then some of the non-living cells also fluoresce orange/yellow, and so a count of living bacteria is not possible, although a total count of both living and dead cells can be obtained. However, this information is not of particular advantage in many food manufacturing operations.

The timing of sample treatments is critical, and although some automation is available, it is a labor-intensive technique, even when image analyzers are used to count the microscope fields, since the fields to be counted are selected by the operator, not automatically.

When milk is being examined a result is available within 20 minutes, so it is a technique giving rapid results. But it is non-selective, although cocci and rods can be differentiated, and it cannot be applied to samples which have low counts.

4. *Flow cytometry* This technique is based on particle counting, and like DEFT uses a UV microscope to count stained organisms. These are suspended in a liquid which flows through a the tube at the rate of 1 ml/min. In order to insure that the particles counted are microorganisms, and not other particles which will inevitably be present in a food or liquid, the sample must be pretreated, and must then be centrifuged to concentrate the microorganisms. As with the other techniques this treatment takes time and the stages of treatment require care in manipulation. The method is more sensitive than DEFT – it can measure counts of approximately 2×10^2/ml of untreated sample – but for counts lower than this the sample must be incubated before examination begins.

A practical application is detecting low numbers of yeast in yogurt. A contamination level of 12 yeasts in a 125 g container of yogurt can be detected provided that 16 hours (that is, overnight) incubation at 37°C is given to the yogurt before testing it by the flow cytometry method.

Use of 'rapid methods' – sterility testing

The four methods described here all require instruments before they can be applied to food samples. These instruments are expensive, and also require microprocessor control either throughout the whole of the examination (the electrical method) or for part of the examination, the counting stage, in the other three methods. The only method which is not labor intensive is the electrical one, since this requires the least pretreatment of the sample before incubation in the instrument. The electrical method is also capable of being selective, by using the appropriate incubation medium to grow a particular microorganism, although the staining applied during the flow cytometer method is selective for yeasts. Because of these factors, very careful thought should be given before introducing any of these methods into a routine laboratory, as their advantages may not be cost-effective.

One of the very successful applications is that of electrical instruments being used to measure sterility in UHT products after pre-incubation at 30°C. There are usually a large number of samples to be examined from each production batch, they all use the same culture medium, and the result is available 18 hours after the start of electrical monitoring, compared with 72 hours if the samples are plated. If the UHT product contains sugar, then the pH of the product can be measured after the pre-incubation period, giving a result in a shorter time than incubating the samples in an electrical instrument. However, when no added sugar is present, as in milk, then the growth of contaminating microorganisms does not always result in the pH being reduced, and so the presence of contamination would not be detected by measuring the pH, although the sample would give a measurable conductance change during incubation in an electrical instrument.

Rapid confirmation/diagnostic methods

Even though electrical instruments can be used for the purpose of detecting certain microorganisms, e.g. *Salmonella*, coliforms or Gram-negative bacteria, the results – like those given when using selective agars in traditional test methods – are only presumptive and further tests must be carried out to confirm the particular microorganism. This confirmation by traditional methods will take at least another 24 to 48 hours, which again can present an unacceptable time delay when microbiological monitoring is being applied as part of the HACCP system to insure correct processing is being achieved within a food plant. To reduce the time required before confirmation is available, rapid diagnostic methods have been developed using Enzyme-linked Immunosorbent Assay (ELISA) kits, DNA probes and latex-agglutination tests. It should be noted that ELISA kits and latex agglutination tests are also used for detection of bacterial toxins in foods, or to determine whether contaminating pathogens such as *Staph. aureus* are, in fact, toxin-producing strains.

Most emphasis on diagnostic rapid methods has focused on *Salmonella* confirmation, and ELISA, DNA probes, latex-agglutination kits and the Vitek Automicrobic System are available for use with this pathogen. All methods are applied following pre-enrichment and selective enrichment of the samples under test, and a further elective enrichment in M broth is given for ELISA and latex-agglutination tests before the diagnostic test is applied. The ELISA method requires 1.5 to 3 hours before the result is obtained, the DNA probe 3 hours, and the latex-agglutination tests give an answer in 3 minutes. Thus the latex-agglutination test gives the quickest answer, and is the least labor intensive of the three methods. However, this method, as with all rapid methods, needs fully trained technicians to apply them, otherwise the results are unreliable and therefore of no use to the production plant.

As might be expected, test kits have been developed for *Listeria monocytogenes*, and an ELISA test, a DNA probe and Vitek Automicrobic System are available. The ELISA and DNA methods identify the genus *Listeria*, not the specific *L. monocytogenes* whereas the Vitek system identifies *L. monocytogenes*. In addition, *Campylobacter jejuni/coli* and enterotoxigenic *E. coli* and *Staph. aureus* have had kits developed for rapid identification.

Evaluation of these methods has been carried out by Leatherhead Research Association (UK) and industrial laboratories, and shows that agreement with traditional confirmatory methods is good when working with laboratory-carried cultures inoculated at the same time as food samples into enrichment broths. However, results are not so consistent when examining naturally contaminated foods, since on occasions the pathogen is not confirmed, although traditional methods have shown the presence of the pathogen under test. The use of these kits must, at present, be treated with caution, and prudently would not be used to replace traditional methods until extensive trials have been carried out on factory-produced samples.

As far as cost is concerned, the minimum cost per identification in 1990 is around $2.50, and the Vitek method uses an instrument which costs $35 000. However, this cost must be weighed against an early result enabling any loss of control during manufacture to be identified, and so corrected as soon as possible.

Growth of microorganisms and culture media

It is axiomatic that the microbiologist only becomes aware of the organism by its growth. Thus growth is a definite indication of its presence, i.e. a positive result. *Remember*, a negative result does not necessarily mean that the organism is absent but simply that it has not been found. The microbiolog-

ist also uses the colonial appearance as an important (often the primary) means of identification of a type or kind of organism. The final growth and appearance of the organism are the interaction of culture media and environmental factors. There are criteria for culture media which Mossel[23] has set forth that clarify thinking about media:

- Productivity – the ability of the medium to support growth
- Selectivity – the ability of the medium to discourage the growth of interfering types
- Electivity – the easy recognition of colonies of different organisms
- Performance with natural food samples – the extent of recovery of sought organisms and the degree of inhibition of interfering organisms
- Taxobias – the extent to which a medium selects or suppresses a particular species or biotype within a selectively enumerated taxo group

Productivity and selectivity of media are confirmed by using reference cultures on the appropriate type or types of media. However, in certain countries it is a regulatory requirement that no stock cultures of pathogens may be held on a factory site. This presents difficulties in implementing Quality Assurance programs, since negative results may be due to poor culture media rather than pathogen-free production samples.

Microbiologists will usually prefer to use prepared dehydrated culture media to make up agars or broths. This is not only convenient and usually cost-effective but ingredients are less likely to have been omitted, which is not unknown especially when a large number of ingredients are required. However, even dehydrated media from large, ethical, international companies can vary. This is discussed by both Mossel[23] and Baird[24] although progress in implementing their ideas is likely to be rather slow.

Other factors which affect performance include the following which are *not* listed in any order of importance or frequency of occurrence:

- Carry-over of inhibitors from the food under examination. Spices are examples of such food – mainly cinnamon, nutmeg, ginger and cloves, but salty dried foods are also inhibitory, as are onions. Inhibition may be overcome by using higher dilutions of the food when plating, or using spread plates when only 0.1 ml of a 10^{-1} here dilution is spread over the agar. A neutralizing agent can be added to the agar, but checks must be made to show that the neutralizing agent itself is not inhibitory. Inhibition can also act in enrichment procedures, as, for example, in *Salmonella* testing, and may need to be overcome by enriching 25 g in 1.25 l of broth (a 1 in 50 dilution), instead of the more usual 25 g in 250 ml (a 1 in 10 dilution).

Carry-over also occurs when plant swabs are taken. Residual disinfectant, or sanitizer, is taken up by the swab and transferred to the diluent. Its presence in the diluent could affect microbial growth. To overcome this, a neutralizing agent is included in the diluent. For example, to neutralize sodium hypochlorite, sodium thiosulphate is used, and to neutralize quaternary ammonium compounds, a mixture of lecithin and Tween 80 may be added to the diluent.

- Media may be insufficiently tempered, i.e. poured too hot (above 45°C), thus 'pasteurizing' the organisms on the plate. Psychrotrophs may be so sensitive that spread plates must be used.
- Spread plates may be over-dried. This results in poor, slow or no growth. There is no universal time/temperature combination for satisfactory drying and experience is required to develop 'laboratory rules'. Factors influencing drying interrelate and include: the number of plates in a stack or how they are 'opened' during the drying period, whether the incubator has a fan, the temperature at which the agar is poured, and characteristics of the agar gel. *Note*: if incubator space is limited closed, inverted plates may be left overnight on the bench to dry. This works if the ambient temperature is below about 25°C.
- The expected growth pattern may be altered if diluent or medium is used directly from refrigerated storage because of 'shortage of time'.
- Any medium has a shelf-life. Over-age medium, whether dehydrated or made up, may not have its designed properties.
- It has been known for an undetected misprint to occur in the laboratory media maker's instructions.
- An ingredient may be omitted in error, or the wrong ingredient may be used. This is more likely where mono-, di-and tri-phosphates are in the recipe or where materials with different waters of crystallization are on the laboratory shelf. To reduce the chance of these errors, it is useful to give in the media instructions additional to the ingredient name, both the chemical formula and the molecular weight.
- A medium may contain an unexpectedly inhibitory component, e.g. Brom-cresol purple (BCP), which is used as a 'screen' to help identify thermophilic flat-sours (*B. stearothermophilus*). Some batches of BCP are highly inhibitory to this organism, so even batches of 'acid free' BCP should be tested before use.
- Selective media by their nature work by discouraging (inhibiting) growth of some organisms to allow the wanted one(s) to grow on. The more highly selective the medium, the more this is so and for some strains of a wanted organism it may be 'over-selective' and they may be suppressed.

This is why two selective broths are used for *Salmonella* enrichment, not just one.

- Too much 'adjustment' of pH because of 'over-correction' may lead to increased salt (NaCl) concentration affecting medium performance.
- Damaged cells may need resuscitation before growing on a medium, but not all strains may respond equally to this process.
- Spread plates work by balancing the restriction of growth (avoiding 'spreaders') against the change in solute concentrations in the drier surface layer. Factors affecting this balance may affect 'typical' growth.
- Incubation temperature is a compromise between the ideal and the available. This may affect growth.
- Incubation time is usually the minimum needed to get results. It may be too short for some strains and/or when cells are stressed.
- Autoclaving by pressure. Since pressure gauges are rarely accurate enough this may result in very different heat treatments to the one intended. This can affect medium performance. Autoclaves should be checked out with thermocouples inside media bottles or tubes, packed both loosely and tightly. This is in addition to those thermocouples in the steam 'atmosphere'. Remember, as with cannery retorts, venting is necessary to achieve a good (even) temperature distribution.
- Leaving media 'tempering' in a water bath (usually about 45° to 50°C) for some hours or overnight is likely to cause hydrolysis and therefore poor setting of acid agars and unknown changes in other media. Good practice sets a maximum holding limit of 4 hours.

Note on the isolation of microorganisms

When a sample is examined for microorganisms, it is most usually examined not just for the total numbers present, but for several types of microorganism. Some of these microorganisms may be present in low numbers, and in several cases the numbers sought are below the sensitivity of the direct plate method – for example, *Salmonella* or *L. monocytogenes* absent in 25 g samples, or *E. coli* absent in 10 g samples; or the microorganisms may be in a stressed condition, so a resuscitation stage is required before the microorganisms can be isolated or enumerated.

Thus, 'isolation' from a sample can take three forms:

1. Direct plating of the sample and dilutions onto several agars; the agars being formulated so that microorganisms other than the required type are either suppressed from growing, or should they grow, their colonies do not give a typical reaction on the agar.

2. Enrichment of a weighed quantity of the sample. The enrichment procedure is designed to allow very low numbers of the particular microorganisms to grow to a level where they may be detected, generally by streaking onto a diagnostic agar.

   For certain pathogens, in which the cells may be stressed, this enrichment is a two-stage procedure. The first enrichment is in non-selective broth, followed by a selective enrichment in a broth which is formulated to suppress the growth of other microorganisms. It must be realized that in the non-selective enrichment the microorganism being looked for is in competition with all the other microorganisms in the sample, or at least those which are capable of growth at the incubation temperature given during enrichment. The dynamics of growth of microorganisms, both those belonging to the same family and those belonging to different genera, are not well known. In 'real food' situations with varying mixtures of organisms, growth dynamics present very real problems to research workers who are investigating this effect with a view to improving methodology as well as to factory microbiologist anxious to know the status of an ingredient or food.

   Another factor to be considered is the effect of the sample itself on the broth. If the sample is acid, then that will affect the pH of the broth. If the resulting broth pH is too low, then the microorganism which is being isolated from the sample may not be able to multiply, so a false negative result would be recorded. This problem can be overcome by adjusting the pH of the broth, so that when the acid sample is added, the pH of the inoculated broth is suitable for the organism(s) sought. An example of this is when acid-whey powder is being tested for total count, coliform, etc. the broth pH is adjusted to 8.4, to give a pH of 6.0 for the mixture prior to incubation.

3. Resuscitation of the sample in a broth, to allow stressed cells to recover from whatever stress they have been subjected to, for example, dehydration in dried samples, or being exposed to a low pH value in acid samples. If resuscitation is not given, these microorganisms often do not multiply to form colonies on agar plates nor do they grow in selective broths. The important consideration in resuscitation procedures is that the incubation of the sample in the broth should not allow the microorganisms to multiply if the resuscitation is to be followed by enumeration of the microorganisms. If multiplication occurs during resuscitation, then a correct enumeration of microorganisms in the sample is not possible. The resuscitation time and temperature must be sufficient for the microorganisms to 'recover' themselves, and so multiply once they are inoculated into agar plates or selective broth, but must not be prolonged in time, otherwise multiplication will start in the resuscitation broth.

Note on the identification/confirmation of microorganisms

This procedure is most often carried out following isolation of microorganisms from a sample. Although microscopic examination of a sample can be usefully applied to known high-count products, e.g. spoiled or temperature abused samples, the information which this gives is general, and refers to types of microorganisms, not to the specific family or species.

However, if the spoiled product is fruit juice, this information may be all that is required, since a microscopic examination will be sufficient to say whether spoilage resulted from yeast growth or bacterial growth, or, possibly, of both types of microorganisms. For the purposes of traceback through the plant, or to raw unprocessed concentrate, to identify where contamination occurred, it may not be necessary for any further identification to be carried out.

In most instances, identification is first done from the appearance of colonies on a selective agar. *Staph. aureus* on Baird-Parker agar and coliforms on Violet Red Bile Agar are examples of this. However, microorganisms other than *Staph. aureus* can grow on Baird-Parker agar, and since all *Staph. aureus* strains do not give the typical, classical reaction, microscopic examination of colonies should be carried out. Some *Bacillus* species give the typical black, shiny colonial appearance on Baird-Parker agar, and in this case a microscopic examination prevents a wrong decision being made regarding the status of the sample with reference to the *Staph. aureus* count. However, should other atypical colonies be seen to be cocci, and to form grape-like clusters, then they are likely to be *Staph. aureus* and the count recorded as such. Further identification may be done to find out if the colonies are coagulase producers, which indicates that they may be capable of producing toxin, and if the numbers are 10^6 per gram or higher, then toxin is presumed likely to be present in the food.

Rapid diagnostic kits are available for confirming colonies of *Staph. aureus*, and there are also kits to detect toxin in foods, and to say whether the toxin is Type A, B, C1, C2, C3, D or E. In foodstuffs, certain foods present difficulties as far as the diagnostic kits are concerned, and there is a loss of sensitivity, for example, in cheese, where the lower quantities of toxin are not detected, although they are detected in meat.

The question that does need to be asked, though, is 'How necessary is identification, and does it always need to be carried through as far as the species?' As far as pathogens are concerned, the species needs to be known for *Clostridium perfringens*, *Staph. aureus*, *B. cereus* and *L. monocytogenes*. This is because other species of the same genus are not pathogenic, yet can give similar reactions as far as colonial appearance is concerned and for several biochemical tests which may be applied. However, for *Salmonella* species, the specific name is not always necessary, as all species are pathogenic, so it is enough to know that there is *Salmonella* present in the sample being examined.

To be sure that the result really is *Salmonella*, it is necessary to confirm the presumptive result which is available either from colonies on a selective agar which gives appearance typical of *Salmonella* species or from the selective enrichment broths which are used in the electrical impedance methods of the Bactometer or of Malthus Instruments (see 'Rapid Methods'). This is because action arising from a confirmed *Salmonella* result is very different from that arising from a presumptive result. A confirmed test results in product recall or product destruction, whereas a presumptive result does not, although, of course, the presence of *Enterobacteriaceae* causing the presumptive result leads to appropriate action, e.g. increased cleaning procedures, within a factory.

Confirmation is also required for the other pathogens mentioned, but any tests made to identify the species will, in fact, act as confirmation tests.

A species may need to be known for epidemiological reasons, e.g. when a specific serotype of *Salmonella* has been isolated from a food poisoning outbreak, or for factory traceback when *Salmonella* has been found, even though the product has not left the factory. Species identification can be of help when the factory plant and environment are undergoing investigation so that the source of the contamination found in the product may be identified. Of course, once a positive *Salmonella* is found, factory clean-up is undertaken, but should a specific item of equipment or part of the environment be located as the source of contamination, then further steps ('root cause corrective (remedial) action') need to be taken to eradicate the contamination source. For example, there may be a crack in a pipeline or a valve diaphragm; or a leak in the roof may be responsible for contamination entering the processing area. Such faults require thorough corrective action, to prevent contamination recurring.

Also, a question to be asked is 'What sort of identification needs to be made? Is it the name of the microorganism or do its particular properties need to be identified?' Within a process, for example, which relies on heat as the major control of microorganism contamination, then the heat resistance of the contaminating microorganism is of significance in determining whether the heat process

was sufficient to reduce the contamination to an acceptable level. Thus, the identification of the heat resistance of the microorganism *in the food in which it was processed* is the kind of information which needs to be known to the processor rather than the particular species, or strain, of the microorganism.

Another example is the rate at which the microorganism grows under defined conditions. If the product depends on pH for its shelf life, then if contaminants are found in numbers greater than expected, it would be wise to identify how rapidly that contaminant can grow at the usual pH value of the food. If in fact the rate of growth at that pH value, and at the temperature at which the food is held during its shelf life, is fast enough to reach high numbers at an early stage of the products shelf life, then a different solution to the problem is required from that when no multiplication is shown to occur. No multiplication means that the high numbers found are due to a high level of contamination during manufacture of the product (possibly arising from an unsatisfactory raw material) whereas multiplication at a rapid rate during storage means that the formulation of the product or its storage conditions require investigation and, quite possibly, changing.

Assessment of results

In making judgments and decisions about environment, equipment, food and processes the wise microbiologist uses all pertinent information. It may be derived from chemical, physical, mathematical, sensory assessment or microbiological techniques. In any situation there is always the question, 'What weight should be given to a particular piece of evidence?' The evidence may be either an individual result or part of an ongoing sequence or trend. To the microbiologist, trends are more important than individual results unless the individual result shows unexpected numbers or kinds of organisms or unusual conditions.

Financial considerations may well put a substantial premium on the speed with which decisions can be made. The microbiologist will, therefore, if it is possible, use evidence which does not require incubation or cultural work. The so-called 'Rapid Methods' may be applicable in these situations although not only capital but application (development) time and running costs need to be considered.

Trends and records

Trends and records are important in accessing the value of an individual result. The significance of trend analysis is that trends can provide either an assurance that a system known to deliver acceptable results is 'in control' or it alerts management that it is not behaving as expected and that action needs to be taken to prevent 'out of specification' running. Thus the management by exception (MBE) principle may be applied. It is useful to remember that in some instances, e.g. with some ingredients, the use of trends means that speed of getting results may not be so critical. The use of control limits in ingredient control charts is discussed by Shapton[12] and illustrates this point.

The form in which records are kept is important because significance of an individual item can be hard to determine in a dense mass of data. What is wanted is a form which allows quick appreciation and hence easy use in decision making. It is useful to remember that the memory of an individual is rarely as accurate and complete as a good record system.

There may be both legal and contractual requirements on the form in which records are kept and also on the time for which they must be retained. It is worth considerable trouble to determine the optimum way of keeping records in the absence of such constraints. The fundamental question to be resolved is 'What must the record system deliver both in content and speed?' The answers to this will shape the record system. Flowers *et al* in their chapter on Laboratory Quality Assurance in Speck[8] indicate the details which may be required:

1. As discussed above, records should be kept documenting the care and disposition of samples during their time in the laboratory. These should show the storage conditions, the personnel with custody of samples, and the final disposition of samples when no longer required.
2. Records are required of all sample analyses including descriptions of analytical methods used, all raw data and observations generated by the analyses, calculations made, and conclusions drawn. The analyst(s) responsible for the testing should be identified. The samples should be described, storage conditions during and after analysis described, and whether a reserve portion was included.

   These records may be in the form of work sheets which become a part of the entire record for each sample, or a notebook which can be referred to in the sample records and correspondence.
3. Analytical records should be reviewed for completeness and accuracy before the results are reported. This review should be at least a two-step operation, with the first review done by another analyst in the laboratory, and a second by the supervisor.

In a food plant it is important to record action taken as a consequence of results. This builds a body of knowledge and experience which greatly helps good decision making. It is helpful if the results of action taken to bring back into control any

procedure or function are readily associated with the appropriate results. By doing this, laboratory results form part of the Total Quality Management (TQM) system and are in accord with the International Standards Organization 1S0 9000 (BS 5750) quality system (see also Chapter 2).

When auditing records it may be important to review them for completeness and accuracy either with another analyst or with supervision. Dependability is an essential feature of laboratory work and therefore examination of records and the record-keeping sytem should form part of any laboratory audit.

Management needs to understand that the term 'accreditation' should not be used primarily as a status symbol but as a demonstration to users and potential users that results are to be trusted.

The question of accuracy and consistency (reproducibility) of results is a difficult and complex one. The key question, which is seldom easy to answer, is 'What can the method deliver when used by careful and competent workers?' As an example, the results of a chemical analysis on skimmed milk powder, which usually are considered more accurate in the sense of being more consistent than microbiological methods, showed interesting differences (see Woodhall[25]). Five laboratories were chosen, two used in-house methods for a nitrate analysis, three used a British Standard method. The results, given as ppm nitrate, are given in Table 9.1.

Table 9.1 Comparison of laboratory results for nitrate in milk powder

Laboratory	Method	Results	
		A	B
1	In-house	50	84
2	B.S.	22	33
3	B.S.	27	28
4	B.S.	18	38
5	In-house colorimetric	1700	1700

Note: A = Normal production; B = after nitric acid clean.

Although nitrate estimations are quite difficult, the extent of the variation illustrates the value of the concept of an error budget linked with the quality assurance of the method.

Accuracy of plate count

The accuracy of counting methods has received a great deal of study. One of the earlier and classic studies which extended over three years was that of Wilson *et al.* in 1935[22]. It is a report of an investigation into ways of assessing the keeping quality of milk but not its safety. They reviewed the plate count and rejected it because of error. The coliform count has a clear rationale in water where it dies off more slowly than enteric pathogens. This is not so in milk, where multiplication means that the count is not a sure indicator of keeping quality or cleanliness. This was therefore rejected. The report provided the technical basis for the modified methylene blue test. This is a simple biochemical test based on reducing properties. It is still of interest because it shows what can be achieved with simple equipment providing that careful attention is given to details of technique together with good experimental design and statistical analysis.

As an example of their attention to detail, since only re-usable glass pipettes were available, the method of cleaning was standardized and errors of delivering 1, 9 and 10 ml quantities of water and milk were determined. It may also truly be regarded as a good HACCP study long before the term was invented.

The report discusses the sources of error in the plate count. When considering these it should be remembered that the plate count remains one of the most widely applicable and useful of microbiological techniques. The principal sources of error are:

- Counting the plates. These include the general errors which affect everyone, e.g. a medium with debris similar in appearance to colonies being counted, poor lighting conditions or absence of a lens.
- A personal bias which is responsible for different counts on the same plate by different persons. There is a tendency for individuals to be consistently 'high', 'medium' or 'low' counters.
- A personal variability shown when the same plate is counted again by the same person. Wilson *et al.* [22] state 'a low error of personal variability is probably the best criterion of a good observer'. It should be noted that errors of personal bias and personal variability vary independently of one another. As an indication of counting errors that might be expected, they suggest that where the plate really contained (i.e. a true count of) 200 colonies, counts by different workers could be expected to vary between 110 and 290. In one experiment under good conditions the variation was about half of this.
- Sampling from the suspension. This is seen when replicate plates are made from the same dilution. Theoretically, the relative error of sampling should diminish as the square root of the number of colonies counted. Experimentally, they found a close correspondence between actual and estimated values for the coefficients of variation up to 200 or 300 colonies per plate.
- Overcrowding of plates reduces the number of colonies found, presumably, because colonies coalesce or because of antagonism between early and late developing colonies. This error showed at about 300 colonies per plate (with milk) and was estimated at about 5 percent with 500 colonies per plate.
- Lack of replication. There is a trade-off between accuracy and cost. It is stated that the magnitude of the standard error of the mean diminishes as the square root of the number of plates counted. Thus to double the accuracy, four times the number of plates must be counted.
- 'Spreaders' which may obscure colonies on the plate.

Wilson *et al* (22) concluded that for milk:

The following figures are suggested as suitable allowance for counts based on different numbers of plates.

Count made on 1 plate	Allow ±90 percent of count
Count made on 2 plates	Allow ±64 percent of count
Count made on 3 plates	Allow ±52 percent of count
Count made on 4 plates	Allow ±45 percent of count
Count made on 5 plates	Allow ±40 percent of count

This is qualified by two assumptions:

1. That a standard technique is used and
2. That the variation in plate count has a normal distribution (i.e. no break-up of clumps of organisms)

Following from this, in order to appreciate the error of the plate count in *your* laboratory with the samples which you have to test, it is necessary to make some systematic experiments. It is essential to get good statistical advice before starting work and to remember that the objective is to get usable information, i.e. sufficiently precise and accurate for *your* purposes.

An excellent up-to-date description of Colony Count Methods is given by Busta *et al.* in Chapter 4 in Speck [8]. This refers to the usage of reporting the count as 'Colony-forming units' (CFU) per g (or ml). They also make the important distinction between 'precision' and 'accuracy'.

Precision is defined as the ability to obtain similar results when repetitive counts are made by the same person or other analysts. Accuracy is the difference between the count obtained and a 'true' count. When considering the entire procedure and the results obtained, both are important. Additionally, this chapter gives very clear descriptions of procedures and rules for counting, which if followed routinely will give confidence both in the assessment of results and in the action(s) taken on the basis of these data.

Note on the limitations and uses of finished (end) product testing

Apart from legal or contractual requirements the real issue is whether or not finished product testing represents the best use of resources. There are, of course, different types of finished product testing; physical, chemical, sensory assessment as well as microbiological, and both product and package may be examined. All processors carry out finished product testing to a great or lesser extent. The statistics of sampling and the extent of variations in microbiological testing mean that the information from such tests is imprecise and, as sometimes undertaken, may easily mislead.

The limitations of finished product testing are clearly expressed by Wooden[26] in a discussion of the early work done to develop 'as close as 100% assurance as possible that the food products we were producing for space [travel] would not be contaminated with pathogens'. He then stated: 'In fact a large part of the production of any particular batch had to be used for testing leaving only a small portion available for space flights.'

The practice of taking a few samples per day, testing them and solely on the basis of these results concluding that 'everything is OK' may comfort those who wish to believe that nothing can ever go wrong in their plant but is scientifically foolish. The assurance of consistent product safety and quality comes from constant application of managerial and technical expertise integrated into a Total Quality Management (TQM) system of which HACCP and Audit are component parts.

Finished product testing cannot, however, be wholly disregarded. There may be legal or contractual requirements why it should be done. As an example, a merchandising customer may wish to have test results on an on-going basis as part of the evidence to show that the quality system is in place and working as expected. Limited finished product testing as part of a monitoring system provides direct verification of microbial status.

Additionally, with a perishable product, it may be important to have knowledge of the product as it was dispatched since this may be important in directing improvements to the stability (shelf life) of the pack, or proving that it was of a certain status when it left the manufacturer.

Note on the use of computers in the laboratory

With the lowering of hardware costs and the development of new software, computers – especially PCs – are becoming increasingly common. Additionally, some computer studies are now a usual part of any technical education.

A computer needs, like any other instrument, to be understood in order to be used effectively. A useful definition to remember is that 'It is a high-speed idiot' in that it will do exactly what it is told – no more and no less – and it will do so very quickly.

In the laboratory computers are used for:

- Control of sophisticated equipment, e.g. chromatographic systems, Bactometer and Malthus instruments used for microbial detection
- Data collection, storage, analysis and display, e.g. graphs, histograms, charts
- Calculations, e.g. heat penetration data analysis and process calculations
- Word processing and report writing

As a principle, always be very clear about the job you want done and how it could develop or change. Thus, first identify the software and then put the hardware together to do the job.

If a single PC is used to control two pieces of equipment, check that they can both be run simultaneously or that alternate running of the equipment will not interfere with the work pattern needed to provide an effective service to production departments.

It is a common experience to acquire a computer for a particular job and then to find that it can or could do much more. At this stage flexibility becomes important as does compatibility (or, too often, lack of it) between systems.

As an indication of widespread use of computers, there is a new journal for the microbiologist, *Binary Computing in Microbiology*.

Use of contract analytical services

This may be done for a variety of reasons, for example:

- To obtain routine Quality Assurance data. In a small company this may be because 'in-house' laboratory facilities are not available. It may also be done to meet a temporary 'overload' situation and enable 'in-house' resources to be used for other purposes, such as urgent development work.
- To get results from specialist methodologies. It may not be economic to buy equipment and hire or train personnel, especially when the volume of testing is believed or known to be limited.
- To get test data from possible new equipment or processes. Usually it is desirable to do this 'in-house' but it may not be practical.
- To get data in response to a challenge by a customer or regulatory agency where a demonstration of independence is useful or necessary. In such cases, if these are 'in-house' facilities, consider testing duplicate samples, as this may enable 'in-house' historical data to be used.

As a rule of thumb, with microbiological testing it pays to contract specialist jobs unless the volume justifies doing them 'in-house' or there is the need to develop 'in-house' expertise.

When considering the use of contract analytical services, begin by:

- Identifying the type of analysis required, e.g. is it for QA, customer service or regulatory purposes?
- Identifying the appropriate methodology, the amount and quality of the work (error budget limits) needed.

Where there is a legal aspect, consider carefully what is appropriate especially where an overseas country or customer is involved. In microbiology there are national preferences for some methods.

- Determining how best to achieve your objectives and the resources which will be required. To obtain value for money, you must be clear as to what you need and why. Do not neglect or underestimate the amount of time needed to manage the contract as well as the total financial cost of using a contract laboratory. Remember that with microbiological testing, special arrangements may have to be made for the transportation of samples.

Having decided, in principle, to use a contract service it is prudent to:

- Visit the laboratory. It is, after all, common practice to visit suppliers and monitor standards. Remember that the money involved is not just the cost of analysis but also that which is affected by decisions based upon the results.
- Check the internal Quality Assurance and Audit procedures used in the laboratory, together with pertinent record keeping.
- Send trial samples to the laboratory. This gives confidence in the results received. With microbiological samples, you should be aware of the likely range of results. Remember, confirmation that the range between laboratories is no more than within a laboratory requires sound statistical design and analysis, together with real commitment of resources.
- Check the qualifications and experience of the personnel actually providing the service. Be particularly careful if there are legal aspects to the testing.
- Not be overly impressed by fancy reporting procedures and computer generated reports. Good presentation of data is always helpful but 'it's the quality of the paint and not just the gloss' which really matters.
- Not make the choice on price alone. An unusually low price may well reflect the quality of the work. Remember that what is sought is value for money.

References

1. Anon (1988) *Control of Substances Hazardous to Health Regulations*, London: HMSO
2. ACDP (1984) *Categorization of Pathogens According to Hazard and Categories of Containment*, Advisory Committee on Dangerous Pathogens, London: HMSO
3. BSI (1979) BS 5726: *Specifications for Microbiological Safety Cabinets*, London: British Standards Institution
4. NSF (1984) *Standard No. 49 Class II (Laminar Flow) Bio-hazard Cabinetry*, National Sanitation Foundation, Ann Arbor, Michigan, USA
5. SAA (1980) *Biological Safety Cabinets*, AS 2551, Standards Association of Australia, Sydney
6. Betts, R. P. and Bankes, P. (eds) (1989) Technical Manual 21. *A Code of Practice for the Microbiology Laboratory Handling Foods*, Campden Food and Drink Research Association, Chipping Campden, Gloucestershire GL55 6LD
7. Collins, C. H., Lyne, P. M. and Grange, J. M. (1989) *Collins and Lyne's Microbiological Methods*, 6th edn. London: Butterworths
8. Speck, M. L. (ed) (1984) *Compendium of Methods for the Microbiological Examination of Foods*, 2nd edn, Washington, DC: American Public Health Association
9. Garfield, F. M. (1989) Sampling in the analytical scheme. *Journal of the Association of Official and Analytical Chemistry*, **72** (3), 405
10. Steiner, E. H. (1984) Statistical methods in quality control. In *Quality Control in the Food Industry*, Vol. 1, edited by S. M. Herschdoerfer, 2nd edn, pp. 169–298. London/Orlando, FLA: Academic Press
11. Jarvis, B. (1989) *Statistical Aspects of the Microbiological Analysis of Foods*, Vol. 21. Progress in Industrial Microbiology. Amsterdam/New York: Elsevier
12. Shapton, D. A. (1986) Canned and bottled fruit products (soups, mayonnaise and sauces). In *Quality Control in the Food Industry*, Vol. 3 edited by S. M. Herschdoerfer, 2nd edn, pp. 261–322. London/Orlando, FLA: Academic Press
13. Steere, N. V. (ed) (1971) *CRC handbook of Laboratory Safety*, 2nd edn. Boca Raton, FLA: CRC Press
14. Collins, C. H. (1988) *Laboratory Acquired Infections*, 2nd edn. London: Butterworths
15. Miller, B. M. (ed) (1986) *Laboratory Safety: Principles and Practices*, Washington, DC: American Society for Microbiology
16. CFR (1987 edn) *Acidified Foods, Title 21 Code of Federal Regulations Part 114*, Washington, DC: US Government Printing Office
17. Mossel, D. A. A. and Corry, J. E. L. (1977) *Detection and Enumeration of Sub-lethally Injured Pathogenic and Index Bacteria in Foods and Water Processed for Safety*, Alimenta, Zurich, 16, Special Issue on Microbiology 19
18. Beckers, H. J., van Leusden, F. M., Roberts, D., Pitezch, O., Price, T. H., van Schothorst, M., Tips, P. P., Vassiliadis, P. and Kampelmacher, E. H. (1985) Collaborative study on the isolation of *Salmonella* from artificially contaminated milk powder. *Journal of Applied Bacteriology*, **59**, 35–40.
19. de Man, J. C. (1975) The probability of most probable numbers. *European Journal of Applied Microbiology*, **1**, 67–78
20. Pettipher, G. L. (1983) *The Direct Epifluorescent Filter Technique for the Rapid Enumeration of Microorganisms*, Latchworth Research Studies Press Ltd
21. Pettipher, G. L. and Rodrigues, U. M. (1982) Rapid enumeration of microorganisms in foods by the direct epifluorescent filter technique. *Applied and Environmental Microbiology*, **44**, 809–813

22. Wilson, G. S. *et al.* (1935) *Spec. Rep. Ser. Med. Res. Coun* London, No. 206, London: HMSO
23. Mossel, D. A. A., (1985) Introduction and prospective to Quality assurance and Quality control of microbiological culture media. *Proceedings of the Third International Symposium*, London, 9–13 January 1984. *International Journal of Food Microbiology*, **2**, (1 + 2), 1–7
24. Baird, R. M. (1985) Pharmacopoeia – step one. *International Journal of Food Microbiology*, **2**, (1 + 2), 9–12
25. Woodhall, M. (1989) The application of hazard analysis and critical control point system to milk powder manufacture. *Journal of the Society of Dairy Technology*, **42**, 4, 102–105
26. Wooden, R. (1988) HACCP approach to product safety. In *Contemporary Quality Assurance – Bulletin of the International Dairy Federation 229*, IDF, Square Vergote 41, Brussels-1040, Belgium, 35–40

9.B Challenge testing (including shelf-life assessment)

Introduction

Challenge testing is a technique which yields information to the manufacturer as to the microbiological status of a product during its 'normal' or 'expected' life before consumption. To the microbiologist it is the most direct evidence of product safety and stability. Built into this 'normal life', however, are conditions which take into consideration non-ideal conditions (abuse) which the product may undergo during distribution, retail storage and when in the consumers' hands.

As generally understood, challenge testing comprises inoculation of a food with microorganisms. When carried out for safety purposes, the microorganisms will be pathogens which are likely to be associated with the product being tested, either as ingredient or post-process contaminants.

When information on stability is required, those spoilage organisms are selected which are known to cause problems in the particular type of food under investigation. For example, an intermediate moisture food (IMF) would be challenged with species of micrococci when spoilage is being investigated and a pickle or acid sauce would be challenged by yeasts – probably by *Zygosaccharomyces bailii*, since this is a yeast which is known to withstand heat processes which destroy other yeasts, and can grow in low pH environments in the presence of preservatives.

It should be noted that, in principle, shelf-life testing is also a form of challenge testing. Microorganisms are not inoculated in much shelf-life testing, and it is primarily carried out to determine the time taken before spoilage of the test product occurs. The information given will not say whether a food is safe microbiologically at the end of shelf life.

However, although not generally inoculated, products undergoing shelf-life trials are challenged by a variety of temperatures during the storage period, and may also be challenged by a variety of packaging types, including modified atmosphere packaging (MAP) at one or more temperatures, or by different preservatives or different levels of the same preservative which may be incorporated into the formulation of the product. The main difference, therefore, is that it is not necessary for microorganisms to be inoculated in a shelf-life test.

In order to insure that the results from shelf-life trials are truly applicable to production, the trials must be carried out using the same ingredients as are to be used, or are being used, in the production plant, using factory equipment in the factory environment. By doing this, then the batches of product used to establish shelf life are likely to carry the microflora which are normally associated with the product in question. Products made in a development kitchen or laboratory from ingredients purchased locally, probably from retail stores, will not yield microbiological data which can be applied directly to the same product being manufactured in a factory. Such development kitchen products will, of course, be perfectly adequate for decisions regarding sensory assessment when developing a new product range.

Setting up a challenge

When considering inoculation challenge testing, the following situations are those in which it should be considered as a useful technique.

- Development of a new product or product range
- Reformulation of an existing product
- Making a processing change during manufacture of a product, e.g. of the heat treatment or changing the packaging conditions

The physico-chemical properties of a product, the process it undergoes, its packaging and intended shelf life, distribution and its use and probable misuse by the consumer are all of importance from the safety standpoint, and all must be thoroughly considered.

Whatever the reason for undertaking a challenge test, useful data will only result when the test takes full account of the following factors:

- The choice of microorganisms for inoculation, their culture and method of inoculation
- The uniformity of the food; whether this is within the pack, between packs or between batches produced of the test material
- The microbiological flora of the food product
- The choice of storage conditions for the test

Campden RA has an excellent 'Guidelines' publication[1] which should be consulted before starting challenge tests. Included in these Guidelines are two examples of challenge tests which have been carried out on products within commercial companies. One product is fresh pasta intended for chill storage, the other is an ambient stable soy sauce. Both of these examples show the care needed in designing these tests and the number of experiments required in order to provide data reliable enough to be used in manufacturing product for sale to the public.

Choice of microorganism(s)

Pathogens

This will depend on the food which is being examined and its likely abuse, as well as the expected occurrence of the pathogen in the ingredients or environment of production or as a post-process contaminant. For example, *L. monocytogenes* is a very widespread microorganism, so may be expected to be present from time to time in factory environments or brought in on raw vegetables or poultry. It can also multiply at temperatures as low as 2°C, so would be an organism to be included in challenge tests on neutral chill distribution products which have an anticipated shelf life exceeding 10 days. *Cl. botulinum* type E would be chosen for vacuum packed fish since it is associated with estuarine and fresh waters and can grow at low temperatures and in anaerobic conditions.

The choice will, therefore, be guided by what is known of the properties of the pathogen related to the type of food product and the conditions under which the food will be distributed.

Even if lawful, a pathogen is not normally used in challenge tests carried out in a food plant laboratory. When testing 'in-plant', non-pathogenic organisms with similar properties are used. For example, a canner may use *Cl. sporogenes* instead of *Cl. botulinum* for heat resistance studies. Organisms used for simulation must have equal or slightly greater resistance to the property being studied, e.g. heat, hydrogen peroxide, pH, etc.

Spoilage organisms

The choice here will be governed much more by the particular spoilage organism or group of spoilage organisms found in a similar product produced either within a company, or those which are isolated from a product already on the market but which is being produced for the first time 'in-house'. It may well be that more than one type of microorganism will be used for a single product, e.g. both pseudomonads and yeasts are spoilage organisms of cottage cheese, and yeasts together with lactobacilli and/or leuconostoc are found in spoiled orange juice.

Organisms isolated from actual spoilage of a particular product often are the best challenge organisms for that product. Screening for resistance to the process is usually the best way to select the organisms-of-choice for the challenge studies.

It is important to culture spoilage microorganisms so that their properties will not change during laboratory transfer. When the organisms are first isolated from a particular food, they are likely to have specific properties, such as resistance to low pH in the sense that they can multiply readily at pH values which other strains of the organism could not; or, for example, they could multiply at temperatures below which most other strains could not. Remember also that during sub-culturing of test organisms these properties can be 'lost' or in some instances be enhanced so that some assurance is needed that they will perform as intended.

Enhancement of properties is known as 'training' and may well be considered as part of the protocol for handling spoilage organisms.

It is also important to inoculate the food so that the inoculum itself does not alter the properties of the food. For example, low- or an intermediate moisture foods (IMF) present problems when an organism is in liquid form. If liquid is added to a low-moisture food, it may alter the a_w and may change it so that growth is allowed which would not occur in the unaltered food. The liquid added can be kept to a minimum, for example by spreading 10 μl of culture inoculum over a measured surface of food. How can you be sure of an even spread? This depends primarily on the skill and experience of the microbiologist. When the spread of inoculum is not even, either on a food surface or throughout the bulk of a food pack, this can cause problems when assessing microbial growth.

The non-homogeneous distribution of naturally contaminating microorganisms in dried foods is recognized by Habraken *et al.*[2]. If it occurs in an IMF undergoing a challenge test and the sampling does not take this into consideration then the resulting data will be incomplete and therefore inaccurate and may lead to mistaken conclusions.

Whether inoculating pathogens or spoilage organisms, consideration must be given to the question of whether only one species is inoculated, whether only one strain within a species is inoculated, or whether a 'cocktail' mixture of strains or species is inoculated. Perhaps for spoilage organisms, the choice most often is whether one or more members of a genus are inoculated, rather than one or more strains of the same species. The difficulty with using a 'cocktail' is that when evaluating results, it is hard to know which strains were the toughest. The problem with using single strains is that it multiplies the work to be done by the number of strains used. In an ideal situation, both single- and multiple-inoculum studies would be made, but this is rarely practicable in most challenge tests.

A further point which needs a decision is whether to adapt the chosen microorganisms to the properties of the food which is being challenged. For example, if the food is acid and is held at chill, should the culturing before inoculation be in an acid broth incubated at a temperature below the optimum for the microorganism? Similarly with foods of lowered a_w, for example, intermediate-moisture foods. Should the growth conditions during culturing of an organism involve the use of a medium with an a_w approaching that of the food?

The most likely source of some pathogens is humans, and so the pathogen has not had to adapt to the environmental conditions in the factory, whereas most spoilage organisms originate from equipment, or are carried over from raw ingredients into finished product when segregation of raw from finished product areas is not being practiced. However, the *Salmonella* in dried milk food for babies was found in the lagging of the milk dryer, and *Staph. aureus* has been recovered from milk concentrate prior to drying.

These two examples show that thought must be given to this question. Even in the case of pathogens, a careful judgment needs to be made as to the likely source of pathogens in the particular product under test.

Uniformity of the food – between packs and between batches

Uniformity applies both to the distribution of particles, e.g. pieces of vegetables or meat in a complex formulated food, and also to the distribution of a property (or added preservative) within what appears to be a relatively uniform food.

Good examples of this are the salt levels in ham and in cheese. In dry cured ham there can be a spread from 1% to 7% brine in the same ham. Similarly in cheddar cheese a 'within-block' spread of 0.62% salt has been found on an average of 2.3% salt, giving a variation from 1.99% to 2.6%. The biggest 'within-vat' spread was 1.06% on an average of 2.47% salt – a variation from 1.94% to 3.0%. It should be pointed out that in general the spread of salt levels within a block of cheese have been found to lie between 0.10% to 0.28% from an average salt percentage of 1.46%.

The importance of attaining uniformity is because of the effect which variation can have on growth of microorganisms.

To be meaningful, therefore, a challenge test requires that each factor has a known range and that values can be determined with known precision in the experimental samples.

As far as the main effect which will result from unexpected non-uniformity within a food product undergoing challenge testing is concerned, it is that misleading results will be given. A local under-concentration, or over-concentration, of a preservative, or differing levels of water activity will affect microbial growth, and may either suppress it or allow more multiplication than should occur during the period of the test. Inaccurate results inevitably mean that sooner or later wrong conclusions are made, and could result in a product launch being abandoned because of spoilage in the market place. Far worse than this, if the lack of uniformity suppresses pathogen growth under test conditions in foods which would permit growth, this could result in an unsafe food being released.

To overcome these problems, it is important in any challenge test to examine sufficient individual samples within a given test, and also to examine more than one batch of product. Additionally, control samples, that is, uninoculated samples, must be examined, so that any difference in the normal microorganism population of the food is detected. A change in this normal flora between different batches could have an affect on the growth pattern of the microorganism inoculated into the food as the 'challenge' organism. When sufficient samples are examined, a statistically valid judgment may be made regarding the distribution of the organism (see Kilsby[3] and Jarvis[4]) and so conclusions regarding the behavior of these microorganisms during the incubation period of a challenge test have a sound foundation.

Growth is dependent on, and affected by, intrinsic or extrinsic factors associated with a food or its packing. Intrinsic factors are those which are found within the food itself and include:

Water activity or a_w (see Chapter 8)
pH and/or total acidity (see Chapter 8)
Type(s) of acid
Redox potential or *Eh* value (see Chapter 8)
The nutrients available for microbial growth
The microflora associated with the food

Extrinsic factors are those factors associated with the processing of the food, the type of packaging and the storage conditions used for the food and include:

Heat treatment
Headspace gas composition
Packaging
Storage temperature – the time/temperature profile is important rather than the nominal temperature
Relative humidity

These factors are listed in the Campden Technical Manual No. 20[1] and this document gives a good discussion of the effect of these factors, with examples of which factors may be associated with certain foods.

The microbiological flora of a food product

The likelihood is that the food product which is being challenged already contains its own microflora, and that during the period of the challenge test, some or all of these microorganisms will in fact multiply. Their growth may influence the growth pattern of the challenge microorganisms, either by stimulating the rate of growth or by exerting an inhibitory effect. It is also possible that, when more than one batch of product is undergoing a challenge test, the composition of the microflora may show a change between batches.

For example, it is not unknown for different batches of product to have predominantly Gram-positive flora on one occasion and a predominantly Gram-negative flora on another. Or the balance may shift from a mesophilic flora to a psychrophilic flora, which again is likely to have a different effect on the behavior of the challenge microorganism.

The use of uninoculated control samples will establish the nature of the background flora in any particular set of challenge tests, but this will not give information as to how the behavior of the challenge microorganism is affected. The ideal way to overcome this problem is to treat some of the product samples so that the natural microflora is killed, yet without altering the physico-chemical properties of the food. This is difficult to achieve in practice, since when heat is applied, and this is the most applicable process to sterilize the food, syneresis of the food may occur which will have a marked affect on the behavior of any inoculated microorganism. As well as releasing liquid from the food, the food substrate itself is changed and is likely to provide a ready source of nutrients for microorganisms which may well not be as readily available in the unheated food. Aseptic filling of product is another solution, although non-aseptic fillers can be the major source of post-process contaminants, and it is these contaminants which may then form the 'natural' microflora of a product. Irradiation is also a possibility, but the effect that it has on fats, which results in unacceptable sensory assessment, must raise questions about the effect of such changes on the behavior of any microorganisms inoculated into these foods. Fat-free foods, however, may be suitable for irradiation in order to destroy the natural microflora.

The choice of storage conditions for the test

Perhaps the first condition which needs a decision is the range of temperature at which the packs will be stored. More than one temperature is necessary for chilled foods, since one of the main reasons for carrying out challenge testing on chilled foods is to know how temperatures above that specified will affect growth of microorganisms.

Despite the fact that for many years the importance of maintaining chill food at temperatures below 5°C (41°F) has been known, it is still possible to find that a rise in temperature occurs once the product has left the factory, and this rise is maintained during retail display, is increased while the consumer takes it home, and is unlikely to return to 5°C (41°F) in the home refrigerator. The manufacturer, therefore, must be aware of the behavior of both pathogens and spoilage organisms in a given product at temperatures which reflect the treatment which the product undergoes after leaving the factory. It is recognized that it is impossible to reproduce exactly what occurs or may occur throughout the whole of the distribution and retail chain. However, the temperatures chosen for the test, including possible 'cycling' at different temperatures, should give information which can be realistically applied to the behavior of the product.

Temperature is not the only condition which will be applied during a challenge test. As given in the list of Extrinsic Factors which will have an affect on microbial growth, the headspace gas composition – for example, in modified atmosphere packs (MAP) or controlled atmosphere packs (CAP) – is a variable, as is the type of packaging, which may be impermeable or semipermeable or have different permeabilities to different gases and so will have an effect on the headspace gas composition.

Relative humidity is another factor interrelated to the packaging and the headspace gas composition, as well as a factor to be considered independently of these.

Heat treatment may also be considered as a variable, since when foods do not have to meet a legal requirement for the heat process which they receive, the manufacturer is able to choose from a range of times and temperatures for the heating process, especially where a preservative is to be incorporated into the product.

Conclusions

These examples of the variables which must be considered when designing a challenge test show that it is not an exercise to be lightly undertaken. It is a complex multifactorial experiment involving the inoculation, incubation and testing of many packs of product, and therefore should only be applied when it is clear that other examinations will not give the information on safety and stability of the product which needs to be known. However, challenge tests may be required in the USA for certain new foods that do not have a prior history of safety and other countries may well follow this lead.

The use and misuse of models

Introduction

Although there is a close link between challenge testing and the development of specific mathematical models, some simple models can be regarded as a generalization with numbers which makes a prediction. '*Cl. botulinum* will not grow below pH 4.6' is an example of a simple model giving an incomplete description of microbial behavior and, as such, is limited. This simple model is, however, successful and is used in establishing sterilization processes for Low Acid Canned Foods. Most usually, though, models are mathematical – involving formulae, diagrams (sometimes three-dimensional), nomograms or computer programs.

Mathematics

For the microbiologist, the common difficulty is in understanding and using the mathematics. This may result in the failure to give critical appraisal to the quality of the microbiological data on which the model is based. Mathematically, the usual problem is to get a 'best fit' of data to a straight line, curve or surface. This is often possible with comparatively straightforward techniques which can make good use of the speed and power of a computer.

To the inexperienced or unwary, the mathematical expressions give an impression of certainty, precision and unquestionable authority which the builder of the model never intended. It is important to appreciate that models are powerful tools when they are properly used to enhance understanding of microbial ecology. This enables work to be concentrated in areas which are likely to give the best return on resources invested. Models are not a species of magic nor a substitute for knowledge and judgment.

There are also rather subtle errors that can be made when using a model, e.g. when extrapolating rather than interpolating data it is easy to underestimate the likely loss of accuracy. The rule most model builders apply is *Do not extrapolate*. Other errors can result from not appreciating the extent of variation about the mean in the results used as data for the model. This results in attributing greater precision than is warranted to the predictions.

It is important when using any model to understand what it will and won't do, and the precision with which it can be used.

Microbiological validity

In order to appreciate the microbiological validity of a model it is helpful to realize that the food microbiologist is concerned with specific aspects of microbial ecology. *Chambers Dictionary of Science and Technology* defines ecology as 'The scientific study of the interrelations between living organisms and their environment'. It is therefore concerned with the physical, chemical and biological factors which relate to population changes. With microorganisms, this concerns changes which environmental factors cause to organisms, e.g. growth and death of different kinds of microorganisms, and also the changes which organisms cause in the environment, e.g. production of acidity or toxins.

Because microorganisms are small, being measured in microns (μ) or 1/1000 of a millimeter, their environment is very small by human standards. A non-homogeneous food, e.g. meat and vegetable in a sauce or even a chicken leg which has been cook-chilled, present, within each food, many different environments with various chemical, physical and microbiological properties. Normally, the microbiological population (or flora) is of mixed kinds all of which, to a greater or lesser degree, adapt rapidly to changing environmental conditions. This is the basis for the sound laboratory 'tradition' that the organism in the 'wild' differs from the laboratory culture. The explanation is that organisms in the 'wild' have adapted to stresses in the environment. In pure laboratory cultures, conditions are near optimal (to which they adapt) so that when put back into the environment they are more susceptible than the 'wild' type to the stresses they encounter. This could mean that the model predicts no or slow growth but in the market place significant growth occurs.

It is clear from these considerations alone that getting good reliable data as the basis for modeling is neither easy nor simple.

Factors to be taken into consideration when designing a model are similar to those in Challenge Testing. The Campden Technical Manual No. 20[1] lists the following critical factors:

Intrinsic Factors	Extrinsic Factors
1. Water activity; a_w	1. Heat treatment
2. pH/total acidity	2. Headspace gas composition
3. Type of acid	3. Packaging
4. Preservatives including salt and spices	4. Storage temperature
5. Nutrients	5. Relative humidity
6. Natural microflora	
7. Redox potential; *Eh*	

All of these factors must be considered and controlled in modeling experiments. In addition to considering the organisms and factors involved, care must be taken with the quality of the data used to build the model. Bratchell *et al.*[5] showed three-dimensional plots of fitted response surfaces for the whole data set, and also plots when this data set was reduced by three (plausible) strategies (scenarios).

This highlighted differences between models, illustrated the consequences of using insufficient data, and demonstrated the risk of developing an erroneous model. They concluded that it is not sufficient to observe that a particular method or model appears to work; it is important to know why it works, otherwise there is no means of judging when and how it will not be applicable.

Analysis

In modeling, the magnitude of the factors may not be known, or known with insufficient accuracy. This can be because the analysis, for example for salt in a ham, is often done on a bulked sample which conceals variations within the muscle tissue which are very significant for microbial growth. It may also be that analysis entirely adequate for control purposes, e.g. regarding all acid in ketchup as acetic acid, does not reflect the true situation when, in reality, perhaps only half the acid is acetic, the rest being citric – which is much less inhibitory to microorganisms. The precision of analysis and the need to know the extent of variation are of critical importance to the validity of the model since these will form the basis for the experimental design.

'Media'

An additional consideration in experimental design is the use of 'media' or simplified substrates. These can be standard laboratory media modified to study a single factor, e.g. pH, or a simplified food system, e.g. a pork slurry modified to be similar to ham. The advantage over using a food is that it is easier to use and hopefully gives more reproducible results, thus giving a quick indication of the magnitude and direction of significant factors. For this reason it is often used at the start of a study, but conclusions always need to be tested on the actual food to substantiate the validity of the model. Difficulties of working with food are illustrated by suggesting that you consider how to model a pizza.

Organisms and their use

Much modeling data are derived from inoculating a pure culture of an organism into a substrate. This is not as simple as it may seem. First, it is no easy matter to be sure that the organism is physiologically comparable with that isolated from food. Second, is the inoculum to be a mixture of strains (a cocktail) or a number of single strains? The dilemma is acute when modeling pathogens when there are a large number of strains of each microbial type derived from outbreaks.

The number of organisms used is likely to affect the result. High inoculum numbers give a better chance of survival under adverse conditions. This effect seems to be more than just a proportional effect but it is a complex situation. Several levels can be used but again this multiplies the work to be done by the number of levels included in the modeling experiments.

In food, mixed cultures, i.e. the presence of several or many kinds of organisms, make up the natural flora. The problem here is to know what are the significant kinds and numbers together with the variability to be found in any specific food brought in from the market. As an example just of changes in number, consider the possible growth of *L. monocytogenes* in chilled products where there has been contamination and temperature abuse during transit.

Models in use

Clearly, much careful thought and work are required to produce an effective model. There is, however, little point when building a model to take factors beyond acceptable sensory limits. The reason models have been built in the past and will be built in the future is that the use of a good model is extremely cost effective in optimizing the use of microbiological resources, the most expensive of which is the time (labor) of microbiologists.

Models may be built for a variety of purposes and a few examples of successful models are:

- Prediction of lag time and growth in response to temperature changes and hence of shelf-life as with the Schoolfield equation and the 'square root rule'.
- Prediction of conditions where inhibition of growth occurs as with ambient stable products. Examples are Bell and Etchells curves for stability of pickles and Microbiological Compositional Analysis (MCA) developed by Tuynenburg Muys[6] for acetic acid sauces. There is also a model by Davey[7] for the combined effects of temperature and water activity on growth.
- Prediction of destruction of spores by heat, e.g. equations by Ball and also Stumbo which can be used for pasteurization.

Currently in the UK the government is placing substantial contracts through the Ministry of Agriculture, Fisheries and Food (MAFF) for model building related to pathogens in food and the program is expected to continue for some years. This development follows the successful models developed by Unilever and an excellent review of the principles of predictive food microbiology by Baird-Parker and Kilsby[8] indicates Unilever's interest.

Spoilage, however, is regarded as the manufacturers' problem and thus is not being funded by MAFF. As an example of world-wide interest in modeling and its importance, a recent paper by Baker and Genigeorgis[9] based on nearly 19 000 samples presents a model for the lag phase of *Cl. botulinum* in fresh fish under modified atmospheres.

Always remember that it is unsafe to base a commercial process on extrapolated, insufficient or unconfirmed data, whatever model is used.

References

1. Campden RA (1987) *Technical Manual No. 20. Guidelines for Microbiological Challenge Testing*, Campden Food Preservation Research Association, Chipping Campden, Gloucestershire GL55 6LD
2. Habraken, C. J. M., Mossel, D. A. A. and van den Reek, S. (1986) Management of *Salmonella* risks in production of powdered milk products. *Netherlands Milk and Dairy Journal*, **40**, 99–116
3. Kilsby, D. C. (1982) Sampling schemes and limits. In *Meat Microbiology*, edited by M. H. Brown, pp. 387–421, London: Applied Science
4. Jarvis, B. (1989) *Statistical Aspects of the Microbiological Analysis of Foods*, Vol. 21. Progress in Industrial Microbiology, Amsterdam/New York: Elsevier
5. Bratchell, N. *et al.* (1989) Predicting microbial growth: the consequences of quantity of data. *International Journal of Food Microbiology*, **8**, 47–58
6. Tuynenburg Muys, G. (1971) Microbial safety in emulsions. *Process Biochemistry*, **6**, (6), 25–28
7. Davey, K. R. (1989) A predictive model for combined temperature and water activity on microbial growth during the growth phase. *Journal of Applied Bacteriology*, **67**, 483–488
8. Baird-Parker, A. C. and Kilsby, D. C. (1987) Principles of predictive food microbiology. In *Changing Perspectives in Applied Microbiology*, edited by C. S. Gutteridge and J. R. Norris, Society for Applied Bacteriology Symposium Series No. 16, pp. 43S–48S. Oxford: Blackwell Scientific
9. Baker, D. A., and Genigeorgis, C. (1990) Predicting the safe storage of fresh fish under modified atmospheres with respect to *Clostridium botulinum* toxigenesis by modeling length of the lag phase of growth. *Journal of Food Protection*, **53**, 131–140 + 153

Chapter 10
Pathogenicity and pathogen profiles

10.A Introduction – pathogens and food processing

Safe food depends upon insuring that it does not contain:

Pathogenic bacteria or their toxic metabolic by-products
Mycotoxins produced by certain fungi
Pathogenic viruses
Pathogenic parasites such as Trematodes, Cestodes, Nematodes, e.g. *Trichinella spiralis* and *Anisakis* spp.; *Giardia lamblia* and *Cryptosporidium parvum*

It is to be expected that most raw materials will carry pathogenic microorganisms, to a greater or lesser extent depending or their origin. Farm as well as wild animals act as hosts to pathogenic bacteria, protozoa and viruses, and crops grown on land which has been grazed by these animals will show some contamination. Additionally, land fertilized by sewage slurries, or crossed by land-ditches carrying sewage, will produce contaminated crops. Some of this pathogen contamination will be carried over on slaughtered carcass meat, including poultry, and on vegetables and other harvested crops, so may enter a plant when unprocessed raw materials are used as food ingredients.

As well as raw foods, processed foods can act as carriers for most pathogens. In the case of toxin-producing bacteria, if they are present only in low numbers, then they are unlikely to be a public health hazard. For some infective pathogens, e.g. salmonellas and viruses, illness can be caused when the total numbers that are eaten or ingested are below 100, and in the case of cheese contaminated with *Salmonella typhimurium* phage-type 10 it was estimated that a single cell per portion of contaminated cheese consumed caused illness[1].

Associations between certain pathogens and certain foods are found when outbreaks of illness are traced to the foods concerned. *Cl. perfringens* in cooked meat and poultry; *C. jejuni* in poultry; *V. parahaemolyticus* in raw seafoods; viruses in cockles, mussels, scallops; and mycotoxigenic molds growing in improperly stored cereals and nuts are examples of links found between pathogens and food types causing illness. However, *Listeria monocytogenes,* salmonellas and *Staphylococcus aureus* are all found in a large range of foods. With *L. monocytogenes*, the reason is thought to be because it is widely found in the environment, including factory (plant) sites; with salmonellas, it is thought to be cross contamination from people and equipment to food as well as from the environment, whereas with *Staph. aureus* it is thought to be people to food contamination which is the main source.

It should be remembered that whatever the food type, finished product contamination will occur when there is not complete separation between raw material areas and areas handling processed food.

Food processing operations are designed to minimize pathogen contamination and growth as follows:

- Undue multiplication of pathogens between harvesting and processing should be prevented. Multiplication to high numbers can result in toxin formation if the pathogen is toxigenic. Since most toxins are heat stable they may be present in the processed foods at levels which cause illness although the pathogen will be killed. For non-toxinogenic pathogens, multiplication results in high numbers which will challenge and may survive the processing stage. Not all processes are wholly bactericidal – for instance, pasteurization of a food may only give a 10^4 kill step with respect to pathogens and less for more resistant spoilage organisms. For a pathogen population of 1 per ml or g, as is found with *L. monocytogenes* in contaminated raw milk in bulk farm tanks, then 1 cell per 10 liters would survive. However, if incorrect storage of raw milk had allowed multiplication to 10^4 per ml or g, then the survivors would number 1 cell per ml or g. This level may not be pathogenic, but it could be a public health risk should subsequent storage conditions permit multiplication, e.g. if there was poor refrigeration control during distribution and storage of a chilled product. For *L. monocytogenes*, a temperature of 8°C allows doubling every 12 hours and a count of 500 per ml or g would result after 5 days from an initial one per ml or g. An increase in temperature of *only* 1.3°C (to 9.3°C) would result in a count of 16 000 per ml or g of food. Foods with these counts are regarded as unsafe.

- The number of vegetative pathogens should be reduced by a factor of at least 10^4, i.e. a 4 log cycle kill. With certain raw materials, e.g. ingredients with high *Enterobacteriaceae* or *Staph. aureus* counts, a process giving a reduction in number of at least 10^6, i.e. a 6 log cycle kill, is more appropriate. However, it should be noted that for shelf-stable low-acid foods, that is, canned foods with a shelf life at ambient temperature of 2 years, the process should give a 12 log cycle kill of *Cl. botulinum* spores as a minimum.
- Multiplication of any survivors of the lethal step in the process should be prevented. This may be achieved by holding the food at a suitable refrigeration temperature; or by maintaining a low a_w in the food; or a low pH; or by a proven 'safe' interaction of two or more factors which will inhibit microbial multiplication. The reason for the high lethality given to low-acid canned foods is that these foods have no 'intrinsic' preservative factor.

Safety assessment of a process

In order to assess whether a process is safe or not, the properties of pathogens must be known in as much detail as possible. Examples of the kind of information which is needed are given for bacteria, mycotoxic molds, viruses and parasites. Be aware that interaction of factors affecting pathogen growth is important, and the necessity for quantifying the effects of these interactions has led to the development of models. *Be sure that any predictive model used in your process establishment is fully applicable.*

Bacterial pathogens

Unless the behavior of a pathogen is known with respect to, for example, heat, acidity, or the a_w of its

substrate, no sound judgment may be made regarding both the hazard presented by a particular pathogen in a food or environment and the proper controls required in processing and handling of food in order to minimize risk of illness.

The information which needs to be known may be presented under the following headings:

Morphology
Oxygen tolerance (aerobe, anaerobe, microaerophilic, etc.)
Habitat
Types of food associated with illness
Significance in food poisoning outbreaks: is it infective or toxinogenic?
Infective dose
Heat resistance
Parameters for development, e.g.
   temperature range
   pH range
   inhibitory salt concentration
   inhibitory nitrite concentration
   any particular resistance to chemical disinfectants

For the main pathogens causing food-borne outbreaks of illness, profiles based upon these headings have been prepared. Information has been gathered from papers in journals, and also from published summaries and reviews, especially Mitscherlich and Marth[2] and Doyle[3]. Limits of growth which are presented for specific properties are based upon 'worst case' values – that is, the lowest pH value at which a particular pathogen, or strain of pathogen, has been shown to develop, or the lowest recorded temperature of growth. It is realized that many of these values have not been given with food as the substrate, but from experiments carried out in laboratory culture media. Thus, a range of values for any property has been reported by various researchers over the years, depending upon the strain of the pathogen, whether the inoculum was in its lag, log, or stationary phase, the number of cells used as the inoculum and, of course, the specific medium used in the experiment. Where wide variation in quoted values has occurred between different workers, then the extremes have been given.

It should be noted that one 'worst case' value has not been included in these profiles, and that is the growth pH of *Cl. botulinum*. It has been reported that this pathogen is able to multiply and produce toxin at pH values below 4.5 to 4.6, the pH generally recognized as being the minimum to allow toxin production. However, growth and toxin production in more acid conditions have only been found when precipitated protein was present in the medium. It is thought that the network of precipitated protein created and maintained acid-quenching zones and only these zones of higher pH (i.e. low acid) allowed growth of *Cl. botulinum*. This special case is not regarded as making current regulations or guidelines invalid[4].

When using these profiles, it must be remembered that it is not always one single factor associated with a food product, either intrinsic or extrinsic, which determines pathogen response. The factors affecting growth may interact. For example, the temperature at which a food is held can affect the response to pH. *L. monocytogenes* will not multiply at pH 5.2 at 4°C, but at 30°C it can multiply at the lower (more acid) pH of 4.4 to 4.6. Thus, holding at a low temperature reinforces the inhibitory affect of acidity.

In other cases, however, a change in only one factor will affect the resistance of a pathogen to adverse conditions and allow it to survive a heating process which is normally lethal. *Salmonella*s are a good example of this. When heated in substrates of low a_w, such as chocolate or milk concentrate, they survive conditions far in excess of those which are lethal at high a_w values. For instance the $D_{57.0}$ of *Salm. typhimurium* increased from 1.4 min in 10% liquid milk to 26.6 min in 50% milk concentrate[5]. It should also be noted that *L. monocytogenes* and some salmonellas both show increased heat resistance if they are held at elevated, but not lethal, temperatures before being heated at 58° to 60°C.

Mycotoxic molds

In earlier times molds were regarded simply as spoilage organisms and as a nuisance. However, the association between illness and molds has been known for many years through ergotism, an illness causing paralysis and which can also interfere with blood circulation. It resulted from eating moldy grain – the mold concerned is *Claviceps purpurea*. But it was not until 1960, when mold-contaminated peanut meal caused liver cancer in turkey poults in the UK and slaughter of 100 000 birds, that widespread investigations were made into toxic metabolites produced by certain mold species. These metabolites were called mycotoxins and the aflatoxins produced by *Aspergillus flavus* and *A. parasiticus* have been the most extensively studied. The peanut meal which caused the illness in the turkey poults was contaminated with *A. flavus*. It is now known that other molds which produce toxins include *Wallemia sebi*, a xerophilic mold which grows in cakes, jams, and dried fruits; some species of *Penicillium*; *Fusarium culmorum* and *F. graminearum*, which grow on cereal crops in the field rather than on stored grain, and *F. moniliforme* which grows on banana fruit.

Generally it is the growth of these molds during storage of nuts, cereals, and dried cereal products which is accompanied by mycotoxin production. However, as well as *Fusarium* spp. growing in field crops, *A. flavus* can grow in peanut plants and *A. parasiticus* in crop plants of maize, cottonseed and tree nuts. The growth in peanut plants appears to be the result of drought stress of the host plants and also depends on the soil moisture levels during plant growth.

Although the food processor can control storage of the harvested, dried crop, growing conditions cannot be readily controlled, although irrigation programs could overcome drought stress. Because mycotoxins may therefore be present in newly harvested crops, and will not disappear even if the crop is correctly stored, it is important that monitoring of mycotoxin levels in cereals and nuts forms part of the HACCP system. Needless to say, the harvested crops must also be stored in conditions which prevent mold growth.

Certain mycotoxins are extremely toxic to animals. Aflatoxin, a carcinogen with the liver as the major target and the kidney and immune system as minor targets, has an LD_{50} of 0.5 mg/kg against the dog. It is also acutely toxic to man, and in India in 1974 more than 100 people died in an outbreak of hepatitis caused by aflatoxin. Because of this acute toxicity, the maximum level of aflatoxin in foods susceptible to mold spoilage should be <10µg/kg (ppb).

Pathogenic viruses

Among these are viruses which cause hepatitis A and gastroenteritis. Foods associated with these viruses are shellfish (including oysters, clams, cockles, mussels and scallops) and watercress. Contamination of these food types is most often caused by water which contains imperfectly treated sewage. It should be noted that water cleansing (depuration) of shellfish does not remove viruses, although it is sufficient to remove bacterial contaminants.

Within a processing plant, viral contamination is only likely to occur if a human, who is either a carrier or who is ill, touches the food with unwashed hands. When GMP is enforced through a personal hygiene program, this should not occur (see Chapter 6).

Pathogenic parasites

Food can act as a carrier of these parasites. Undercooked pork, lamb, beef or poultry may contain *Toxoplasma gondii* and undercooked pork *Trichinella spiralis*. Another source of parasites is water, and human infection may be caused through three possible routes. The first is when contaminated water is drunk directly; the second is by vegetables, e.g. watercress, being irrigated by contaminated water; and a third route is by the use of contaminated water in a processing plant. Certain water-borne parasites, such as *Cryptosporidium* spp., are resistant to chemical sanitizers, so a contaminated water supply will itself contaminate equipment surfaces during cleaning. In manufacturing, contamination of cottage cheese could occur, since washing of curd with cold water is part of the cheese-making process. *Cryptosporidium* cysts are removed by filtration of the water supply, so use of correctly filtered water should overcome this contamination hazard.

10.B Pathogens and illness

Illness is caused either by infection or intoxication. Infection results from consuming a food which contains living pathogenic microorganisms. In some cases, only very few pathogens need to be ingested for illness to occur.

Intoxication results from consuming a food which contains bacterial toxin. The toxin is formed in the food by the pathogen, and when bacterial numbers reach at least 10^6 per ml or g, there is sufficient toxin in the food to cause illness of the consumer. Toxins are heat stable, so will remain in food which has been heat processed and the pathogenic bacteria destroyed.

Pathogens may be regarded as belonging to two groups:

Usual pathogens – which are well documented as causing outbreaks of food poisoning, and are responsible (or have been in the past) for many cases of illness.

Emerging pathogens – which are associated with food poisoning, but which are not responsible for many known cases of illness. It should be noted that among the usual pathogens, *Cl. botulinum* and *Staph. aureus* cause illness of a few people each year, although many more people were made ill just 30 or 40 years ago. This reduction in illness results from the effective application of control measures in food processing, based on the knowledge of the properties of each of those pathogens. However, failure to apply correct controls can swiftly change the illness figures, as was shown in the UK in 1989. Because a botulinum cook was not given to an ingredient of yogurt, an outbreak of botulism resulted, with the death of one of the affected people.

Profiles of usual pathogens

Bacillus cereus

- Morphology

Gram-positive rods with central spores 1.0–1.2 by 3–7μm
Aerobe, and facultative anaerobe

- Habitat

Found in the majority of farm and in all bulk milks. However, its source may not be the straw and soil, which is the usual entry for spore-forming bacteria, but it may contaminate milk from farm and dairy equipment. The primary source of this contamination is not clear.

- Types of food

Found in vanilla sauce, meat dishes, chicken soup, mashed potatoes, vegetables, rice dishes and dried mixes, spices. Commonly found in pasteurized milk or cream held at 15°C or more:

- 129 out of 486 milks in Scandinavia
- 150 out of 161 cream samples
- 86.7% of pasteurized milk in Rumania, at levels ranging from 1×10^{-1} to 1×10^4 per ml

- Significance

It has been implicated in food poisoning since 1906, when meat balls caused an outbreak. Vanilla sauce has also caused illness, following storage overnight at room temperature. The sauce was found to contain between 10^7 and 10^8 *B. cereus* per ml. In the years 1960–1966, it was the third most common pathogen in Hungary, often in meat dishes and the level of spicing was thought to be responsible. In the UK the majority of food poisoning cases have originated from holding cooked rice overnight before using. Only one milk-associated outbreak has been reported. This was in 1975 and involved pasteurized cream – only 2 people became ill.

- Infective dose

The minimum level found in foods implicated in outbreaks in the UK is 1×10^3 per ml, the highest count is 5×10^{10} per ml, and the median count is 1×10^7 per ml; 76% of the foods had 1×10^6 per g.

B. cereus has two toxins; one is pre-formed in the food and causes illness (diarrhea and/or vomiting) after approximately 4 hours. The other toxin may be pre-formed in food, but may be produced *in vivo* after ingesting large numbers of viable *B. cereus* cells.

Preformed toxin is:

- Thermostable
- Formed during stationary phase
- Peptide (MW 5000)
- Found only in serotype 1.

In vivo toxin is:

- Thermolabile
- Produced during logarithmic phase
- Protein (MW 45 000)

• Heat resistance

The vegetative cells are no more heat resistant than vegetative cells from non-spore formers.

Temperature(°C)	Spore heat resistance D values (min)
85	220 in phosphate buffer at pH 7.0
90	71 in phosphate buffer at pH 7.0
95	13 in phosphate buffer at pH 7.0
100	5.5 to 8 in phosphate buffer at pH 7.0 2.7 to 3.1 in skim milk 5 in low acid food
121	0.0065; 0.03; 2.37 at pH 7.0 30 in soy bean oil 17.5 in olive oil
135	Reported in the UK that 1 in 10^5 to 10^6 spores survived for 4 hours. The strain was selected from a spoiled pack produced during a UHT run at 135°C.

• Parameters for development

	Minimum	Optimum	Maximum
Temperature	Typically 10°C*	28°C–35°C	50°C
pH	4.9 Some strains 4.35		9.3
Salt	Growth inhibited by 10% NaCl		
a_w	0.912		

* Some strains can multiply at 4°C: 2 log cycles in 7 days and 3 log cycles in 10 days

Campylobacter jejuni

• Morphology

Gram-negative vibroid shaped rods; one or many turns per cell.
0.2–0.8 by 0.5–5 µm (spiral forms 8 µm long). Coccoid or spherical when old.
Motile – characteristic corkscrew-like darting movement; single flagellum at each pole.
Obligate micro-aerophile: may be cultured in atmospheres with 6–10% oxygen supplemented with 5% carbon dioxide.

• Habitat

C. jejuni is found in the intestine of farm animals and poultry, domestic animals, and in farm wastes.

It is also found in sewage and in river water, since sewage treatment allows survival of a small proportion (<0.1%) of campylobacters. There is evidence that low (10–10^2 per liter) numbers in water supply can infect poultry flocks.

• Types of food

The foods most commonly contaminated with *C. jejuni* are poultry carcasses (ranging from 50 to 60% in the USA, Canada and UK to 94% in Australia). Offal (from pig, cows and sheep) showed a 47% contamination rate, and retail beef, pork and lamb showed 15 to 23% contaminated samples. More than a 100 samples of each foodstuff were examined in this survey, although another survey of red meat retail samples showed a 1% contamination rate, based on a sample size of 4933 meats.

Milk has been reported as carrying *C. jejuni* in 0.9% of raw milk in a US survey; in 4.3% of UK raw milk; and in 0.2% of Dutch milk. Cooked meats in the UK showed a 2.3% contamination rate.

• Significance

The number of cases of food poisoning due to *Campylobacter* is equal to those caused by *Salmonella* in the USA, and in the UK exceeds those caused by *Salmonella*. Most of the cases are not outbreaks; 90% are sporadic or family associated. However, in both the USA and the UK there have been milk associated outbreaks: five in the USA in 1982 and 13 in the UK between 1978 and 1981.

There is a strong association between poultry and food poisoning due to *Campylobacter*, and since there is some evidence to show that illness is caused by a small number of *Campylobacter* being ingested, contaminated chicken carcasses may act as vectors carrying *Campylobacter* into kitchens. Illness may be caused by cross contamination of other foods, or direct transfer to the mouth when handling contaminated raw chicken (or meat). Freezing of meat can cause a reduction in numbers, as can storage at chill temperatures, but presence of the pathogen in a large percentage of foods in retail stores shows that frozen or chilled storage cannot be used to eliminate *Campylobacter* from contaminated foods.

• Infective dose

This is thought to be <10^6 cells, and may be as low as 5×10^2 per person affected by illness. Symptoms of illness are acute colitis – abdominal pain with bloody diarrhea.

Heat resistance:

Temperature(°C)	D values (min) in skim milk
50	1.3–4.5 (2 strains); 3.5–5.4 (5 strains)
55	0.74–1 (5 strains)

It is also reported that 10^6–10^7 cells per ml failed to survive when heated at 60°C for 1 minute in skim milk, i.e. 10^6–10^7 reduced to <1 per ml.

- Parameters for development

	Minimum	Optimum	Maximum
Temperature	32°C	42°–45°C	45°C
pH	5.3*	6.0–8.0	9–9.5
	*1 of 3 strains grew at 4.9		
Salt	Growth in presence of 1.5% NaCl at 42°C. No growth at 2% salt level at 42°C, 35°C and 30°C.		

Clostridium botulinum

- Morphology

Gram-positive rods which can form sub-terminal spores.
0.5–1.0 by 1–2 μm in size
Motile at 20°–25°C
Anaerobe

- Habitat

Irregular occurrence of types A, B, E, F and G in soil, fresh water sediments and possibly also in the sea.

In the UK, has been found in lake muds (types B, C, D and E found and some samples had more than one type) and in 10% of soil samples (type B only).

In the USSR, incidence ranges from 11% of soil samples in Siberia, to 12.7% in European zone to 14.2% in middle Asia.

Type E was found in 62% of positive samples
Type B was found in 28% of positive samples
Type A was found in 8.3% of positive samples

Types C and D are seen as obligate parasites and are seldom found except where animals and birds congregate.

- Types of food

Strain type	Food	Disease caused
A	Vegetables Meat conserve Fish conserve	Human botulism in western USA
B	Prepared meat Silage Feed	Human botulism in eastern USA and Europe – less predominant than type A
C	Spoiled vegetables Maggots	Thought not to be implicated in human botulism

Strain type	Food	Disease caused
C	Silage Wastes	Cattle, sheep, mink. Thought not to be implicated in human botulism
D	Wastes	Cattle in South Africa
E	Fish Fish products	Human beings, mink and wild birds in North America, North Europe and Japan
F	Meat Meat products	Human botulism in USA and Denmark
G	Soil (Argentina)	Not implicated in human botulism

- Significance

Botulism is caused by toxin(s) produced by *Cl. botulinum*, and high numbers of cells are necessary for toxin to be present in the food. The toxin is extremely potent. The spores are able to survive extremes of heat, drying and chemical exposure. Germination of spores in low acid–high moisture foods will be followed by outgrowth at ambient temperature of both proteolytic and non proteolytic strains. At chill temperatures (below 10°C) only non-proteolytic strains will develop.

Symptoms. These arise from consumption of food in which *Cl. botulinum* spores have germinated, then grown and produced toxin. The toxin interferes with transmission of nerve stimuli and causes paralysis, starting with blurred vision, then difficulty in talking, swallowing and finally breathing.

The toxin will cause illness at low concentrations in food. It is relatively heat stable and is only destroyed after heating at 79°C for 20 minutes or 85°C for 5 minutes.

Infection of infants, where illness is caused by toxin production in the intestine as distinct from illness caused by food-borne toxin, can occur. It also appears possible that adults may be infected by *Cl. botulinum* – in the USA the number of infections has been estimated at 15 out of a total of 300 cases during the years 1973 to 1985.

- Heat resistance

Vegetative cells are not particularly heat resistant, but since the spores are heat resistant, processes must be designed to destroy spores. Proteolytic and non-proteolytic strains have spores of different heat resistance so separate data will be given.

Proteolytic strains comprise: all strains of Type A
some strains of Type B
some strains of Type F

Temperature (°C)	D value (min) in Buffer at pH 7		
	Type A	Type B	Type F
95	12–38	23	
104			3.5–6.3*
110			1.45–1.82
112.8	1.09–1.23	0.15–1.32	
115.6	0.74		
121	0.20	0.11–0.17	

*Equivalent heat resistance was found in crab meat

Non-proteolytic strains comprise:	some strains of Type B all strains of Type E some strains of Type F

Temperature (°C)	D value (min) in Buffer at pH 7		
	Type B	Type E	Type F
76.6			1.66–6.64
77		0.77–1.95	
80		0.6–1.9	
82.2	1.49-32.3		0.25–0.84 (1.16 in crabmeat)
95	0.1	0.1–0.6	
100		0.003–0.017	

HTST pasteurization will not destroy spores of either proteolytic or non-proteolytic strains.

A botulinum cook is defined as that giving a 12 log cycle kill, i.e. a destruction of 10^{12} spores, of the most resistant proteolytic strain. In commercial terms this is a cook of 3 minutes at 121°C and is the minimum to insure a product which will not present a botulinum risk when stored at ambient temperature.

For non-proteolytic strains Type E, a heat treatment of 0.2 minutes (12 seconds) at 100°C is necessary to give a 12D kill.

Parameters for development

Proteolytic strains

	Minimum	Optimum	Maximum
Temperature for growth and toxin production	12°C	30°–40°C	48°C
pH	4.6	7.0	9.0

Salt: tolerance is affected by temperature and pH
10% NaCl inhibits growth and toxin production at 35°C and pH 7.2
5% NaCl inhibits growth and toxin production at 35°C and pH 5.2
Lowering the temperature increases the inhibitory effect of salt

e.g. 3% salt allows toxin production in 2 days at 30°C
3% salt allows toxin production in 15 days at 20°C

Sodium nitrite and salt (cured meats)

Growth and toxin production is inhibited by:
100 ppm NaNO$_2$ and 5% salt at 35°C and pH 6.2
250 ppm NaNO$_2$ and 4% salt at 35°C and pH 6.2

Non-Proteolytic strains

	Minimum	Optimum	Maximum
Temperature	Type E 3.3°C Type F 4°C Type B 5°C	25°–37°C	45°C
pH	5.0	6.5–7.0	9.0

Salt: tolerance is affected by temperature and pH
6.5% NaCl inhibits growth and toxin production at 30°C and pH 7.2
3.5% NaCl inhibits growth and toxin production at 30°C and pH 5.2
5% NaCl inhibits growth and toxin production at 15°C and pH 7.0
3.5% NaCl delays growth and toxin production at 6°C and pH 7.0 for up to 120 days.

Sodium nitrite and salt (cured meats)

Growth and toxin production is inhibited by:
100 ppm NaNO$_2$ and 3% salt at 35°C and pH 6.2
250 ppm NaNO$_2$ and 2% salt at 35°C and pH 6.2

Clostridium perfringens

● Morphology

Gram positive rods with sub-terminal spores 09–1.3 by 3.0–9.0 μm
Anaerobe, but can grow in presence of oxygen in the log phase.

● Habitat

Distribution is widespread. Type A is found in soils, faeces, marine sediments, and dusts, and it is this type which is the most common cause of illness in man. It also comprised 94% of isolates from meat, crab, and cured sausages; a total of 339 samples were examined.

Types B, C, D and E appear to be obligate parasites of animals, although they may be occasionally found in man.

● Types of foods

Isolated from meats: 16.5% of samples

Ground beef: 47% of samples, with counts ranging from 5 to 100 per g

Chicken: in relatively small numbers

Turkey: 62% of samples

Dairy products: in small numbers in raw milk, and so in pasteurized milk, milk powders, etc.

• Significance

Cl. perfringens is one of the main causes of human food poisoning and the symptoms are caused by a heat-labile enterotoxin. To cause illness, high numbers of vegetative cells need to be ingested, and as these sporulate in the intestine the toxin is formed. Small amounts of toxin can be produced by vegetative cells, but the ratio of production of toxin in vegetative cells to sporulating cells is 1:1000.

The symptoms of illness are stomach cramps and prolific diarrhea.

• Infective dose

The causative food generally contains high numbers of cells, and the median count is 7×10^5 per g in implicated foods in the UK

Note: Powder for calf feeding may have a limit set of 5 cfu per 2 g of powder, as it appears that young calves are susceptible to very low numbers of *Cl. perfringens*.

Powder for use in pot snacks has a limit of 100 per g maximum, as under conditions of customer abuse, i.e. snacks held warm for some time after hot water is added, spore germination and outgrowth might occur.

The germination time at temperatures between 30°C to 43°C is about 10–12 minutes, so an infective dose can be reached in a few hours in food which is held warm, as in reconstituted snacks or cooked meat.

• Heat resistance

The vegetative cells are no more heat resistant than vegetative cells of non-spore formers.

	Spore heat resistance
Temperature(°C)	*D* values (min)
100	Strains range from 0.31 to 17.6
105	Strains range from 5 to 27

• Parameters for development

	Minimum	Optimum	Maximum
Temperature	12°C*	43°–45°C	50°C
pH	5.0	6–7.5	8.3
Salt	Growth Inhibited by 6% NaCl		

*At 15°C, growth occurs in 7 days at pH 7.2, but none occurred in 7 days at pH 5.6 or pH 6.5

Listeria monocytogenes

• Morphology

Gram-positive coccoid rods 0.4–0.5 by 0.5–2µm in size
Motile at 20°–25°C
Aerobe, or micro-aerophilic
Psychrotroph

• Habitat

Ubiquitous in nature, and frequently in poorly made silage, sewage sludge, vegetable matter, river water and soils.

• Types of food

Found in a range of foods, including raw vegetables, raw ground meats, raw chicken, fermented sausages, pate, raw milk, ice cream mix, ice cream bars.

Surveys of raw milk made in the USA showed 3.7% of milks contained *L. monocytogenes*. A Canadian survey in 1988 found 1.3% of milks contained this pathogen. It is suggested that infected milk typically contains one bacterium per ml.

• Significance

Immunocompromised people are the most susceptible, contracting mainly meningitis and septacaemia. Abortions can occur when pregnant women become ill.

Coleslaw, Mexican-style cheese and Vacherin Mont d'Or cheese have caused outbreaks of illness. Approximately 30% of confirmed cases of Listeriosis in these outbreaks have died. Pasteurized milk was associated with an outbreak in Massachusetts in 1983.

• Infective dose

This is unknown. It should be noted that outbreaks have extended over a period of time, and the implicated food has been examined for presence or absence of *L. monocytogenes* in 25 g or ml samples rather than for direct counts of the the bacteria. It should also be noted that in surveys of soft mold-ripened cheese when counts have been made, although the majority of cheeses have counts of less than 100 *L. monocytogenes* per g, approximately 15% have counts ranging from 10^4 to 10^6 per g.

A recent study on pathogenicity in mice by Conner *et al.*[6] has shown that for immunocompromised mice, the LD_{50} was 53 bacteria or less for 12 bacterial strains, and 480 for 1 bacterial strain. The LD_{50} was between 7.2×10^5 and 8.5×10^7 for the 13 strains examined in non-immunocompromised mice. These results appear to indicate that as far as immunocompromised humans are concerned, relatively low numbers of *L. monocytogenes* in foods would cause illness, whereas non-immuno-compromised people would be unaffected by high numbers of *L. monocytogenes*. For food, the policy of enforcing absence in 25 g or ml of food is correct to insure the safety of immunocompromised consumers.

● Heat resistance

Temperature(°C)	D values (seconds) in milk	
	Freely suspended	Intracellular
57.8	331	429
62.8	38	55.2
66.1 – sealed tubes	16.9	16.7
– 'slugflow' plant	19.1	18.4
68.9 – sealed tubes	8.6	3.9
– 'slugflow' plant	5.1	9.1
71.7	3.1	5.0
– 'slugflow' plant	0.6–2.0	Results similar to freely suspended cells
74.4	1.1	1.5

Legal HTST pasteurization (71.7°C for 15 seconds) should, therefore, give:

approximately five decimal reductions of freely suspended *L. monocytogenes*
approximately three decimal reductions of intracellular *L. monocytogenes*
or 1 per ml in raw milk would reduce to:
1 per 100 liters if freely suspended
1 per liter if intracellular

Temperature (°C)	D values (min) in food slurries					
	Chicken		Beef		Carrot	
	Strain Scott A	Strain 11994	Strain Scott A	Strain 11994	Strain Scott A	Strain 11994
60	5.29	5.02	8.32	6.27	5.02	7.76
62	2.51	2.21	4.20	2.90	3.10	3.29
64	1.56	1.84	2.19	2.21	1.47	2.50
66	0.68	0.95	0.94	0.93	0.90	0.73
68	0.38	0.41	0.35	0.27	0.30	0.44
70	0.16	0.20	0.20	0.14	0.23	0.27

● Generation times and lag periods at temperatures between 0°C and 13°C

Temperature (°C)	Generation time (hours)	Lag period (days)
0	62–131	3–33
2.5	24–45	3–8
4	18–30	
5	13–25	1–3
8	12–13	
9.3	5–9	<1–2
10	4–10	
13	3–4	

The range of values is a result of either differing growth media, or because of strain variation in different experiments.

For example, at 10°C the generation time is:
4 hours in whole milk
5 hours in skim milk
10 hours in 11% cream

At 0°C
– one strain had a generation time of 77 hours
– one strain had a generation time of 98 hours
– one strain had a generation time varying from 62–131 hours.

(Culturing of the inoculum and growth medium at 0°C appeared to affect the response.)

Work with one strain (NCTC 11994) in milk showed that at 4°C, the generation time was unaffected down to pH 5.5, but pH 5.4 gave doubled generation time and pH 5.2 prevented growth.
The lag period was also extended at pH values below 5.5.

● Parameters for development

Temperature	Minimum 0°C	Optimum 25°–30°C	Maximum 45°C
pH 30°C+	4.4–4.6	6.5–8.0	9.5
25°C	4.5		
20°C	4.5		
7°C	4.8		
4°C*	5.2		

+ In orange serum broth, minimum pH was 4.8
* In orange serum broth, minimum pH was 5.0. The lag was 15 days, then 1.5 log increase over 25 days, then a decrease at 50 days.

NaCl   *c.*10% limits growth at 30°C and 25°C over a 24 hour period. In broth, 10% at 25°C permits growth after 24 hours at pH values between 5.0 and 8.0.
10.5% allows survival for 15 days at 37°C and 30 days at 22°C
20–30% allows survival for 5 days at 37°C and 10 days at 22°C.
Between 10.5% and 30% permits survival for 100 days at 4°C.

Nitrite   100 ppm inhibits growth in presence of 3% NaCl with a pH ≤5.5. 50 and 120 ppm in salami resulted in counts decreasing 1 log cycle in 21 days (residual nitrite level was 0.5 ppm)

Salmonella

● Morphology

Gram-negative rods with peritrichous flagella 0.5–0.7 by 1–3 μm

Motile, except *Salm. gallinarum*
Aerobe, and faculative anaerobe

● Habitat

Salmonella, of which there are more than 2000 serovars, is associated with:

Domestic and wild animals
Poultry and non-domestic birds
Insects
Man

The usual habitat is the intestinal tract of the host, so water contaminated with sewage or farm manure, or manure carried off the farm by vehicles, are also sources.

Levels of approximately 100 000 per 100 g have been found in sludge used for spraying grassland. Survivors have been recovered from land for up to 30 to 40 days after spraying (UK data). In Holland, *Salmonellae* have been found in 94% of effluent samples and in 27% of gulls around sewage plants.

● Types of food

Commonly found in:

Poultry	36% of chicken carcasses 89% of turkey carcasses
Pork sausages	12%
Raw milk	70% of US milk plants, ranging from 1.5% to 25% of milk sampled at these plants.

Outbreaks have been caused by:
  Pasteurized milk in USA (1985)
  Pasteurized milk in UK (1986)
  Cheddar cheese in Canada (1984)
  Dried milk in UK (1985)
  Salami sticks in UK (1988)
  Eggs in Spain, USA and UK (1988 onwards)

These products have all caused notifiable food poisoning outbreaks, and 7 deaths were associated with the US pasteurized milk outbreak which affected 16 000 people over a 4-week period. One infant death was associated with the UK dried milk outbreak.

It should be noted that processed animal food is often contaminated with *Salmonella*, so will introduce *Salmonella* to otherwise clean farms and livestock.

● Significance

Salmonellosis is caused by ingesting viable *Salmonella*. These attack the intestinal walls, causing the symptoms of nausea, vomiting, pain, diarrhea, headache, which are the most usual found. Generally, only 5% of people affected need hospital treatment.

In the UK, there are approximately 16 000 notified cases of Salmonellosis per year, with a mortality of 40. However, certain groups of people are more vulnerable, for instance elderly people, young children and those in hospital; mortality is higher among these groups. *Salm. dublin* is more invasive than other types, and when illness results from this type, 80% of people are admitted to hospital and the mortality rate is 25%.

● Infective dose

In the US pasteurized milk outbreak, the milk contamination level ranged from 5 to 24 per ml.

In the Canadian cheddar cheese outbreak, levels of between 1.5 and 9.1 per 100 g were found in cheese recovered from patients' homes. Samples from warehouse cheese contained between 0.36 and 9.3 per 100 g.

Salmonella can survive in cheddar cheese and in powder for periods ranging from 8 months to 10 months. These findings are from naturally contaminated products.

● Heat resistance

This is not high in liquids.

The following *D* values have been found in milk using a range of *Salmonella* species:

Temperature (°C)	*D* Values (min)
60	0.058–0.098
61.5	0.043–0.070
63	0.037–0.061
64.5	0.034–0.040

The following values have been found with:
Salm. typhimurium – recovered from the US 1985 milk outbreak
Salm. senftenberg 775 W – one of the most heat resistant of the salmonellas at high a_w

Temperature (°C)	*D* Values (minutes)	
	Salm. typhimurium	*Salm. senftenberg 775 W*
60		6.3
62.8	0.11	
71.7	0.003	0.09

From these values, HTST pasteurization should give 68 log cycle destruction of most *Salmonella* species.

Salm. senftenberg would give 2.7 log cycle, i.e.
  10^6 per ml would be reduced to *c.* 10^3 per ml
  10^3 per ml would be reduced to *c.* 1 per ml

● Enhanced heat resistance

D values can be increased in the following two instances:

1. If heat shock is given to a suspension of *Salmonella*. A heat shock of 48°C for 30 minutes resulted in the survivors having the D_{59}°C value increased from 0.9 minutes to 3.9 minutes.
2. If the a_w of the food is lowered. This has been demonstrated in both chocolate and milk concentrate.

For example, in chocolate *Salm. typhimurium* has a:

D_{70}°C value of 678–1050 minutes
D_{80}°C value of 222 minutes
D_{90}°C value of 72–78 minutes

● Parameters for development

	Minimum	Optimum	Maximum
Temperature	5.1°C*	37°C	45°–47°C+

* 65/109 strains developed within 4-weeks.
+ 80/81 grew at 43°C within 18 h; the remaining culture within 2 days.

pH	4.0 (HCl and citric acid) 4.4 (lactic acid) 5.4 (acetic acid)	6.5–7.5	Can survive pH 9.0
Salt	Inhibited at >8% NaCl		
a_w	Limited at 0.95 and below		
Sodium nitrite	Inhibited at >160ppm at pH 6.0		

Note. Survival will occur in dried foods. Also, in nutrient broth with 12% salt, there was survival for 22 days at 20°C and 55 days at 5°C.

Staphylococcus aureus

● Morphology

Gram-positive cocci, spherical or ovoid in shape.
0.7–0.9 μm in diameter.
Non-motile.
When examined microscopically, characteristic grape-like cell clusters are seen.
Aerobe, and facultative anaerobe.

● Habitat

Ubiquitous in man's environment. The primary habitat is on the skin, and in the nose and throat of man and animals. A large proportion of healthy people carry *Staph. aureus*. Nasal carriers form 40–44% of the population; hand carriers vary from 14% to 40%. It is also present in cow's udder, so can be found in raw milk supplies.

● Types of food

Staph. aureus is found in a variety of foods. These include fish, fresh meat, tinned meat, cream confectionery, milk and cheese. Although much of the occurrence is due to contamination from food handlers, *Staph. aureus* can also establish itself on food processing equipment, and so heat-treated milk or food can become contaminated during manufacture if the source of *Staph. aureus* is not removed during CIP treatment of equipment.

● Significance

Approximately 60% of the strains of *Staph. aureus* produce toxin and illness is caused by ingesting food which already contains *Staph. aureus* in the food. The population of *Staph. aureus* needs to have reached a level of 5×10^6 per gram before sufficient toxin is produced. However, since the toxin is heat stable, and growth may have occurred before any heat processing stage, the food may not contain any living *Staph. aureus* cells. Ingestion of 1 ng per g is thought to be sufficient to cause illness.

Staph. aureus produces five toxins – A, B, C, D, and E

In the UK:
    toxin A is responsible for 52% of outbreaks
    toxin D is responsible for 6% of outbreaks
    A and D combined are responsible for 19% of outbreaks
    C and D combined are responsible for 9% of outbreaks

The number of outbreaks and cases of food poisoning attributable to *Staph. aureus* toxin has become less significant in recent years when compared with illness caused by *Salmonella* or *Campylobacter*. However, in 1982 there were 699 cases in the USA caused by *Staph. aureus*, and two outbreaks in the UK in 1953 were traced to milk powder containing toxin which caused over 1000 cases of illness among schoolchildren.

● Heat resistance

This shows great variation as reported by different workers. Little work has been done at 71°C, the legal HTST pasteurization temperature.

Temperature (°C)	D values in milk
61.7 (143°F)	20 minutes
62	20 to 65 seconds (a mixture of strains was heated)
71.7	4.1 seconds

When cultures were aged, i.e. in the stationary phase, the D values increased threefold, i.e. 20 seconds to 60 seconds – 65 seconds to 195 seconds.

Toxins. These are more heat stable than *Staph. aureus*, and are not destroyed by milk pasteurization. However, in general the times and temperatures applied in normal canning processing are sufficient to inactivate the amount of enterotoxin which, if present in foods, would cause an outbreak (0.5–10µg per 100 g food).

- Parameters for development

	Minimum	Optimum	Maximum
Temperature	11°C (also reported at 7°C)	37°C	48°C
pH	4.0	6–7	9.8–10
a_w	0.86 (0.83) (Generation time is 300 minutes)	0.98	0.99
Eh	−220 mV	+200 mV	+200 mV

- Parameters for toxin production

	Minimum	Optimum	Maximum
Temperature	10°C	40°–45°C	48°C
pH	4 (limited at 5)	7–8	9.6
a_w	0.85	0.98	0.99

The following data illustrate how temperature affects the time taken for detectable toxin production, both when *Staph. aureus* is grown in low-count media (without competition) and when grown in high-count milk (with competition from other microorganisms).

Temperature (°C)	Low-count milk and broth	High-count milk
35	in 12 hours	Toxin detected – no time given
30	in 18 hours	none detected
25	in 24–36 hours	none detected
20	in 48–96 hours	none detected
17.8	in 8 days	
16	in 21 days	

Vibrio parahaemolyticus

- Morphology

Gram-negative pleomorphic rod – straight or curved with flagellum (occasionally peritrichous flagella) 0.65 by 1.3µm
Aerobe and facultative anaerobe
Halophile

- Habitat

Found in coastal sea waters when the water temperature is 15°C and above. It cannot survive at the pressures found in deep water. It may be isolated from sediments near the coast even when the water temperature is below 10°C. Distribution in coastal waters appears to be world wide.

- Types of food

Raw fish and shellfish are commonly associated with *V. parahaemolyticus* outbreaks in Japan. Counts are low on freshly caught fish and shellfish, but multiplication can occur on seafood at temperatures between 12°C and 30°C.

- Significance

This pathogen is only likely to cause illness when raw fish or shellfish are eaten following storage at temperatures above 4°C. Pathogenicity of *V. parahaemolyticus* strains is linked to their being hemolytic and producing a thermostable extra cellular hemolysin. Such strains are known as Kanagawa positive.

Illness is caused by infection, and in Japan this pathogen caused between 50% and 70% of bacterial food poisoning cases from the early 1960s to the mid-1970s. Outbreaks also occurred in Europe and the USA in the mid-1970s.

- Infective dose

This has not been established satisfactorily, although trials with volunteers indicated that 10^{10} cells per ml of Kanagawa-negative strains did not cause illness among volunteers. Among isolates from seafood and marine environmental samples, Kanagawa-positive strains formed <5% of those examined. Since it is assumed that only Kanagawa-positive strains will cause illness, it seems probable that an infective dose may be no more than 10^4 cells.

- Heat resistance

Detailed work is scant. A 3 to 4 log cycle reduction in numbers occurred in peptone – 3% NaCl medium after 5 minutes at 60°C. At 100°C, a 6 log cycle reduction in numbers occurred in a shrimp homogenate – 3% NaCl slurry after 1 minute. The heat resistance is increased when salt is present in the broth in which the inoculating cells are grown – levels of 3 or 7.4% NaCl increase heat resistance compared with cells grown in 0.5% NaCl.

- Parameters for development

	Minimum	Optimum	Maximum
Temperature	12.8°C	37°C	42°–43°C
pH	4.5–5	7.5–8.5	11
Salt	0.5%	3%	8%
a_w	0.948 (with salt as solute)	0.992	

The generation times in seafood range from 13 minutes at 37°C to 15 to 18 minutes at 30°C.

Viruses

Viruses are very small living particles weighing 10^{-17} g, which are unable to reproduce themselves. They can therefore only be carried by food, and will not replicate in it. It is only when the viruses carried by food are pathogenic to man that illness may result.

Historically, viral illnesses that have been shown to be food-borne include:

Tickborne encephalitis: in goat's milk

Poliomyelitis: in cow's milk

Hepatitis A: in water, milk, oysters, clams, potato salad, orange juice, whipped cream cake, watercress, cold meats, salads, cockles, ice cubes, sandwiches

Gastroenteritis: in shellfish, butter cream icing, green salads

Hepatitis A and gastroenteritis are still of importance as far as the UK and the USA are concerned. In the UK, there are approximately 400 cases per year of Hepatitis A caused by viral infection; gastroenteritis cases total approximately 11 000 per year, having increased from approximately 4500 cases per year in the early 1980s.

The type of food implicated usually dictates the route of viral contamination, as the following examples illustrate.

With shellfish it is likely to be waterborne, the virus getting into the water in the first place from sewer outlets, and the shellfish then retains the virus in its digestive apparatus. There is no virus replication in shellfish.

With watercress and green salad material, contamination is likely to be from where the crop grows, or from irrigation if this is done with sewage or polluted water. There is no evidence that viruses active against man have any action on plants. With other foods, it is likely to be from an infected human handling the food, and transmitting the virus via unwashed hands. This possibility of human transmission is less likely to occur in food factories than in catering establishments, but must be guarded against in factories where a lot of handling of finished product occurs, e.g. frozen gateau manufacture, sandwich assembly and small-scale speciality salads preparation.

Insects and contaminated food contact surfaces have also been shown to cause virus contamination of foods.

● Viruses associated with food poisoning

Hepatitis A	20–30 nm in size, icosahedral form and has no envelope. Single strand of RNA.
Gastroenteritis	
Small round structured virus (includes Norwalk virus)	25–30 nm in size, round, with cup-shaped depressions on surface. Undetermined nucleic acid type. Has been described as 'like Parvo virus', with exception of Norwalk.
Parvo virus	20–22 nm in size. Single-stranded DNA. Needs special conditions to replicate in human cells, but reported as present in diarrheal stools.
Rotavirus	65–75 nm in size, double coat of protein surrounds segmented double-stranded RNA. Several known serotypes.
Calicivirus	Approximately 30 nm, cup-shaped depressions on surface. Single-stranded RNA. Reported to have been observed in human diarrheal stools.
Adenovirus	70–75 nm, icosahedral, with knobbed fibers projecting from vertices. Single large molecule of double-stranded DNA. Associated with respiratory and conjunctival infections but also isolated from diarrheal stools. 41 known serotypes.
Astrovirus	28 nm, spherical, but shaped figures are seen on the surface. Undetermined nucleic acid type, but an astrovirus from lamb contains single-stranded RNA. Five serotypes reported. Detected in human diarrheal stools, and illness associated with children under one year of age.

● Infective dose

This is unknown, but thought by some authorities to be low, possibly of the order of 100 virus particles per person infected. It is known that susceptibility to infection varies from individual to individual, and also varies between groups of people.

● Inactivation of viruses

Heat
There is very little reliable quantitative data on heat inactivation. In milk, when four viruses were

inoculated to give 10^4 plaque-forming units per ml, heating for 30 minutes at 55°C gave complete inactivation.

> Cliver, an authority on Food Viruses, says: generalizations are difficult, but necessary . . . in the absence of extremely high or low levels of solute and of extreme values, the time for 90% inactivation of viruses that emanate from the human intestines are likely to be measured in:

Seconds   at temperatures above 60°C
Minutes   at temperatures above 50°C
Hours   at temperatures above 40°C
Days   at temperatures above 30°C
Weeks   at temperatures above 10°C

However, it appears that Hepatitis-A virus is more resistant, and at 60°C in serum it was not inactivated after 60 minutes. In water, a reduction occurred at this temperature/time combination, and inactivation at 100°C after 5 minutes. The initial titer of virus was not given.

- Chemical disinfectants

Viruses are regarded as being more resistant than bacteria, but detailed information is sparse. Chlorine, when used in water treatment, is effective against viruses, as is chlorine dioxide. Quaternary ammonium compounds are not very effective.

- Irradiation

Cobalt–60 gamma rays of approximately 4 kilograys (0.4 Mrad) will cause 90% inactivation of viruses.

Yersinia enterocolitica

- Morphology

Gram-negative rod
0.5–1.0 by 1–2µm in size
Motile at 22°C
Aerobe
Psycrotroph

- Habitat

Found in water (thought to need the presence of organic matter to survive), pigs, small rodents, and house pets.

- Types of food

Milk – both raw and pasteurized
Milk products
Egg products
Raw meat – especially ground pork
Poultry
Vegetables

- Incidence in milk

	Raw	Pasteurized
Northern Ireland (1986)	11.3% (150 samples)	6% (100 samples)
Denmark (1982)	10% (251 samples)	
USA (1982)	12% (100 samples)	1% (100 samples)
Canada (1978)	18.2% (unknown number of samples)	

- Significance

Many of the serotypes found in environmental and food habitats are either not those serotypes associated with *Y. enterocolitica* infections or are not the virulent strains of the serotypes isolated from outbreaks. However, epidemiological studies in Belgium of sporadic cases found in the community show an association between consumption of raw pork and illness, especially among young children and elderly people.

In the USA there have been four food associated outbreaks since 1976. The largest implicated pasteurized milk, and some thousands of people were affected but illness was thought to be caused by *Y. enterocolitica* surviving and multiplying on the outside of milk cartons in chill storage. Milk contamination was presumed to occur as the cartons were opened and poured.

- Serotypes associated with illness

Serotypes 0:3 and 0:9 are those implicated in Europe, Eastern Canada, Japan and South Africa.

0:8 is implicated in the USA and western Canada.

- Symptoms

Gastroenteritis is the main symptom often accompanied by acute abdominal disorders and arthritis.

- Infective dose

Unknown, but infections peak in autumn and winter months in Europe.

- Heat resistance

Temperature (°C)	*D* values
62.8	0.7–17.8 seconds; 0.18–0.96 minutes (both these sets of data in whole milk)
68.3	5.4 seconds (milk)

- Parameters for development

	Minimum	Optimum	Maximum
Temperature	0°–1°C	32°–34°C	44°C
pH	4.6 at 25°C*	7.0–8.0	9.0+

* No growth at this pH at 3°C.
+ 2 log cycles growth

NaCl (broth experiments)

5% allowed a maximum 2 log cycle increase after 10 days at 3°C
7% resulted in death at 3°C

- Growth rate at 3°C (broth experiments)

2 log cycles in 4 days, i.e. 1 to 100 bacteria per ml.
5 log cycles in 7 days, i.e. 1 to 100 000 bacteria per ml.

Emerging pathogens

The following pathogens are known to cause food or waterborne illness, although not all of them have caused large, well-defined outbreaks.

Bacteria

Shigella Comprising *Shigella dysenteriae*, *S. flexneri*, *S. boydii* and *S. sonnei*. Although some outbreaks of Shigellosis are food-borne, investigation of US outbreaks found that the main contributing factor was poor personal hygiene of food handlers at the place of consumption. The number of recorded cases arising from food-borne outbreaks in the USA and the UK is low.

Escherichia coli Although invasive *E. coli* can cause an illness resembling Shigellosis, it is toxigenic *E. coli* which are mainly responsible for food poisoning. Enterotoxinogenic *E. coli* (ETEC) have been found in Brie cheese which caused food poisoning; they can also be waterborne. Entero-hemorrhagic *E. coli* 0157:H7 can cause hemorrhagic colitis, and outbreaks arising from food have occurred in the US, Canada and the UK. The pathogen has been isolated from beef, pork, lamb and poultry.

Aeromonas hydrophila A psychrotroph associated with sporadic cases of diarrhea in the very young, or in older adults. Two US outbreaks were linked with raw oysters originating from the same growing area. In both the US and the UK, retail foods including seafoods, red meats and poultry all carried *A. hydrophila*. *A. hydrophila* and *A. sobria* have been found on retail lamb and offal in Australia.

Both species produce haemolysin and toxin, indicating their link to pathogenicity.

Bacillus subtilis and *Bacillus licheniformis*. Both these bacteria have been implicated in outbreaks over the last 10 years, and a range of foods were involved. For *B. subtilis* in the UK, there have been 49 episodes, with at least 175 cases between 1975 and 1986. The counts in 30 foods examined ranged from 10^5 to 10^9 per g. For *B. licheniformis* over the same period, 24 episodes have occurred with at least 218 cases. The counts in 10 foods examined ranged from 1.7×10^6 to 2×10^9 per g. An outbreak in the US in the early 1960s reported *B. subtilis* as responsible, although test results indicated that *B. licheniformis* may have been a more likely identification. 161 cases were involved in this outbreak.

Protozoa

Giardia lamblia is mainly water-borne, and cases of illness in England and Wales have risen from approximately 1000 in 1969 to approximately 6000 in 1987. In 1985 there was a UK outbreak traced to a water main repair, and in 1986 in the US there was an outbreak in which fruit salad acted as the transmitter. Its origin was from a woman who changed the diaper of a child who was a symptomless excretor.

Cryptosporidium parvum may be water-borne or carried in raw milk. It causes diarrhea and vomiting and the first human case was reported in 1976. In 1989 there was a widespread outbreak caused by the oocysts being present in a regional water supply. Comparative figures for illness caused by *Cryptosporidium* in England and Wales were 5000 in 1989 and 1600 in 1988. The oocysts are resistant to chemical disinfection, and they can survive in the laboratory for up to one year in aqueous solutions though they are likely to be killed by desiccation.

The food manufacturer needs to know these pathogens, and also to be aware that there are others including what are regarded as 'suspect' pathogens, e.g. *Plesiomonas shigelloides* or certain enterococci, or an invasive and potentially lethal pathogen, e.g. *Vibrio vulnificus*. At the moment, these pathogens appear to be responsible for only a few cases of illness, but with better epidemiological information and diagnostic methods it can be expected that some of these pathogens will become usual, rather than less usual. As far as is known from investigations already made, the properties of growth or survival are such that established processing and GMPs should be effective in controlling these pathogens.

References

1. D'Aoust, J.-Y., Warburton, D. W. and Sewell, A. M. (1985) *Salmonella typhimurium* phage – type 10 from cheddar cheese implicated in a major Canadian foodborne outbreak. *Journal of Food Protection*, **48**, 1062–1066

2. Mitscherlich, E. and Marth, E. H. (1984) *Microbial Survival in the Environment*. Berlin: Springer-Verlag
3. Doyle, M. P. (ed) (1989) *Foodborne Bacterial Pathogens*. New York: Marcel Dekker
4. Hauschild, A. H. W. (1989) *Clostridium botulinum*. In *Foodborne Bacterial Pathogens*, edited by M. P. Doyle, pp. 111–189. New York: Marcel Dekker
5. Dega, C. A., Goepfert, J. M. and Amundson, C. H. (1972) Heat resistance of *Salmonellae* in concentrated milk. *Applied Microbiology*, **23**, (2), 415–420
6. Conner, D. E., Scott, V. N., Sumner, S. S. and Bernard, D. T. (1989) Pathogenicity of foodborne, environmental and clinical isolates of *Listeria monocytogenes* in mice. *Journal of Food Science*, **54**, (6), 1553–1556

Further reading

Christensen, M. L. (1989) Human viral gastro enteritis. *Clinical Microbiology Reviews*, **2**, 51–89

Cliver, D. O. (1988) Virus transmission via foods; a scientific status summary by the Institute of Food Technologists' expert panel on food safety and nutrition. *Food Technology*, **42**, (10), 241–248
Cliver, D. O. (ed) (1990) *Foodborne Disease*. San Diego, CAL, London: Academic Press
Food Focus No. 11 (1990) *The Emerging Pathogens Yersinia, Aeromonas and Vertoxigenic E. coli. A* literature survey. British Food Manufacturing Industries Research Association, Leatherhead, UK
Halligan, A. C. (1989) *Food Focus: Salmonella and Listeria – Factors Affecting Their Growth and Survival in Foods*, British Food Manufacturing Industries Research Association, Leatherhead, UK
Jackson, G. J. (1990) Public health and research perspectives on the microbial contamination of foods. *Journal of Animal Science,* **68**, (3), 884–891
Miller, A. J., Smith, J. L. and Somkuti, G. A. (eds) (1990) *Foodborne Listeriosis*, (Topics in Industrial Microbiology). Amsterdam: Elsevier, on behalf of the Society for Industrial Microbiology.

Chapter 11
Spoilage – including enzymes and their importance

11.A Notes on spoilage

Introduction

Spoilage may be defined as changes in sensory qualities of a product which make it unacceptable or unfit for its intended purpose. Spoiled food is not necessarily injurious or pathogenic, it may simply be unappetizing or unpleasant. Sensory attributes – the 'Quality' aspects or 'Elegance factors' – are the proper concern of the processor whereas regulatory authorities whose duty is to safeguard Public Health have a direct interest in issues concerning food poisoning. Spoilage which has no relation to food safety includes yeast fermentation of fruit juice, fruit yogurt or acetic acid pickles. Furthermore, spoilage may not necessarily result from a failure to apply recognized Good Manufacturing Practices (GMPs) which are general principles determined by the 'state-of-the art'. Spoilage will occur because of process/package/distribution design and/or control failures when the food was under the direct control of the processor; and/or product abuse or misuse when it has left the direct control of the processor.

Be aware that spoilage may result from one or more causes, i.e. two or more things must 'go wrong' together for it to happen. Be aware also that spoilage is not always the result of microbial changes, e.g. softening of pickles or gelation of UHT milk can be enzymatic changes. Ordinary chemical reactions may cause spoilage, e.g. color changes in sensitive fruit products which react with atmospheric oxygen. However, microbiological spoilage forms a substantial portion of the total.

It is important to recognize that the Longitudinally Integrated Safety Assurance (LISA) strategy applies just as much to spoilage organisms. The reason for this is that safety and spoilage are two closely similar aspects of microbial ecology. The difference is that the number of types or kinds of organisms which can cause food spoilage is greater by orders of magnitude compared with those which cause food poisoning. Additionally, spoilage organisms can be remarkably robust and it is quite usual for adaption or 'training' to occur. A plant (factory) develops its own microbiological flora (or population) which can have an unusually high resistance to ingredients, e.g. salt, sugar or vinegar used in the processes. Be aware that not all strains of all organisms have been recognized or described and that experienced microbiologists know of spoilage incidents where the causative organism was growing actively at or beyond the published limits.

In any realistic commercial process, there is a 'margin of safety' to insure that for any individual lot or batch the chance of spoilage is very low. Such 'margins of safety' depend substantially on the quality of judgment and on the experience of the microbiologist, together with the willingness of management to apply this judgment.

Intuitive responses to spoilage

What follows is a gross oversimplification and to some extent something of an exaggeration or caricature. However, it is important to recognize that there are legitimate differences in approach to a spoilage incident which can be expected to happen from time to time. Consider the responses of Production Management and the Microbiologist.

- Production management
  Aim – put it right as soon as possible.

Methods – try anything or everything that might work (or help).

Afterwards – when it is 'cured' – aim to reduce, as quickly as is feasible, any increase in standard production costs.

- Microbiologist

Aim – find the cause (or causes) of the spoilage – how did it happen; was it internal, attributable to process/packaging/distribution design failure or failure in application, or was it externally caused by abuse or misuse?

Methods – take the HACCP route to root-cause corrective (remedial) action.

Afterwards – record results of investigation as guide to future action. If a number of changes were made simultaneously, investigate which were the most effective, so that a temporary or permanent change may be made to 'margins of safety' as appropriate.

An approach for the microbiologist involved in an incident

The most important rule is *Don't panic* but remember that what is wanted is usable information that enables the right corrective (remedial) actions to be taken quickly. Obtain data and assess the extent and nature of the problem. It has been known for lot markings or codes to be misread. Actual numbers of affected containers or packages are important – it is rarely true that 'they are all spoiled'. The distribution of affected containers within a shipment may well be an early clue to the cause of spoilage.

Obtain and assess data on processing and distribution of affected code lots. This is when the ability to pinpoint small intervals of time between code changes is really appreciated. Estimate how likely it is that other lots are affected and determine appropriate management action. Remember, there is good advice on pages 4 to 7 of Campden Technical Manual 18[1] which ends 'Look for patterns, trends and common data but do not jump to conclusions. Try to establish the best direction in which the investigation is to proceed.'

Microbiological data are important *but must* be put into the overall context. To do this *all* pertinent data are needed. This includes but is not limited to:

- The pattern of reported spoilage – how it was discovered, who made the discovery, the known extent of the problem, any possible associated illness.
- Product composition of spoiled and adjacent codes of unspoiled product.
- Packaging, composition and integrity of spoiled packs and adjacent codes of unspoiled product.
- Processing records including ingredient, processing, filling and/or packaging data. (These

should be available from previously identified critical control points.)
- Storage and distribution records

Collecting information for microbiological assessment usually begins with an examination of a spoiled pack or packs. To begin with, it may be very limited, e.g. abnormally high numbers of organisms seen on microscopic examination. These can be significant even if they are non-viable. At this point in the investigation if is prudent to consider the realistic options and to be aware of the implications of the time it will take to get test results. Always remember that what is needed is usable information for management to take rapid corrective action. It may therefore not be necessary, for example, to continue identification beyond the presumptive level. The focus of interest in a spoilage investigation is on the behavior of any suspected organism, i.e. its physiology, rather than in its identity. Remember, identification can always be done at a later date if cultures are kept.

As information becomes available there will be a need to review HACCP analyses. At this point advice from those with experience with the causative organism(s), e.g. in Research Institutes or Associations, can prove valuable. It is unfortunate, but those with the most experience are likely to be in direct competition, so the extent of available help is limited. This illustrates the importance of good laboratory reports so that a body of 'in-house' information is built up over the years.

The appropriate microbiological methodology will vary greatly according to circumstances. However, Campden Technical Manual No. 18[1], Speck[2], Samson and Van Reenen-Hoekstra[3], King *et al.*[4], and Skinner *et al.*[5] may form a suitable starting point for your laboratory's essential collection of books and reprints. Remember that when under the pressure of a spoilage investigation, it is hard to evaluate calmly the limitations of new methodologies. It is also helpful to have made previous contact with suitable contract-work laboratories since time is valuable and you will want to make arrangements as quickly as possible.

The prudent microbiologist keeps alert for warning signs that spoilage may be developing and takes rapid action to contain an incident.

References

1. Brown, K. L. (1987) Technical Manual No. 18. *Examination of Suspect Spoiled Cans and Aseptically Filled Containers*, Campden Food Preservation Research Association, Chipping Campden, Gloucestershire GL55 6LD
2. Speck, M. L. (ed) (1984) Compendium of Microbiological Methods for the Examination of Foods, 2nd edn, Washington, DC: American Public Health Association

3. Samson, R. A. and Van Reenen-Hoekstra, E. S. (1988) *Introduction to Food-borne Fungi*, 3rd edn. Baarn, The Netherlands: Central bureau voor Schimmelcultures
4. King, A. D. Jr., Pitt, J. I., Beuchat, L. R. and Corry, J. E. L. (eds) (1986) *Methods for the Mycological Examination of Food*. New York/London: Plenum Press
5. Skinner, F. A., Passmore, S. M. and Davenport, R. R. (eds) (1980) *Biology and Activities of Yeasts*, Society for Applied Bacteriology Symposium Series No. 9. London: Academic Press

11.B Notes on enzymes and their importance

These are intended primarily as 'background' or 'training' material.

Introduction

Enzymes are complex molecules present in all living cells and are usually regarded as biological catalysts, because they speed up reactions without being permanently changed themselves. Enzymes are proteins which may be simple or conjugated. The activity of simple protein enzymes depends only on the protein structure whereas conjugated enzymes requires the presence of a non-protein molecule called a cofactor. The cofactor may be a single metallic ion or a complex organic molecule called a coenzyme. Some enzymes are active only in the presence of both types of cofactors. Enzymes and cofactors are linked together with different degrees of affinity. They can be separated easily, by dialysis for instance, or linked covalently in which case they are called prosthetic groups. The entire enzyme complex is called an oloenzyme; after removing the cofactor, the protein moiety (or part) is called an apoenzyme. Cofactors are usually heat stable but the protein moiety of the molecule is usually heat labile.

Enzymes accelerate both forward and reverse reactions thus achieving equilibrium more quickly through formation of an unstable intermediate having higher energy than either the initial substrate or the final product. The reaction constant can be expressed by the Arrhenius equation:

$$k = A * exp\,(-E_a/(R * T)) \qquad (1)$$

where:

k = reaction rate constant
A = frequency factor (a thermodynamic constant)
E_a = Arrhenius activation energy
R = universal gas constant
T = absolute temperature in degrees Kelvin

(Note, in this section the * sign means multiplied by.)

It is therefore possible to compute the acceleration brought about by an enzyme, when the reaction rate in the presence and absence of the particular enzyme is known. Taking as an example the decomposition of hydrogen peroxide, which without catalysis (k) has an activation energy (E_a) of 70 kJ/mol, but which in the presence of the enzyme catalase (k') has an E_a value of 7 kJ/mol, a very high value is obtained.

$$k'/k = exp\,((70 - 7)/RT)$$
$$= exp\,(63/2.5)$$
$$= 8.8 * 10^{10}$$

where $R = 8.314$ J/mol $* T$ which is the absolute temperature in degrees Kelvin.

The turnover number, i.e. the number of substrate molecules transformed in one minute by a single enzyme molecule, usually ranges between 10^3 and 10^7.

Specificity

Single enzymes have different degrees of specificity and they may act on different but closely related substrates. In most cases they catalyze one reaction only. It follows that living cells must contain a vast number of enzymes in order to carry out the very great number of vital metabolic activities, and it can be inferred that each living cell contains about a thousand molecules each of about 3000 types of enzymes, i.e. about three million enzyme molecules per cell.

The specificity of enzymes may be:

- Low, if the enzyme activity is directed only towards the linkage to be split, e.g. an ester linkage, irrespective of the type of acid and/or alcohol.
- Group specific, when the enzyme may act upon a substrate having a specific linkage and a specific group on one side of the linkage, e.g. trypsin cleaves only peptide linkages involving the carboxyl group of lysine or arginine.
- Stereochemical, where enzymes show a strong preference for one member of a pair of stereoisomers, e.g. lactic dehydrogenase oxidizes L(+)−lactic acid to pyruvic acid, but has no effect upon D(−)−lactic acid.
- Absolute, where only one substrate is modified by each enzyme, e.g. urease splits urea into water and ammonia; or maltase hydrolyzes maltose to form two molecules of alpha-D-glucose. This is the most exclusive type of specificity and the most widespread among enzymes.

Effect of temperature and pH

Most enzymes are active in the temperature range 20° to 50°C, with the maximum rate at temperatures close to 30° to 35°C. Reaction rate decreases at either side of the optimum value. At lower temperatures the $Q(10)$ is about 2, i.e. the rate of reaction doubles for an increase of 10°C. At higher temperatures the protein moiety of enzymes is denatured with a rate increasing with temperature but the $Q(10)$ value is again about 2. Some enzymes are still active at 70°C, but most are inactivated at temperatures just about 60°C. Enzymes treated at very high temperatures (greater than 125°C) for very short times may lose activity only temporarily. They may regain activity (reactivation or regeneration) after storage at a suitable temperature, often around ambient.

Most enzymes have an optimum pH at which the reaction rate has the highest value, although some are equally active irrespective of pH. For most enzymes the optimum pH lies between 4.5 and 8. There are exceptions, e.g. pepsin has an optimum pH at 1.8; arginase at 10. Sometimes, the optimum pH coincides with the isoelectric point of the protein. Typically, significant enzyme activity usually occurs within a narrow range of pH (optimum pH ±2 pH units); outside this range reaction rate decreases until it stops.

Enzyme kinetics

The interrelationship of enzyme and substrate concentrations is of paramount relevance in order to achieve a defined reaction velocity. In suitable environmental conditions and at a given high substrate concentration the rate of enzymatic activity increases as the enzyme concentration rises, with the substrate being in excess. Beyond a certain point there is no increase in reaction rate for a rise in enzyme concentration since the concentration of the substrate becomes the limiting factor. Also, for any given enzyme concentration, the enzymatic activity increases as the substrate concentration rises until a limiting substrate concentration is reached, beyond which the reaction rate of the system does not increase further (Figure 11.1).

These effects are explained in terms of the formation of an enzyme-substrate complex acting as a limiting factor in the enzyme reaction:

$$S + E \rightleftharpoons ES \rightleftharpoons E + P$$

where S is the substrate, E is the enzyme and P is the product. On this basis, the kinetic approach leads to the fundamental Michaelis–Menten equation

$$V = (V_{max} * [S])/(K_m + [S])$$

where:

V is the reaction velocity
V_{max} is the maximum velocity that can be reached with the available amount of enzyme
S is the substrate concentration
K_m is the Michaelis–Menten constant measured in moles per liter.

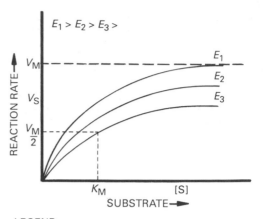

LEGEND
REACTION RATE V_S
MAXIMAL RATE V_M
HALF VALUE OF THE MAXIMAL RATE $V_M/2$ WHERE E_1, E_2 AND E_3 ARE ENZYME CONCENTRATIONS

Figure 11.1 Enzyme kinetics

We can use this equation to show that when $V = V_{max}/2$, it becomes:

$$V_{max}/2 = (V_{max} * [S])/(K_m + [S])$$

and $\frac{1}{2} = [S]/(K_m + [S])$

so that $K_m + [S] = 2 * [S]$

so $K_m = [S]$

i.e. K_m is equal to the substrate concentration at which half of the enzyme maximum reaction velocity is developed.

Different enzymes have different K_m values which change with pH, temperature and substrate. The degree of affinity between an enzyme and a substrate is inversely related to the K_m value.

Functions of enzymes

In living cells, enzyme activity carries out all reactions required for both metabolic and catabolic purposes, e.g. they are involved in transport mechanisms of nutrients inside the cell, catabolism of nutrients, anabolism of new cellular materials, cell multiplication, movement and energy production.

Food products may be affected by enzymic degradation of carbohydrates, lipids and proteins. The specific degradation involved characterizes the spoilage and mainly derives from the microbial flora found in the product. For example, polysaccharides are broken down into low molecular weight compounds (oligosaccharides, disaccharides, monosaccharides), and in turn monosaccharides are oxidized to organic acids (lactic, acetic, formic), ethanol, carbon dioxide and water. Lipids are hydrolyzed to free fatty acids and glycerol. Proteins are firstly degraded into oligopeptides, peptides and free amino-acids, and then into final degradation products, e.g. hydrogen sulfide, ammonia, indole, scatole, etc.

Physico-chemical factors affecting microbial metabolism and growth also influence the activity of degradative enzymes.

The use of enzymes in food processing

The use of enzymes is of growing importance in food processing. An important reaction is the hydrolysis of biopolymers to a degree which may be slight or complete, e.g. the preparation of crystalline glucose from starch.

The use of enzymes in food processing is, however, very ancient. For example, the hanging of meat results in autolytic changes which improve texture and flavor before spoilage occurs. Rennet has been used traditionally in cheesemaking. This is a saline extract of the abomasum or fourth stomach of calves usually slaughtered before they are 30 days old. It is a mixture of rennin (or more correctly chymosin) and some pepsin. More recently, 'vegetable rennet' – which comprises enzymes derived from fungi – has been used to prepare vegetarian cheeses. The uses of enzymes in yeast to ferment grape juice, produce beer from grain and to raise bread are also very ancient.

Enzymes may be used in a number of forms, e.g. as liquids or as microfine powders agglomerated by a number of encapsulating or aggregating techniques. For instance, detergent enzymes are supplied to washing powder manufacturers in agglomerated form. Immobilized enzymes are also used. There are two forms available, in one the protein molecules are attached to an inert support, e.g. glass; titanium dioxide; resin particles; or on the inner surface of narrow bore nylon tubes. Alternatively, the enzyme may be held within a cell in which the cell wall is treated chemically to improve both mechanical strength and substrate permeability. The cells may then be packed in a column and given a controlled substrate flow.

An example of this is immobilized glucose isomerase used to produce 'High Fructose Corn Syrup'. This syrup has up to 90% fructose, is very sweet and a large tonnage is produced annually in the USA. Other important enzyme applications include pectic enzymes in fruit and vegetable juice production. They are used to produce both cloudy and clear fruit drinks.

The 'Hot-break' process for tomato juice and paste, pioneered by Heinz, depends on inactivation of the very fast acting pectic enzymes in the tomato fruit in order to retain the pectin which is important in ketchup production.

Conclusions

These examples indicate that enzymes have widespread and important applications in food processing. Because they are such potent materials they must be used with care, and with the appreciation that there may be significant 'Health and Safety' aspects which need management attention.

Because enzymes are a 'foreign protein' in the human body, plant employees (operators) can be exposed either to a smoke of microfine powder or an aerosol or splashes from liquid enzyme preparations. Generally, the exposure has no obvious effect but some people are, or may become, highly sensitive to these materials because of an immune response. These effects are mainly respiratory, ophthalmic and/or dermatological.

Further reading

1. Fersht, A. (1985) *Enzyme Structure and Mechanism*, 2nd edn. New York: W. H. Freeman
2. Cornish-Bowden, A. and Wharton, C. W. (1988) *Enzyme Kinetics*. Oxford/Washington, DC: IRL Press
3. Birch, G. G., Blakebrough, N. and Parker, K. J. (eds) 1981 *Enzymes and Food Processing*. London: Applied Science

Chapter 12
Aspects of microbiological safety in food preservation technologies

Introduction

The notes in this chapter are primarily for product developers and microbiologists who are moving out of familiar product/processing/packaging/distribution systems. When doing this it is *essential* to remember that microbial behavior is not open to negotiation, unlike 'costings' or 'marketing concepts'. Microbes do not see, hear or respond to human pleas. They have no respect of persons or hierarchical status. What they *can and will* do is to breed rapidly unless positive preventive action is taken. Pathogens should *always* be assumed to be present as well as spoilage organisms and appropriate control measures taken.

These brief notes outline the background and draw attention to some points which experience in the technologies has shown to be significant in the control of microorganisms. Because markets and technologies are always changing, the notes are one of a number of starting points in the development of safe, wholesome food manufactured and distributed under TQM.

Thus, when beginning product development, the technical reasons determining safety of the product must be known and systems (often called Quality Systems) developed and implemented to deliver a positive assurance of product safety. Remember, a *positive assurance is required* which should then be successfully demonstrated to a technically knowledgeable and impartial auditor.

This calls for the use of the Longitudinally Integrated Safety (LISA) strategy, implemented as in the 'equation':

$$\text{LISA} = \text{HACCP} \rightleftharpoons \text{AUDIT}$$

where HACCP is Hazard Analysis Critical Control Point and AUDIT includes compliance and systems audits.

When developing systems for safety (and quality), you will need to know and implement pertinent legal requirements and technically based Good Manufacturing Practices (GMPs). As a general GMP guide, there is a useful Institute of Food Science and Technology (UK) publication *Food and Drink Manufacture – Good Manufacturing Practice: A Guide to its Responsible Management*). The 2nd edition of 1989 includes a list of Codes of Practice, Guides and Standards as an Appendix. A recent (1990) US publication, *Current Good Manufacturing Practices (CGMP's)/Food Plant Sanitation* by Wilbur A. Gould includes helpful extracts from the current Code of Federal Regulations in force in the US and an overview of sanitation which would be valuable as training material for plant management and supervision. It is published by CTI Publications Inc., Baltimore, Maryland 21218-4576. *Mechanisms of Action of Food Preservation Procedures*, edited by G. W. Gould, provides valuable background for the technologist. It was published in 1989, by Elsevier Applied Science, London and New York.

12.A Chilled foods

Introduction

Over the last five years the market in chill, or refrigerated, foods has shown a marked increase in Europe, the USA and Japan. The increase has been both in terms of value, including shelf space in the supermarkets, and in the range of products available. These include those sold directly to the consumer to take home, and Foodservice (catering) sales, where chilled products are sold in delicatessens, 'work place' cafeterias, restaurants or institutions for consumption on the premises. Perhaps the groups showing the greatest increase are ready meals ranging from pizzas and pasta dishes to complete meals with separate vegetable and meat or fish components; salads, with both mayonnaise or vinaigrette dressings and unprepared vegetables; sandwiches with fillings which include fish, meat, egg or vegetables with or without a mayonnaise dressing; and desserts ranging from more traditional fermented milks to fromage frais (often with fruit included) to multicomponent, multilayer desserts with cream or custard as a major ingredient. This variety contrasts with the meat, fresh fish and dairy products (milk, butter, cheese and cream) which were the traditional produce in the chill cabinet.

Product range expansion satisfies two expectations – that of the consumer for convenience foods, which save time when preparing meals in a household where the adults are likely to be employed outside the home, and that of the manufacturer for more added-value products. As well as convenience, consumers also expect 'freshness' in their foods. Freshness is regarded as being healthier than, for example, traditional heat processed foods and the sensory quality will be more 'natural'. The trend is therefore away from overtly processed foods; consumers often do not appreciate that chill foods are in fact processed. Healthier foods, in the consumer's view, should not contain 'chemical' preservatives, and this has resulted in lower levels of salt and nitrites in preserved meats. Another trend in the chill product market has been towards extending the shelf life. This combination of multicomponent products, less severe heat processing, reduction or removal of preservatives and extended shelf life has resulted in a large number of chill products being categorized as high risk. These are defined in the *Guidelines for Good Hygienic Practice in the Manufacture of Dairy Based Products*[1] as those foods that:

- allow growth or survival of contaminating micro-organisms which may jeopardize the safety or quality of the food.
- are handled or exposed to the factory environment after heat treatment
- are likely to be consumed with no further heating.

The manufacture of these 'high risk' products should take place using stringent control measures, including:

- strict segregation of raw materials;
- 'high care' areas (see 'Prevention of cross-contamination' in Chapter 6.A) for the handling of product immediately after processing and during assembly and packaging of multicomponent foods;
- correct temperatures being maintained throughout storage and distribution.

Ideally, the 'high care' area should be maintained at a temperature of 12° to 14° and the air supply should be filtered to Class I. The purpose of these measures is to reduce the microbial load, both of pathogens and spoilage organisms, and to minimize any microbial growth in product. In view of these changes in chilled foods manufacturing, the control of pathogens is primarily dependent upon maintaining a low temperature throughout the *entire* shelf life of the food. However, for certain pathogens, notably *Listeria monocytogenes,* non-proteolytic *Clostridium botulinum,* and *Yersinia enterocolitica* (and enterotoxigenic *Esherichia coli* and *Aeromonas hydrophila* among the emerging pathogens) growth will occur at temperatures below 5°C. Some *Salmonella* strains can grow at temperatures around 7°C, and it should be remembered that micro-organisms can sustain growth at temperatures below their minimum growth temperature under certain conditions. Thus, if growth has already started during a period when the temperature was above the minimum, e.g. in a food subjected to the hazard of temperature abuse, then when returned to the correct chill temperature microbial growth would be expected to continue.

For these reasons product formulation is increasingly designed to include factors which inhibit pathogen growth. Because of the sensory qualities expected of chilled foods, the use of a single factor at a level sufficient to cause pathogen death is often excluded. For example, if pH was the only limiting factor, the type of acid and the concentration which would cause death would probably mask a mild or neutral flavor. The strategy, therefore, is to use several factors in combination. Separately, they have insufficient effect at the use levels but together they prevent multiplication of pathogens. Factors generally considered to achieve this purpose are an acid pH, a lowered water activity and a mild (sub-lethal) heat. Currently, the use of competitive microflora is being investigated for chilled foods, having been used traditionally in such products as fermented dried sausages.

It should be pointed out that Controlled Atmosphere Packaging (CAP) or Modified Atmosphere Packaging (MAP) is a technology now being used for fresh meat, poultry, fish and certain ready meals, as well as for fresh vegetables. The use of CAP and MAP is primarily to extend the shelf life as applied to the sensory qualities of the foods. The atmospheres used are known to inhibit spoilage microflora, but the effect on pathogens needs to be known when the safety of the food is being assessed. Although some work has been carried out on pathogen development in CAP meats[2] and MAP fishes[3], the published data base at present available is not adequate for well-founded decisions on safety. Not until more investigations are made on pathogen growth in a range of chill foods packed under CAP or MAP will the safety limits of this technology be known.

To insure the safety of chill products, therefore, the HACCP system must be rigorously applied, with special attention to those pathogens having low minimum growth temperatures. The type of food and the proposed packaging for each product must also be taken into account. *Remember* that for pathogens not able to grow at chill temperatures, refrigeration is not a kill step – survival can be prolonged beyond the end of shelf life at those lower temperatures.

Ingredients

Any unprocessed (raw) ingredient may be expected from time to time to contain pathogens, e.g. raw milk, vegetables, dried herbs, etc. For certain ingredients, e.g. dried or frozen food which will receive no 'kill' step during processing, microbiological monitoring of incoming batches (lots) may be carried out to confirm that they conform to the pathogen specifications for that ingredient. Should pathogens be found then the batch may be rejected. However, for most ingredients, monitoring is impractical, and control is effected through Good Manufacturing Practices (GMPs). Thus, ingredients on a site will be segregated from finished product; employees responsible for storage and transport of raw ingredients should not enter finished product areas without change of clothing and hand-washing; and any processing given to raw ingredients should take place outside the 'high care' area.

Supplier Quality Assurance (SQA) should form part of the quality system which insures that ingredients are acceptable for their intended use. The ingredient specifications should require the supplier to notify the user of *any* change made to the ingredient, as this change could affect the safety of the final product. The user (not the supplier) must decide if the change compromises the safety of an ingredient.

Ingredients must be stored so that microbiological growth is either prevented or minimized. For chill ingredients, storage should be at no more than 5°C and preferably less than 3°C, and the storage time

needs to be limited. Dried ingredients must be stored in ways which maintain the dry condition.

Stock rotation is an important part of GMP of a quality system, and all ingredients must be used within their 'use by' date or shelf life. Integrity of ingredient containers must not be broken, otherwise contamination may take place, whether from the container material, pests, water or the general environment, e.g. dust. Container integrity should be checked on arrival, and damage should not occur during unloading or stacking. If there is damage, spilled material must be removed immediately and the affected area cleaned and sanitized.

Ingredient pallets should be stacked away from walls or pillars to allow effective placing of pest control bait at these points. There should be enough space between stacked pallets for easy, and therefore safe, access of trucks for loading and unloading.

Processing methods

Since so many processing methods may be applied to chill foods it would be impracticable to detail each one in this section. Processing may be divided into two main categories: primary and secondary. Table 12.1 gives the main examples in these categories. Some ingredients may have received their processing before receipt, whereas others are under the direct control of the chill food manufacturer. However, the following points must be

Table 12.1 Processing of chilled products

Primary processing	*Secondary processing*
Drying – meats, fish	De-boning – cooked meats, poultry, fish
Smoking – meats, fish, cheese	Brining – cheese
Pasteurizing (cooking) – meats, poultry, fish, liquids (milks, soups, juices, sauces), baked goods, vegetables	Dicing/slicing – cooked meats, poultry, vegetables, cheese, bread, cakes, confectionery decoration
Curing – meats	Grating – cheese, chocolate, nuts
Fermentation – meats, fish, milks, vegetables (sauerkraut)	Whipping – cream, dessert topping
Acidification – salads, fish, vegetables (pickles)	Assembling – ready meals, pizzas, sandwiches, gateaux confectionery, desserts, pies
Salting – fish, vegetables	
Sanitizing (disinfecting) washes–vegetables, herbs	

Final manufacturing stage: packing and chilling

considered and controlled wherever the processing is carried out.

Primary processing

This is a stage which causes either microbial death or inhibits growth, and a chilled food receives at least one primary process during manufacture. Some foods, e.g. salami sausages, may receive three primary processes – drying, smoking and fermentation. Each process must have the control options to insure its effectiveness. For example, in meat the temperature at the center of the foodstuff must reach the specified temperature before cooking is complete. Remember that temperature and time are linked – the lower the temperature of cooking, the longer the meat must be held at that temperature. The cooking process should achieve at least a 4D kill of *Listeria monocytogenes*[4], or other authorities cite a 6D kill of *L. monocytogenes*[5]. This organism is selected as it is regarded as the most heat resistant of non-sporing pathogens.

Cooling must also be strictly controlled, as germination and outgrowth of surviving spores should be prevented, especially those of *Clostridium perfringens,* a pathogen closely associated with meat-borne food poisoning. The US Code of Federal Regulations[6] and the USDA/FSIS directive 7110.3[7] both clearly state the rate of cooling and the final temperature to be applied when cooling various meats. The USDA/FSIS directive requires that the internal temperature of the product must reduce from 54.4°C (130°F) to 26.7°C (80°F) in 1½ hours, then fall to 4.4°C (40°F) in 5 hours – a total cooling time of 6½ hours. These requirements insure that cooling is a rapid process, to prevent spore germination. A chiller of fully adequate capacity should be used to achieve the correct cooling rate, employing either chilled air or water. Contamination from either of these sources must be prevented.

Note: For certain products, e.g. double (48% fat) cream, the rate of chilling should not be too rapid, otherwise the cream becomes solid and is unacceptable to the consumer.

Sanitization (disinfection) is a process which has become necessary for vegetables or herbs included in certain salads, where the pH is higher than the traditional pH 3 to 3.5. Less acid salads may have a final pH of 5; a pH which may inhibit pathogens, but will not cause the numbers to decline, unlike salads at pH 4 with acetic acid as the acidifying agent in which vegetative pathogens show a significant decline under chill conditions[8]. Because of pathogen survival in the higher pH salads ingredients expected to carry pathogens must be sanitized before being added to a salad. Hypochlorite solutions are often used, but blanching in steam is acceptable, e.g. for onions.

It should be noted that preparation and composition of mayonnaise and salad dressing must conform to regulatory requirements in the USA [9, 10] and in many other countries. Remember, whatever primary process or processes are used in the manufacture of chilled food, safety must be confirmed by carrying out challenge testing (see Chapter 9.B). As a matter of policy a full record of this work should be kept permanently on file, for reasons similar to the keeping of Scheduled Process Establishment records for canned foods.

Secondary processing

This processing does not cause microbial death and, in fact, is more likely to recontaminate products following their primary processing treatment. Therefore, the objective during manufacture is to maintain the safety of a food by preventing microbial contamination during secondary processing. All equipment must be thoroughly cleaned and sanitized before use, which means rigorous application of cleaning schedules as part of GMP. Remember, graters, dicers and whippers are difficult pieces of equipment to clean and sanitize effectively and are well known as contamination sources.

De-boning of cooked meats and fish, and assembling of ready-meals, sandwiches, desserts, etc. is often a manual, and labor-intensive, process. To minimize contamination good personnel hygienic practices must be enforced, but environmental contamination must not be forgotten and these operations should take place in a 'high care' area (see Chapter 6.A).

Treatment of ingredients for multi-component foods

These ingredients are used when assembling salads, ready meals, pizzas, sandwiches, desserts, etc. Usually they are themselves held in chill store then brought to the manufacturing area prior to use. Once in this area they would be held, ideally, at a temperature of 12°C to 14°C (54°F to 57°F), although the temperature may be as high as 20°C to 25°C (68°F to 77°F). These high temperatures are ideal for growth of many spoilage microorganisms, and permit growth of some pathogens. Microbial growth could begin during the time that an ingredient is held in the manufacturing area, if this period is extended, and might continue when returned to the cold store if all the ingredient was not used. It is part of GMP, therefore, that only enough ingredient for use in a single batch of final product should be taken from the cold store. Returning any unused ingredient to cold store is not good practice. The time that an ingredient batch can be held in the manufacturing area will depend on the temperature of the area – it can be longer when the temperature is 12°C to 14°C (54°F to 57°F) than when it is 20°C to 25°C (68°F to 77°F). To confirm GMP, the temperature of an ingredient should be monitored within its container. Effective control is therefore achieved by accurately forecasting the ingredient quantity needed for a single batch of product, removing that quantity from the cold store, and knowing its temperature profile while it is held in the manufacturing area.

Process control and validation

Each of the many primary and secondary processes used for chilled food production must have its appropriate controls applied during manufacture. For example, drying of salami will need the temperature and time of drying to be specified, and the allowed tolerances (deviations) for these specifications. Temperature monitoring, relative humidity measurement in the drying room, date and time labelling for each batch as it enters the drying room are all part of effective control, and each batch must conform to the product specification when drying is completed. Alarm systems linked to temperature and relative humidity monitors will alert the processor to any loss of control in these systems.

Pasteurization of liquids will be regulated in the usual way by the specified temperature being achieved, correct flow-rate insuring that the liquid is held for the specified time; and a flow diversion valve fitted to reject under-processed liquid.

Pasteurization (cooking) of meat, poultry, pies, etc. requires temperature monitoring of the slowest heating portion of the product batch to insure that the pasteurization temperature is reached, and positioning the products within the oven to allow heat penetration to *all* the separate pieces. Both cooking and cooling of products must be monitored for temperature and time[7]. All other processes must be similarly controlled. Whatever control is applied, instruments used for measurement must be accurately calibrated and maintained to insure that they are functioning correctly within specified limits, e.g. thermometers to ±0.5°C (1°F) or pH meters to ±0.1 pH unit.

Validation is by the use of automatic temperature and time recording (as with milk pasteurization) or through written records (as with batch drying). Verification may be either chemical (pH measurement of mayonnaise, vinaigrette dressings and final product; chlorine levels in vegetable sanitizing (disinfecting) solutions; moisture levels in dried salamis), physical (temperature), or microbiological (counts on products, or on equipment to confirm effective cleaning and sanitizing). All processes should also be validated by adequately detailed safety and quality audits.

The final process applied by the chilled food manufacturer is, of course, refrigeration. Processed, packaged foods must be chilled and maintained at a temperature which, for most foods, should not exceed 5°C (41°F[11]. Certain foods may be stored at lower temperatures, down to 2°C (35.6°F). These are the 'high risk' foods with reference to *L. monocytogenes* and non-proteolytic *Cl. botulinum*. Other foods, e.g. fermented meats, hard cheese and butter, may be stored at up to 8°C (46.4°F).

The refrigerated store should be equipped with temperature recorders as well as monitors and GMP needs to include what action to take should the specified temperature be exceeded. Preferably, audio-visual alarms should be used to alert operators when the temperature has risen.

Manufacturers should know how long products take to reach the temperature of the chill store after pallets enter the store. Monitoring of inner packs of pallets should be done, so that the usual temperature profile for different pack sizes and stacking patterns are known, and appropriate GMP schedules prepared based on these data. The chill store refrigeration equipment should be adequate, thus insuring that product reaches its specified storage temperature as soon as possible after it enters the store.

Packaging materials

These are of many types, but whichever material is used for a particular product the final pack must exclude microbial contamination throughout shelf life. This applies particularly to any seams or joins in the pack, and also to the lids. Secondary packaging may, in fact, insure that 'snap-on' lids remain in place, even though the main purpose of this packaging is for labelling and display. Similarly, 'tamper-evident' packaging – often a plastic ring around the edge of the lid and the carton – will prevent the lids being loosened, and this in turn excludes contamination. When the lid is foil bonded to a plastic container, special care may be needed to maintain integrity of the bonding. Product splashes into the lip of the carton will interfere with the bonding and increase the chances on contamination at the point of splash, and an 'easy-peel' lid bond must be of sufficient strength not to become loose during the handling it receives before it reaches the consumer.

Since packaging materials can carry microorganisms, they must be manufactured hygienically to minimize their microbial load. They should be adequately protected after manufacture, and throughout their storage, which itself should be in a clean, dry area. Dust in the storage area should be controlled by vacuuming, not by brushing. The protective outer wrapping should not be removed until immediately before the packaging material is taken into the processing or packing area. No outer coverings should enter this area because of microbial contamination and foreign material risks to the final chill product.

When packaging is wound on a spool, the first 4 meters should be discarded before the packaging is fed into the forming equipment. GMPs should insure that 'ready-formed' containers are handled in a hygienic manner before filling.

Chilled foods packaged in CAPs or MAPs require that the packaging material conforms to specified permeability or impermeability. The specification must insure that the correct atmosphere is maintained in the pack throughout product shelf life. Permeable packaging material must also maintain integrity, as with any other material.

Coding and labeling

Both coding and labeling must conform to regulatory requirements in the country or state of manufacture and in those where the food is to be marketed. Coding may be described as information to the user regarding the date by which the food should be consumed. In the UK, the words 'best before' or 'use by' are followed by a date, which is the last day that the food will maintain its sensory properties, and remain safe microbiologically. The manufacturer's decision on the 'best before' date is based on quality assurance procedures during manufacture and distribution, including both challenge and shelf-life testing. In the USA, 'use by' and 'sell by' dates may be used, but in European countries within the European Economic Community 'sell by' dates are being phased out, and will no longer be allowed by the end of 1992.

Labelling covers information to the consumer including storage instructions, cooking instructions where appropriate, and nutritional data. Storage instructions should include the words 'keep refrigerated' on each pack[5] and for high risk products the temperature limit of +5°C should be stated. It should be noted that some products with secondary (outer) packaging surrounding the container(s) do not have any label or code on the product container itself. This information is only printed on the outer packaging. Such packaging can constitute a hazard, in that the user (whether in retail sales or in the home) may remove the outer packaging on a re-usable product, e.g. a salad, a 'dip', or a pate, and *not* replace it when the container is returned to the refrigerator. There is then no means of knowing the 'use by' date.

Cooking instructions, e.g. for meals with uncooked meats, fish or other sensitive ingredients, should be validated by challenge testing, just as when determining shelf-life. The cooking, or heating, must insure safety. Chicken is known to carry *L. monocytogenes, Salmonella* spp. and *Campylo-*

bacter, and the cooking process should insure at least a 4 log cycle destruction of these pathogens. Microwave cooking in the home is still suspect for safe heating as the heat distribution within the equipment is not uniform. Two Technical Memoranda from Campden FDRA give important data on the effect of microwave heating[12,13].

Cooking instructions given to the user must be based on validated heating procedures which insure that the whole of the product or ready meal achieves the specified temperature for the correct time.

Storage and distribution of product

The refrigerated store must be constructed to maintain chill temperatures and to meet all the hygienic requirements for storage premises (see Chapter 4) including fully adequate pest control. Product pallets must be separated to allow free air circulation and easy access for product removal. Pallets should be clearly labelled so that date codes are easily seen, as well as date coding of each pack as previously described. Stock rotation must insure that product is dispatched in sequence, as determined by the date coding labels. At dispatch, product temperature as well as the date should be recorded.

Loading should be into a previously chilled vehicle which has its own refrigeration unit and is well insulated, and should be directly from the cold store into the vehicle and, wherever possible, made using an enclosed loading bay. This prevents temperature increase during the loading operation. The vehicle temperature during transit should be monitored and recorded.

When the vehicle is unloaded, either into a transit cold store or at its final delivery point, product temperature should be checked before unloading, and transfer to the chilled storage should be immediate.

Special considerations

These considerations all arise from the fact that in the current marketplace many chilled foods receive less processing than was applied in the past. Although some retailers in the UK regard a short shelf life as a marketing plus, the emphasis (as can be seen in many articles in the trade press) is on extending the shelf life of chilled foods. Each manufacturer must give full consideration to what length of shelf life should be set for chill foods with reference to the safety of the food when it is consumed. Every aspect of chill food manufacture is linked to shelf life – from product formulation; ingredient choice and handling; selection and control of process and packaging; assembling of ready meals, sandwiches and confectionery; to the packing, storing and distribution stages. The Campden Technical Manual No. 28[14] gives an outline structure of the logical approach which should be followed when determining the shelf life of your product.

For safe manufacture, GMPs based on a HACCP plan must be established, enforced at all stages of manufacture and confirmed by audit. However, even when working within such a quality system the manufacturer must not over-extend product shelf life. Obviously, economics dictate that shelf life must be sufficient to give the distributor, retailer and consumer some margin when handling these products, but this margin must take full account of the behavior of psychrotrophic pathogens, such as *L. monocytogenes, Cl. botulinum* types B, E, and F, *Y. enterocolitica* and pathogenic strains of *E. coli.* It has been reported in a useful review paper by Lechowich[15] that *Y. enterocolitica* 'will increase in population from 100 to 10 million cells per g when held at 7°C (45°F) for 10 days'. Even when the primary processing is sufficient to reduce pathogen numbers to a safe level, post-process contamination may occur. Initial low numbers in or on the food, because of their ability to grow at chill temperatures, may result in an unsafe food if the shelf-life is over extended.

Since refrigeration is not a 'kill' step, product formulation should always consider, and be likely to include, the use of several factors (hurdles or barriers) such as acidity, use of preservatives and reduction in water activity to inhibit microbial growth. The main emphasis is to limit pathogen growth, but spoilage microorganisms will also be inhibited – conferring a quality benefit as well as insuring food safety. Product formulations should be validated by challenge testing as well as shelf life assessment. Remember, however, that the kitchen stage of product development is not always suitable to use for such experiments. Ingredients used for kitchen-scale products are not necessarily those used in production trials, and may well have different microorganisms or physico-chemical (growth-inhibiting) characteristics. It could be misleading at best, and unsafe at worst, to set a production product shelf life on kitchen-scale testing.

Shelf life must also be the main factor in evaluating the safety of CAPs or MAPs for products; 'Sous vide' technology may be included here. The atmosphere in these packaged foods (a vacuum in the case of 'sous vide') is likely to allow the growth of anaerobic pathogens, particularly surviving spores of *Cl. botulinum* types B, E and F. In the case of 'sous vide', where heat processing follows packing of the product, the food will lack any competitive post-processing microorganisms. Until the published data base on microbial behavior in these packs is more comprehensive, these types of packaging should not be regarded in themselves as a safe technology to give extended shelf life.

Obviously, rapid chilling of product to the correct temperature must be a major objective when manufacturing chilled foods. Guidelines which give the processor valuable, practical help in achieving their objective are those produced by the NFPA[5] and the UK Institute of Food Technology[11]. Remember that it is the (slowest cooling) product temperature, not the air temperature, which determines the prevention or inhibition of microbial growth. Achieving rapid cooling depends on correct handling of processed and packed product, and correct stacking of pallets as well as the performance of chill store refrigeration plant. The processor should establish and record the temperature profiles of all products manufactured on-site, including different container sizes and stacking patterns. Since it is not practical to monitor the product temperature profiles of each production batch, the profiles should be used to determine handling procedures which insure rapid chilling. Temperature monitoring should start at the time of packing, continue in the cold store and through the distribution system, including retail outlets. This has a twofold purpose: (1) it provides the data to establish good GMPs and to improve any weak points in the chain; (2) it gives an accurate measure of the inevitable temperature abuse (rise) which occurs following manufacture. This last information enables challenge or shelf life testing to be based on practicalities, so enabling a realistic assessment of safe shelf life to be made.

References

1. DTF (1989) *Guidelines for Good Hygienic Practice in the Manufacture of Dairy Based Products,* Dairy Trade Federation, 19 Cornwall Terrace, London NW1 4QP
2. Silliker, J. M. and Wolfe, S. K. (1980) Microbiological safety considerations in controlled-atmosphere storage of meats. *Food Technology,* **34**, (3), 59–63
3. Baker, D. A., Genigeorgis, C., Glover, J. and Razavilar, V. (1990) Growth and toxigenesis of *Clostridium botulinum* type E in fishes packed under modified atmospheres. *International Journal of Food Microbiology,* **10**, 269–290
4. NAC (1990) *Hazard Analysis and Critical Control Point System,* HACCP Working Group of the National Advisory Committee on Microbiological Criteria for Foods, USDA/FSIS, Room 3175, South Ag. Building, Washington, DC 20250
5. NFPA (1989) *NFPA Bulletin 42-L Guidelines for the Development, Production, Distribution and Handling of Refrigerated Foods,* Microbiology and Food Safety Committee, National Food Processors Association, 1401 New York Avenue, NW, Washington, DC, 20005
6. USDA (1990) 9 CF3 Ch 111 (1-1-90 edn) 318.17 Requirements for the production of cooked beef, roast beef and cooked corned beef. Food Safety and Inspection Service (Meat and Poultry), United States Department of Agriculture, Washington, DC 20250
7. USDA/FSIS (1988) *Time/Temperature Guidelines for Cooling Heated Products,* Food Safety and Inspection

Service Directive 7110.3, United States Department of Agriculture, Washington, DC
8. Van Schothorst, M. (1980) Miscellaneous Foods III Salads. In *Microbial Ecology of Foods,* Vol. 2, Food Commodities, pp. 824–827. International Commission on Microbiological Specifications for Foods, New York: Academic Press
9. FDA (1990) 21 CFR, Ch 1 (4-1-90 edn) 169.140 Mayonnaise. Food & Drug Administration, Health and Human Services, Washington, DC
10. FDA (1990) 21 CFR Ch. 1 (4-1-90 edn) 169.150 Salad Dressing. Food & Drug Administration, Health and Human Services, Washington, DC
11. IFST (1990) *Guidelines for the Handling of Chilled Foods,* 2nd edn, Institute of Food Science and Technology (UK), 5 Cambridge Court, 210, Shepherds Bush Road, London W6 7NL
12. Bows, J. R. (1989) Technical Memorandum No. 527. *The Influence of the Thermal, Electrical and Physical Properties of Food Heated by Microwaves,* Campden Food and Drink Research Association, Chipping Campden, Gloucestershire, GL55 6LD
13. Walker, S. J., Bows, J. Richardson, P. and Banks, J. G. (1989) Technical Memorandum No. 548. *Effect of Recommended Microwave Cooking on the Survival of Listeria monocytogenes in Chilled Retail Products,* Campden Food and Drink Research Association, Chipping Campden, Gloucestershire GL55 6LD
14. Campden R. A. (1990) Technical Manual No. 28. *Evaluation of Shelf Life for Chilled Foods,* Campden Food and Drink Research Association, Chipping Campden, Gloucestershire GL55 6LD
15. Lechowich, R. V. (1988) Microbiological challenges of refrigerated foods. *Food Technology* **42**, (12), 84–85, 89

Further reading

Campden R. A. (1990) *International Conference on Modified Atmosphere Packaging 15th–17th October 1990*: *Conference Proceedings Parts 1 and 2.* Campden Food and Drink Research Association, Chipping Campden, Gloucestershire GL55 6LD
Lachica, R. V. (1987) Factors for estimating the safety of temperature-abused chilled and frozen foods. *Activities Report of the Research and Development Associates,* **39**, (1), 138–144
Lioutas, T. S. (1988) Challenges of controlled and modified atmosphere packaging: a food company's perspective. *Food Technology,* **42**, (9), 78–86
Moberg, L. (1989) Good manufacturing practices for refrigerated foods. *Journal of Food Protection,* **52**, 363–367
Palumbo, S. A. (1986) Is refrigeration enough to restrain foodborne pathogens? *Journal of Food Protection,* **49**, 1003–1009

Guidelines and codes of practice

CFA (1989) *Guidelines for Good Hygienic Practice in the Manufacture of Chilled Foods,* Chilled Food Association, 6 Catherine Street, London WC2B 5BB
DTF (1989) *Guidelines for Good Hygienic Practice in the Manufacture of Dairy Based Products,* Dairy Trade Federation, 19 Cornwall Terrace, London NW1 4QP
IFST (1990) *Guidelines for the Handling of Chilled Foods,* 2nd edn, Institute of Food Science and Technology (UK), 5 Cambridge Court, 210, Shepherds Bush Road, London W6 7NL
NFPA (1989) *NFPA Bulletin 42-L Guidelines for the Development, Production, Distribution and Handling of Refrigerated Foods,* Microbiology and Food Safety Committee, National Food Processors Association, 1401, New York Avenue, NW, Washington, DC 20005

Chapter 12.B Frozen foods

Background and introduction

Freezing is one of the oldest forms of food preservation. Nature was used to accomplish the lowering of temperatures until the second half of the last century, when mechanized freezing came about. Food products maintained under frozen temperatures were able to be stored almost indefinitely. The process does not drastically alter the shape, color, size or texture of the food and allows for convenient preparation. In the USA by the mid-1980s, frozen food accounted for approximately 40% of the food consumed. In Europe the percentage ranges from 10 to 25%[1].

Freezing as a process is utilized to protect perishable foods from both enzymatic and microbial degradation. As freezing occurs ice is formed. Solutes found in the liquid phase are concentrated. This change alters the water activity (a_w) as well as the pH of the product. In most cases, the biochemical equilibrium is changed sufficiently to reduce or stop the enzymatic activity which begins taking place once the raw material is harvested. The measurement of the enzymatic reaction times will vary, depending on the food type and the processing treatment used prior to freezing. These changes and processes will also influence the growth limits of microorganisms found in the product.

Several factors influence the effectiveness of the freezing process as a preservation technique, including the initial microbial load, the physical dimensions of the product being frozen, the packaging material, the freezing rate, the time and temperature of storage, the consistency of the storage temperature, and the time/temperature conditions of thawing. The freezing process can be broken into three stages:

1. Cooling down from the initial temperature of the product to the temperature at which freezing begins.
2. Chilling at the latent heat 'plateau' during which there is a phase change of water into ice.
3. Further cooling to the ultimate temperature for storage.

The initial cooling may kill or injure a proportion of the microflora through cold shock. The decrease in temperature will reduce the growth rate of those cold-tolerant organisms still able to grow. The conversion of water into ice may mechanically injure or kill a proportion of the microorganisms. The third stage of cooling to the storage temperature further inhibits growth of the most cold-tolerant microorganisms until it ceases at about −8°C[2]. Slow freezing to a final temperature of −10°C is more lethal than rapid freezing to a final one of −20°C. During slow freezing ice crystals form to concentrate the soluble solids that affect the stability of the cellular proteins. During rapid freezing the temperature passes through the freezing point to inactivate the solutes[3]. Even though slow freezing may be the method of choice to reduce the numbers of microorganisms, this process may have a negative impact on the quality of the food product being frozen. A balance between food quality and microbial quality will have to be established.

Freezing causes the apparent death of 10 to 60% of the viable population. This number may be gradually reduced during frozen storage depending on the storage conditions. Freeze-resistant bacteria have been observed to persist through 10 years of frozen storage[3]. Generally, Gram-negative organisms are the most susceptible to death by freezing. Gram-positive nonsporulating rods and spherical bacteria are the most resistant vegetative cells. Bacterial spores are not affected by freezing. The toxins of *Clostridium botulinum* and *Staphylococcus aureus* are not destroyed by freezing; thus it is possible for food poisoning to occur from ingesting frozen product containing pre-formed toxin. Higher organisms are more sensitive to low temperatures than are bacteria. Thus, freezing and frozen storage may be used to destroy parasitic protozoa, cestodes, and nematodes in various foods[2].

Most frozen food will contain a mixture of microflora. The type and number that are recovered in analysis will depend on the type which were the dominant organisms at the time of freezing and on their relative rate of death during both freezing and frozen storage. Bacteria that undergo abrupt transition from 37°C to 0°C in the exponential phase of growth lose viability. Alteration in the permeability of the cytoplasmic membrane causes leakage of amino acids and adenosine triphosphate. This effect is the most noticeable among the Gram-negative organisms. Many cells that are damaged during freezing do not die, but are unable to initiate growth on minimally nutritive media or media containing inhibiting agents. However, if given an energy source in the absence of inhibiting agents, the cells rapidly repair the damage and become able to grow. The recovery period persists for 1 to 6 hours[3].

Non-lethal injury can be important from the standpoint of food safety and public health. Bacteria that are cryo-injured may not be detected using conventional methods and product could be released for consumption containing injured pathogens. These organisms have the potential for recovery during thawing and may then be able to grow or produce toxin at normal temperatures. Because different organisms respond differently to freezing, in certain foods there could be a selective effect for specific types of microflora. Several pathogens could be favored by these selective effects and become major portions of the population. Therefore, when determining methods to be used for enumeration, recovery of injured pathogens should be considered.

Preparation and storage of ingredients

Fruits

Changes affecting texture, flavor, and color of the fruit can occur at any of several stages of handling. Varietal selection, cultural practices, climate, method of harvesting (mechanical versus hand), transport and storage of raw material all play an important part in the ultimate quality of the finished product. A poor variety for freezing cannot be made into a high-quality product; on the other hand, the best raw material can be quickly downgraded by poor processing or other handling practices[4].

Most fruits for freezing are picked as near eating-ripe maturity as possible, since they do not ripen appreciably after picking. Exceptions are apples, which ripen slowly in storage, and Bartlett pears, which are always picked at a hard-green but sweet stage. Some tropical fruits are similar to pears in this respect. Both tomatoes and strawberries have been successfully ripened by the use of gaspacking. If fruit is ripe on one side but green on the other, it sometimes can be used in a sliced product or jam after trimming, or it can be used for puree or juice after passing it through a finisher to remove the unripe portion along with skin, seeds, etc. [4].

Since ripe fruits in general are subject to infection with molds and other microorganisms, a constant program of control by the processor is required. This includes chlorination of wash water, protection of fruit from bruising during handling, and sorting out of damaged and moldy fruit, as well as frequent and thorough cleaning of equipment. Spraying or dipping the fruit to cover the surface with a mold inhibitor has been tried with varying success[4].

Meat

To preserve the quality of meats, e.g. tenderness, color, juiciness, flavor, and aroma, raw material handling is important. Stress during the interval preceding slaughter has a direct physiological effect on such qualities as color and texture. Post-mortem chilling is extremely important. It slows the drop in pH (which slows the loss of water-holding capacity) and inhibits proteolysis and growth of microorganisms. This sequence of events impacts the texture of the final product.

Meat is a highly perishable product, with spoilage microorganisms growing quickly on it. Freezing enables meat to be held for some time without appreciable change in properties. With the exception of the external surface and the gastrointestinal and respiratory tracts, the tissues of normal healthy animals contain few microorganisms. The organisms that comprise contamination of meats will vary with the location of the farm the animal was raised on, the method used to transport the animals to the slaughterhouse, the holding conditions at the slaughterhouse, and the diet and health of the animal. The organisms present on frozen meats will then be a reflection of those present after slaughter[2].

Poultry

Phases of poultry processing essential to the economical production of optimum-quality frozen poultry begin with the selection of high-quality live birds, and include slaughtering, defeathering, evisceration, chilling, cutting-up, further processing, packaging and freezing. The USA has standards for quality of individual live birds. Assignment of grade is based on health, feathering, and absence of excessive numbers of pin feathers, conformation, fleshing, fat covering or finish, and freedom from defects such as skin and flesh bruises, and breast blisters.

The chronological age of the bird influences palatability primarily through its well-known effect on tenderness. Diet generally does not affect the quality of poultry although fish oils in the diet will change flavor and aroma. A brief period of fasting prior to slaughter has no influence on the cooked aroma or tenderness of the poultry product[4].

Seafood and fishery products

Chilling delays and minimizes spoilage and the ideal chilling system cools fish rapidly to wet-ice temperatures. It is essential to minimize bacterial contamination of the fish during all stages of handling. Dirt should be washed off the fish as landed and the surfaces with which the fish come in contact be maintained in a clean condition. Care must also be taken to wash the fish after gutting it, and to use clean ice. Fish properly iced will cool rapidly and will retain quality for 1 to 2 weeks, depending on the species. Storage of fish in refrigerated seawater has been shown to be as fast and effective as chilling in ice for certain species. Ease of handling, reduction of weight losses, and elimination of bruising are advantages to be considered in short-term storage in refrigerated seawater. In all cases it is important that adequate seawater circulation be maintained and that the temperature be kept at $-1°C$ (30°F). Since bacterial build-up in tanks and in connecting pipelines can be a problem, it is essential that a refrigerated seawater system be designed for easy cleaning. Chlorine should be used to sanitize the entire system.

Technical requirements have been established for the freezing of fish at sea and for thawing these fish for reprocessing ashore. Methods in use include moving air, contact-plate, and immersion. Shellfish such as crabs and lobsters, must be kept alive after capture, and in healthy condition up to the time of processing.

Since shrimp live only a few minutes after removal from their natural habitat, microbial spoilage starts immediately through marine bacteria on the surface and in the digestive system, and through micro-organisms which happen to contaminate the shrimp on the ship's deck, in handling, and from ice used during their storage. Fish and other marine organisms caught with the shrimp may also, chiefly through slime and exuded intestinal contents, contaminate them. Removal of the heads reduces the bacterial count somewhat because the head carries approximately 75% of the bacteria. With expeditious handling and thorough washing under commercial conditions, headless shrimp can evidently be placed in ice storage on board trawlers and carry only a relatively low microbial load. In commercial practice shrimp are packed in alternate layers of ice and shrimp. Since the melting ice from the upper layers of shrimp washers down over the lower layers, position in the bin influences bacterial count.

The prevention of deterioration in the quality of fresh and ice-stored shrimp involves not only maintaining low microbial counts, but prevention of oxidation, chiefly of phenols, into melanins. This condition is known as 'black spot' or melanosis.

Vegetables

Freezing preserves 'garden-fresh' quality for the consumer which is maximized when the time between harvest and cooking pot is only that required for washing, cutting, sorting, or other handling that makes the vegetable ready for serving. Only those who raise vegetables within easy reach of the kitchen achieve this degree of freshness. To reduce the delays between harvest and processing and to reduce the effects of delays, careful scheduling of harvests to minimize build-up of raw materials should be the rule and so should the precise dispatching of hauling trucks. Cooling vegetables by cold water, air blast, or ice will often reduce the rate of post-harvest quality losses sufficiently to provide extra hours of high-quality retention for transporting raw material considerable distances from the fields to the processing plant. However, inadvertent freezing during transportation of the vegetables can have a deleterious effect on the raw material.

For those vegetables stored for long periods of time, bruising, mold and other fungus infection and insect infestation must be prevented or kept to a minimum.

Dairy dessert products

Frozen dairy desserts include ice cream, ice milk, sherbet, ices, and various modifications such as yogurt, parfait, mousse, custard, bisque, and many novelty products. The microflora of frozen dessert mixes before pasteurization is that of the ingredients: milk and cream, condensed milk and dried milk, buttermilk, butter, sugar, chocolate, fruits and nuts, and eggs.

The same dangers of illness caused by drinking raw milk are inherent in ice cream either made from raw milk and cream, or handled under unsanitary conditions following heat processing[2]. In this product if pathogens are present, they may survive for many months. During the past, raw or improperly pasteurized eggs containing *Salmonella* were inadvertently added to the mix. However, in recent years, pasteurization of liquid eggs has been required to eliminate this problem.

More recently, ice cream products have been recalled because of the presence of *Listeria monocytogenes* contamination. Staphylococcal food poisoning has occurred due to contamination of the mix with *Staph. aureus* and subsequent temperature abuse that permitted growth sufficient to yield toxic levels of enterotoxin. Also use of dried whey as a source of milk solids may present a special hazard, for failure of starter cultures to grow normally during cheese manufacturing may permit uninhibited staphylococcal growth and enterotoxin formation. Enterotoxin will be in both the whey and the cheese and being heat stable will carry over into dried whey, which then may be responsible for the presence of enterotoxin in ice cream.

Any ingredient or additive added after pasteurization is a significant contamination risk. However, applying a Supplier Quality Assurance (SQA) program will minimize this problem, providing they are carefully stored and handled in the plant before use.

Processing methods
Fruit

The processing of fruits for freezing may include peeling, cutting, treating to control enzymatic browning, adding of syrups or sugar, see Fig. 12.1. Most fruits to be preserved by freezing are not blanched because heating causes softening and loss of moisture. An exception is fruit destined for bakery products that are to be heated during preparation. Therefore, the organisms normally found on frozen fruit are the same as those present on the raw ingredient[2].

Microbial populations on fruits to be frozen are best controlled by adequate washing, removal of obviously diseased fruit, careful handling to prevent bruises, frequent cleaning and sanitation of handling

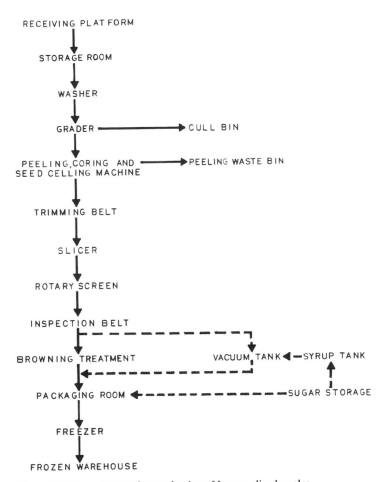

RECEIVING PLATFORM

STORAGE ROOM

WASHER

GRADER ———————→ CULL BIN

PEELING, CORING AND ———→ PEELING WASTE BIN
SEED CELLING MACHINE

TRIMMING BELT

SLICER

ROTARY SCREEN

INSPECTION BELT

BROWNING TREATMENT   VACUUM TANK ◄—SYRUP TANK

PACKAGING ROOM ◄— — — — — — — —SUGAR STORAGE

FREEZER

FROZEN WAREHOUSE

Figure 12.1 Flow diagram for production of frozen, sliced apples

and conveying equipment, and prompt freezing of the prepared fruit[2].

Quick freezing, which was developed to retard the enzymatic changes in vegetables, may not be the best method to be used for all fruit types. In fruits, enzymes producing off-flavors are not especially active; darkening of color by oxidative enzymes such as polyphenol oxidase is more important. Where enzyme-catalyzed changes in color are not pronounced, speed of freezing is of lessened importance[4]. The rate of freezing may then impact texture and other quality factors.

Meats

The primary processing of meat consists of slaughtering, dressing, chilling, and cutting. The initial flora is mainly determined by surface contamination received during these operations. Although meats processed before freezing will contain organisms picked up from equipment and air of the processing area, the normal flora of unprocessed meats will resemble that of raw chilled meat[2].

The number of organisms within the tissue is normally very small, approximately 1/10–100 g[2]. Microbial contamination from the skin or hide includes the normal flora of the skin (*Staphylococci, Micrococci,* pseudomonads, yeasts, and molds) as well as organisms of fecal and soil origin. The total number of organisms on the skin may exceed $10^9/cm^2$[2]. Other sources of contamination during dressing includes the knives, hands, and clothing of the workers and wash water. With good sanitation, contamination from these sources is likely to be negligible compared with that from the animal itself. Attempts to clean carcasses usually have little effect on microbiological contamination[2].

Fast chilling at low temperatures with high air speeds and low humidities may reduce bacterial numbers, but under less severe conditions, growth of psychrotrophic organisms can alter the proportion of psychrotrophs to mesophiles. The time the carcass is held in the chiller may have a more significant effect on the microbial population than the chill temperature. Carcasses cooled at ambient temperatures of 15° to 20°C (58° to 78°F) must be expected to show growth of mesophiles, including pathogens. As edible offals often are more heavily contaminated than carcass meat they should be chilled immediately[2]. Meat packers do not usually 'age' or 'hang' meat.

Contamination received by meat during cutting, boning and packaging depends on local sanitary conditions. During the considerable handling, fresh surfaces are exposed, thus increasing the risk of contamination. The temperature of the boning room, the time meat is held, and the cleanliness of the cutting tables, conveyor belts, saws, knives and other equipment all affect the microbial flora. If the temperature of the room is ≤10°C (50°F) then microorganisms on the outside of carcasses cause most contamination. This will be spread as the carcasses are cut, so equipment hygiene must be enforced. However, products that undergo extensive handling are likely to become contaminated with bacteria of human origin. For example, the proportion of human derived strains of staphylococci present on meat has been shown to increase during processing[2].

Poultry

Poultry processing is made up of the following steps:

1. Slaughtering – the standard method for slaughtering is to bleed the bird. The effects of inadequate bleeding results in an objectionable red appearance of the skin layer giving a lower-grade carcass.
2. Scalding – feathers are loosened by immersion in or spraying of water at controlled times and temperature. Excessive scalding can result in increased toughness in the cooked meat and loss of moisture.
3. Picking – removal of feathers from scalded birds is done by mechanical means. Excessive processing during picking may also lead to toughened meat.
4. Evisceration – poultry is normally eviscerated as soon as possible following feather removal to inhibit bacterial growth and to prevent the transfer of visceral flavors to the meat.
5. Chilling – after warm evisceration, it is desirable to lower the temperature rapidly in order to inhibit bacterial growth and other deteriorative changes. Over one-third of the total heat to be removed to attain the desired −18°C (0°F) frozen storage temperature is removed during the washing and chilling process. Most poultry is chilled by combinations of cold water and ice, the

particular procedures and types of equipment vary widely. In the USA the chilling process must conform to USDA regulations.
6. Aging – poultry aging prior to freezing directly impacts the toughness of the meat. Aging can take from 1 to 16 hours for different types of poultry. Processors optimize the aging process between slaughter and freezing.
7. Freezing – the effect of freezing and frozen storage on poultry appearance and palatability is influenced by rate of freezing, temperature and duration of storage packaging, and handling during and after thawing.

Seafood and fishery products

The usual raw seafood destined for freezing is a fillet, or steak, or a shucked oyster. Shrimp, lobster, and crab are usually subjected to blanching, boiling, or steaming under pressure prior to freezing. The duration of the cook is usually short to minimize quality loss. Temperatures range from 95 to 100°C. Meat from cooked crabs and lobsters is picked by hand prior to freezing. Shrimp may be cooked either before or after shelling. The cooked meat is packaged and frozen without further treatment, or after being battered and breaded[2].

Cooking crab and lobster reduces the bacterial count significantly, especially that of heat-labile organisms, e.g. *Salmonella* and *Pseudomonas*. However, recontamination occurs with picking and shell separation, which may involve brine flotation, giving total counts around 10^5/g of Gram-positive rods and cocci, Gram-negative rods, and yeasts. Shrimp is handled differently from crab and lobster, being landed dead. Varying periods of shipboard storage result in bacterial growth and spoilage prior to the land-based processing. Bacterial counts on shrimp received at shore plants may vary from 10^3 to 10^7 per g. After cooking the shrimp by boiling or steaming for a few minutes to release the shell, the count is reduced approximately a hundredfold but may rapidly return to the former level after peeling and sorting. Because of handling, low levels of coliforms, *Escherichia coli*, and *Staphylococci* may be present in finished products[2].

Hand picking of crabmeat means that the possibilities for transfer of potentially pathogenic bacteria from humans are high. Flotation in brine can allow contamination by *Staphylococci*, because they survive well in salt solutions. Since *Staphylococci* do not compete well with the normal spoilage flora of crabmeat they die off slowly during storage[2]. Although *Vibrio parahaemolyticus* is heat labile it occurs naturally on crustacea, and outbreaks of *Vibrio* food poisoning from cross contamination have occurred with shrimp, crab, and lobster.

Vegetables

Vegetables are prepared for freezing by cleaning and washing to remove field dirt, debris, and surface residues, which provides the first step in control of microbial contamination. Inedible parts are removed by trimming and vegetables are cut, sorted and size-grade as appropriate[4].

Physically moving large quantities of vegetables, both before and after processing, is one of the big operations in the freezing plant. This is done in a variety of ways: belts, air conveyors, flumes, screw conveyors, tote bin, etc. Moving vegetables by water in a flume or by pumping has special attractions to the freezer because it simultaneously moves, cools, and washes. Unfortunately, immersion in water can also leach sugars and flavors, and cause water disposal problems. In recent years air conveyor systems have been used for moving such vegetables as peas, corn, diced carrots, string beans, and lima beans[4].

Inspection and hand sorting of vegetables is a continuous process, from the moment raw material is received at the plant until the processed product is ready for the final packaging. Mechanical harvesting greatly increases the necessity for careful inspection. Sticks, stones, twigs, leaves, miscellaneous trash, etc. are always mixed in with the load from the field. Although metal-detecting devices and various ingenious sorting, screening and vibrating equipment are used, vigilant inspection and hand sorting is still required for most vegetables[4].

The usual processes of washing, trimming, size reduction, blanching, etc. are applied as appropriate to the product, with obvious microbiological consequences.

Some vegetables require blanching before being frozen. This stabilizes them by inactivating certain enzymes, and it also reduces the numbers of microorganisms. Although color and texture of the vegetables will be affected by blanching, underblanching can cause changes during storage that are even less desirable than those that occur with unblanched vegetables. To measure the effectiveness of blanching, whether by steam (for cut or sliced vegetables) or by hot water (for large vegetables) catalase and peroxidase monitoring of the vegetables is carried out, since these enzymes lose their reactivity in the range of importance for stability of frozen vegetables. Specific tests have been developed for these enzymes, and correctly blanched vegetables show a negative reaction.

Vegetables are cooled immediately following blanching; cold water flumes are easy and convenient for this, although spraying with cold water or blowing cold air are alternatives sometimes used. Microbiological quality of the cooling water is of obvious importance.

Dairy dessert products

The basic steps of production in manufacturing frozen dairy dessert products are composing (weighing or making) and blending the mix, pasteurization, homogenization, cooking, aging, flavoring, freezing, packaging, hardening (if required), and storage. These steps are shown in Figure 12.2. for a flavored product. Other processing procedures may be employed depending on the requirements of the specific product such as the culturing of yogurt prior to freezing or whipping of ingredients for mousse.

The first step of processing is composing the mix. This procedure may range in scope from the small batch operation where each ingredient is weighed or measured and added, to the large pushbutton operation where many of the ingredients are metered into the batch. The common procedure is to: (a) add liquid materials (cream, milk, or other liquid milk products) to mix vat or pasteurizer; (b) apply heat (optional) and then add dry solids such as egg yolk, gelatin, etc. (mixing dry products with three parts of sugar and adding to the mix will aid in their dispersion); (c) add sugar when the mix reaches approximately 49°C (120°F); (d) use caution to insure that all materials are dissolved before pasteurization temperature is reached[4].

Pasteurization is either the Low-Temperature-Holding procedure (LTH) at 68.3°C (155°F) for 30 minutes; the High-Temperature-Short-Time method (HTST) at 79.5°C (175°F) for 25 seconds; or the Ultra-High-Temperature method (UHT) at 138°C (280°F) or higher (US regulations) or 148.9°C (300°F) for at least 2 seconds (UK regulations). Fruits and nuts may be added before or after pasteurization depending upon whether these processes are detrimental to them. Color is generally added after pasteurization. The minimum LTH and HTST heat treatments permit survival of thermoduric non-spore-formers and spores of many aerobic and anaerobic spore-formers. Only the more resistant spores survive UHT processes. Fruit-flavored (water) ices because of their low pH (3.5 or less) are not pasteurized[2].

Homogenization: (1) thoroughly blends the ingredients of the mix; (2) breaks up and disperses the fat globules; (3) improves the texture and palatability of the ice cream; (4) makes possible the use of different ingredients; (5) reduces aging and aids in obtaining overrun; (6) produces a more uniform product[4].

Aging the mix before freezing is responsible for the following changes: (1) the fat is solidified; (2) if gelatin has been used as a stabilizer it swells and combines with water; (3) the proteins of the mix may change slightly; (4) the viscosity is increased, largely due to the other changes. Smoothness of body and texture, resistance to melding, and ease of whipping are also improved by aging[5].

The aging process may range from 4 to 24 hours

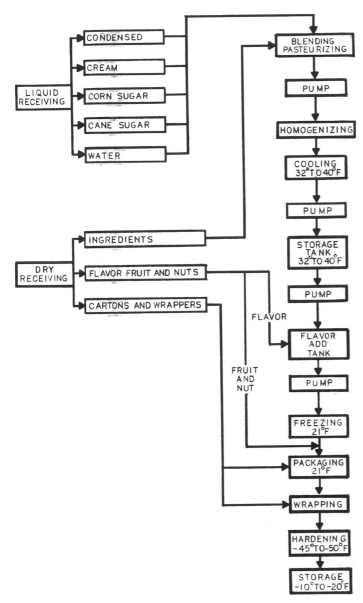

Figure 12.2 Flow diagram for production of flavored frozen dairy dessert product. Note approximate temperature conversions are as follows:

−50°F = −46°C; −45°F = −43°C;
−20°F = −29°C; −10°F = −23°C;
+21°F = −6°C; +32°F = 0°C;
+40°F = +4°C

with temperatures not to exceed 10°C (50°F). When the mix has been properly aged it is ready for the freezing process, which generally follows immediately[5].

Freezing the mix is one of the most important steps in the making of ice cream. The freezing process should be accomplished as rapidly as possible to insure small ice crystals and smooth texture in either the batch freezer or the continuous freezer. The function of the freezing process is: (1) to freeze a portion of the water of the mix; and (2) to incorporate air into the mix. There are four phases of the freezing process: (1) lowering the temperature from the aging temperature to the freezing point of the mix; (2) freezing a portion of the water of the mix; (3) incorporating air into the mix; (4) hardening the ice cream after it is drawn from the freezer.

Process control and validation

Microbiological quality control involves development and use of processes designed to keep microbial numbers low by reducing contamination and preventing microbial proliferation. The establishment and use of HACCP, as described in Chapter 3, is an essential part of process control. The process should include: (1) selection of low-count ingredients; (2) use of processing techniques as described previously in which the heat treatments are severe enough to reduce bacterial populations; (3) avoidance of post-heating contamination and constant maintenance of low temperature during storage. The effectiveness of these methods is monitored by microbiological examination of plant, equipment, materials, and products. Post-processing contamination and cross contamination from raw sources are the major areas of concern for processed frozen foods[2].

Inspection of facilities and equipment and observations of hygienic practices of personnel, which are often required by regulatory authorities, are commonly used to check adherence to good food-handling practices[6]. For frozen foods, this type of inspection can identify areas of concern for environmental contamination and breakdown in sanitation practices.

Along with physical inspections, ingredients, samples from lineflow points, and finished product are examined for microorganisms. Normally this examination will be performed to assess sanitation. However, in some instances foods are examined for specific pathogens or their toxins. Indicator organisms may be used in a quality control program rather than detection of specific pathogens. Although the presence of indicator organisms alerts to the possible presence of pathogens their absence does not prove the absence of pathogens. Microbial counts are also used to assess the effectiveness of cooking processes to control the microbial population including the pathogens. Limits have been established for some microorganisms, but are not warranted for all food products. In the USA, a joint committee has been established utilizing experts from regulatory agencies, academia, and industry to assess the need for microbiological criteria in foods based on individual categories of foods.

For individual frozen food categories, the following specific microbial analysis may be warranted:

1. For frozen fruits, yeasts and lactic acid bacteria can be used to assess manufacturing practices. Most samples of fruits would not need to be analyzed for pathogens. Microscopic mold counts can be useful for assessing the wholesomeness of raw fruit prior to processing.
2. Microbiological analysis of raw meats is important to assess the quality of the raw material to be used. However, application of criteria to the raw product would not be feasible. The condition of the raw product will have a direct effect on the quality of the finished product and the overall shelf stability.

The pathogens of frozen meat are the same as those of the meat before freezing, but their numbers may be very much changed. Official records indicate that the most frequent causes of food poisoning associated with meat are *Clostridium perfringens*, *Salmonella*, and *Staph. aureus*. Vegetative cells of *Cl. perfringens* rapidly decrease in number in frozen meats, but spores remain unchanged and multiplication can be rapid during thawing under warm conditions. Present indications are that *Cl. perfringens* occur normally at a level of one or fewer per 100 g deep within the muscle tissues immediately after slaughter. *Salmonellae* occur with variable frequency. Liquid enrichment procedures are necessary to recover the organisms.

Cooked meats should be analyzed for pathogenic organisms such as *Salmonella* to determine the effectiveness of the process and for *Staph. aureus* and Enterobacteriaceae as indicators of post-processing contamination.

3. Raw frozen poultry may contain large numbers of organisms, some of which may be pathogens. Microbiological criteria for these products would be impractical. However, screening for aerobic plate counts and coliforms would yield information on processing practices.

Processed poultry products should be evaluated for the presence of *Salmonella*, *Staph. aureus*, and coliforms. *Yersinia* and *Campylobacter* have also been associated with poultry. In products such as pot-pies, analysis for *Clostridia* might be beneficial to insure product safety.

4. Toxins that might be present in the raw product or are produced as a result of bacterial growth in seafoods prior to freezing would, of course, not be inactivated by the freezing process. Conditions for bacterial toxigenesis would also permit development of large populations of other bacteria, and this would be detected in most frozen thawed products as an excessively high aerobic plate count and obvious spoilage. The use of vacuum and modified atmosphere packaging will reduce the numbers of spoilage organisms in fish held chill after it has been thawed. Should temperature abuse occur in this type of package, then conditions for outgrowth of *Cl. botulinum* and toxin production would be established.

For frozen cooked seafoods, microbiological line sampling to determine adequate sanitation of surfaces and to measure contamination and growth in product will help evaluate post-cooking manufacturing operations. Regular tests on final product are advisable. Coliforms, *E. coli*, *Staph.*

aureus, and aerobic plate count are good indicators. On frozen raw fish, microbial analysis gives some information about the quality of the fish before it was frozen. Fecal *Streptococci* determined on the frozen product may give a more useful measure of processing sanitation than *E. coli*, in view of their greater ability to survive freezing. *Staphylococci* are reasonably resistant to the effects of freezing and provide some index of the degree of human contact with the product. Paralytic shellfish poisoning (PSP) toxin in shellfish meats and/or scombroid poisons can be detected by laboratory analysis[2].

5. Frozen vegetables, like fruits, have very little problem with microbial control. Mold counts on the raw vegetable can give evidence of the product quality, as will aerobic plate counts. Studies have shown that coliforms have been detected at many stages of frozen vegetable manufacturing. The correlation of the presence of these organisms with any unsanitary practice was not found; it appears that neither the coliform group nor *E. coli* were suitable indicators of poor sanitation during the processing of frozen vegetables. Frozen vegetables are rarely involved in food poisoning incidents because (a) non-spore-forming pathogenic bacteria do not survive blanching, (b) any pathogens contaminating the product post-blanching cannot grow at the temperature of the frozen food, and (c) most are cooked before being consumed. In limited studies, pathogens were found in raw product at extremely low levels. Allegations of food poisoning from frozen vegetables are inherently unlikely and the possibility of contamination and time–temperature abuse during preparation for serving should be thoroughly investigated[2].

6. For the most part, sampling and testing of the usual hardened ice cream or of ice cream mix for aerobic plate counts and coliform group at various stages of preparation and handling is routine to check plant sanitation. In addition, a test for *Staph. aureus* is a useful control of soft-serve ice cream. Variations from counts normally experienced point to the need for corrective (remedial) measures. Obviously, an investigation of a food-borne disease outbreak must include sampling and analysis for the appropriate pathogens, such as *Salmonella*, *Staph. aureus*, and *L. monocytogenes*. To insure product safety, environmental as well as equipment monitoring of post-processing areas should be part of GMP in frozen dairy product plants. Of particular concern with ice cream specialty products is the blast freezer. *L. monocytogenes* growing on the inside of the tunnels during shutdown periods has contaminated product during blast freezing and resulted in recalls. This equipment must be included in an environmental monitoring program.

Evaluation of fruits, nuts, flavors and colors prior to addition will aid in establishing the microbial quality of the finished products. Remember, pathogens can survive in frozen dairy products.

Packaging materials

The packaging of frozen foods imposes certain special requirements. Because of the tendency of water vapor to sublime from frozen food surfaces to colder surfaces in freezers and storage rooms, packaging materials for frozen foods should have a high degree of water vapor impermeability. Most foods expand on freezing, some to the extent of 10% of their volume. Therefore packages in which food is frozen should be strong, have a degree of flexibility, or should not be completely filled. As with all foods that may be stored for months or years, packages should be protective against light and air. Some frozen foods will be thawed at the time of use in their containers, so packages should be liquid-tight to prevent leaking upon thawing.

Many packages and packaging materials such as cans, metal foils, plastic-coated cardboards, and plastic films are all satisfactory for frozen foods[7]. Barrier properties found today in films, foils, laminates and paper can be selected to minimize the effect of moisture loss. The addition of vacuum packaging or gas flushing for some products will also retard deterioration of the product. Selection of packaging materials should include consideration of intended product use and storage time[2].

Storage of finished product and distribution

Background

Proper freezing alone will not insure the successful distribution of frozen foods – successful storage is at least as important as freezing itself. Unless frozen foods are warehoused under proper conditions of storage humidity and temperature, all of the careful production and quality control methods that a packer may have used could easily be lost before the product is delivered to the final consumer.

In the earlier days of frozen storage, there was a tendency to recommend the lowest possible temperature for frozen foods, e.g. $-30°C$ ($-22°F$) at the producer and wholesale level, and $-18°C$ ($-0.4°F$) for retail stores and home freezers. This has been somewhat modified in recent years because of energy considerations. Different storage temperatures may also be recommended for different product types. However, many of these products are held at the same time within the same facility, making product-specific storage temperatures impossible to accomplish. Compromise temperatures to cover multiple products would be $-22°C$ ($-7.6°F$) for the wholesale level, reasonable precautions against excessive temperature rise during transport, and retail cabinets operating at no warmer than −10°C (14.0°F) in any one point below the load line[1].

Sanitation of frozen storage will be maintained with attention to good manufacturing practices. These include the prevention of infestation of insects and rodents within the structure walls; prevention of corrosion on surfaces; prevention of condensation around pipes and openings; and attention to air purity.

Some freezing rooms are very likely to have odors that come from the unfrozen product when it was originally placed in the freezer. Some products continue to give off odors even at −18°C (0.4°F). These odors should be removed or modified and equipment is available to handle this problem. Absorbers may be placed in the room to purify and deodorize by circulating the room air through activated carbon filters. Another type of air purifier is the germicidal ultra-violet light lamp.

From a theoretical standpoint, frozen foods should be stacked in solid piles in such a way as to reduce to a minimum air circulation around the products. Such a system is practical only when the storage has a cold floor and the products are not piled against warm walls, and when all of the foods moved into the storage are as cold or colder than the storage itself. Under such conditions, desiccation and oxidation will be less than when considerable air circulation is permitted.

In actual practice, palletized frozen foods are often unloaded from trucks or refrigerated cars in which the products have warmed above the storage temperature. In this case, it is important that the food temperature be quickly reduced to that of storage[4].

Fruit

The normal microflora of frozen fruit consists mainly of yeasts and molds. Growth of these organisms and consequent spoilage of the product is influenced by the temperature of storage. Conditions which lead to partial or complete thawing will predispose the fruit to spoilage organisms, notably yeasts. Properly handled frozen fruit will not contain pathogenic bacteria. The acidic environment normal to fruit tissue does not allow bacterial pathogens to survive[2].

Meat

Meat which has been properly frozen and stored is not spoiled by microorganisms, but two malpractices make spoilage possible. In the first, temperatures in the range −5° to −10°C (23 to 14°F) permit slow development of molds, such as *Cladosporium herbarum* which forms small black colonies. Usually

there is little off-odor or off-flavor. In the second, a high microbial load before freezing may result in slow spoilage of meat during frozen storage. Although these microorganisms cannot multiply when frozen and gradually 'die off', their enzyme systems especially lipases and lipoxidases may remain active down to about −30°C (−22°F). Abuse by thawing and refreezing under uncontrolled conditions must be expected to result in high microbial populations[2].

Poultry

Poultry stored at or near −18°C (0.4°F) will not spoil from microbial activity; it will only spoil before freezing and after thawing. Some yeasts and molds, however, can grow on frozen carcasses at temperatures as low as −7°C (19.4°F)[2]. Organisms commonly implicated are *Cladosporium herbarum,* causing black spots; *Thamnidium elegans* and *Thamnidium chaetocladioides,* causing whiskerlike growth; and *Sporotrichum carnis,* causing white spots.

Freezing and frozen storage can reduce the number of viable microorganisms on poultry; some are killed while others are only damaged sublethally. Aerobic plate counts on poultry skin may decrease by 10 to 95% as a result of freezing; further death then occurs during frozen storage but at a slower rate. When frozen poultry is thawed and held at chill temperatures, then the rate of spoilage is the same as that of chilled poultry.

Fish

Fish in frozen storage undergo changes in flavor, odor, texture and color. The speed of these changes varies with the handling and processing techniques employed before, during and after freezing. Microbial growth is not a problem, because at temperatures below −9°C (15°F) the activity of marine bacteria is largely inhibited. Enzymes secreted by marine bacteria may, however, adversely affect the stability of the frozen product[4].

Vegetables

In most frozen vegetables, the predominant microorganisms are the lactic acid bacteria. Microbial spoilage is rare. In the frozen states, spoilage is essentially precluded by the low temperature and reduced a_w. Aerobic colony counts ranging from 10^1 to over 10^5 per g are normal and should not be cause for alarm[2].

Dairy dessert products

Hard ice cream manufacture should not present microbiological problems during storage and distribution, provided that GMPs are followed throughout processing. However, soft-serve ice cream is not frozen immediately after heat processing but is distributed to retail outlets as a mix. It is then held liquid until soft-frozen when dispensed to the consumer. Microbial abuse is likely to occur under these conditions, since some outlets have poor refrigeration facilities and inadquate cleaning and sanitizing schedules for storage and freezer equipment. For these reasons, soft-serve ice cream must be regarded as a hazardous product.

Codes of practice and GMP guidelines

The commercial producer of frozen foods needs to know not only the mechanics of getting his produce ready for market but also all the trading requirements such as standards, specifications, regulations, and codes. Quality standards and specifications, among other things, provide a common language for use in purchase and sale negotiations and other business transactions. Based on commercial quality levels, they serve as useful guides in packing the frozen product.

Regulations and codes primarily are rules governing the packing and marketing of frozen foods. Fundamentally developed in the interest of the consumer, they also encourage uniformity in packing and handling practices[4].

Microbiological criteria are applicable at the international, federal, state (both health and agriculture departments), and local levels, as well as by the food industry. The types of criteria at each of these levels are influenced by the missions and responsibilities of the agency or organization involved. International groups include: the Joint FAO/WHO Food Standards Program as implemented by the Codex Alimentarius Commission, the European Economic Community (EEC), and the International Commission on Microbiological Specifications for Foods (ICMSF).

In the USA The Food and Drug Administration (FDA) is charged with establishing: (1) a reasonable definition and standard of identity, (2) a reasonable minimum of quality, and (3) a minimum standard of fill of container. The operational program of FDA may specify microbiological criteria. The US Department of Agriculture (USDA) has jurisdiction over the production of egg, dairy, meat and poultry products as well as agricultural commodities. Specific regulations for frozen foods will be found published in the Code of Federal Regulations and the Federal Register.

References

1. Jul, M. (1984) *The Quality of Frozen Foods.* New York: Academic Press

2. ICMSF (1980) *Microbial Ecology of Foods, Vol 2, Food Commodities.* International Commission on Microbiological Specifications for Foods. New York: Academic Press
3. Sandine, W. E., Ayres, J. C. and Mundt, J. O. (1980) Prevention of food spoilage. In *Microbiology of Foods,* edited by B. S. Schweigert, pp. 61–70. New York: W. H. Freeman
4. Derosier, N. W. and Tressler, D. K. (1977) *Fundamentals of Food Freezing.* An AVI book published by AVI acquired by Van Nostrand Reinhold, New York
5. Arbuckle, W. S. (1986) *Ice Cream,* 4th edn. An AVI book published by Van Nostrand Reinhold, New York
6. ICMSF (1986) *Sampling for Microbiological Analysis: Principles and Specific Applications,* 2nd edn. International Commission on Microbiological Specifications for Foods, Toronto, Canada: University of Toronto Press.
7. Potter, N. C. (1976) Food Packaging. In *Food Science,* pp. 258–259. An AVI book published by AVI acquired by Van Nostrand Reinhold, New York

Further reading

APHA (1985) Standard Methods for the Examination of Dairy Products, 15th edn, edited by G. H. Richardson. Washington, DC: American Public Health Association
BAM (1986) *Bacteriological Analytical Manual of the US Food and Drug Administration,* 6th edn. Association of Analytical Chemists (for the FDA), Washington, DC
Speck, M. L. (ed) (1984) *Compendium of Methods for the Microbiological Examination of Foods,* 2nd edn. Washington, DC: American Public Health Association
Bryan, F. L. (1982) *Diseases Transmitted by Foods* (a classification and summary), 2nd edn. Atlanta, GA: Centers for Disease Control
NAS (1985) *An Evaluation of the Role of Microbiological Criteria for Foods and Food Ingredients.* National Academy of Sciences, Washington, DC: National Academy Press
ICMSF (1978) *Microorganisms in Foods. Vol. 1. Their Significance and Methods of Enumeration,* 2nd edn. Toronto, Canada: University of Toronto Press
International Insitutute of Refrigeration (1986) *Recommendations for the processing and handling of frozen foods,* 3rd edn. Paris: Insitutute international du froid
Arthey, D. and Dennis, C. (eds) (1990) *Vegetable Processing.* Glasgow: Blackie. Also useful for sections 12.A and 12.D

Codes of Practice

Codex Alimentarius Commission (1976) *Recommended Code of Practice for the Processing and Handling of Quick Frozen Foods,* CAC/RCP8-1976, Codex Alimentarius Vol. E, Food and Agriculture Organization, United Nations, Rome: World Health Organization
FFRT (1987) *Frozen Food Handling and Merchandizing: a Code of Recommended Practices Endorsed by the Frozen Food Roundtable,* prepared by the Frozen Food Roundtable, Washington, DC

12.C Dried foods

Introduction

Although dried foods are not usually regarded as a rapidly expanding segment of the market apart from the 'traditional' baby foods, there has been in recent years a considerable increase in the types of products available to the consumer. These are mainly dried 'instant' snacks, soups or drinks and their manufacture includes new technologies as well as different ingredients from those used in the past. Extrusion technology is among one of these new processes used for dried foods, but it lies outside the main scope of this section. Remember, however, that even though high temperatures are used, and high shear applied to the products during extrusion, it must not be assumed that these will sufficiently reduce microbial numbers to render the food safe. As for any process, the effect on microbial load must be confirmed experimentally and sufficient data generated to confirm product safety.

Any dried food has safety aspects, particularly since the market for some foods is among the more vulnerable members of the population, such as infants and young children, the elderly and the ill. When foods need to be reconstituted, as many of them do, then abuse by holding the reconstituted food for some time before it is consumed can result in low numbers of microorganisms multiplying to levels which can cause food poisoning if they are pathogens.

Since 1950 there have been 17 major food poisoning outbreaks due to milk powders[1], 15 of which have been caused by *Salmonella* spp. Two were caused by staphylococcal toxin in the powder concerned, arising from *Staphylococcus aureus* growth in the concentrate before it was dried. Chocolate has also been responsible for *Salmonella* outbreaks, at least one of which arose from *Salmonella*-contaminated cocoa powder being used to manufacture the chocolate. Dried egg, when first introduced commercially, was also a cause of *Salmonella* food poisoning, and an outbreak originated from an egg-based pasta containing *Staphylococcal* toxin formed in the dough prior to drying.

Dried foods may be prepared from liquids, e.g. milk, whey, eggs, flavorings, or from high-moisture solid foods, e.g. fruits, vegetables, meats, pasta doughs, etc. Whatever the source, the dried food will have a low moisture content, ranging from ⩽4% for milk powders to ⩽14% for cereals. Therefore, the a_w will be low which, combined with increased osmotic pressure, inhibits microbial growth. However, dried foods will act as carriers of microorganisms, and survival can be prolonged.

Different products will need differing drying regimes to reach the required moisture level; a good review of these is given (under the particular foodstuff) in ICMSF, *Microbial Ecology of Foods,* Volume 2[2]. Included in these regimes are air drying (both at ambient and after heating the air), drum drying, fluidized bed drying and freeze drying. For 'instant' products the drying regime is less severe than for non-instant ones and uses fluidized beds, which can be a significant cause of microbial contamination in the finished product.

Certain dried products, e.g. muesli-type breakfast cereals, are made by blending dry ingredients, and the product may be consumed without a further microbial kill-step. The ingredients must therefore be part of a Supplier Quality Assurance (SQA) program to insure that they do not carry pathogens. When dried products are manufactured from liquids, pasteurization of the liquid must be carried out, immediately before the concentration stage. If concentrate is shipped into a drying plant, it should be diluted then pasteurized and re-concentrated, unless the concentrate tanker and the equipment used to transfer the concentrate into the plant can be *guaranteed* not to contaminate the concentrate. For most plants, this is not practical without considerable investment in both fabric, equipment and procedures.

Manufacturing area environment is of critical importance for all dried products, including blended ones. Spray-drying equipment is physically very large, and uses large volumes of air both for drying and, more importantly, for transporting the cooled powder to storage and filling equipment. Real practical problems must be solved in order to overcome environmental contamination (see example 1: Powdered Formula Milk in Chapter 3), and these include hygienic equipment design and operation.

Ingredients

As with all products, ingredients for dried foods should be of good microbiological quality, they should be stored to minimize microbial growth before drying, and microbial contamination during storage should be prevented. For liquids and high-moisture solids storage should be in chill and storage time limited. Grains and cereals (whole or powdered) are already dried to a low moisture content but they are liable to contamination from pests. Container integrity must therefore be maintained throughout storage and pest control measures enforced. Certain dry ingredients, such as vegetables, herbs and spices, can carry high numbers of bacteria, including spores which will survive through to the finished product. Some dry flavorings, particularly if they were reconstituted into a slurry before drying, may have high spore counts. Careful SQA is the most effective way of preventing these unnecessarily high microbial loads.

Dry ingredients must also be kept dry, as any moisture getting onto a cereal, powder or flour may allow mold growth, with the possibility of aflatoxin production. Storage time should be limited to one month. If longer than this, appropriate precautions should be taken, which include reducing the product temperature to about 15°C (59°F), blowing in nitrogen or carbon dioxide gas as well as extra pest disinfestation measures.

Vegetables should be clean, undamaged, correctly stacked and storage time controlled.

Processing methods

Processing should achieve controlled drying of the liquid or particulate under conditions which prevent microbial growth and also contamination. The resulting product should be uniform in moisture content, with a final moisture level low enough to prevent microbial growth. Some 'instant' products, i.e. those readily soluble in cold liquid when reconstituted, undergo processing (instantizing) which increases the likelihood of microbial survival compared to less soluble products. In addition, fluidized drying beds are known sources of microbial contamination, so special attention needs to be paid to cleaning procedures and to the quality of cooling water when this is used for cooling.

Pasteurization of liquids should be part of processing, as the heating which the liquid receives while it is being concentrated is not a legal pasteurization treatment. There is no flow-diversion device for underheated liquid in evaporators, so pasteurization should be carried out before concentration (evaporation) of a liquid. Different products will each have their own scheduled concentration procedure, with specified temperatures and levels of total solids for each stage of concentration. Since concentration is under vacuum, thermophile growth is possible in certain stages of the evaporator, and the final concentrate will permit growth of *Staph. aureus*. The holding temperature before drying must be above 55°C (131°F) and holding time should be ⩽ 4 hours. Two concentrate balance (stock) tanks should be used and full CIP and sanitizing treatments be given alternately each 4 hours.

Even when correctly processed regarding time and temperature of drying, the problem of contamination remains. Environmental contamination of product is a particular hazard with dried foods. Many plants are large which makes effective cleaning of the environment, as well as the plant, difficult to achieve. A recent paper[1] illustrates this clearly with reference to pathogens in milk powder. However, other products are also open to environmental contamination and procedures applied to a rice-based infant food are used to illustrate how environmental and equipment contamination may be minimized. Figure 12.3 shows the flow diagram

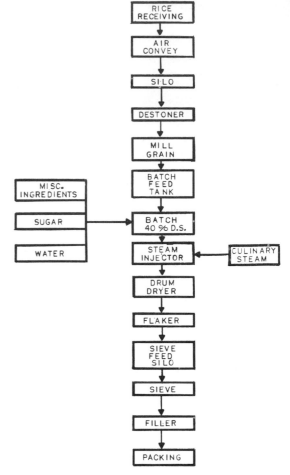

Figure 12.3 Flow diagram of drum-dried rice-based infant food

of this process, in which the product is drum dried, and the text gives the procedures under the following headings:

Equipment for cleaning manufacturing areas
Procedures for cleaning manufacturing areas
   Introduction
   Cleaning schedule for fabric of:
   Reception area for rice in sacks
   Ingredient loading area
   Rice-milling room
   Processing department
   Bulk finished-product silos
Cleaning schedules for machinery and equipment in:
   Reception area for rice in sacks
   Bulk finished-product silos
   Packaging department

Equipment for cleaning manufacturing areas

Brushes

- For brushes used in wet areas, washing and rinsing in hot water and detergent each time after they are used
- For brushes used in dry areas, strong beating of the brushes after use

Mops

Mops can be used to wash floors where a mechanical washer/dryer cannot be used or to finish off corners and areas not covered by this machine. A practical device consists of a metal frame on wheels that can be moved anywhere, with a fiber mop, a device for squeezing and wringing the mop, and two buckets – one for the detergent solution and one for water to rinse the mop as required.

Keeping in mind the need to minimize humidity, when the mop is used care should be taken to:

- Spread only a very fine layer of liquid over the floor so that this will dry quickly in the air
- Wash and rinse the mop and buckets thoroughly after use, with hot water and detergent
- Wring the mop as dry as possible

Vacuum cleaners

In the case of vacuum cleaners, two types must be available:

- For dry areas
- For wet areas (i.e. where washing with water is necessary)

The water vacuum cleaners must *on no account* be used for cleaning powder and dust in dry areas.

Dry vacuum cleaners are carefully emptied in the 'high care' area. They should not be routinely taken outside the high care area, as the whole cleaner would then need to be cleaned and sanitized before its return. Provided that the correct cleaning and operating procedures are followed, the powder residues from the vacuum cleaner should not be a contamination problem, and careful emptying should minimize dust.

Wet vacuum cleaners must be thoroughly washed – both tank and accessories – in hot water and detergent. In addition, they should be thoroughly dried if they are not used within the next 12 hours.

Note that the corrugated expanding flexible tubes often fitted to vacuum cleaners present a cleaning problem as dirt will collect at the bottom of each corrugation.

Two sets of accessories must be provided for the cleaners and should be marked with different colors (red and blue, for example):

- One for walls and floors
- One for machinery and points potentially in contact with food products

All vacuum cleaners must be equipped with a filter system for outgoing air leaving the vacuum point.

In the case of cleaners destined to be used in manufacturing areas, these must be equipped with a double air filter, i.e.

- Prefilter = 1μ approximately
- Filter = 0.3μ

Washer/dryers

The washer/dryer is ideal for fast washing and drying of floors. It has the big advantage of collecting the dry liquid immediately after the brushes have been used and saves a significant amount of manual work. The brushes clean any hard filth very energetically, since the pressure can be adjusted as required.

After each cleaning operation, the washer/dryer tanks containing the dirty liquid must be emptied and thoroughly rinsed. After rinsing a disinfectant solution must be put in the tanks.

Procedures for cleaning manufacturing areas

Introduction

To facilitate discussion of what is a major topic – cleaning and sanitation of manufacturing departments – we will highlight three separate points:

- Cleaning of manufacturing premises
- Cleaning of machinery and equipment
- Disinfestation

Our descriptions of the type and frequency of operations are only a summary, aimed to give an overall picture of the logic and sequence of procedures.

We have not attempted to go into any further detail since we firmly believe that:

- Carefully supervised on-the-job training of personnel in specific step-by-step procedures is the key to setting up good cleaning and sanitation procedures.
- Procedures must be constantly monitored and adjusted as necessary to improve performance. They must be constantly followed, even though repetitive.
- Success in applying the first two guidelines is vital for insuring on-going product safety – as well as insuring good overall product quality.
- Cleaning schedules for fabric of:

Reception area for rice in sacks

Remove all residue from floor using brushes and vacuum cleaner. Remove empty sacks which on no account must be left in area overnight. Ideally sacks should be removed on an on-going basis.	Daily at end of run

Wash floor, after brushing, with washer/dryer	At end of week
Dust walls, window sills and skirting boards using vacuum cleaner. Wash windows	Every 4 months
Clean wireways, light bulb holders, pipes, etc. using dry cloths	Every 4 months

Ingredient loading area (soya, etc.)

Vacuum clean and brush all residue from floor. Remove empty sacks which on no account must be left in area overnight. Ideally, sacks should be removed on an ongoing basis.	Daily at end of run
Wash floor, after brushing, with washer/dryer	At end of week
Dust walls, window sills and skirting boards using vacuum cleaner. Wash windows	Every 4 months
Clean wireways, light bulb holders, pipes, etc. using dry cloths	Every 4 months

Rice-milling room

Vacuum clean and brush all residue from floor. Remove empty sacks which on no account must be left in area overnight. Ideally, sacks should be removed on an ongoing basis.	Daily at end of run
Wash floor, after brushing, with washer/dryer	At end of week
Dust walls, window sills and skirting boards using vacuum cleaner. Wash windows	Every 4 months
Clean wireways, light bulb holders, pipes, etc. using dry cloths	Every 4 months

Processing department

Collect main residue from floor. Then wash floor with washer/dryer and finish off with mop.	Daily at end of run after cleaning of machinery
Clean tiled walls with moist rags or sponges	Daily at end of run after cleaning of machinery
Wash windows.	At end of week

When cleaning has been finished, spray floors and walls with a nebulized disinfectant solution of quaternary ammonium salts using a shoulder pump. Pour about 10-20 ml of concentrated disinfectant into discharge outlets.

Bulk finished-product silos

Vacuum clean residue from floors	Daily
Wash floors using washer/dryer and finish off with mop	At end of week
Vacuum clean dust and powder from walls, window sills, skirting boards, etc.	Fortnightly
Clean and wash windows, wireways, lamp holders, etc.	Every 3 months
Brush floors and grilles	Every 3 months

Packaging department

Constantly clean manufacturing residue from floor using brushes and vacuum cleaner. At end of production, complete cleaning, taking care to remove all residue from under machines, corners, etc., once again using brushes and vacuum cleaners.	Daily
Wash floors with washer/dryer and finish off with mop.	At end of week
Vacuum clean dust and powder from walls, wireways, lamp holders, window sills, etc.	At end of week
Wash windows	Every 3 months

- Cleaning schedules for machinery and equipment in:

Reception area for rice in sacks

| Sacks. Careful attention should be given to pallets holding rice or other raw materials before they enter the processing departments. If they are very dusty, they should be cleaned using a jet of compressed air. If they are very dirty, sacks should not be allowed to enter the processing departments. | |
| Platform scales. Remove all residue from platform and dust framework with dry rags. | Daily at end of run |

Flexible screw feeder. Run the screw feeder until pipes have been completely emptied.   Daily at end of run

Sack dumps. Clean the inside of the hopper using compressed air. Discharge residue from below. Dust the outside of the hopper using dry rags. Remove resistant dirt (stuck to surfaces) using moist rags.   Daily at end of run

Clean magnet using plastic scrapers   Every week

Screener (sieve). Open up equipment and remove all residue present on net. Check that sieves are intact; replace those that are damaged. Clean outer part of equipment with dry rags (moist if necessary).   Every week

Aspirator-destoner. Dismantle and remove all residue contained inside. Clean outer part of equipment using dry rags – moist if dirt is stuck to surfaces.   Every week

Air filter. Dismantle and clean filter brushes with compressed air. This should be done outside production departments.   When necessary

Drums. Empty receptacles and clean them roughly   Daily at end of run

Wash with water and detergent, rinse well, and blow dry   At end of every week

Bulk finished-product silos

Product collector filter. Dust outer surfaces   Weekly

Open up and clean interior using compressed air. Discharge residue into silos (which should be empty and ready for subsequent cleaning)   Every 6 months

Silos. Disconnect terminal portion of silos from below and clean inner surfaces using a brush   Every 6 months

Screener (sieve). Dust outer walls. Open up and inspect. Remove residue using a vacuum cleaner.   Daily, end of run

Note: This is a critical control point in terms of contaminants. Inspect all discarded material (both large and small) carefully for foreign bodies, insects, etc.

Packaging department

Note: We recommend that most of the route covered by conveyors be protected with sheets of plexiglas. This stops both the product and the surfaces on which it is placed from being hit by dust and hence by bacteria.

Cyclones. Beat outer surfaces using a rubber hammer. Clean inner walls using compressed air working from above. Discharge residue from below   Daily at end of run

Surge hopper. Clean inner surfaces using compressed air. Clean outer surfaces using dry rags (slightly moist if necessary)   Daily at end of run

Metal detector. Open up and inspect. Remove all residue carefully, using plastic scrapers   Daily at end of run

Packaging machine. Dismantle protective walls. Remove surface residue using a vacuum cleaner. Finish off with short bursts of compressed air   Daily at end of run

Conveyors. Blow belt surfaces using compressed air and, if necessary, wash away any splashes of glue   Daily at end of run

Wash with detergents and sponges. Next rinse and disinfect with a bactericide in spirit solution   Every 3 months

Overwrap machine. Dust outer surfaces using dry rags   Daily at end of run

Fill scales. Remove all residue from platform and dust the entire piece of equipment with dry rags   Daily at end of run

Dust collectors. See 'Cyclones' above.

Outline HACCP plan for a rice-based infant food

A different and rather more complex rice-based infant food is use to illustrate a HACCP plan applied to the manufacture of a dried product. Figure 12.4 gives the process flow diagram; it should be noted that a drum drying stage is followed by spray drying after addition of ingredients, some of which would be unsuitable for drum drying due to their texture and the fact that drum drying would cause loss of flavor. The outline HACCP plan for the process is given in Table 12.2. As well as listing the critical control points (CCPs), with the accompanying options and limits, improvements are noted. These should be introduced during scheduled plant refurbishment. CCPs relating to construction features and management procedures to minimize environmental contamination are given in Chapter 3, as Example 1 of HACCP applications. Remember that the drying and filling areas are often the principal contamination source for dried products, and although Example 1 refers to milk powder, the information given there may be extended to the environment of other products.

Process control and validation

The principal control of the drying process is by the temperature of drying and the speed at which the product is dried. The size of the particle is also important. Whether it is droplets as in liquids, or pieces of, for example, meat or vegetables when drying solid foods, the particle sizes should be similar. If they are not, the moisture content of the finished product may not meet the specified level throughout a product batch.

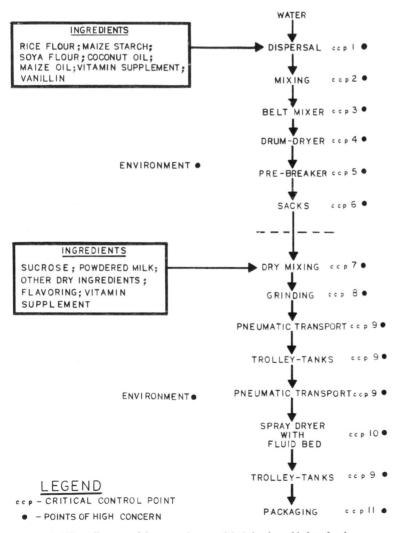

Figure 12.4 Flow diagram of drum- and spray-dried rice-based infant food

Table 12.2 Analysis chart of process and control options: pre-cooked dried rice infant food

Critical control	Description	Hazard	Controlled by	Control limits	Improvements
1. Dispersal	Addition of hot water	Bacterial proliferation occurs over time	Daily cleaning	Maximum of two production runs before cleaning.	Lines should be doubled up to permit more frequent cleaning
			Temperature control	Temperature adjusted over __°C	
2. Mixing	The solution is mixed in stirred tank	As above	As above	As above	
3. Belt mixer	For achieving the required mixing of all ingredients	As above	As above	As above	Inclusion of an in-line cooker
4. Drum dryer	Evaporation of water by a Drum Dryer (DD) The dryer works at 150°C The product reaches ____°C	The product leaving DD meets current of air caused by extractor. Microorganisms may be transferred to the product. Bacterial proliferation occurs over time in plant and environment	Daily cleaning and inspection. Temperature control by steam pressure. Procedures for personnel	Repeat the cleaning when parts of the plant are not approved	Tunnel with flows of 'sterile' air directed toward sheet of product leaving DD
5. Pre-breaker	The sheet form of the product is reduced in pieces				
6. Sacks	The product is put in sacks	Contamination of sacks	Only new sacks (checked by personnel) are utilized	Inspection Sacks rejected if not clean and intact	Trolley-tanks with over pressure filtered air
7. Dry-mixing	Addition of other ingredients to the semi-finished product prepared	Pathogens in milk and fruit powders (raw materials)	Supplier assurance including factory audit	Freedom from pathogens when agreed testing scheduled followed	
8. Grinding	The dry mixture is ground by mill	Contamination from mill	Regular cleaning		
9. Pneumatic transport – trolley tanks – pneumatic transport	Pneumatic transport of product, then stored in trolley tanks until transferred to spray dryer by pneumatic transport	During these operations the product can be contaminated by trolley-tanks as these tanks are transferred from one building to another. Contact with air can introduce microorganisms	Regular cleaning of trolley-tanks with hot water and sanitizing solutions. Drying by hot air. Filtration of ducted air. Inspection of ducting for cracks.	Inspection The filter improperly cleaned must be retreated. Class 1 filter on inlet air. Production halted until any cracks or leaks are repaired	Keep all trolley tanks over pressure with filtered air during the discharge.
10. Spray dryer Fluid bed	The dry mixture is added with water and treated in spray dryer with a fluid bed	Atomizer rotor is subject to bacterial growth	Daily dismantling washing and sanitizing	If the rotor isn't OK on inspection the cleaning operations must be repeated	
11. Packaging	Product placed in sachet of impermeable nitrogen flushed laminate before heat sealing. Sealed sachet placed in a box	Packaging material could be contaminated. Contamination (pathogens) from the environment. Product residues in filler could contaminate fresh product as it is filled	First meters of spool discarded Prevent build up of residues by dry cleaning the filler		Filtration of air in the packaging area

Deficiencies and cross references should be noted against the appropriate CCP no. and recorded on this matrix.

Moisture content of product must be measured to validate that the specified level is achieved. The choice of method is important, and so is the way it is carried out. Small changes in methodology can result in misleading results, even with the traditional 'hot air' oven test. Automated instrumental methods, such as Near Infrared Reflectance, require correct calibration of the instrument at least once a day. Accurate control of moisture content is needed to prevent microbial growth; localized high moisture content in powder or other dried material may allow mold growth, causing spoilage at best and aflatoxin production at worst.

As with all processes, thermometers must be accurate and calibrated as must instruments measuring total solids of concentrated material.

In the environment, incoming air (both to the room and to the dryer and powder transport system) is filtered[3]. The air filtration system must be of sufficient capacity to handle the large volume of air required, and filters must be maintained (or replaced) regularly. Movement of personnel must be controlled and hygienic procedures must be enforced to prevent contamination entering the drying/filling areas, and cleaning must be thorough.

Validation of the environment regarding pathogen contamination is through microbiological monitoring of vacuum cleaner residues and swabbing of selected areas. These environmental samples are enriched and examined for *Salmonella*. *Guidelines for Good Hygienic Practice in the Manufacture of Milk-based Powders*[3] gives useful advice on where to take environmental samples and indicates the frequency of sampling. It also includes a recommended sampling level of product for *Salmonella* testing, as does *An Evaluation of the Role of Microbiological Criteria for Foods and Food Ingredients*[4], which gives the current US sampling plans. Product testing for *Salmonella* is also a validation of safe processing, but must not be regarded as a substitute for environmental testing.

Microbial distribution is not homogeneous in dry material, and the sampling method as well as sample size must take account of this. The paper by Habraken *et al.*[5] states that continuous sampling gives the most accurate assessment of microorganisms in milk powder. Certainly, within-bag as well as bag-to-bag variation has been demonstrated with a powder containing approximately 5 *Salmonella* per kg. Commercial availability of continuous sampling devices is currently (1990) limited, but they are obtainable although with a long delivery date. As with all sampling points inadvertent contamination must be avoided in use.

When examining products other than milk or egg, remember that some herbs and spices, e.g. mustard, cinnamon, mace, nutmeg, etc., have anti-microbial properties. These must be taken into account when determining microbial counts; it may be necessary to use a higher dilution when plating, or to use a higher ratio of enrichment broth to sample when testing for *Salmonella*, e.g. 50 to 1 rather than the usual 10 to 1.

Many powders, including milk as well as spices and herbs, will need a stage in their examination which allows damaged microorganisms to resuscitate or repair the damage. Different products need different procedures to allow resuscitation, but the main points to consider include the way the powder and diluent are mixed, and the time and temperature of holding the diluted powder. Remember, microbial multiplication should not occur during resuscitation. It should only enable the microorganisms to repair any damage which occurred while being held in a dry environment. If resuscitation is not done, then false negative results will occur, and the true safety of the product is not being assessed and unsafe product could be allowed to leave the manufacturing plant.

Packaging materials

These are not usually sterilized before use but should not add contamination to the product. They should therefore be manufactured hygienically (SQA should be required), be protected during storage, and stacking should allow easy access for cleaning and laying pest bait. Cleaning should be by vacuum cleaner, not with a brush. The 'outers' should be discarded just before the packaging materials enter the 'high care' filling and packing area, to minimize contamination of the area.

Coding and labeling

Coding is important for identification of factory, batch, filling line and time of filling as well as for stock rotation. Coding should be clearly visible; this may mean that pallets must be marked if the sacks or packs of product are stacked horizontally.

Labeling on each sack or pack must give clear, unambiguous instructions regarding storage in dry conditions and product reconstitution where this applies. When reconstituted product is held at ambient temperatures, or above as with infant foods, microbial multiplication is likely to occur. This may result in an unsafe food, so the instructions should state the correct method of use for each product. Since customer abuse is possible, the manufacturer should determine the behavior of products when held at temperatures allowing growth of pathogen or spoilage microorganisms (the abuse step). This will provide accurate information for the instructions as well as assessing product safety.

Storage and distribution of product

Finished product should not be stored in the same area as ingredients. The product must be kept dry, so pack damage must be prevented both during loading of pallets and stacking by fork lift trucks. Should pack damage occur, the pack must be removed and the contents only re-used following adequate re-processing. Any spillage of dried product on the floor must never be re-used, but promptly removed with a vacuum cleaner. Distribution must maintain pack integrity.

Special considerations

The most important consideration relating to dried products is the hazard of contamination from the environment in the drying, intermediate storage and filling areas. The dried material is exposed to air, and being transported by air, to a greater extent than most processed foods. Because many powders are fine the environment is dusty. This will cause cleaning problems and, if the dust becomes damp, it can support microbial growth, which then becomes an additional contamination source. Remember, too, that fine dusts present both fire and explosion risks[6],[7].

Fluidized beds, which complete the drying process, are also sources of contamination of dried product. They are not easy to clean and are operated at temperatures which allow microbial growth.

Control of environment must overcome these problems, starting with the design and fabric of the building and equipment, continuing with planned maintenance, adequate cleaning schedules and, not least, management of personnel. Although end product testing is part of the control system, microbiological monitoring of environmental samples with reference to *Salmonella* or other pertinent pathogens is the main verification that effective control has been achieved by the Quality Assurance program.

References

1. Mettler, A. (1989) Pathogens in milk powder – have we learned the lessons? *Journal of the Society of Dairy Technology,* **42**, (4), 48–55
2. ICMSF (1980) *Microbial Ecology of Foods,* Volume 2, Food Commodities. International Commission on Microbiological Specifications for Foods, New York: Academic Press
3. ABPMM (1987) *Guidelines for Good Hygienic Practice in the Manufacture of Milk-based Powders.* London: Association of British Preserved Milk Manufacturers
4. NAS (1985) *An Evaluation of the Role of Microbiological Criteria for Foods and Food Ingredients,* pp. 187–190 for dried milk, 248 for dried egg. National Academy of Sciences, Washington DC: National Academy Press
5. Habraken, C. J. M., Mossel, D. A. A. and van den Reek, S. (1986) Management of *Salmonella* risks in the production of powdered milk products. *Netherlands Milk and Dairy Journal,* **40**, 99-116
6. Beever, P. F. and Crowhurst, D. (1989) Fire and explosion hazards association with milk spray drying operations. *Journal of the Society of Dairy Technology,* **42**, (3), 65–70
7. Abbott, J. A. (ed) (1989) *Users Guide to the Prevention of Fires and Explosions in Driers.* Rugby: Institution of Chemical Engineers

Further reading

Gibbs, P. A. (1986) *Scientific and Technical Survey No. 154. Microbiological Quality of Dried Foods,* The British Food Manufacturing Industries Research Association, Randalls Road, Leatherhead, Surrey, England
Luh, B.S. and Bhumiratana, A. (1980) Breakfast rice cereals and baby foods. In *Rice: Production and Utilization,* edited by B. S. Luh. AVI Publishing Company, Inc., acquired by Van Nostrand Reinhold, New York

Codes of practice or GMP guidelines

ABPMM (1987) *Guidelines for Good Hygienic Practice in the Manufacture of Milk-based Powders,* Association of British Preserved Milk Manufacturers, 19 Cornwall Terrace, London NW1 4QP
3-A Accepted Practices and Standards relating to dry milk and dry milk products are published by the International Association of Milk, Food & Environmental Sanitarians (IAMFES) in the *Journal of Food Protection.* These practices and standards are revised on an on-going basis, and current *Journal of Food Protection* issues should be consulted, or the IAMFES contacted at PO Box 701, Ames, Iowa, 50010, USA for the up-to-date documents.

12.D Conventionally canned foods

Background

The technology of conventional canning may be summarized as filling food into containers, e.g. of metal, glass or plastic, followed by sealing the containers and heat treating them. This heat treatment is sometimes called 'terminal sterilization' and is designed to eliminate extremely large numbers of spores of the pathogen *Clostridium botulinum,* and reduce the chances of survival of the much more heat resistant spores of spoilage organisms. This condition is usually called 'commercial sterility'. After heating, the containers must be cooled and subsequently handled in a way which maintains container integrity and avoids re-infection caused by leakage.

The amount of heat required to give safety from *Cl. botulinum* varies enormously and depends primarily on the composition of the food. One of the most important factors is the acidity as measured in pH units. A 'low-acid' food has a pH greater than 4.6 and a water activity greater than 0.85 (US regulations) or a pH greater than 4.5 (UK Code of Practice).

The minimum sterilization process for *Cl. botulinum* is usually expressed as 12 log cycle or decimal reductions assuming all organisms are spores, and is called a '12D reduction'. This is often understood as requiring an $F_o = 3$. One F_o is equivalent to one minute at 121°C assuming instantaneous heating and cooling; so $F_o = 3$ means 3 minutes. Meats usually receive a minimum of $F_o = 6$.

The technology of aseptic packaging, sometimes called aseptic canning, is very different. In the aseptic process the food and containers (which may be metal cans) are 'commercially sterilized' separately. The food is cooled to a suitable filling temperature, then the containers are filled and sealed, all under aseptic conditions.

In conventional canning, the sterilization operation is often described in technical literature or conversation in terms which may cause some confusion. It is sometimes called the 'cook' although this includes the 'come up', 'holding' and cooling of the containers. A 'botulinum cook' means the amount of heat needed to deal adequately with *Cl. botulinum* spores but not necessarily those of the more heat resistant spoilage organisms. Another term for the sterilization operation is the 'process' which may be confused with the total preparation and/or manufacturing process unless the context is clear.

Because the 'kill step' of terminal sterilization is given to sealed containers and because of the excellent control which is technically possible over container integrity, conventional canning is inherently a very safe technology. The technology is also suitable for the HACCP approach, which can provide the required positive assurance of safety.

Preparation and storage of ingredients

To insure that food is safe and wholesome at the time the sterilization process is applied, Good Manufacturing Practices (GMP) must be applied before and during the preparation and storage of ingredients. The reason for this is that while the sterilization process will kill organisms it will not destroy heat resistant toxins, such as staphylococcal enterotoxin. Furthermore, sterilization should not be expected to remove taints caused by previous spoilage; indeed some taints may only develop after heating in the sterilization process. Apart from attention to GMP it is worth remembering that careful examination of materials held as a result of line breakdown or stoppage(s) is important in preventing tainted finished product.

Processing methods

These will be considered under the following headings:

- Manufacturing processes
- Sterilization process establishment
- Container integrity
- Post (sterilization) process container handling

Manufacturing processes

These include standard preparation processes of ingredients for which the usual (GMPs) apply. For example, size reduction (dicing, mincing); blanching; mixing; pre-or part-cooking. Since cooking also occurs during 'commercial sterilization' allowance is made for this in product development. However, ingredients containing air or gases which are driven off by heating, e.g. vegetables, are blanched before filling. This avoids the presence of air in the container which during 'commercial sterilization' could cause additional stresses that might affect container integrity as well as causing unwanted longer-term oxidative changes to the product.

One manufacturing process which has very important safety implications is the acidification of low-acid foods. Where this can be done, and its application is limited by flavor changes, it enables a 'pasteurization' process to be used instead of 'sterilization'. Some products such as palm hearts, artichoke hearts, pimentos and many pickles can be acidified. Because the safety of the product under these conditions depends on each and every container having a pH below 4.6, acidification *must* be properly achieved. Further details are given in an excellent text by the Food Processors Institute[1], which lists five methods of acidification. These are (a) blanching foods in acid solution, (b) immersion of foods in acid solution, (c) direct batch acidification, (d) direct addition of a predetermined amount of acid to individual containers, and (e) addition of acid foods to low-acid foods in controlled proportions. Be aware that when one component is above pH 4.6, especially if it is in large dices or pieces, it can take a significant time to reach equilibrium.

Sterilization process establishment

Responsibility for process establishment should be given only to those with expert knowledge of the subject and experience of the product range together with the sterilization equipment and procedures that will be used. Remember, many judgments are made in the detailed HACCP that is required to insure product safety, and the quality of the judgments is critically important. Botulism from your products could cost you your business. This is because the toxin is very potent, the disease is rare but readily identifiable and the fatality rate high. Expect any cases to be publicized with, historically, extremely serious consequences to the packer of the implicated food. Remember that commercially canned foods have a high reputation for safety.

The objective of process establishment is to achieve 'commercial sterility' and to optimize sensory and nutritional attributes while meeting the microbiological requirements.

Those with experience are aware of the considerable effect of the product on heat resistance of organisms. Acidity as measured by pH is an obvious factor but it is not the only one. Other materials such as salt, sugar, nitrate and fat may also have an effect. It is worth remembering that it is not only the effect of heat on spores in the product but also that of the recovery medium or product which determine survival.

Further consideration of this topic is beyond the scope of this chapter and readers are referred to Stumbo[2] and Campden Technical Manual 3[3] for further information.

'Commercial sterility' involves both a pathogen, *Cl. botulinum*, and spoilage organisms. The minimum process for *Cl. botulinum* spores is given in Campden TM3[3] as sufficient to reduce the probability of survival to not more than 1 in 10^{12} cans (containers). This is a safe definition which takes into account the volume of the containers, and requires rather more lethal heat than the original $12D$ concept. It is also used in the UK Codes of Practice No. 10[4].

With regard to spoilage organisms, the position is more complex. Campden TM3[3] recommends as a minimum that the survival rate of spores should not be more than 1 in 10^4 cans with the comment that most processes in use would give even lower chances of survival, e.g. not more than 1 in 10^6 cans. Stumbo et al. [5], suggest as a minimum, assume a 4 log cycle ($4D$) reduction of *Clostridium sporogenes* with a D_{121} of 1.5 minutes. Thermophilic spore formers are not usually a problem in temperate climates providing the cans are cooled quickly (say, within 1 hour of completion of cook) to about 40°C (104°F). For packs for hot can vending machines, the process must insure acceptable thermophilic spoilage rates under appropriate incubation conditions, e.g. 55°C (131°F).

With regard to sensory and nutritional attributes, this is largely a matter for the individual canner. However, be sure that any material added for nutritional reasons does not carry an unacceptable microbial load or violates regulations.

The result of process establishment is summarized in a scheduled process document (or report) which is specific for each product, container type and size, and type of sterilization equipment. It is obtained by bringing together an estimate of the lethal heat needed for microbial destruction and the lethal heat derived from product heating data in the proposed sterilization equipment.

The amount of lethal heat needed to deal with an expected microbiological load may be determined experimentally using methods described by Stumbo[2]. It is more usual to use process evaluation data from similar products which are known to give satisfactory results. However, this should only be done when there is an ample margin of safety with respect to *Cl. botulinum*.

The product heating data required may come either from heat penetration tests which measure the temperature/time profile (usually using thermocouples) or from direct microbiological methods. Sterilization equipment simulators are often used for these tests but be sure that there is good evidence that the simulator compares to the full-scale equipment with known accuracy. It has been known, and this can be seriously misleading, for inherent design features of a simulator to give better, more favorable, i.e. higher, F values than the full-size equipment.

Heat penetration tests are preferred because they are quick, relatively easy and give data which can be treated mathematically to calculate equivalent processes at other times or temperatures and for other container sizes. Further information on these tests is given in Campden Technical Manual 3[3] and by Hersom and Holland[6].

Be aware that because of variation between individual containers, rates of heat penetration may vary appreciably for the same product. It is therefore good practice to take at least three replicate containers for each experiment and perform sufficient experiments at each stage of recipe development to be sure an adequate process is achieved under the most adverse conditions likely to be encountered.

It is well established that the geometric center is not always the slowest heating point in a container. For this reason, a center F_o value of 3 should not be assumed to provide safety against botulism. The location of the slowest heating point should be determined during the initial stages of process establishment.

It cannot be emphasized too strongly that scaling up a recipe from kitchen batch to factory production may result in significant changes in heat penetration rates. The magnitude and direction of these changes cannot be readily predicted. While heat penetration tests should be performed at each stage of recipe development, final confirmation should aiways be made on product prepared under normal factory operating conditions.

Direct microbiological methods are used if the process cannot be assessed by heat penetration tests, e.g. if a rotary sterilizer does not allow thermocouple wires to be installed.

The principle of these tests is that the product is inoculated with organisms of known (high) heat resistance and filled into containers. After sterilizing, the survivors are counted. This is also called the inoculated pack or count reduction method[6]. A variant of this is to put the spores in a suspending medium in a glass bulb or capillary tube[6]. Another variant is to incorporate the spores in an alginate ball or block as described by Perkins *et al.* [7] and by Campden workers[8,9].

Since it is important to know the limits of accuracy this means using the appropriate number of replicates. Direct microbiological methods are expensive as they are time consuming and use large volumes of valuable and sometimes hard-to-make spore suspension of the required heat resistance.

It is useful to remember that product heating data is influenced by both intrinsic, i.e. product dependent, and extrinsic, i.e. equipment dependent, factors. Both must be considered when establishing a process because any factor which alters the lethality of the process is a 'critical factor' which must be specified in the 'scheduled process'.

As examples of intrinsic factors, consider the need to rehydrate pasta fully, or the effect of changes in the type of starch or the method of cooking starches. Another example is a change in dice size of vegetable garnish which not alters only the rate of heat penetration into individual dice but may affect the density of packing in the container and thus alter the product heating data.

Extrinsic factors are related to the performance of filling and sterilizing equipment. Filling is important, not only to get the correct ratio of solids to liquid, but where the headspace bubble is responsible for increasing the rate of heat transfer, it becomes a 'critical' factor.

Although not usual, the type of closure may affect the process. It has been known for the use of tin-free steel ends in place of enameled ends to affect the rotation of the can and hence the process.

Sterilizers of many different types are used in conventional canning. They usually use steam or water as the heating medium or, less usually, a controlled mixture of steam and air or direct heating of the (metal) container. Containers may be held stationary (non-agitated) or rotated, e.g. end-over-end, axial rotation or intermittent axial rotation used in 'cooker-coolers' of the reel and spiral type.

Each type and size of sterilizing equipment has its particular characteristics which must be known both for product heating rates and for the accuracy of sterilizing times and temperatures which they deliver. This information is always required for the scheduled process and may be demanded by regulatory authorities.

It is *absolutely essential* to know the limits of accuracy of both time and temperature given to each and every container. Do not fall into the trap of thinking that this is the same as the accuracy of an individual instrument, e.g. a thermometer. The knowledge required comes from knowledge of the total system accuracies of equipment used for measuring, controlling and recording time and temperatures under 'worst case' conditions. This knowledge is needed for both process establishment and validation (audit).

It is also important to consider how equipment malfunction and line stoppages could affect the sterilization process. It has been known for some mechanical fillers to put a 'mat' of product into a container if intermittent short stoppages occur. The HACCP approach – with questions such as what if? how would we know if it happened? what difference could it make? – is the one which the experienced professional takes.

Process establishment using 'outside' data is sometimes practiced. When a new product is proposed, in order to reduce experimental work, it may be legitimate for preliminary trials to use a process obtained from a research association publication or another responsible source. In these circumstances, it is extremely important to insure that the new product is essentially the same as the one for which the process was designed. It must be clearly understood that products with the same name or title may have very different formulations. This is especially true of garnished products with a sauce thickened with starch. Even simple products such as vegetables in brine will have different process requirements depending upon the size, shape and amount of vegetables in the can.

Therefore, when 'outside' data are used, heat penetration or other tests *must* be made to confirm that any sterilization process for low-acid packs is at least sufficient to insure safety with respect to *Cl.*

botulinum before the product is eaten by *any* person.

Having decided on a process for a particular product and container, it is usual to perform incubation tests for the more heat resistant spoilage organisms. It is wise to regard these as 'supporting evidence' only unless they are part of an 'inoculated pack' experiment. This is because the results depend on the initial microbial load, the effect of the sterilization process and the statistics of sampling.

Incubation tests are used to predict probable rates of spoilage in given circumstances. Because of this, it is essential that the purpose of the test and its statistical basis are clearly understood before predictions are made from the results. It is not widely appreciated that even when no spoilage is observed after incubation the probability of spoilage in the lot from which the samples were drawn may still be quite high. This is especially true when low numbers, e.g. 12 cans, are considered. As a rule-of-thumb 3000 cans would have to be incubated with no spoiled cans found in order to be reasonably sure that the true spoilage rate did not exceed 1 in 1000. For this reason, although incubation tests provide valuable information, the results must *never* be used as the sole criterion for assessing the safety of a process or the bacteriological status of a quantity (a 'lot') of production.

Providing that their limitations are understood incubation tests provide useful information and are widely used.

Although there are many ways of performing incubation tests, they may be considered in two main categories. Firstly, laboratory incubation tests which are destructive and designed to assess the probability that organisms surviving the process will spoil the product under very severe conditions. Their value is in providing some confirmatory evidence of satisfactory process establishment and partial supporting evidence for the disposal of goods produced where a deviation from the correct process has occurred.

Secondly, bulk incubation tests, which are non-destructive and designed to assess the post process leaker spoilage rate. Their main value is that they enable corrective action to be taken to avoid expensive spoilage outbreaks.

From time to time it is necessary to re-establish a process. Changes in recipe, manufacturing method, filling method or location may significantly and adversely alter the effectiveness of the sterilization process. In these circumstances, heat penetration or other tests must be made to re-establish a satisfactory sterilization process. This evidence must be obtained and kept permanently on file by the department responsible for process establishment to show that the pack continues to have a satisfactory process.

Since a positive assurance of safety is required,

the proper approach is to check unless there is good and sufficient evidence to show that the change would not alter the process.

Container integrity

Achievement and maintenance of container integrity are essential for microbiological safety and to minimize spoilage. The details of what needs to be done depend on the container and closing (sealing) equipment. For cans and glass containers see Section 9 of Canned Foods[1] and for flexible packages see NFPA Bulletin 41-L[10].

Reputable manufactures of containers and closing (sealing) equipment provide detailed recommendations for their systems. What the processor must do is to manage the operations of the equipment so that container integrity is achieved. This requires the establishment and audit of detailed procedures appropriate to the individual needs. When establishing a system remember that commercial customers as well as regulatory agencies want assurance that these operations remain under control.

Post-process ('sterilization') container handling

Although early work from the 1950s was done with cans, the principles apply to other container types. It was established that, under certain conditions cans could leak. Leakage may be either air or water.

If water enters, it is much more likely than air to carry microorganisms although if the volume leaked is small enough it may not carry an organism. Shapton and Hindes[11] showed this difference between air and water clearly by comparing the blown can rate for oxtail soup. When wet cans were run over wet runways, during a 7-week period 87 666 incubated cans gave a spoilage rate of 1.17 cans per 1000. This was considered excessive and the practice stopped. In the following 3 months, with dry cans run over dry runways, 60 000 cans were incubated with zero blown cans.

It is important to realize that leaker infection or spoilage does not always result in blown cans. Thorpe and Everton[12] state that it is well established that for certain packs the non-blown spoilage rate considerably exceeds the blown rate. It is also well established that some packs such as 'garden peas' are more susceptible to leaker spoilage and blowing than others, e.g. carrots.

By taking appropriate measures the leaker infection or spoilage rate can be dramatically reduced, and with good manufacturing practices, rates of 1 in 5000 to 1 in 10 000 or substantially better may be achieved. Note that these rates are for cans 'blown' during incubation at 30°C (86°F) for 10 to 14 days.

Thorpe and Barker[13] point out that leaker infection is the most common form of spoilage encountered in the canning industry. However, it should be clearly understood, as Shapton and Hindes[11] state, that the rate of spoilage depends on the method of assessment, e.g. sales credits for spoilage give very different figures from those obtained with bulk incubation tests.

It is important to remember that pathogenic organisms may also leak into cans. *Staphylococcus aureus, Salmonella* spp. (including *S. typhi*) and *Cl. botulinum* type E have all been involved. Four key factors in effective control are:

1. Sound container and seam or seal formation. Thorpe and Everton[12] state 'If seams are of poor quality or if seams are subjected to abuse during passage of the can handling equipment, then the percentage of leaking cans will be greatly increased'.
2. Disinfection of water used to cool the containers. The first chance of post-process infection is during cooling. The use of a disinfectant, e.g. chlorine, not only insures a low microbial count but also acts as a positive disinfectant for any organisms washed out from protecting rust or scale in a retort or sterilizer.
3. Dry container, e.g. can, handling. If is important to remember that the liability for infection remains as long as the can is wet. Thorpe and Everton[12] state 'However, if dry cans are re-wetted or conveyed over wet runways, the spoilage hazard is immediately re-introduced'. This point was developed by Thorpe and Barker[13], who state that

   Wet surfaces, therefore, need to be disinfected. If surfaces are inadequately cleaned, subsequent treatment with disinfectants is ineffective. As a result, both large numbers of viable bacteria and ample nutrients remain on the surfaces and the original high degree of contamination is soon re-established.

   It has been found that if surfaces of can handling equipment which are wet, or may become wet, have less than 500 bacteria per $26 \, cm^2$ ($4 \, in^2$) the occurrence of leaker infection is reduced to a commercially acceptable level. This standard or better (100 to 200 bacteria per $26 \, cm^2$) may be achieved consistently if the wet surfaces are regularly and properly cleaned and disinfected.

   Surfaces of equipment that remain dry throughout production periods need not conform to this standard but will still need to be cleaned and disinfected, although at less frequent intervals.
4. Avoidance of abuse. Food containers, e.g. cans, do not need to be treated like eggs, but changes of speed and direction put stresses on the cans

which may momentarily 'spring' the seam. Clearly, the stronger the forces, the more likely this is to happen. There is an excellent treatment of both impact and pressure abuse by Thorpe and Barker[13] and of the importance of hygienic (sanitary) design in minimizing its effect. The review on leaker spoilage by Thorpe[14] is required reading for those concerned with leaker spoilage and also gives detailed GMPs.

Process control and validation

This has two main aspects; first, equipment – including instrumentation – should be suitable for its intended purpose and second, suitable control and validation procedures must be devised and validated (or audited). It is axiomatic that equipment must be properly designed, carefully installed, adequately cleaned and maintained and correctly operated in order consistently to deliver what is required of it.

For the purposes of this section, consideration will be limited to control and validation of 'cook-room' (sterilization area) operations under the following headings:

- Equipment
- Cook-room facilities
- Procedures
- Record keeping
- Cook-room audits

For notes on cook-room personnel see the later sub-section 'Special considerations'.

Equipment

- General

Sterilizing equipment performance and temperature distribution will inevitably be affected by ancillary equipment and the way in which utilities (services) are supplied. For example, vents and bleeds (bleeders) should be suitably sized and sited and discharge freely to atmosphere; air and water supplies must not leak into the retort during sterilizing; build-up of condensate in steam retorts must be prevented; drains must be adequately sized, sited and protected against blockage by fallen containers or other debris.

It is always helpful to remember that a job is more likely to be done correctly if it is made straightforward for the operator. For example, it is helpful if there is uniform and clear color coding of utilities on valves as well as lines. As another example, handles which turn 90° between 'on' and 'off' should be consistently arranged to work in the same direction relative to the operating position.

Particular attention should be given to older equipment and/or to equipment used for seasonal production to insure that it meets the required standard of performance. For further information, see Section 8 of *Canned Foods* by the Food Processors Institute[1].

Every sterilization process is defined by at least two parameters, i.e. time and temperature, and in some instances by others, e.g. can rotation or over-pressure. It is therefore essential that these parameters are measured, controlled and recorded within known, close limits of accuracy.

● Determination of time.

Process time should be measured, controlled and recorded so that the process time given to the containers is known and can be shown to be not less than the scheduled process time. This apparently simple requirement has quite far-reaching implications. It is obvious that an accurate clock should be easily readable from all operating positions in the cook-room. However, when two or more clocks are used they must be of a similar type and synchronized to avoid confusion and consequent processing and recording errors. Ideally, each retort would have its own timer.

Continuous sterilizers are of very different designs and the cook-time is adjusted in different ways. For example, with a cooker-cooler it is the reel speed which affects cook time and speed of can rotation; in hydrostatics it is the carrier chain speed and, if the position of the water seal can be altered for the same steam dome (or section, or chest) pressure, it is the position of the water level chosen for the hydrostatic legs which determines the process time as well as chain speed.

In an agitating hydrostatic retort, rotation is indirectly transmitted to the containers by the relative speeds of the main carrier chain and the chain which rotates the 'clusters' of carrier bars.

Although an automatic timing device is usually installed to indicate and sometimes record the machine speeds which determine cook time it should not be used as the only means of determining machine speed. The simplest way to obtain an independent check may well be to measure machine speed against a stopwatch once per shift. As a sound principle, all automatic devices should be designed and installed to 'fail safe' with appropriate 'out of tolerance' alarm signal or signals.

Process time should be recorded so that the cook time may subsequently be determined to the required degree of accuracy which may be ±1 minute. With short cooks, greater accuracy may be required; with long static processes the cook time may be 'rounded' to a 5-minute interval although the thermograph can be read to greater accuracy.

● Determination of temperature

Since the lethal effect of temperature changes substantially over only a few degrees Centigrade, it is obvious that accurate temperature measurement, control and recording is needed. To do this in practice means that a 'standard' or 'master' thermometer must be available with a known accuracy calibration over the range with which the user is concerned. These are usually available from or relate to a National Standards institution. The original standard for technical reasons is likely to be a platinum resistance thermometer although mercury-in-glass 'master' thermometers are available. When using a 'standard' or 'master' instrument be sure that you fully appreciate *all* the factors which affect its accuracy.

Platinum resistance thermometers (PRTs) have the advantage of being easily calibrated and their electrical output can be linked directly to reliable solid state indicating, controlling and recording equipment.

● Retort thermometers

Protected mercury-in-glass thermometers are the traditional cannery retort thermometers. They are 'well damped', i.e. slow responding to changes compared with platinum resistance thermometers, and so appear stable and 'accurate'. This 'accuracy' – which makes the performance of control equipment look good – needs to be carefully scrutinized in order to know the real accuracy limits.

It must be appreciated that all thermometers measure temperature at a single point and this is assumed to be representative of the large volume of the retort or sterilizer. Installation is therefore important and account must be taken of factors which could lead to inaccuracies, e.g. the proximity of metal at a lower temperature. For these reasons, thermometers in steam retorts are usually mounted in a blister or stub tube which is continuously bled and the thermometer itself is provided with a 1.6 mm (¹⁄₁₆ in) bleeder or bleed. These 'bleeds' are designed to keep steam moving over the sensing element and thus make the measured temperature more representative of the whole. Do not confuse a bleeder (bleed) with a vent – they serve very different purposes.

A typical mercury-in-glass retort thermometer specification would include the following:

– The scale length should not be less than 180 mm (7 in) between the highest and lowest graduation.
– The scale should be graduated at intervals of not more than 1°C (2°F) over a range of not more than 50°C (100°F).
– The bulb length should not be less than 80 mm (3 in) from the base of the mounting thread to the end of the bulb sheath.

For most purposes a scale of 90 to 130°C (170 to 270°F) will be found to be convenient. Black scale markings on a white background and the 'red mercury' feature make it much easier for temperature reading.

To minimize the risk of processing errors, it is important that all thermometers and controller/recorders on each bank of retorts are graduated in the same units and it is preferable that all the instruments on a 'retort bank' are of the same design.

● PRT criteria

Appropriate criteria for platinum resistance thermometers (PRT) are necessarily detailed. A UK standard has been published by Richardson and Bown[15]. This includes reference to the British Standards Institution BS 1904: 1984 'Specification for industrial platinum resistance thermometer sensors'. It is good practice for the sensor to have a resistance accuracy at least $\frac{1}{3} \times$ DIN 43760: 1968 over the normal operating range of 0 to 130°C (32 to 266°F) for most sterilizer operations. Remember, however, that it is the overall accuracy of the complete system which is important.

Retort thermometers need to be checked for accuracy and this is not the simple matter it may appear. Because the protective metal casing expands on heating, do not test mercury-in-glass retort thermometers in an oil bath as this has been shown to cause errors. Use steam and be sure that the thermometer is in the correct working position. Retort thermometers which do not agree with the master thermometer to within ±0.5°C (1°F) should be withdrawn from service.

Although legal requirements may be for thermometers to be checked on installation and once per year thereafter, it is good practice to check thermometers together with controller/recorders on installation and at intervals of not more than three months thereafter.

● Controller/recorders

Unless local legislation requires stricter tolerances, it is good practice for the controller accuracy to give for steam cooks a temperature tolerance of +1, −0.5°C (+2, −1°F). For water retorts, using a sterilizing process of at least 40 minutes, a cook temperature tolerance of ±1°C (±2°F) within 15 minutes of the start of the cook should be achieved and held for the remainder of the cook. For shorter times, a temperature tolerance of +1, −0.5°C (+2, −1°F) should be achieved at the start of the cook and be maintained throughout the cook.

The recorder accuracy should be such that at cook temperatures the record may be clearly and distinctly read to the nearest 0.5°C (1°F). Note that local legislation may require minimum chart and scale dimensions. With a digital printout (output) it is important to highlight deviations from the pre-set tolerance otherwise key information can be lost in a block of figures. For corrective action it is important to have figures recorded with the required frequency, e.g. at one minute intervals.

Controller/recorders should be calibrated to agree with the retort thermometer to within 0.5°C (1°F) at cook temperatures. This means that, at worst, the record will be 1°C away from the 'standard' thermometer, i.e. retort thermometer 0.5° plus controller/recorder 0.5°. Remember that when reconstructing a sterilization process from records, the 'worst case' assumptions about tolerances must be made (i.e. what was the lowest possible temperature and the shortest time of the cook).

● Sterilization equipment design and fittings

This is too extensive a topic to be considered here. Unless local legislation requires more stringent criteria it is sound practice to use FPI/NFPA recommendations in *Canned Foods*, Section 8 – Thermal processing systems[1] as minimal requirements. Remember, their soundness has been widely proven over the years.

● Venting of steam retorts

This important topic is not always clearly understood. It is essential to appreciate that it is a gas-flow phenomenon, *not* a question of pressure differentials. Venting is an essential operation in retorts which use saturated steam as the heat transfer medium. Venting or 'purging' may be defined as the displacement of air from the retort by saturated steam before the cook commences.

The closed retort initially contains cans and air. When steam is admitted and the vent is open to the atmosphere, mainly air escapes at first because of expansion as it is warmed by the steam. Much of the steam in the early stages of venting is condensed on the retort walls and on the cans, but as these surfaces become hot the flow from the vent increases in velocity and becomes a mixture of steam and air. This flow must continue until the air has been swept from the retort, because air is an insulator, and air pockets prevent uniform heat distribution.

Air in the steam has two effects. Firstly, it tends to form films on the surfaces of the cans; a thin film of air insulates a can and may be equivalent to increasing its dimensions by 6 mm (¼ in) all round. Secondly, the effect of air is to lower the temperature of steam at any particular pressure. An air pocket among the cans will diffuse very slowly, but in the instrument blister or pocket the steam will be virtually pure, the temperature and pressure will agree and there will be no indication at all of the presence of air elsewhere. Operators should therefore be made aware that agreement between temperature and pressure does not necessarily confirm that the retort is fully vented.

The problem in establishing a venting schedule is to know when the air has been removed from the retort so that the vent may be closed. There is no simple method of detecting this fully vented state, therefore reliance must be placed on rules of procedure which have been shown, by temperature distribution tests, to lead to efficient venting of the particular retort type.

When venting schedules are established, retort type and piping arrangements are obvious factors to be considered. Less obvious are the effects of retort load. A large number of closely packed, small, cold cans will require more venting than a much smaller number of loosely packed, large, hot ones. The method of packing baskets, e.g. scrambling or close packing, and the perforation pattern of the divider plates are other factors which significantly affect the flow of steam through the load.

In order to avoid confusion and consequent risks of mistakes in retort operation, a single venting schedule should be established, for each type of retort, which has been shown by temperature distribution tests to be satisfactory under the most adverse conditions likely to be encountered.

In good practice, venting schedules should be expressed in the form: 'Vents must be fully open during the entire venting period. Vent for × minutes and for such additional time necessary to reach a temperature of $t°$', together with any additional instructions that may be necessary for any particular retort installation.

Temperature distribution test data used to establish the venting schedules under the most adverse conditions likely to be encountered should be kept permanently on file. These data should be available in the department establishing processes and to the regulatory authorities.

Cook-room facilities

• General

Proper process establishment together with confirmation that the sterilization equipment will function within known and acceptable limits of accuracy is the first step that the canner must take to insure safe processing. What is then required is a positive assurance either that the process has been correctly administered to each and every container or that the affected containers have either been segregated for further examination or removed for destruction under supervision. The required assurance can only be satisfactorily given after a detailed and systematic HACCP of all aspects of cook-room operation backed up by adequate, careful safety and quality audits. Both HACCP and audit should include consideration of cook-room facilities and layout; process control procedures; routine auditing; and training of cook-room personnel.

• Layout

Experience shows that cramped layout and poor working conditions increase the chance of errors being made. It is a considerable advantage in controlling, and keeping separate, the movements of sterilized and unsterilized containers when their routes can be kept physically separate. It is also obvious that poor lighting decreases the efficiency of operations and supervision. It is particularly important that all instruments including clocks can be easily read from the operating position during the entire working period. It is also worth remembering that cook-rooms tend to be hot and steamy places and that high temperatures and humidities throughout a shift may lead to operator discomfort which results in loss of efficiency.

Remember that cook-room floors should be impervious to water, resistant to wear, smooth where baskets are moved on trolleys or trucks and well drained. This is because good cook-room floors promote safe working and minimize spillage of containers during the movement of baskets.

Microbiological safety of the cook-room operation depends partly on physical factors, e.g. good design and maintenance, and partly on effective control procedures.

• Separation of unsterilized and sterilized containers

With continuous sterilizers it is considered sufficient if the in-feed and out-feed runways and mechanisms are adequately separated by physical or spatial barriers. Instances have been known where malfunction of the systems has resulted in cans jumping from one runway to another. Further evidence that containers have been sterilized may be obtained by marking the containers with heat sensitive ink or paint, but this is not adequate as the only means of differentiation.

With batch sterilizers, ideally there should be physical barriers such that the only route for filled containers from filling to packing is through the sterilizing retorts. Because of the increased chance of human error, two additional means of differentiating sterilized from unsterilized baskets are required, e.g. use of a tape and disc procedure.

In continuous sterilization processes good engineering provides for the complete segregation of unsterilized and sterilized containers. When, in an emergency, unsterilized or sterilized containers have to be discharged from the continuous system they should be collected in baskets or other suitable containers and identified in a manner similar to that described for batch processes.

It is good practice to insure that the lines and runways are clear of containers before and after a period of production. It is particularly important that unsterilized containers should not be left on the line long enough for a drop in initial temperature or for consistency changes which may affect the lethality of the sterilization process. It is also important that containers do not get 'hung up' on a line when they may re-enter the system and be incorrectly sterilized (and wrongly labeled). The rule is therefore to always check that *all containers*, especially unsterilized ones are removed from the lines and runways *before and after* each period of normal or experimental production of a scheduled variety/size.

Containers occasionally fall from baskets or jump from runways and may come to rest at some considerable distance from where they fell. As there is usually no way of being absolutely certain of the status of such containers, the only safe course of action is to destroy them. The temptation to 'save money', e.g. on the part of 'keen' operators, must be resisted and there should be a very strictly enforced rule as follows:

All sealed containers in the cook-room which are not in baskets, on runways or held under quarantine must be regarded as unsafe and should be punctured on the code end before removal from the cook-room for destruction. It is expressly forbidden to place such containers in baskets, on runways or to otherwise withhold them from destruction.

An actual case of botulism in the US canned meat industry was almost certainly due to such an occurrence.

When an emergency discharge of sterilized containers from a continuous sterilizer has to be made into baskets it is the correct practice to treat these procedurally as quarantined basket stock, and immediately to mark them as such.

Procedures

● General cook-room

These procedures should achieve as a minimum the following goals:

1. Separation of sterilized and unsterilized baskets for batch operations;
2. 'Reconciliation' of numbers of containers entering and leaving the cook-room – this is on a basket basis for batch retorting;
3. Recording sterilization operations, i.e., by log sheets and recorder charts;
4. Quarantining 'out of tolerance' production.

These procedures must take into full account the fallibility of human nature and equipment malfunction. The system of control procedures is designed, by multiple checking, to minimize the chances of mistakes being made. Such a system begins when the containers have been sealed and ends when they have been proved suitable for shipment. There is an important distinction between batch and continuous sterilization processes since mistakes are more frequently and easily made in batch process operations.

1. Separation of sterilized and unsterilized baskets. The first requirement of a system for batch processes is to insure that unsterilized and sterilized baskets are easily distinguished and that unsterilized baskets are taken to the correct retorts. The second requirement is that the retort loader shall know that he is loading the retort with unsterilized baskets from the correct line. The third requirement is that, when the sterilization and cooling cycle is complete, the retort unloader shall know that he is removing sterilized baskets from the retort. The fourth requirement is to insure that sterilized baskets are taken to the correct packing line. A simple system uses identification plates which, by a combination of material, shape, marking or color, differentiate between unsterilized and sterilized baskets and between different products or lines. In conjunction with the use of heat sensitive tape, ink or paint and reading the can code such a system meets these basic requirements. An acceptable procedure uses at least two means of differentiating sterilized from unsterilized baskets in the cook-room. To avoid errors, confusion, and bad housekeeping, do not allow the use of cardboard or paper to identify or indicate the status of any basket unless forming part of a specially designed heat sensitive indicator.

The next requirement is to be able to re-trace the movement of any basket or baskets through the cook-room at a later date.

In any one production period for each filling line, the baskets must be serially numbered when the basket is being filled. This number identifies the basket until it has been unpacked and a tally must be kept by the basket filler, the retort operator and the basket unpacker. At the end of the production shift, these tallies must be inspected and reconciled.

2. Reconciliation of numbers of baskets entering and leaving the cook-room. For each filling line and line or bank of retorts, the baskets should be serially numbered as they are filled and a record maintained – usually known as the loader's log sheet. The retort operator needs a log sheet to show basket numbers loaded into each retort, the actual sterilization process given to each individual retort load, and the basket numbers unloaded and available as floor stock or for labeling. The basket unloader must also keep a tally and log or record. Comparison of these logs is the means by which proper reconciliation may be made and this can then be checked by a physical stock check.

3. Recording sterilization operations. The retort operator's log is a key document. Entries must therefore be made indelibly, erasures strictly forbidden and any alterations clearly shown and initialed with the time marked.

With continuous sterilizers, a log must also be kept recording the actual sterilization of specific periods of production. These also must be made indelibly, erasures strictly forbidden and any alterations clearly shown and initialed with the time marked.

Unless local legislation requires a longer period, retort and sterilizer operator's log sheets should be kept on file with the corresponding sterilizing recorder charts for a minimum of 3 years. Where appropriate, electronic records are acceptable providing that proper precautions are taken against corruption of data.

Experience shows that the minimum information which should be written on each sterilizer recorder chart before use is: retort or sterilizer number, date, variety or product code (or full name), and container size. Additionally, the retort operation supervisor or foreman should, at the time of removal, initial and record the time on the charts for which he has been responsible.

It is important that a responsible person with experience in the appropriate sterilization technology but who does not have direct sterilizing department responsibilities should examine, reconcile and approve by initialing the sterilizer record charts, retort operator's log sheets and other pertinent production records and tallies. This audit of each day's records should be completed by the end of the next normal working day.

4. Quarantining 'out of tolerance' production. A system for identification, segregating, recording and proper, controlled disposal is essential if quarantined material is to be handled in an acceptable manner. It is vitally important to ensure that goods under quarantine are not inadvertently shipped. Remember, in the USA this is the most common cause of product recall. For this reason, the goods must be clearly identified and stored so that they are physically separated from 'normal' production. For maximum security, this should be in a designated, locked area. Following normal good practice, when the disposal of quarantined goods is being considered, the quarantine form together with all the relevant information should be available. Quarantines should always be raised for 'out of tolerance' sterilization conditions. Co-packers are expected to inform the appropriate Heinz department of this happening within one working day and to retain the quarantined goods until Heinz agreement on disposal has been obtained in writing. Information required will include, as appropriate, sterilizer or recorder charts, basket tally sheets, sterilizer or retort operator's log sheets and the results of laboratory tests, etc. A supervisor should sign a quarantine log that he personally observed the quarantined product going out of the designated quarantine area.

● Hold before shipping

It takes time in respect of any production lot to check-out that: the container integrity is sound; the sterilization has been correctly administered; there is no reason to expect excessive spoilage and that sensory and other criteria have been met. GMP requires that this be done before the goods are shipped beyond the direct control of the canner. In the absence of a legal or contractual requirement, a 10- to 14-day hold is regarded by Heinz as a reasonable period. If early release is undertaken, it must be in accordance with a written procedure which clearly defines the responsibility (individual accountability) for insuring the microbiological safety of the lot. This means that critical factors, including container integrity, sterilization conditions, chlorine or other specified disinfectant levels in cooling water, have all been checked out and found to be within specified tolerances.

Canners should remember that this holding period is the last practicable opportunity to avoid shipment of unsafe or unsatisfactory production and that there may be legal or contractual obligations to perform tests, including incubation tests, before production is shipped.

Warning Under no circumstances should the results of an incubation test be used as the sole or major criterion of the safety of production.

● Routine confirmation of sterilizing processes

It should be recognized that small, unplanned changes do occur as a result of normal ingredient variations and equipment wear. Additional changes occur because of minor modifications and sometimes maintenance operations. In order to provide a continuing assurance of safety, it is necessary to have routine experimental confirmation of the established sterilization process. This may properly be regarded as part of the audit program. The frequency should be such as to deliver the necessary assurance and as a minimum guide for low-acid foods, check once during each 30-day period in which each variety/size is produced. Records of the routine confirmation of sterilizing processes should be kept on file for a minimum of 3 years unless local legislation requires a longer period. This information should be available to the department which established the processes.

● 'Recall'

This is needed because it may be legally required, and is an essential part of GMP. It is a recognition of the fallability of human beings and systems. It is good practice to test the recall system's ability to locate a particular 'code lot' of production at least annually. It is important to test the ability of the recall procedure to function effectively at week-ends and peak vacation times. As a minimum, a recall booklet such as that produced by the NFPA (Bulletin 34–L, *Successfully Managing Product Recalls and Withdrawals*), should be in the office safe of each canner, updated with any change, including home telephone numbers of key personnel, and checked every 3 months.

● Record keeping

Full, sufficient and accurate records are required so that the sterilization history of any 'code lot' of production can be reconstructed. Such records include not only process times and temperatures but records of production and other factors critical to the thermal process and container integrity. There is excellent material on records for product protection in Section 6 of the NFPA/FPI book *Canned Foods*[1]. As an example of the care to be taken, if temperature-sensitive tags (cook checks) are used, they should be attached to the recorder chart record for each retort in a batch system.

● Cook-room audits

The purpose of these is to provide an independent confirmation that there is an on-going positive assurance of safety. It supplements the assurance of safety provided by the competence and diligence of the operators, the quality of the supervision and inspection together with the care taken in reconciliation of records. The periodic survey or compliance audit should be made of equipment, procedures, and personnel training, using applicable GMP and local legislation. In the absence of detailed local legislation the NFPA/FPI book *Canned Foods*[1] provides excellent material and Shapton[16] gives an example of a hypothetical checklist which might be used as a basis for vertical steam retorts.

Such audits should be made at at intervals of not more than 12 or less than 3 months. It is important that the timeliness of remedial action and its adequacy receive senior management attention and be recorded. In addition to compliance audits, systems audits should be made to scrutinize the ability of all parts of the sterilization system together with associated systems and procedures to deliver what is expected of them. The priority and timing of these systems audits are a matter of expert judgment of local needs. However, a significant part of the system should be audited on an approximately annual basis. By means of such audits a proper assurance of safety can be given and timely changes to systems and procedures made.

Packaging materials

These may be regarded as 'ingredients' with appropriate specifications and sampling inspection schemes. Because they are manufactured items, a Suppliers' Quality Assurance system should provide the principal assurance that the goods supplied are to the agreed specification. Remember, problems can arise when the specification is inadequate, i.e. is met but does not deliver what is intended.

Coding and labeling

The systems used must comply with legislation in the country in which the goods are sold. Additionally, the canner may wish to identify small 'code lots' as part of a Quarantine system. Ink-jet coding which can be easily and frequently changed has an advantage in this.

Storage of finished product and distribution

Canned foods are usually intended for long storage at ambient temperatures, the exception being canned hams intended for chill storage. There are therefore no particular problems providing that the cans are kept dry. Occasionally, warm humid air in contact with cold cans condenses a film of moisture which in extreme cases may cause slight rusting but with an extremely low risk of causing perforation of the can.

One persistent problem connected with distribution is the use of sharp knives to open shrink-wrapping and fiber cases in stores. This may cut a can, which if it contains a thick product may dry and seal the cut. The result may not be noticed until the customer opens the can when mold growth or discoloration may be seen. Canners should therefore warn distributors and store keepers against this practice.

Special consideration – cook-room personnel

Although the training of personnel to insure container integrity is no less important that that for cook-room personnel, because of the wide variety of container types consideration will only be given here to cook-room personnel.

Unless the principles of sterlization and the need for careful adherence to scheduled processes are appreciated by those who work in the cook-room, serious and expensive mistakes must be expected. Comprehensive training and re-training must therefore be provided at all levels for *all* concerned with sterilization, e.g. retort or sterilizer operators, sterilizing inspectors, cook-room supervision and additionally for ancillary personnel such as basket loaders and unloaders.

The following comments are intended to be a useful brief guide unless local legislation requires more stringent standards. They are given under the headings of 'Retort operators', 'Sterilizing inspectors' and 'Cook-room supervisors'.

● Retort operators

Experience has shown that the best results are obtained when retort operators are given an understanding of the elementary principles of sterilization and trained on the equipment which they will use. Excellent material is available from the NFPA and Campden FDRA, but remember that the presentation needs to be given in the operator's own culture and language. The record of such training should be entered in the employee's personnel file. As a matter of policy, no one should be permitted to sterilize foods for commercial distribution unless they have been suitably trained to operate the equipment and follow all appropriate procedures.

● Sterilizing inspectors

In every production unit there is a need for at least one sterilizing inspector who checks the application of the scheduled processes. It is essential that such an inspector should be independent of cook-room supervision and have the authority to quarantine goods which he considers *may* not have been sterilized within the specified limits. This provides the first independent check on the sterilizing operation and minimizes the consequences of errors.

Experienced retort operators often make good sterilizing inspectors since the primary requirement for this job is a thorough knowledge of the sterilizing operation and equipment.

● Cook-room supervisors

In every production unit, it is necessary to have at least one person available during the sterilization operation who is directly responsible for the close supervision of the operation. The 'cook-room supervisor' is therefore a description of a task and not necessarily a grade of management.

A cook-room supervisor should have a sound understanding of the basic principles of correct sterilization and the consequences of under-sterilization, a detailed knowledge of all types of retort and sterilizer in his charge and be fully conversant with all relevant sterilization control procedures. His main function is to ensure the smooth running of the sterilizing operation, which requires that he is available to deal with any difficulties as they arise.

He will be unable to do this effectively if he is assigned other duties which conflict with his primary responsibility, particularly if they take him away from the cook-room.

As an important matter of policy therefore, it is not acceptable for the cook-room supervisor to be assigned any duties which conflict with his primary responsibility for the sterilizing operation.

References

1. NFPA/FPI (1989) *Canned Foods – Principles of Thermal Process Control, Acidification and Container Closure Evaluation*, 5th edn, The Food Processors Institute, 1401 New York Avenue, Washington, DC
2. Stumbo, C. R. (1973) *Thermobacteriology in Food Processing*, 2nd edn. New York: Academic Press
3. Campden RA (1977) Technical Manual No. 3. *Guidelines for the establishment of scheduled heat processes for low-acid foods*, The Campden Food Preservation Research Association, Chipping Campden, Gloucestershire GL55 6LD
4. DHSS (1981) *Food Hygiene Codes of Practice No. 10. The Canning of Low Acid Foods – a guide to good manufacturing practice*. London: HMSO. (*Note:* This is currently being revised and is out of print)
5. Stumbo, C. R., Parohit, K. S., Ramakrishnan, T. V., Evans, D. A. and Francis, F. J. (1983) *Handbook of Lethality Guides for Low Acid Canned Foods. Vol. 1, Conduction-Heating, Vol. 2, Convection-Heating*. CRC Press: Boca Raton, FLA
6. Hersom, A. C. and Holland, E. D. (1980) *Canned Foods – Thermal processing and microbiology*, 7th edn. Edinburgh/London/New York: Churchill Livingstone
7. Perkin, A. G., Davies, F. L., Neaves, P., Jarvis, B., Ayres, C. A., Brown, K. L., Fallon, W. C., Dallyn, H. and Bean, P. G. (1980) Determination of bacterial spore inactivation at high temperatures. In *Microbial Growth and Survival in Extremes of Environment*, edited by G. W. Gould and J. E. L. Corry, pp. 173–188, Society of Applied Bacteriology Technical Series No. 15. London: Academic Press
8. Richardson, P. and Gaze, J. (1986) Technical Memorandum No. 429. *Application of an Alginate Particle Technique to the Study of Particle Sterilization under Dynamic Flow*, Campden Food Preservation Research Association, Chipping Campden, Gloucestershire GL55 6LD
9. Gaze, J. E., Carter, J., Brown, G. D. and Thomas, J. G. (1988) Technical Memorandum No. 508. *Application of an Alginate Particle Technique to the Study of Particle Sterilization under Dynamic Flow*, Campden Food and Drink Research Association, Chipping Campden, Gloucestershire GL55 6LD
10. NFPA (1989) NFPA *Bulletin 41–L, Flexible Package Integrity Bulletin*, Flexible Package Integrity Committee, National Food Processors Association, 1401 New York Avenue, NW, Washington, DC
11. Shapton, D. A. and Hindes, W. R. (1965) Some aspects of post processing infection. In *Proceedings of the 1st International Congress of Food Science and Technology 1962*, Vol. 4, edited by J. M. Leitch, pp.

205–211. London: Gordon and Breach Science Publishers

12. Thorpe, R. H. and Everton, J. R. (1968) Technical Manual No. 1. *Post-Process Sanitation in Canneries,* reprinted 1986, Campden Food Preservation Research Association, Chipping Campden, Gloucestershire GL55 6LD

13. Thorpe, R. H. and Barker, P. M. (1985) Technical Manual No. 8. *Hygienic Design of Post-Process Can Handling Equipment,* Campden Food Preservation Research Association, Chipping Campden, Gloucestershire GL55 6LD

14. Thorpe, R. H. (1987) Technical Bulletin No. 56. *Leaker Spoilage of Foods Heat Processed in Hermetically Sealed Containers,* Campden Food Preservation Research Association, Chipping Campden, Gloucestershire GL55 6LD

15. Richardson, P. S. and Bown, G. (1987) Technical Bulletin No. 61. *A Standard for Platinum Resistance Thermometers, for use on Food Industry Sterilizers and Pasteurizers,* Campden Food Preservation Research Association, Chipping Campden, Gloucestershire GL55 6LD

16. Shapton, D. A. (1986) Canned and bottled food products (soups, mayonnaise and sauces). In *Quality Control in the Food Industry,* Vol. 3, edited by S. M. Herschdoerfer, pp. 261–322. London/Orlando, FLA: Academic Press

Further reading

Pflug, I. J. (1987) *Textbook for an Introductory Course in the Microbiology and Engineering of Sterilization Processes,* 6th edn, Environmental Sterilization Laboratory, 100 Union Street, Minneapolis, Minneapolis 55455

Lopez, A. (ed) (1988) *A Complete Course in Canning and Related Processes,* 12th edn, 3 vols, Baltimore MD: CTI Publications

Rees, J. A. G. and Bethson, J. (1990) *Processing and Packaging of Heat Preserved Foods*. Glasgow: Blackie. Includes aseptic processing and packaging

Codes of practice or GMP Guidelines

Reference [1] (which gives key USA regulations in an appendix), and reference [4] above; together with: Codex Alimentarus (in press) *Code of Hygienic Practice for Low-Acid and Acidified Low Acid Canned Foods,* Draft document ALINORM 87/13A, FAO/WHO, Rome, Italy

Chapter 12.E Aseptically packaged foods

Background

In aseptic packaging, a product is sterilized, cooled and transported under sterile conditions to a container which has also been sterilized. The product is filled and sealed under sterile conditions into the container. The end product is a hermetically sealed container holding sterile food, which can be stored for prolonged periods of time at ambient conditions. The hermetic closure and the ability of the container to maintain its hermetic nature throughout distribution of the product are essential to maintain commercial sterility of the product. Commerical sterility essentially means absence of microorganisms, including those of Public Health significance, capable of growing in the food at temperatures at which is likely to be held during manufacture, distribution and storage. Product sterilization is accomplished thermally, with a degree of heating related to product composition and acidity. Product viscosity and presence of particulates influence the heating and filling rates. Depending upon the material, packaging sterilization may be accomplished by heating, use of chemical sterilants (H_2O_2 presently) and ionizing radiation or combinations thereof. The sterility of filling can be maintained by air filters, positive pressure, vaporized H_2O_2 and heat.

These aseptic equipment systems can be grouped based on proximity and mechanism of filling operations to package-forming units. Many of the units presently being used or offered for sale are form/fill and seal operations. Other systems incorporate thermoforming operations to form the package. Filling a sterile product into a sterile package while in the mold, or forming a sterile package (rigid or flexible), sealing and transporting it to the point of use, opening, filling and resealing in a sterile environment are other options.

Package forms and systems are also quite varied. Dole aseptic canning operations have been and are used for several products. Bag-in-box systems are used for a variety of acid products, primarily for institutional and product re-manufacture markets. Composite first-generation package forms such as those from Brik Pak, Combliboc, International Paper, etc. are being used today for an increasing number of both high- and low-acid products. Second-generation formed plastic containers, thermoformed and blown, have not yet made a substantial impact. Many of these systems are presently available with a great deal of development work currently being done. It would appear that such package forms have a substantial future.

Since this is a relatively new technology, development of test methods is an area of major import-ance. Package integrity and maintenance of sterility, package performance in distribution, package sterilization techniques, and package residual and environmental monitoring of H_2O_2 are active areas of development and concern. Several methods exist for measurement of concentration levels of H_2O_2 solutions. An on-line continuous monitoring method is needed.

Several tests are being used by product manufacturers to assist in assessing the quality of their aseptic packages. These include visual and package rip-down tests which are being used to indicate package alignment and seal integrity. Electrolytic testing of packages is being proposed to determine leakage through aseptic packages or damage to the inner layers of plastic. Plastic prohibits current flow unless the film has been damaged. If leakage is indicated, it can be traced using an ink penetration test. Sterility testing can be accomplished using traditional methods modified for aseptic packages. Package performance in distribution using traditional or modified techniques must be evaluated. This is an important area for test method development.

It is really important to recognize that aseptic packaging systems demand high standards of technical and management expertise. Those with experience of an aseptic system will fully endorse the words of Thomas W. Holzinger, Corporate Director of Borden Dairy, 'This system leaves no room for error'.

Preparation and storage of ingredients

Acidic fruit juices, sliced fruits and fruit fillings, and low-acid milk and pudding products are the major food items that have been commercially aseptically packed. Obviously, all reasonable measures should taken to assure ingredient freshness – this is especially true where 'High Temperature Short Time' or 'Ultra-High Temperature' (HTST/UHT) processing precedes the aseptic fill, since certain thermostable microbial enzymes will survive the heat process. In conventional processing at temperatures up to about 120°C (248°F) the time–temperature relationships are determined mainly by the heat resistance of spoilage bacteria and are such that enzymes in the food are destroyed. With foods processed by the HTST/UHT methods, however, the heating may be insufficient to inactivate enzymes although quite adequate to destroy the spoilage organisms. This is due to the differing slopes of the enzyme inactivation curve and the thermal death–time curve so that HTST processes favor enzyme survival.

While not of public health concern, the survival of these thermostable enzymes implies that premature loss of shelf life can occur. Therefore microbial proliferation – 'incipient spoilage' – in the ingredient material must be held to a minimum if such shelf-life shortening is to be avoided. The literature indicates that psychrotrophic bacteria, e.g. *Pseudomonas* spp. and *Aeromonas* spp., can be a source of undesirable amylases, lipases, and proteases.

For a quality product you obviously must start out with high-quality ingredients – you must check incoming ingredients for compliance with specifications, and reject shipments that fall short. Order ingredients (and packaging materials) only from suppliers that have been approved in order to make sure they are capable of meeting your specifications and are reliable. Suppliers should furnish written guarantees that the ingredients are not adulterated and, where appropriate, meet FDA or other regulatory requirements. Accurate receiving records must be maintained showing batch codes or lot numbers. No changes in specifications should be made unless their effects on the finished product have been thoroughly investigated and found satisfactory.

Processing methods

Because of the wide variety of products and packaging systems, detailed consideration of specific applications is beyond the scope of this section. It is essential that the potential user of any aseptic system makes a detailed, expert and rigorous HACCP analysis of the proposed product, manufacturing operations, filling, sealing and subsequent handling and distribution systems together with a realistic assessment of customer use and abuse. It is also important to be aware that foods which have particulates of larger size, say over 6 mm^3 (¼ in^3) are likely to present particular difficulties both in process establishment and in handling. What can be done here is to outline briefly a few of the lessons of experience with High Acid and Low Acid packs.

In practice, an aseptic packaging line is considered to have two distinct parts. The first is the aseptic product line which delivers sterile product to the filling and packaging system and the second is the filling and packaging equipment. Burton[1] in his Chapter 4 describes and illustrates a range of aseptic product processing equipment. Consideration of any aseptic processing line reveals the *continuous* mode of operation. The implication of this is that aseptic operation involves a certain complexity and the necessity of an attendant high degree of control and monitoring in order to assure a safe product.

The basic concept of producing aseptically packaged shelf-stable products can be developed in two parts. First, the product must be processed to inactivate microbiological activity and enzymes that cause spoilage. This is normally done by heating the product to an elevated temperature and holding it there for a certain period of time. The time and temperature vary, depending upon the product and

on the microorganisms to be inactivated. Once heat treated in this manner, the product is then cooled under aseptic conditions, i.e. microorganisms are kept from re-entering the product.

The second part is the aseptic filler, which takes the aseptically processed product, fills it into sterilized containers, and seals the containers. This method of processing and packaging allows the use of paperboard and plastic containers. The need for cans and energy-inefficient retorts is eliminated.

There are many aseptic fillers available that the processor may choose from, but before a product can be aseptically filled it must be sterilized. Sterilization systems have been designed to suit particular product characteristics.

In their strictest sense, 'sterile' and 'aseptic' mean 'the absence of any living microorganisms'. These terms and the term 'commercial sterility' are often used interchangeably in the food industry to describe the condition that must be achieved for food, equipment and containers.

In aseptic processing systems, product is sterilized before filling into the container. The product is cooled to near-ambient temperature and transported (pumped or gravity/pressure transferred) to a sterile container. The product's commercial sterility is maintained from the heat processing to the packaging operations. The end product is a hermetically sealed container holding commercially sterile product. It may then be stored at ambient temperatures for prolonged periods of time.

With the approval by the FDA of hydrogen peroxide as a sterilant for packaging materials, aseptic processing and packaging of liquid foods has emerged as an important technique among American and other food processors. Initially, aseptic processing was limited to products of a fluid nature such as juices, juice drinks and milk, but improved systems can now accommodate juice concentrates, puddings, sauces and products which contain particulate matter.

Simply stated, the objective of aseptic processing is to provide products which will retain desirable sensory qualities for an acceptable period of shelf time without the need for refrigeration or freezing. To do this, product is quickly heated and cooled in commercially sterile equipment under closely controlled time/temperature relationships before being packaged in aseptic containers.

The critically important time/temperature relationship depends very much on the characteristics of the product being processed and on such factors as chemical inhibitors, preservatives, water activity and, most important, pH. For most high-acid fruit juices which have a pH value of 4.6 or less, 80°C (176°F) for 30 seconds holding is sufficient to kill spoilage organisms. Since fresh juice, however, requires a slightly higher thermal treatment to achieve pectinesterase inactivation, 93 to 96°C (200 to 205°F) for 15 to 30 seconds is the practical compromise. Additionally, specific organisms such as *Bacillus coagulans* may cause difficulties at relatively low pH levels. In the case of tomato juice with a pH close to 4.6, high thermal treatment at 124 to 126.7°C (255 to 260°F) for 30 to 45 seconds is required.

Non-acid foods with a pH in excess of 4.6 need higher heat and less time; the sterilization of milk products is best accomplished at temperatures of 137.8 to 149°C (280 to 300°F) using holding times of 2 to 5 seconds.

Successful aseptic processing of foods requires at least the following conditions:

1. Mechanical and thermal ability of the system to process the product
2. Sterilizable equipment
3. Sterile packages
4. Repeatable ability to bring sterile product and sterile packages together in sterile filling and closing areas

The variety of equipment and systems available as classified by the type of heat exchange include the following:

1. Direct steam infusion or injection
2. Tubular heat exchangers
3. Scraped-surface heat exchangers
4. Plate heat exchangers

Direct steam infusion or injection

Infusion sterilizing is a direct heating/cooling method that utilizes a steam chamber and vacuum vessel. Raw product flows into a controlled atmosphere of culinary steam. Inverse (inverted) distributor cups cause the product to flow in free-falling umbrella-shaped films which are immediately intermingled with the steam. The level of product plus condensate is controlled so that the infuser operates essentially empty except for enough fluid to maintain a liquid seal at the outlet. The infusion heater operates over an extended range of flow rates and temperatures.

Operating pressure in the infusion heater is held at 0.35 to 0.70 kg/cm² (5 to 10 psi) above the final product temperature saturation point. This allows instantaneous condensation of the steam yet keeps a relatively small temperature difference between product and steam.

Infusion heaters are frequently combined with flash chambers, plate, or scraped surface heat exchangers to create a custom designed sterilizing system. These may be pre-piped packaged systems or field-assembled components arranged to meet specific plant space requirements.

For some applications, steam injection is preferred. In steam injection, steam is dispersed into the product stream by using an injector; steam infusion, on the other hand, relies on dispersing the product into a steam atmosphere.

Direct steam sterilizing is a versatile aseptic processing method designed primarily to heat and cool fluid foods through their critical heat ranges in a matter of seconds. Infusion/injection heating produces the fastest heating of the four heat exchange systems considered here. This rapid processing minimizes flavor changes and product damage normally associated with high processing temperatures. It is especially important for low-acid products which require sterilizing up to 149°C (300°F). As an example, in the case of milk and milk products, the temperature is raised from approximately 77°C (170°F) to approximately 149°C (300°F). When limited to holding times of 2 to 4 seconds the result is a clean pasteurized flavor and continuous operating runs of 20 hours or longer are possible.

Acquisition costs for direct steam heating systems are low when high flow rates are being processed. Since there are few moving parts, service costs are also low.

Tubular heaters

Tubular aseptic sterilizing is an indirect heating/cooling method that uses stainless steel coiled or trombone tubular heat exchangers. The tubing diameter is relatively small compared to product flow. As a result, extremely high flow velocities within the tubing maximizes turbulence which induces very rapid heat transfer. This allows the sterilizing to take place in a very short time, minimizing harmful thermal effects.

The tubing diameter is suited to the product flow and viscosity. Tubes are fabricated into coils or bundles and placed, along with special media baffles, into stainless steel jackets. Hot water, steam or cold water pass through these jackets to heat or cool the product flowing within the tubes.

A high-pressure reciprocating pump is used to force the product through the system. The pump can operate at pressures up to 350 kg/cm² (5000 psi). The actual pressure required will vary according to capacity requirements, the viscosity of the product, and homogenization needed, if any.

CTA (Circular Tubular Aseptic) systems provide high heat transfer rates and a scrubbing action that reduces 'burn on' or fouling in the tubes. This results in a very short processing time. Generally, it takes less than 30 seconds for the product to travel through the entire system which helps to preserve the natural flavor of the product. This is especially important with low-acid products requiring sterilizing treatment up to 149°C (300°F).

Scraped-surface heat exchangers

Scraped-surface sterilizing is an indirect heating/cooling method incorporating a dasher (shaft) with scraper blades that rotates inside a heat exchange cylinder. Product is pumped through the cylinder and heat exchange medium is circulated through a jacket on the outside of the cylinder. The centrifugal force of the rotating dasher holds the scraper blades against the inside of the cylinder wall. The blades continuously scrape the cylinder wall clean of product, simultaneously producing a desirable level of product mixing. With the continual removal of product from the cylinder wall the product film is reduced to an absolute minimum, permitting long processing runs without product build-up on the heat exchanger wall.

Different dasher diameters may be selected to give the desired annular space between the dasher and cylinder wall. This space is dictated by the type of product or size of particulate to be processed.

Scraped-surface heat exchangers process product over a broad temperature range. Product viscosity may vary from light fluid to heavy viscous pastes, with or without particulates. Small systems may be prepackaged on a common base along with water and product tanks, pumps, holding tubes and related piping and controls.

The various horizontal and vertical configurations allow this form of heat exchanger to be adapted to specific systems or plant requirements.

Plate heat exchangers

Plate heat exchange is an indirect heating/cooling method. It incorporates a number of stainless steel gasketed plates which are compressed and locked together in a rugged frame or press. Narrow spacing between the plates accommodates the product and the heating or cooling media in adjacent streams. Porting within the plates directs the medium to one side and the product to the other and heat exchange takes place through the stainless steel plate.

As product is pumped through the plate heat exchanger, the flow is distributed as a thin film which moves over the irregular plate surface, producing a level of turbulence desirable for uniform heating and maximum run lengths.

The addition of terminals allows several independent sections to be located in one press or frame. These sections can be used for preheating, regeneration, sterilizing, holding, and cooling.

If 'regeneration' is employed, the hot sterilized product may be used to preheat incoming cool product. This is accomplished by substituting the hot product for the heating medium in that particular section of the assembly. When a product-to-product regenerator is used, it should be designed, operated and controlled so that the pressure of the sterilized product is at least $0.07 \, \text{kg/cm}^2$ (1 psi) greater than a non-sterilized product. This is to insure that if a leak occurred it is likely to be from 'sterile' to 'unsterile' product.

Process control and validation

Introduction

Five conditions must be met in order to succeed to produce a satisfactory product:

1. Sterile equipment
2. Sterile product
3. Sterile container
4. Sterile environment during filling
5. Seal and container integrity maintenance

If any of these five conditions are not met, you are in serious trouble! Quality assurance is therefore of the utmost importance in aseptic systems and must be taken into account from the inception of the idea for a new product, through all planning stages, including the selection of equipment, process, and packaging. All quality assurance programs must be tailored to the specifics of a given product, process, package, and plant.

If you are considering aseptic packaging, it is important to understand the technology and its rationale and to be very rigorous and thorough in all HACCP studies. As a start, it is a good investment of time to read widely. To be included on the reading list is a 1985 review by Hersom[2], who also deals briefly with spoilage rates; the US Food and Drug Administration regulations[3], which even if you are not bound by them are detailed GMPs; the NFPA *Principles of Aseptic Processing and Packaging*[4] and Campden Technical Manual 11, which is in two parts, 'Principles' and 'Test Methods'[5]. When the Codex draft is finalized and published a 'Code of Hygienic Practice for Aseptically Packaged Low-Acid Canned Foods' will be available.

Because any surviving spores may grow in aseptically packaged foods and some may develop toxins, *Clostridium botulinum* is the pathogen against which sterilization processes must be developed and a scheduled process established. The scheduled process must also be adequate for more heat-resistant spore-formers. Heat-stable enzymes may be inactivated between 76.7 and 87.8°C (170 and 190°F) for appropriate periods of time.

Because *Cl. botulinum* is the target pathogen, the categorization of foods into low-acid with a pH above 4.6 (USA) or 4.5 (UK), and acid is a valid one, just as it is for conventional canning. For products such as fruit juices and sauces, e.g. ketchup, a 'high-acid' category is recognized, in which the target will be acid-tolerant spoilage organisms, especially the ascospore formers including the yeast *Zygosaccharomyces bailii*.

In seeking FDA acceptance of a proposed low-acid aseptic installation, three separate experiments are expected to be performed. First, the raw product is inoculated with a specific number and type of organisms, and their absence must be demonstrated in the finished product. The processing temperature is gradually lowered until some organisms are found to survive. This enables the safety factor of the process to be calculated. Next, the packaging material paper stock or plastic film needs to be inoculated with an organism to demonstrate that the peroxide treatment, or whatever sterilization system is used, destroys all contaminants present. The third test consists of inoculating the air space in the sterile zone of the filler to demonstrate the effectiveness of the pre-sterilization procedures in destroying these organisms. Different organisms are to be used for the different tests[6].

After the sterilization processes for product and equipment have been established, and validated by inoculation experiments, NFPA recommends that at least four commercial runs of uninoculated product be packed and incubated followed by examination for evidence of spoilage and for proper functioning of the sealing equipment. After commercial production is underway, evaluation of equipment should be an on-going function. For those aseptic packaging systems that rely on H_2O_2 as the package sterilant, you must initially demonstrate (and continually monitor to assure) that the H_2O_2 residue in the final container does not exceed the 0.5 ppm residue level stipulated by FDA.

Packaging materials – rigid containers

Metal cans have long been the mainstay for containment of shelf-stable perishable products. Evaluation of container integrity is, and has been, an area of vital importance. A variety of tests have been utilized in this area including:

● External defects

Visual examination of metal containers plays a vital role in assessment of potential can failure. The severity of external defects such as denting, perforations, buckling, presence of corrosion and stains, and rough edges must be determined. The presence of flippers or swollen cans must be identified with appropriate disposition of affected containers.

● Double seam evaluation

Can ripdown provides necessary information relating to proper closure of can end to body. Measurements such as body hook, cover hook (overlap), seam width, countersink depth, and seam tightness are routinely made.

● Level of vacuum

Measurement of headspace vacuum is accomplished using a device which punctures through the top and which is connected to a vacuum gauge. Level of vacuum provides useful information which can relate to container integrity.

● Biotesting

Biotesting can be done on rigid, semi-rigid, and flexible containers. It involves filling the container with a nutrient broth and sealing in a manner normal to that container. The container is then maintained in contact with a medium infected with a test organism. After contact for a specific period of time, the container is incubated, followed by assessment of microbial growth.

Packaging materials – semi-rigid and flexible containers

Seal and/or container integrity can be assessed using many different techniques including:

● Vacuum methods

Vacuum methods can be used with many types of products. Both standard American Society for Testing Materials (ASTM) and commercial techniques exist. Packaged products are placed in a vessel connected to a vacuum pump. Vacuum is applied to the vessel and observation made for expressed product or bubbles. Fluorescent dyes can be substituted for product to make visual observation less difficult and to pinpoint leakage sites. This procedure can be modified so that dye or colored product is drawn into the package through leak points. Smoke particles can be used in addition to dye.

● Pressure methods

Containers can be pressurized by puncturing with a tube which injects compressed gas. The packages are then submerged under water and observation for bubbles made. Dyes can also be used with this type of test. Simple techniques also exist for mechanically stressing packages. A known weight is placed on the package. After a short period of time, observation is made for expressed product.

● Helium, carbon dioxide, oxygen, organic solvent and water vapor leak detection methods

All of these gases can be utilized to determine if leakage exists and the rate of that leakage in sealed packages. Leakage can be monitored inside-out or outside-in using detection systems specific to the gas in use. Leakage rates can be determined by the difference between gas transmission of the package with and without leakage sites masked off.

Container integrity – issues for semi-rigid and flexible containers

Flexible and semi-rigid package forms are often thought of as being less structurally sound than their rigid counterparts, although this may not be true. Under certain situations, flexibles may be subject to flex cracking, abrasion, pinholing, seal failure, puncture, etc. Damage resulting from these events could result in loss of integrity and microbial contamination. Pinholes occurring as a result of physical damage or poor material manufacturing processes provide potential access for microorganisms. Abuse prior to filling and filling itself can expose non-sterilized material to the product, and microbial contamination could result from contact with such material. Package performance in handling, warehousing and transportation must be assessed because of the potential impact on container integrity. Vertical shock from various drop heights and onto a variety of surfaces can be potentially damaging. The role of the product in transmitting shock will differ with each type of product. Flat drops, side drops onto corners, seals, tops, etc. may all induce different types and amounts of damage. Vibrationally induced damage can result in loss of container integrity. Potential for damage can be affected by primary container, pack distribution, and mode of transportation.

Compression-induced damage due to vertical loading could result in rupture or seal damage. The adequacy of containers depends upon pallet patterns and heights, environmental conditions and load-unitization methods.

Seal integrity is essential if container integrity is to be maintained. Foreign matter or product in the seal area could interfere with proper sealing. Channelling, folds, wrinkles, or occluded matter can all cause seal failure. Sealing equipment must be routinely maintained and inspected. Incoming material must be inspected to insure that the sealing layer is intact. Even when sealed properly, flexing, shock and/or vibration-induced damage can cause seal damage or disrupt areas adjacent to seals.

Quality assurance

Quality assurance at all stages of operation is necessary to insure safety in the finished product. Inspection of incoming material for contamination, specifications, and integrity are necessary to maintain quality of raw material. Improper handling prior to a package's formation can cause pinholing, damage to the potential seal area, and to the structure-bearing components of the material. Lack of uniformity in the make-up of board stock or plastic contact film could lead to problems in machine operation. The package sterilizing process, maintenance of sterility during filling and sealing, and the operation of automatic machine shutdown

devices are all investigated when sterility is not achieved because of a suspected malfunction.

Coding and labeling

It is obvious that a legible coding job – spelling out critical production data such as date, hour, machine number, etc. is of vital importance should there be a future need for a recall. These codes must not be impressed on the container with such force as to rupture can enamel or otherwise compromise package integrity. The 'expiration' or 'best used by' date should be prominently displayed on the package. Retailers will need the UPC (Universal Product Code) bar-code to appear on the container.

With the exception of UHT milk, the labeling of aseptically packaged food products does not significantly differ from that of other types of non-refrigerated containers. As pointed out in reference[7], UHT milk labeling requirements (in the USA) are somewhat more rigidly prescribed. In the USA, each panel on the container requires certain information in a specific order to assure compliance with the PMO (Pasteurized Milk Ordinance). Packages larger than single-service should bear a warning that any unused portion of the product should be refrigerated after opening.

Storage and distribution

Semi-rigid packs are less robust than rigid containers such as cans, and therefore they require extra care during storage and distribution. The somewhat shorter shelf life of certain aseptic packs implies that due heed must be paid to the 'first-in, first-out' principle of stock rotation. Obviously, excessive storage temperatures will compromise product shelf life.

These aseptic containers should be packed into cartons and overwraps designed to protect them from mechanical damage that may result in loss of integrity; overstacking should be avoided. Personnel handling this type of product should be informed of the inherent risks of rough handling. The relative frailty of some aseptic containers leads to the need to stress good pest control in the warehouses.

Special considerations

Personnel

Aseptic processing and packaging and the retorting of flexible and semi-rigid containers is a different technology than used with rigid containers. It is more complex and requires higher levels of skill than with conventional canning operations. Sanitation, cleaning, equipment maintenance, inspection, quality assurance, quality audit, record keeping and interpretation must be integrated. Longditudinally Integrated Safety Assurance (LISA) is essential for on-going satisfactory operation. Employees must understand not only their own particular job but have an appreciation of the entire system and, more importantly, be motivated to produce a safe, high-quality product. Plant employees and supervisors will need to have intensive training in several disciplines. Management, especially at high levels, must have an understanding of the total operation as regards product safety and quality. Full implementation of Total Quality Management (TQM) is a *must*.

It can happen that management makes decisions which cheapen the product, process or packaging without making a rigorous HACCP analysis. The 'cutting corners' philosophy will not work with this new technology. Those organizations that are not sufficiently knowledgeable or motivated at all levels of the company or who attempt to cut corners will be likely to create problems for the rest of the industry to deal with. Currently, there are limited numbers of highly trained, capable individuals. The demand for these individuals makes them mobile. This may create problems in daily operations as people move from company to company. Industry must therefore work with educational institutions to insure that potential new employees receive the necessary education and training.

High-acid packs

Experience in this field has grown from our pioneering work on aseptic bulk storage in tanks of capacity up to 461 8201 (122 000 US gall) and railroad tank-car transportation of tomato and fruit purees and pastes. This has led to 'clean-room' or 'semi-aseptic' packaging of 'resistant' products, i.e. not sensitive to spoilage. In such a system, for example, filled but open containers may be briefly exposed to 'clean-room' but not 'sterile' air conditions. It is critically important for management to appreciate that a 'clean-room' or 'semi-aseptic' system is *totally unsuitable for low-acid packs* because of hazards from pathogenic microorganisms, even if some of the equipment has passed review by FDA for low-acid use. It is also *unsuitable* for 'sensitive' high-acid packs because of unacceptably high spoilage risks or for packaging systems which would become a hazard if gas-producing fermentation occurred.

It is critically important for management to appreciate that 'clean-room' systems are not tolerant of errors nor of slack or careless operation. Problems should be expected, e.g. from cracked sight glasses or defective sealing on aseptic vessels, as well as from extending production runs without scheduled cleaning and resterilization of equipment. Less obvious, perhaps, are problems which can occur if tandem product-handling pumps get out of synchrony causing a localized high-vacuum condition.

Because of the great variety of equipment and products it is prudent to treat each high-acid aseptic system on an individual basis for HACCP of the total product line and the development of the 'Quality System'. Development in aseptic packaging systems can be very rapid and it is essential that Quality Systems are adequately detailed and kept updated. Always remember that apparently minor changes may undo or at least jeopardize the asepsis of the system.

At this point it may be useful to give a few illustrations of items not to be overlooked in the consideration of a product line.

Obviously the detailed sanitary design must be to a high standard with CIP used as widely as possible so that cleaning is under automatic pre-set control. Poor installation may negate good design, e.g. by causing liquid to be 'trapped' such that it will not drain from a pipeline. Experience shows that particular care needs to be taken in sterile air supply to balance or aseptic storage tanks.

As with any aseptic line, the operating sequence starting from a clean line is:

1. Pre-sterilize
2. Change-to-product (if hot water sterilization or cooling has been used)
3. Run product
4. Change-to-water
5. Clean

Remember that all aseptic operations depend on accurate instrumentation for measurement, control and recording so fully adequate calibration and maintenance routines are essential. During operation, an example of the kind of detail not to be overlooked is that during pre-sterilization a significant amount of time is needed for large pieces of metal to reach 'sterilization' temperatures. For instance, large valves must reach the scheduled temperature in all places where microorganisms may lurk, since if they are not killed they may enter the product on the cold 'sterile' side. Similarly, be certain that all steam bleeds are free (unblocked) and fully operational and beware of pressure surges ('hammer') which may result in momentary loss of a designed positive pressure.

In many respects, high-acid systems need to be treated as if they were low-acid ones although the penalty for failure to maintain sterility is spoilage rather than the food poisoning and spoilage penalty for low-acid systems.

Low-acid packs

Although UHT milk can be regarded as a low-acid pack, it has been developed over the years and may be considered to be a well-established technology usually subject to local regulatory requirements. (*Note*: regulation in the USA is somewhat more complex, and will be discussed later in this section.)

Other low-acid packs, especially particulate packs, require the application of considerable engineering, microbiological and management expertise. As a matter of policy these must be developed and run with the determination to provide a positive assurance of safety equivalent to that of conventionally canned foods. It is important for management in all countries to implement this policy even if local regulatory requirements are less strict.

Non-technical management needs to appreciate that there are considerably more problems and research needs with low-acid aseptic packs than with acid aseptic or conventionally canned packs.

High-temperature process establishment

As an illustration of technical difficulties, consider the implications of a 12D *Cl. botulinum* cook. To get direct experimental evidence of death rates in the region of 115° to 125°C (239 to 257°F) is not easy, but is achievable given correct operation of good equipment. The problem with the higher temperatures used in aseptic processing is that the lethal effect is so great that very large numbers of spores are destroyed in very short times. Additionally, the heating and cooling effects (or contribution) become a significant part of the whole. This is discussed by Lund[8], who points out that if the *D* value of *Cl. botulinum* is 0.2 *minutes* at 120°C with a z value of 10°C, then at 150°C, 12D would take 0.144 *seconds*.

This simple calculation illustrates the extent of the technical problems. Because of these difficulties there is a lack of published data at temperatures above 130°C, yet 130° to 140°C is not an unusual aseptic processing temperature range.

One way of dealing with these problems practically is to build in a substantial margin of safety. However, the validity of extrapolation of data from lower temperatures depends on the accuracy of the model. For example, does the *z* value remain constant or does it change with increasing temperature, or does the Arrhenius activation energy model fit better? These are not just academic niceties but matters affecting vital safety margins. Campden TM 11 Part 1[5] gives a table of what at the time (1986) was considered the best estimate of 12D requirements for *Cl. botulinum*.

Temperature (°C)	Time (seconds)
130	54
135	30
140	18

High-temperature process administration

Having made a decision on the amount of lethal heat required, how are the high temperatures and short times to be measured? Accuracy of measurement is of prime importance because of the way in which

small changes in these values have a large numerical effect on the evaluation of the process.

Even for a simple liquid-phase product, process evaluation is not straightforward. Should one use only the exposure in the holding section (tube), as is required by the US Food and Drug Administration, or can the contribution of heating and cooling be estimated reliably enough to be included? The types of flow in the heating and holding sections are critical factors since the type of flow determines the relationship of the minimum residence time to the average or mean residence time. GMP requires that this flow pattern should be determined experimentally for each product type and line configuration used and the dated records kept permanently on file in the department establishing the scheduled process. *Any* changes to the heating and holding sections should be tested and the process confirmed. Note that the holding section must slope upwards so that it is filled at all times. It is essential that 'worst case' conditions are known and used to establish that the scheduled process is adequate.

Administration of aseptic processes to particulates is even more complex since they are usually carried or suspended in a liquid phase. Size, composition and integrity of particle are critically important, together with factors such as solid-to-liquid ratio. This is because the liquid acts as a heat transfer medium with all that this implies for rates of heat transfer and temperatures actually achieved.

Different approaches may be taken to estimating the lethality of the particulate process, for example:

- Heat the particle to the required temperature then count only the lethality given in the unheated holding section (tube). If this is used, how is the temperature at the slowest heating point of the particle to be determined?
- Assume that the temperature of the particle is equal to that when it entered the heat exchanger prior to the holding section (tube). Count only the time of the particle in the holding section (tube). This approach is a more conservative assumption.

Both these approaches involve the determination of residence time of the fastest particle, i.e. the one with the shortest 'heating and holding' times. This is not easy.

- A third approach is to determine the combined thermal contribution of heating system and holding section (tube). The cooling contribution is disregarded because particle integrity cannot be controlled.

These approaches are discussed by the NFPA authors Chandarana, Gavin and Bernard[9, 10] in two chapters, 18 and 40, in *Food Protection Technology*. When considering margins of safety needed, remember a key factor is the accuracy of measurement and the ability to control and record critical factors in the scheduled process.

These brief notes show the need for multi-disciplinary expertise in the development of safe aseptic processing, and similar expertise is needed in the application of safe packaging systems.

Packaging systems

To illustrate the detail which needs to be considered with packaging systems, the Dole system uses open-top sanitary cans and sterilizes them with superheated steam. Should a film of moisture condense on the surface of the cans, e.g. in storage just before use, this film of water must be evaporated before the cans will heat at the designed rate.

The Dole system is unusual in using superheated steam as a sterilizing medium. Many use hydrogen peroxide (H_2O_2) to sterilize a plastic-coated laminate which can be sealed (welded) by heat and pressure. For further discussion see Campden TM 11[5] and NFPA, Principles [4]. The underlying basis of the packaging sterilization process depends on the fact that microbial numbers on the outer surface should be very low with no opportunities for growth. This is quite different from the situation in the unprocessed product. Application of H_2O_2, a strong oxidizing agent, to a chemically and physically clean surface is designed to reduce spore survival to that of a 'botulinum cook'. Any surface dirt, oil, finger marks, etc. reduce the efficiency of the process, so the condition of packaging material storage and handling are critical factors. Hot peroxide is more effective than at ambient temperatures, so temperature at the packaging surface and, obviously, concentration of peroxide are also critical factors. If a spray system is used, the design of the spray pattern and the operational need to avoid blockage of the nozzles and insure correct feed rate are important considerations. Whether a spray or a peroxide immersion system is employed, the peroxide will be activated and removed by heat and air. Peroxide is unacceptable in the product so it must be removed before filling. The FDA limits for residual peroxide are strict (0.5 ppm) but should also be met under GMP for those operating outside FDA jurisdiction. It is normally important that filling is clean, otherwise product on the seal surfaces will interfere with the sealing and thus increase the chances of leaker infection. This may not be so necessary where the product is a thin liquid without fibers and the system is designed to seal through the product. Because heat and pressure are used to weld two plastic surfaces, good seal jaw design and careful adjustment and maintenance are needed.

Location of packaging equipment

Because of the operating need to presterilize package presentation, filling and sealing areas in order to maintain asepsis, the equipment should be located in a dust-free and reasonably dry area.

Sterile air

Apart from the obvious hazard of damaged or incorrectly fitted air filters, consider the effect of reduced flow, e.g. caused by running pre-filters too long, on the pattern of sterile air flow. Correct air-flow patterns not only act as a 'blanket' but also prevent ingress of contaminated air.

Product monitoring

Before any lot is shipped a positive assurance is required that the sterilization processes for product and package are fully adequate against *Cl. botulinum* and that spoilage should not occur. Incubation tests are widely regarded as part of this assurance, although they must never be used as a substitute for the examination of records in a well-designed Quality System. It is necessary to be aware of the limitation of incubation tests (see Section 12.D). No practical routine test before shipment, for example, has been adequate to reveal the existence of micrococci in UHT supplemented milk after 3 months' storage at ambient temperatures. For low-acid packs, it is usual to consider a half or full shift output from one machine as a 'lot'. In taking samples, since the object is to detect excessive spoilage, it may be considered preferable to take more samples at start-up and shut-down with fewer during the run rather than sample at a steady rate. If contamination occurs early in the run on the cool sterile side, samples taken near shut-down can show almost 100% spoilage.

Incubation temperatures of 30 to 32°C (85 to 90°F) are usual, but times of incubation vary and in the absence of legislation they need to be considered carefully. One view, and this is regulatory in the UK, is to incubate low-acid packs for 15 days (compare with 7 days' incubation of acid packs for gas production by yeasts). Alternatively, some packs may be pre-incubated at 30 to 32°C for 3 to 5 days before being streaked on plates or using electrical methods to detect growth. In such a regime, packs are also incubated for 14 or 15 days before testing. Some additional packs should always be kept for the full shelf life at ambient temperatures as reference samples.

Whatever system is used, it must deliver a positive assurance that, except for an occasional 'leaker', spoilage is not expected. Failure to do this may result in a costly recall operation. GMP requires a positive release procedure, in writing, by an authorized individual and should include records of case or tray numbers and lot codes.

Remember the words of T. N. Holzinger used earlier: 'This system leaves no room for errors!'

Consumer use

Consumers must be educated in how to use, handle and inspect for damage and/or tampering of these containers. Most people realize that dropping a glass jar is not the thing to do, but they do not realize that abusing a flexible package could result in loss of container integrity. It may be common sense but it must be stressed that, once opened, these containers must be consumed or refrigerated. Also, intentional tampering can occur, so 'tamper-evidenting' is an important feature.

References

1. Burton, H. (1988) *Ultra-High-Temperature Processing of Milk and Milk Products*. London/New York: Elsevier Applied Science
2. Hersom, A. C. (1985) Aseptic processing and packaging of food. *Food Reviews International*, **1**, (2), 215–270
3. FDA (1988) *US Code of Federal Regulations (CFR) – Food and Drugs*, Ch. 21, parts 100 to 169, revised (April 1, 1988), obtainable from Superintendent of Documents, US Government Printing Office, Washington, DC 20402
4. Nelson, P. E., Chambers, J. W. and Rodriquez, J. H. (eds) (1987) *Principles of Aseptic Processing and Packaging*, The Food Processors Institute, 1401 New York Avenue, NW, Suite 400, Washington, DC 20005
5. Rose, D. (1986) (pt 1) and (1987) (pt 2) Technical Manual No. 11. *Guidelines for the Processing and Packaging of Low-acid Foods*, Pt 1, Principles of design, installation and commissioning; Pt 2, Test methods in design and commissioning; Pt 2B, Test methods in production, Campden Food Preservation Association, Chipping Campden, Gloucestershire, GL55 6LD
6. Food Processors Institute (1983) *Capitalizing on Aseptic – Proceedings of NFPA Conference October 11–12, 1983, Washington, DC* Food Processors Institute, 1401 New York Avenue, NW Suite 400, Washington, DC 20005
7. Food Processors Institute (1985) *Capitalizing on Aseptic II – Proceedings of NFPA Conference April 11–12, 1985, Washington, DC*, Food Processors Institute, 1401 New York Avenue, NW, Suite 400, Washington, DC 20005
8. Lund, D. (1987) Aseptic processing of particulates – technical advances and industrial applications. In *Food Protection Technology*, edited by C. W. Felix, pp. 377–385. Chelsea, MICH: Lewis Publishers
9. Chandarana, D. I., Gavin, A. III and Bernard, D. T. (1987) Aseptic processing of low-acid heterogeneous foods in relation to current good manufacturing practices. In *Food Protection Technology*, edited by C. W. Felix, pp. 201–206. Chelsea, MICH: Lewis Publishers
10. Chandarana, D. I., Gavin, A. III and Bernard, D. T.(1987) Aseptic processing of particulates – approval procedures. In *Food Protection Technology*, edited by C. W. Felix, pp. 387–391. Chelsea, MICH: Lewis Publishers

Further reading

Reuter, H. (ed) (1989) *Aseptic Packaging of Food*. Lancaster, PA: Technomic Publishing

Campden RA (1989) Technical Manual 24. *The Microbiological Aspects of Commissioning and Operating Production Processes*, Campden Food and Drink Research Association, Chipping Campden, Gloucestershire GL55 6LD. This manual is a follow-up to Parts 1 and 2 of Technical Manual 11[5]

FPI (1988) *Canned Foods – Principles of Thermal Process Control, Acidification, and Container Closure Evaluation*, The Food Processors Institute, 1401 New York Avenue, NW, Suite 400, Washington, DC 20005. Particularly the chapter entitled 'Aseptic processing and packaging systems', pp. 107–116; and the Appendix section entitled '21 CFR, Part 113 – Thermally processed low-acid foods packaged in hermetically sealed containers, pp. 171–190

van den Berg, M. G. (1990) Gamma and electron energy treatment of packaging materials. In *Newsletters of the International Dairy Federation No. 113* (Dairy packaging newsletter 18), pp. 5–11, International Dairy Federation, 41 Square Vergote, B – 1040 Brussels, Belgium

NFPA (1989) *NFPA Bulletin 41 – L. Flexible Package Integrity Bulletin,* National Food Processors Association, 1401 New York Avenue, NW, Washington, DC 20005

USDA (1984) *Guidelines for Aseptic Processing and Packaging Systems in Meat and Poultry Plants*, Meat and Poultry Inspection Technical Services, Food Safety and Inspection Service, US Department of Agriculture, Washington, DC 20250

Codes of practice or GMP guidelines

Milk and other aseptically packaged low-acid foods are regulated under the Code of Federal Regulations (CFR) low-acid food provisions; manufacturers wishing to aseptically process low-acid foods in hermetically sealed containers must register their plants and file their heat treatment and packaging processes with FDA under provisions of 21 CFR 108 and 113 of the low-acid food regulations. The specific FDA regulations for aseptic processing and packaging systems are covered under 21 CFR 113.40 g, see reference[3]. We recommend that the interested reader consult this document directly. It is also copied verbatim in the FPI *Canned Foods* given as further reading.

Some required process control devices are spelled out in the regulations, others are not. This is because most UHT processing systems and aseptic packaging systems differ sufficiently in design that process safety controls must be tailored to a specific installation. Cited here are a number of requirements as set forth in the above CFR:

• Operators of low-acid food processing systems must have attended a special short course developed by FDA in conjunction with the NFPA, or they must be under the supervision of an individual who has attended such a course.

- In plate-type heat exchangers (and in all indirect heating systems), the sterile product must be at all times under greater pressure than the unsterile product. This is so that in case of a pinhole leak or crack in a seal, the sterile product does not become contaminated with unsterile one.
- In direct heating systems, there is no such required pressure differential. In a direct system, the product is heated through the injection or infusion of steam. In order to remove from the product an amount of water equal to the condensed steam added, the sterile product is flash-cooled in a reduced pressure or vacuum chamber.
- FDA does not allow the heat administered during the come up or retained during the come down to be included in the calculation of the F_0 value for a given scheduled process – the F_0 must be derived solely from the time and temperature in the holding tube.
- Aseptic processing systems for low-acid products must be provided with a flow diversion valve actuated automatically when the product temperature at the end of the holding tube falls below the one listed on the filed scheduled process.
- The timing pump must be a positive pump, and must be sealed so that its throughput cannot exceed the capacity filed in the scheduled process.
- Recording and indicating thermometers must be located at specific prescribed locations.
- Specific procedures must be followed in case of a process deviation. Usually the product will be placed on hold until the deviation has been fully evaluated by a processing authority.
- FDA has very specific requirements for record keeping. Critical process control points must be monitored at given frequencies. Findings must be recorded and made available to FDA upon request.

A different branch of FDA concerns itself with indirect food additives. These regulations apply equally to high- and low-acid foods. FDA permits only a hydrogen peroxide residue of 0.5 ppm in containers.

If milk or milk products are processed an additional FDA group, the Milk Safety Branch, has to review the sanitary design of the equipment and verify its compliance with the Pasteurized Milk Ordinance (PMO). This group maintains jurisdiction over milk, even if it is sterilized, since some microorganisms associated with raw milk can produce toxins which are not destroyed by the heat treatments that destroy *Cl. botulinum*. Milk must be handled with proper sanitary practices from the cow to the UHT sterilizer to prevent excessive bacterial growth during that period. This agency is also concerned with the potential dilution of milk during processing. Ratio-controllers are required for all direct heating systems to make sure that the flash cooling in the vacuum chamber removes an amount of water equivalent to the amount of steam added in the heating process. If the temperature of the product going into the steam infuser or injector is equal to the temperature of the product going out of the vacuum chamber, there is neither product dilution nor concentration.

The Milk Safety Branch of FDA concerns itself also with the nomenclature of milk products. Currently sterilized milk must be labeled as 'Long Shelf Life Milk' and must have the statement 'Refrigerate After Opening'.

The 3A Sanitary Standards Committee is an association of industry, Public Health organizations, and the International Association of Milk, Food, and Environmental Sanitarians which sets sanitary standards that equipment for milk and egg processing must meet. The 3A group concerns itself with such things as the finish of the stainless steel, the ability to clean threads and grooves on fittings, and design features to insure that the equipment can be readily and properly cleaned.

Other guidelines include the Campden Technical Manual 11, given as reference [5], and the USDA Guidelines for meat and poultry products listed under Further Reading.

12.F Physico–chemically preserved foods

Introduction

This section of the chapter deals with a group of technologies, some traditional, e.g. fermented foods, some newer, e.g. canned carbonated beverages, which share the characteristic of being multi-factor preservation systems that are successful when dealing with a low initial microbial loading. Editorially, because of the differences between the technologies, most of the material is given under the heading of 'Specific considerations' and for the convenience of the user, references follow the individual technologies.

The technologies considered here are:

- Foods stabilized with gases
- Fermented foods (other than milk based)
- Sauces, dressings and pickles

Preparation and storage of ingredients

Because of the variety of ingredients, the only generalization that can be made is that appropriate GMPs should be applied. This needs to be considered in pertinent detail, e.g. the distribution of salt and/or acidity levels in a barrel is a Critical Control Point (CCP) for prepared pickle ingredients to be used out-of-season or to extend the season.

Processing methods

These vary greatly according to the process, product, and packaging/distribution system but include the usual operations of washing or cleaning ingredients, grading, size reduction, blanching, heat treatment, composing (assembly) of batch, filling, packaging, labelling and overwrapping.

Process control and validation

These may be simple or complex as necessary. Importantly, they should deliver a positive assurance of safety (and quality) as part of an integrated safety and quality system.

Packaging materials

The specification of these needs to be fully adequate for the intended purpose, which may not be easy with new and complex packaging systems.

Coding and labeling

These will need to be in accord with good practice and the regulatory requirement in the country of sale. Attention to detail is important, so that if a pack, e.g. a dressing, should be refrigerated after opening, then this should be clearly stated. This is usually done either on the label or closure.

Storage of finished product and distribution

This should be appropriate for the pack. Where chilled conditions are part of the system, e.g. CAP/MAP packs, it is important to realize that temperature is probably the most important single safety 'hurdle'. The integrity of the cold chain under these conditions is a critical factor in assuring both safety and quality.

Specific considerations

Foods stabilized with gases

Foods may be treated with gases to enhance or preserve sensory qualities, to prevent microbial growth or to kill organisms. Carbon dioxide, sulphur dioxide, ethylene and propylene oxides, oxygen and nitrogen are among the gases that have been used for these purposes.

For those wanting further information, there is an excellent overview given by ICMSF[1] of the action on microorganisms. Applications for specific products are dealt with in ICMSF[2].

Carbon dioxide – mode of action

Carbon dioxide (CO_2) is a colorless gas although in a vat the surface may be seen to shimmer. It is not a poison but will not support life, and concentrations

Table 12.3 Flow chart of stabilizing a canned beverage with CO_2

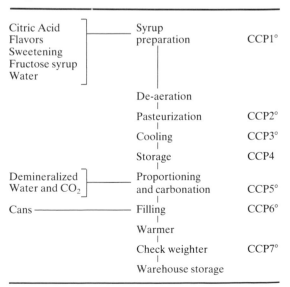

Citric Acid ⎤	Syrup	
Flavors	preparation	CCP1°
Sweetening		
Fructose syrup		
Water ⎦		
	De-aeration	
	Pasteurization	CCP2°
	Cooling	CCP3°
	Storage	CCP4
Demineralized ⎤	Proportioning	
Water and CO_2 ⎦	and carbonation	CCP5°
Cans ——————	Filling	CCP6°
	Warmer	
	Check weighter	CCP7°
	Warehouse storage	

°Points of high concern

Table 12.4 Analysis chart of process and control options: carbonated beverage

Process stage	Description	Hazard	Controlled by	Control limit	Improvements	Reference
Reference 1 – Syrup preparation	a – Prepare a weighed mix of ingredients: – fructose syrup – drinking water – flavors – citric acid – colors (only for cola) b – Blend and eventually dilute with drinking water until _____	°Spoilage organism from raw materials °Product residues in the plant could contaminate fresh syrup	°Microbiological control of water and other raw materials (particularly those have a high) spoilage spoilage potential, such as sugar syrup). °Suppliers assurance and audit °Correct cleaning and sanitizing of plants	°Control of the efficiency of water treatment plant °Reject contaminated raw materials		
2 – De-aeration of syrup	Allow air entrapped in the syrup to escape during the dissolving/ mixing process with vacuum (STD _____ Torr, minimum _____ Torr).	Product residues in the plant	Correct cleaning and sanitizing of plant			
3 – Pasteuriz- ation	Heating syrup to ___°C for ___ seconds in a plate heat exchanger	Incorrect temperature and/or holding time	Correct intrumentation of heating plant	If temperature falls to ___°C product is quarantined. °If temperature falls below ___°C production is stopped		
4 – Cooling	Cooling in plate heat exchanger to 5°C maximum	Incorrect temperature	Correct intrumentation of cooling plant	If temperature exceeds 8°C production is stopped		
5 – Storage	Syrup stored in tank at 5°C (maximum 10°C) for up to 24 hours before filling	°Temperature rise may allow bacterial growth °Product residues in the plant could contaminate fresh syrup	°Correct temperature (and time) during storage. °Correct cleaning and sanitizing of plant	Maximum storage time 16 h at max 10°C. If timeexceeds 16 h but not 24 h product is quarantined. If time exceeds 24 h the syrup is controlled organoleptically and microscopically by _____ before any possible utilization: product is quarantined		
6 – Proportioning and carbonation	a – Demineralized water and syrup are cooled in a plate heat exchanger to ___°C to increase the quantity of dissolved CO_2 b – During proportioning precise quantities of syrup and water are mixed to produce product of correct composition. c – In 'Premix system' water is carbonated prior to mixing with syrup and the resultant carbonated product held in a CO_2 atmosphere prior to filling	°Incorrect carbonation °Temperature rise may prevent a correct carbonation °Air in the water or beverage inhibits effective carbonation °Spoilage microorganisms from treated water °Product residues in the plant could contaminate fresh preparation	°Correct instrumentation of carbonation plant and plate heat exchanger °Correct cleaning and sanitizing of plant	Pressure CO_2 in premix must be ATM _____, temperature of the product not more ___°C. If there are variations of these parameters, production is stopped		

Table 12.4 Continued

Process stage	Description	Hazard	Controlled by	Control limit	Improvements	Reference
7 – Filling	a – Cans are given a short rinse b – Finished beverage is introduced into cans c – Cans are sealed with double-seaming	°Spoilage organisms from water. Cans may lose carbonation through microleakage of the seam or of the can	°Correct filler instrumentation	The quantity in finished product may be constantly not less than _____ vol CO_2/vol soft drink. Product below vol CO_2 but not less than ____ is quarantined. If carbonation falls below _____ vol CO_2, production is stopped		
8 – Warmer	Inverted cans are heated to ___°C (max ___)	Cans may swell and lose carbonation through microleakages due to overpressure	Correct warmer intrumentation	The warmer temperature may be not more than _____°C		
9 – Check weigher	Re-inverted cans are all weighed to reject leaking cans	Uncalibrated or incorrectly calibrated automatic weigher	Correct calibration of automatic weigher	°Control hourly weigher efficiency. If it doesn't discard underweights less than −15 g, production is quarantined °Check once a day the kind of defects causing underweights		
10 – Warehouse storage	The finished product is palletized then it's transported to warehouse	Cans may be damaged and lose CO_2 (and product) through the leak	Correct handling of finished product	Periodic inspection of product in storage		

Note. Some values shown by a _____, e.g. temperature, pressure, are deliberately omitted from this analysis for commercial reasons.

in air of 10% or more will cause unconsciousness. Care must therefore be taken when entering spaces in which CO_2 has been used.

When CO_2 is dissolved in water carbonic acid is formed which lowers the pH, giving an effect additional to that of the CO_2 alone. Microorganisms vary in their response to CO_2. Some, like yeasts, are unaffected, some, like Lactobacilli in a candled-jar, are stimulated and others are inhibited. Usually, concentrations of 5% upwards show a linear inhibition with concentrations of over 25% showing maximal effect. Lower temperatures show a synergestic effect on sensitive organisms.

Controlled Atmosphere Packaging (CAP), Modified Atmosphere Packaging (MAP) and Vacuum Packing are used for a wide variety of foods including meat, fish, fruit, vegetables and ready meals. The primary aim is to improve sensory properties during chilled or ambient storage. The packaging protects the contents from contamination but microbiological safety *must* be demonstrated and not just be assumed. The safe assumption is that any extension of shelf life requires positive proof of microbiological safety. Papers by Hotchkiss[3] and Lioutas[4] give useful reviews of microbiological aspects of CAP and MAP.

Carbonated beverages

Carbonation provides both desirable sensory properties and kills or inhibits spoilage organisms. The stabilization of a canned beverage is shown in the process flow chart in Table 12.3 This is followed in Table 12.4 by a HACCP analysis of each process stage. It will be noticed that figures for control points have been omitted as have the columns 'Improvements' and 'Reference'. This is because they will vary for each line. It will be remembered that 'Improvements' mean the next planned steps to improve line control or performance and that 'Reference' means the cross-reference to other HACCP analyses. Table 12.5 indicates the level of concern at each process stage.

Table 12.5 Level of concern for each process stage

1 – Raw materials	
Sugar syrup(1)	high
Water (tap and demineralized water)	medium
Sweetening	low
Flavors	low
Citric acid	low
CO_2	low
2 – De-aeration	low
3 – Pasteurization	high
4 – Cooling	high
5 – Storage	medium
6 – Proportioning and carbonation (2)	very high
7 – Filling	high
8 – Warmer	medium
9 – Check weigher	high (not a microbiological concern)
10 – Warehouse storage	medium

Notes: (1) High fructose corn syrup is used as a liquid with a Brix of 70 to 72°.
(2) It is very important to standardize an accurate procedure for determining the amount of carbonation.

References

1. ICMSF (1980) *Microbial Ecology of Foods, Volume 1. Factors affecting life and death of microorganisms*, International Commission on Microbiological Specifications for Foods. New York/London: Academic Press
2. ICMSF (1980) *Microbial Ecology of Foods, Volume 2. Food commodities*. International Commission on Microbiological Specifications for Foods, New York/London: Academic Press
3. Hotchkiss, J. H. (1987) Microbiological hazards of controlled/modified atmosphere packaging. In *Proceedings of the Third International Conference on Controlled/Modified Atmosphere/Vacuum Packaging*, Itasca, Illinois, September 16–18 1987. Princeton, NJ: Schotland Business Research, Inc.
4. Lioutas, T.S. (1988) Challenges of controlled and modified atmosphere packaging: a food company's perspective. *Food Technology*, **42**, 78–86.

Fermented foods (other than milk-based)

Manufacturers of fermented and acidified foods rely on the presence of edible acids in the finished product to aid in reducing the pH below 4.6, thus assuring that the product will not be a public health problem. The important factors in the preservation of fermented, pickled and acidified foods are the high salt and/or sugar concentration, low pH, (high acidities) together with the presence of preservatives.

Reliance on one of the above-mentioned factors alone would be foolhardy because there are many microorganisms that are capable of adaptation to their environment and are able to bring about spoilage. When used together in various combinations, dependent upon the food system, these factors can be extremely effective in insuring safety and prevention of spoilage. Each factor should be examined to determine what their role would be in influencing the effectiveness of each other.

Salt

This is one of the oldest materials used in preserving foods. However, salt alone will not prevent the growth of all microorganisms although it will stop the outgrowth of microorganisms of Public Health significance if used in concentrations of 10% and higher in liquids with a pH of 4.6 and below. There are certain halophilic bacteria and fungi that will grow in saturated salt solutions, and many oxidative and fermentative yeasts will grow at concentrations of salt over 10%. When the sugar supply is exhausted the fermentative yeasts will stop growing and when the oxidative yeasts are subjected to sunlight or brought to an anaerobic condition they cease to be a problem.

Salt inhibits microorganisms by reducing the available water to the organism. This deprivation of water is termed lowering the water activity or a_w. The higher the salt concentrations, the lower the a_w, but many microorganisms are prevented from growing at salt concentrations much lower than those necessary to reduce the water activity to an inhibition point. It is quite probable that salt solutions are inhibitory because of high osmotic pressures that cause cell dehydration. The chloride ion is also toxic to many bacterial cells.

In salt stock pickle manufacture, sub-surface yeasts ferment all the available carbohydrate that has not been utilized by lactic acid bacteria and then they stop growing. Salt stock tanks that are directly exposed to sunlight do not have scum formed on their surfaces but if covered by boards or canvas quickly become covered by a film of surface yeasts.

Sugar

This is another material that inhibits microorganisms by bringing about a reduction of water activity. However, much higher concentrations of sugar are necessary to inhibit most organisms when compared to salt. As a rule, bacteria are more sensitive to sugar than are fungi. Yeasts will grow in saturated sugar solutions but not when the pH is below 4.6. Molds will grow on the surface of saturated sugar solutions unless the surfaces are protected from air.

Preservation prediction charts, e.g. those of Bell and Etchells[1], have been developed for different levels of sugar, salt and acetic acid with and without preservatives. These are available to those interested in this preservation technology. The charts demonstrate that sugar concentrations can be decreased as the acetic acid concentration increases and vice versa. Both sugar and acid amounts can be decreased if preservatives such as sodium benzoate are employed. If potassium sorbate is used instead of sodium benzoate then sugar and salt can be decreased even further.

As an alternative to the Bell and Etchells approach, Tuynenburg Muys[2] used a formula method based on concentration of acetic acid, salt and sugar in the aqueous phase to assess safety and stability.

pH

This is generally defined as a measure of hydrogen ion activity. We simply define it as a 'measure' of hydrogen ion concentration in a solution. This is quite different to the total acid concentration as determined by titration.

The pH scale is inverse, that is, if the pH is low then the concentration of hydrogen ions (H+) in solution is higher. The scale extends from 0 to 14 and is not linear, thus a change of one unit reflects a tenfold change in hydrogen ion concentration. Pure water is considered to have a pH of 7.0 which is the middle of the scale and is neither acid nor basic but neutral.

Each microorganism will generally respond differently to ranges of pH; the range for optimum growth is usually quite narrow, e.g. 3 to 4 units. Most microorganisms grow best at pH values near neutrality but there are a relatively few bacteria that will grow below a pH of 4.6 although few of these are of known public health significance. The lactics are the most common organisms that will grow at low pH and during growth the pH falls even lower.

Yeasts and molds will generally grow best in an acid environment (pH 4.5 to 6.0) and may grow actively at a pH of 3.0. However, if sufficient oxygen is present, molds and yeasts will use the acid for growth and so increase the pH. Thus, it can be seen that in processing acid foods, the processor would do well to prevent air from being in contact with the food for long periods of time unless adequate precautions are taken to prevent yeast and mold growth. Remember also that fermentative yeasts will grow well at low pH in the absence of air if sugar is present.

This pH effect on microorganisms is enhanced as the temperature increases. Microorganisms are much easier to kill with heat at a low pH than they are at a high pH. This phenomenon holds true with bacteria, yeasts and molds. Bacterial spores also become less heat resistant as the pH is decreased. Pasteurization temperatures and heating times have to be adjusted with changes in pH. However, one must be careful because pasteurization temperatures for fermented foods, pickles or sauces cannot be relied upon to kill spores.

In low-acid foods above pH 4.6, spores such as those of *Clostridium botulinum* require heating well above pasteurization or boiling temperatures. Even with boiling the spores will survive the process. Fortunately at pH 4.6 or below surviving *Cl. botulinum* spores will not grow and produce toxins. However, there are some spore forming bacteria which can grow and cause severe problems in some acid foods (pH 4.0 to 4.6). The butyric anaerobes will cause vile-smelling pickles to be produced. Pasteurization temperatures are not high enough to destroy these spores and they will grow in pickles if the acid level is too low. Pickles are high-acid foods and have been treated as such for years and their preservation methods have developed along these lines. In response to marketing trends to 'milder' products, i.e. if the pickles approach pH 4.6, the severity of pasteurization should be greatly increased.

Acidity

This term describes a compound that will release one or more hydrogen ions (H+) by dissociation into solution. Acids are either mineral or organic. Mineral acids containing inorganic elements include hydrochloric and sulfuric. These are strong acids. Organic acids such as lactic, acetic and citric acids are carbon-containing compounds and are weak acids. The strength of acids are a measure of how quickly they will give up a hydrogen ion (H+). In dilute solutions mineral acids will dissociate completely, whereas the dissociation of organic acids depends upon the individual acids and other components of the system. Vinegar, for instance, which contains acetic acid is the most commonly used acid to acidify foods and it is relatively weak. Lactic acid, the predominant acid in fermented foods, is considerably stronger than is vinegar and citric acid found in citrus fruits is stronger yet. The anti-microbial effect of the acid, however, is caused both by the hydrogen ions and by the toxicity of the undissociated acid molecules. In this respect, acetic acid is significantly more effective than lactic or citric acids.

Acidity refers to the total amount of acid present in solution. It is routinely measured by titration with a base and is expressed as a percentage of some specific acid – the titratable acidity. The acidity of foods packed in liquids is stated as grams of a named acid per 100 ml of the packing medium. As an example, vinegar contains acetic acid and is commonly expressed in the USA in grains. This refers to the percentage of acetic acid times ten (10), e.g. 100 grain vinegar contains 10% acetic acid.

Cells of microorganisms are well equipped with protective mechanisms. Only certain compounds can get in or out. The cell surface is usually negatively charged and this repels negatively charged molecules. Still more complex systems keep out positively charged molecules like hydrogen ions (H+). The internal pH of a healthy cell is around neutrality (pH 7). Cells are normally impermeable to dissociated acids (acids that cannot diffuse into the cells) but are permeable to undissociated ones. Cells will concentrate organic acids inside the cell. Once an undissociated acid enters a cell it may become mostly dissociated in the internal environment at a high pH and will be unable to get back out of the cell. This is possibly the mechanism by which common food acids, such as acetic and lactic, kill cells. They are probably concentrated enough to so alter the pH that the cell cannot continue its metabolism.

There are differences in the effectiveness of various organic acids and this effectiveness will be quite different with different organisms. For example, the decreasing order of germicidal activity of acids on *Staphylococci* is acetic > citric > lactic > malic > tartaric > hydrochloric; whereas the decreasing order of these acids when inhibiting yeasts is acetic > hydrochloric > lactic > malic > tartaric. Be aware that the results obtained vary with the organism and conditions of growth[3] and the choice of acid(s) in your products should be validated before commercial use.

Preservatives

These are chemicals permitted for use in the preservation of food. Generally they have highly restricted use. Apart from any questions of toxicity, regulatory authorities, very properly, wish to prevent possible misuse to disguise poor sanitation. Nitrates and nitrite have been used for many years in cured meats but their present use has come under intense scrutiny in certain countries, particularly the USA. The amounts of nitrite in foods are normally limited by law. Sulfur dioxide and sulfites have been deleted from a number of foods. Calcium propionate is still the compound of choice with baked goods. Sodium benzoate and potassium sorbate are the two most widely used chemicals in food. Methyl and propylparabens are similar to benzoate in effectiveness when added to acid foods and have an advantage of retaining their antimicrobial activity at higher pH levels than benzoate.

The activity against microorganisms by sodium benzoate, potassium sorbate and calcium propionate is very dependent on the pH of the food system. These preservatives are the salts of organic acids and dissociate completely in solution and the negatively charged ions (anions) are in equilibrium with the undissociated acids. The lower the pH, the more undissociated acid present. Since inhibition of microorganisms is directly related to the concentration of undissociated acid it is apparent that benzoic and sorbic acids have little activity at pH 6.0 and above. They are most effective at about pH 3.5.

Benzoate is considered less than one-half and sorbate about two-thirds as effective at pH 4.5 as at pH 3.5.

Methyl and propylparabens, while chemically related to benzoate, are not directly affected by pH because they are not acids.

Food preservatives function primarily by an inhibition of the microorganism rather than by killing them. In general, potassium sorbate is more effective than sodium benzoate in protecting acid foods from yeasts and molds, whereas the opposite is true when bacteria present the major problem. Be aware that lactic acid bacteria are almost completely insensitive to the sorbates and certain anaerobic bacteria, such as *Cl. botulinum*, will even use sorbate as an energy source.

Although many producers are turning to the use of preservatives to obtain mild flavor and longer open shelf life, preservatives are not an instant technological fix. Regulatory authorities will permit levels which require low initial levels of spoilage organisms if preservatives are to be effective. Organisms can develop resistance to preservatives, e.g. *Zygosaccharomyces bailii*. These organisms, after becoming a natural inhabitant of the environment in a pickle plant, can become extremely troublesome. It has been shown that *Z. bailii* isolates from a pickle plant, where they had never been exposed to sorbates, nevertheless had a significant resistance to sorbates. Worth[3] and Eyles and Worth[4] discuss mechanisms of resistance to preservatives. The processor does not rely on preservatives as a single preservation source, but they can be used together with low pH, the presence of other organic acids and other factors to enhance the keeping quality.

Microorganisms for fermentation

Having said all this, food technologists are aware that the natural fermentation of fruit juices for wine, the lactic fermentation of milk to produce cheese and the rising of bread by microorganisms was done for centuries before there was any knowledge of what brought it all about. It was successful because of a natural selection of desirable organisms which occur naturally in certain foods. These fermentations are now controlled under carefully monitored conditions, mainly using pure cultures of the most desirable organisms for each type of product. However, sauerkraut, pickle and olive fermentations are still mostly dependent upon the natural selection of lactic acid bacteria.

In sauerkraut, the natural fermentation of salted shredded cabbage involves all lactic acid bacteria that are found in processing pickles and a few extra. These bacteria produce the acid necessary for preservation as well as the characteristic flavor. It is essential that the surface of the sauerkraut brines be protected from air. The salt concentration is low when compared to salt stock pickle brine and a variety of oxidative molds and yeasts can grow on brine surfaces if oxygen is available. These organisms can produce serious defects in color, odor, taste and texture.

When cucumbers are placed in salt brine at a reasonable temperature, sugars are rapidly extracted from the tissue and a natural fermentation by lactic acid bacteria develops very quickly. How quickly depends upon temperature and the initial salt concentration. If the salt is too high or the temperature too low the fermentation will be slow in starting. Under normal conditions, the lactic acid bacteria consistently predominate and reach a high population in the brine in 24 to 48 hours. This period is a critical one and it should be monitored to assure that the production of lactic acid is initiated quickly and the pH decreases rapidly to pH 4.0 or below.

Most cucumbers which are fermented are fermented for 'salt stock' pickles. The three species of lactic acid bacteria that consistently predominate in the brines are *Pediococcus cervisiae, Lactobacillus plantarum* and *Lactobacillus brevis*. The first two species are very desirable because they produce primarily lactic acid from sugar, i.e. are homofermentative, while the latter species produces considerable amounts of carbon dioxide and alcohol in addition to lactic acids.

The accumulation of large amounts of carbon dioxide in brines frequently cause serious losses due to hollow pickles (or 'bloaters'). Purging procedures to rid the tanks of this gas must be used in these fermentations to prevent 'bloaters' when large diameter cucumbers are used.

Modern processors have been using controlled processes for the fermentation of pickles. This involves killing most of the microorganisms on the skin of the cucumber and then inoculating the brine with a pure culture of *Lactobacillus plantarum*. The key element of the process is the acidification of the brine to insure that undesirable and potential spoilage bacteria have no opportunity to grow, and the use of a buffer to allow the homofermentative lactic acid bacteria to ferment most of the sugar that is present.

In natural fermentations of salt stock, sugar not used by the lactic acid bacteria is fermented by the yeasts which produce large amounts of carbon dioxide. There are large numbers of these yeasts present and they are all more resistant to acid and salt than are the bacteria. These organisms do not cause defects in pickles if the tanks are properly purged.

Since salt stock brines are normally anaerobic except at the surface, molds and filamentous yeasts can only develop at the surface. Most fermentation tanks are not covered as exposure to sunlight prevents the growth of these organisms. However, if the surface is protected from sunlight it will become quickly covered with filamentous yeasts which are extremely salt tolerant and will oxidize the lactic acid in the brine. They should be either discouraged from forming or periodically skimmed off.

Molds are not nearly as salt tolerant as yeasts and have not been a problem in cucumber fermentations. They become a problem if permitted to grow on brine surfaces because molds produce tissue-softening enzymes such as pectinases and cellulases. They are strictly aerobic and grow only on the surface of the brine. The procedures used to control yeasts will also control the molds.

References

1. Bell, T. A. and Etchells, J. L. (1952) Sugar and acid tolerances of spoilage yeasts from sweet cucumber pickles. *Food Technology*, **6**, (12), 468–472
2. Tuynenburg Muys, G. (1971) Microbial safety in emulsions. *Process Biochemistry*, **6**, (6), 25–28
3. Worth, A. D. (1989) Relationships between the resistance of yeasts to acetic, propionic and benzoic acids, to methyl paraben and pH. *International Journal of Food Microbiology*, **8**, 343–349
4. Eyles, M.J. and Worth, A.D. (1989) The response of *Gluconobacter oxydans* to sorbic and benzoic acids. *International Journal of Food Microbiology*, **8**, 335–342

Further reading

Costilow, R. N. (1979) *Acidified and Fermented Foods: Principles of Handling*. St. Charles, ILL: Pickle Packers International
Campbell-Platt, G. and Anderson, K. G. (1988) Pickles, sauces and salad products. In *Food Industries Manual*, 22nd edn, edited by M. D. Ranken, pp. 285–333. Published with the authority of BFMIRA (the Leatherhead Food Research Association) by Blackie, Glasgow and London, and in the USA and Canada by AVI, an imprint of Van Nostrand Reinhold Company, New York

Sauces, dressings and pickles

The consumer in the USA is familiar with a large variety of sauces and dressings. Sauces based on tomatoes as the main ingredient include chili sauce, seafood cocktail sauce, hot sauce (salsa), pizza sauce, 57 sauce, spaghetti sauce and taco sauce. Some of these sauces depend upon sterilization at an F_0 value sufficient to inactivate *Bacillus coagulans* var. *thermoacidurans* spores and are filled at temperatures over 190°F. Others depend upon the salt, acidity and solids level to provide them with their preservation characteristics. Another group of sauces are dealt with insofar as the preservation parameters are like those used in pickles and relishes. In the UK the term 'pickle' is more usually used for pieces of vegetable in an acid sauce. These sauces would include tartar sauce and horseradish sauce, and as both sauces are susceptible to spoilage by yeasts and lactobacilli upon opening it is usual to recommend that once the product is opened, it be placed under refrigeration. Sauces such as steak sauce, 57 sauce and Worcester sauce depend upon their acidity and solids ratio together with their low pH so that they are normally resistant to spoilage and do not require refrigeration after opening.

Salad dressings have low pH's (3.1 to 4.0) with vinegar (acetic acid) as the principal ingredient which maintains the preserving quality of the product. However, they sometimes must employ preserving acids such as sorbic and benzoic to aid in preventing spoilage from occurring. The limitation for the use of benzoate in the USA is 0.1% (1000 mg per kg) but there are no such limitations with sorbic acid. Other countries have regulatory limits for both these preservatives. If the two are combined in the product the combined percentage rarely exceeds 0.13% because at this point you can generally expect the flavor of the product to be unduly influenced. If oil is present, partitioning of the preservative may occur, resulting in a significantly lower concentration in the aqueous phase where the microorganisms occur.

The reason preservatives are pressed into service to avoid spoilage is to obtain varieties of salad dressings that are light (low) in salt and sugar. The manufacturer may have to rely on chemical means for preservation when the acidity and salt/sugar ratio is less than a traditional recipe.

The normal recipes of salad dressings would employ 3.0 to 4.0% salt, 20 to 30% sugar and 1.2% acetic acid to protect against all except the yeasts, molds and a few bacteria. *Saccharomyces* spp. and *Bacillus vulgatus* have been implicated in a few spoilage incidents[1]. There has also been spoilage involving high lactobacillus populations[2]. Other spoilage organisms which have been found include *Z. bailii*, *Pichia* spp. and lactobacilli growing at an unusually low pH range. Salad dressings as well as mayonnaise also rely on their high oil content to concentrate solutes in the aqueous phase, thus lowering the water activity (A_w) to help protect from food-borne pathogenic bacteria, according to Smittle[3]. The principles of microbial safety (and stability) given by Tuynenburg Muys[4] for emulsions apply equally to the aqueous phase of sauces, dressings and pickles.

In the UK sauces are usually referred to as either thick or thin. Examples of thick sauces would be fruit sauces, tomato sauce and apple sauce. They are mixtures of sugar, salt, spices and vinegar with the addition of gums and/or modified starches to thicken them. Examples of thin sauces are Worcester sauce or mushroom sauce. Further information on UK-style products is given by Dakin *et al.*[5] and by Campbell-Platt and Anderson[6]. Prior to the widespread use of pasteurization, the primary preservation factor was the concentration of acetic acid (as vinegar). The rule quoted by Dakin[7] was that pickles should contain not less than 3.6% acetic acid calculated as a percentage of the volatile constituents. This method of calculation is rather similar to that of Bell and Etchells[8]. Pickles and sauces usually have a pH of around 3.5 or below.

The spoilage organisms capable of attacking products with vinegar as a main preservative ingredient are limited and sauces in the UK are subject to spoilage much like those in the USA. The individual organisms reported in spoilage incidents vary somewhat from those reported in the United States; *Saccharomyces* spp. and *Pichia* spp. have been involved as have *L. brevis* and *Lactobacillus buchneri* as outlined by Dakin and Day[9].

It must be remembered that the key to producing products such as these depends primarily on excellent sanitation of the equipment and environment in the production area, the ingredient material quality and the manner in which personnel handle the product. The greater the incidence of contamination the more stress is placed on the preservation system. This applies equally to 'Safety' and 'Spoilage'.

References

1. Appleman, M. D., Hess, E. P. and Rittenberg, S. C. (1949) An investigation of mayonnaise spoilage. *Food Technology*, **3**, 201–203
2. Kurtzman, C. P., Rogers, R. and Hesseltine, C. W. (1971) Microbiological spoilage of mayonnaise and salad dressings. *Applied Microbiology*, **21**, 870–874
3. Smittle, R. B. (1977) Microbiology of mayonnaise and salad dressings: A review. *Journal of Food Protection*, **40**, 415–422
4. Tuynenburg Muys, G. (1971) Microbiological safety in emulsions. *Process Biochemistry*, **6**, (6), 25–28
5. Dakin, J. C., Binstead, R. and Devey, J. D. (1971) *Pickle and Sauce Making*. London: Food Trade Press
6. Campbell-Platt, G. and Anderson, K. G. (1988) Pickles, sauces and salad products. In *Food Industries Manual*, 22nd edn, edited M. D. Ranken, pp. 285–333. Published with the authority of BFMIRA (the Leatherhead Food Research Association) by Blackie, Glasgow and London, and in the USA and Canada by AVI an imprint of Van Nostrand Reinhold Company, New York.
7. Dakin, J. C., (1962) Pasteurization of acetic acid preserves. In *Recent Advances in Food Science*, Vol II, edited by J. Hawthorn and J. M. Leitch, pp. 128–141. London: Butterworths
8. Bell, T. A. and Etchells, J. L. (1952) Sugar and acid tolerance of spoilage yeasts from sweet cucumber pickles. *Food Technology*, **6**, (12) 468–472
9. Dakin, J. C. and Day, P. M. (1958) Yeasts and lactobacilli causing spoilage in acetic acid preserves. *Journal of Applied Bacteriology*, **21**, 94–107

Further reading

Harris, B. R. and Epstein, A. K. (1937) *Mayonnaise and Salad Dressings*. Chicago, ILL: The Emulsol Corporation

12.G a_w Controlled foods

Background

Water activity is a factor with a major selective influence on microbial growth and is therefore of importance in the microbial ecology of foods (see also Chapter 8.B.1). There is an excellent introduction to this subject by Christian[1], although thirty years earlier Scott[2] introduced the concept of water activity, and demonstrated that reduction in growth rates and cell yields occurred when water activity was reduced in various foods and laboratory media. Leistner and Rodel[3] also related microbial inhibition to water activity.

Growth of microorganisms in food can be slowed down or entirely prevented either by reducing the water content, e.g. as in traditional sun-dried foods, or by binding water to make it inaccessible to microorganisms, e.g. as when salt and sugar are mixed with the food. The extent to which water is 'bound' in foods is expressed in terms of water activity (a_w). This is a measure of the relative vapor pressure of water above the food at equilibrium (see 'Process control and validation' in this section for the measurement of a_w).

Fresh foods with a_w values above 0.95 rapidly spoil if not refrigerated. In this a_w range bacteria outgrow yeasts and molds as their metabolic rates are generally higher and they compete successfully for available nutrients and space. At water activities in the range 0.95 to 0.90, bacteria still dominate the spoilage flora but nutritionally fastidious types such as Gram-negative rods and spore-forming bacteria progressively give way to lower a_w-tolerant cocci and lactobacilli. At still lower water activities, yeasts and then molds become the main spoilage organisms with growth minima of approximately 0.88 and 0.75 a_w respectively (see Table 8.1).

Exceptional species of bacteria, yeasts and molds can grow well below the minima stated here and these exceptions are invariably responsible for the food microbiologists' problems. The terms halophilic and xerophilic are broad descriptions used to include those exceptional organisms that either tolerate or require relatively high concentrations of solutes in the growth medium.

Tables 8.5 and 8.6 in Chapter 8 show the water activities of foods in decreasing order and the water activities of foods and spoilage organisms.

Microbial growth above $0.90a_w$

The minimum a_w values for microbial growth quoted here are based on test systems where sodium chloride was used as the a_w controlling solute, unless otherwise stated.

- Gram-negative bacteria

The majority of food spoilage organisms and those of Public Health significance grow at an optimum rate close to an a_w of 1.0, which is the a_w of most fresh foods. Important members of the Families *Enterobacteriaceae* and *Pseudomonodaceae* are responsible for both food poisoning and food spoilage at high a_w.

Enterobacteriaceae are primary environmental saprophytes and scavengers often found in the intestinal tract of man and lower animals. They are very important to man as both primary and secondary pathogens, and because of their association with the intestinal tract they easily become introduced into food either during slaughter of animals or during subsequent processing and food handling in conditions of poor sanitation.

Salmonella spp. grow well in mixed flora at high a_w. Their growth minima was shown to be around 0.94 to 0.95 a_w by Christian[4, 5], who used both nutritionally simple and complex media together with a variety of a_w controlling solutes. Horner and Anagnostopoulus[6] studied the effect of a_w on heat survival of two species of *Salmonella* and found the same a_w minima for growth, but this was tested only in laboratory media. Although *Salmonella* spp. are unable to grow in Intermediate Moisture (IM) foods they have been shown by Corry[7] to survive at low a_w for over a year in sugar and polyol solutions. This is significant as, since they can have a very low infectious dose of around 1 to 10 organisms per g of food, infection after ingestion could occur without any growth of *Salmonella* in the food.

For many years *Salmonella* spp. have been responsible for most reported food poisoning outbreaks in the UK but recently their place has been taken by *Campylobacter* spp. which now account for the majority of reported acute bacterial diarrheal infections. It is likely, however, that cases involving *Salmonella* spp. are under-reported more compared to cases of *Campylobacter* spp. because of the latter's severe bloody-diarrhea symptoms. The subject of *Salmonella* spp. in IM foods has been thoroughly reviewed by Corry[8].

Escherichia, *Shigella* and *Klebsiella* are members of the *Enterobacteriaceae* that are present in the commensal flora of the gastro-intestinal tract of vertebrates. The majority are non-pathogenic in the gut but can cause infections elsewhere in the body. *Shigella* is only rarely transmitted in food and has a limiting a_w of 0.96.

The most significant organism among these is *Escherichia coli*, of which six serotypes are important in infections of man. All can be transmitted in food, associated especially with poultry and poultry products. The limiting a_w for these serotypes is approximately 0.95 a_w [9].

Proteinaceous foods such as fresh meats are

spoiled often by pseudomonads which are proteolytic psychrotrophs with a limiting a_w of 0.95. Surface drying of carcases immediately following slaughter of animals can cause the a_w of the meat to fall to a level that inhibits bacterial development but modern post-slaughter techniques are designed to insure maximum water retention in the meat for economic reasons and this practice aids their proliferation[10].

Many species in the family *Pseudomonodaceae* are of marine origin and some can tolerate high salt concentrations, e.g. saturated salt solutions of $0.75 a_w$. The growth of vibrios is stimulated by salt and some (halophiles) require it for growth. They are Gram-negative rods and include species such as *Vibrio cholerae*, which can be spread via food and infective humans, and *V. parahaemolyticus,* associated with warm water shellfish.

V. parahaemolyticus is the causative agent of an infectious food poisoning syndrome associated with warm summer months and the consumption of raw seafood. It is 'top of the food poisoning league' in Japan and has a limiting growth a_w of approximately 0.92, equivalent to 10% NaCl[11]. Nickelson and Vanderzant[12] have written a useful review containing a table of biochemical characteristics based on reported work from numerous sources. As the most a_w-tolerant Gram-negative pathogen, this species could be significant in IM foods with an a_w near the upper limit of 0.90, particularly if seafood ingredients or crude solar salt were used in its manufacture.

● Gram-positive non-spore-forming bacteria

The more tolerant gram-positive non-spore-forming types such as the lactic acid bacteria and *Micrococcaceae* and spore-forming *Bacillaceae* (see next section) are able to grow in the range 0.90 to $0.95 a_w$.

The relative a_w tolerance of some of these organisms is exploited in the production of fermented meat products. Starter cultures may contain a mixture of *Micrococcus* spp., which reduce nitrate to nitrite, giving the products a red appearance, and *Lactobocillus* spp. or *Pediococcus* spp., which produce lactic acid. These organisms are also capable of causing spoilage in moderately low a_w foods such as bacon and some lactic acid bacteria form slimes on spoiling products, as do pseudomonads. Members of the genera *Lactobocillus* and *Pediococcus* have a limiting growth a_w about 0.94[13], micrococci are slightly more tolerant of low a_w and are limited at 0.93[14].

● Spore-forming bacteria

The genera *Bacillus* and *Clostridium* have been extensively studied in relation to IM foods because their spores may remain viable indefinitely in these foods. Damaged packaging and storage abuse such as holding in humid conditions could lead to localized areas of increased a_w allowing viable spores to germinate.

Marshall[9] found the minimum a_w for growth of *Bacillus* spp. to be in the range 0.90 to 0.94. *B. cereus* causes two distinct forms of food-borne illness; a diarrheal-syndrome and an emetic-syndrome which last for about 24 hours. The growth-limiting a_w for *B. cereus* is approximately 0.91.

The genus *Clostridium* has been extensively studied by food microbiologists with considerable attention given to *Cl. botulinum*. This species is relatively tolerant of low a_w and its name is derived from 'botulus', Latin for sausage, since it is said the organism was first isolated from sausages containing curing salts, following a food poisoning incident. There are at least seven serological types of botulinal toxin (all are heat labile and can be destroyed by heating to boiling for 10 minutes).

Jakobsen and Trolle[15] studied 55 strains representing 18 different species of *Clostridia* isolated from spore populations acting as food contaminants. Growth from vegetative cell inocula occurred at $0.945 a_w$ for four strains, including *Cl. perfringens*. The next lowest a_w tested, 0.93, showed no growth. The same growth-limiting value was found for three of the four strains that grew when heat-activated spore inocula were used. Investigations show that the lower limits of a_w for growth and toxin production in laboratory media are 0.95 for type A; 0.94 for type B; and 0.97 for type E (Ohye and Christian[16]) and Baird-Parker and Freame[17]).

Microbial growth below $0.90 a_w$

Organisms that are able to grow below $0.90 a_w$ are potential food poisoning or spoilage organisms of IM foods.

● Yeasts and molds

The contribution that Pitt[18] made to the International Symposium on Water Relations in Foods regarding, as he defined them, xerophilic fungi will undoubtedly remain a classic source paper. The main updating will probably only concern mycotoxins, which are being actively investigated world-wide as their full significance becomes known. Additional a_w minima for the growth of some species may also have to be added as spoilage is encountered in the growing number of IM foods being developed.

Pitt provided a list of all the known xerophilic fungi, that is, yeasts and molds capable of growth at a water activity below 0.85, which included only eleven genera of a conservatively estimated 5000 genera of known fungi. The eleven are all ascomycetes or asexual forms of ascomycetes. There are no pathogenic yeasts among those listed, so yeast growth in IM foods would cause spoilage and not food poisoning. However, 16 species of molds listed are known to produce mycotoxins.

Mycotoxins are secondary metabolites of filamentous fungi which can be formed during their growth on food and animal feeds. Since their discovery in England in 1959, a wide range of molds representing every group of filamentous fungi have been shown to produce them. The first compound studied was named Aflatoxin B1 and was found in the feed of turkeys farmed in East Anglia. About 100 000 young turkeys and tens of thousands of ducklings and young pheasants had died from what came to be known as 'Turkey X Disease', and by 1964 the compounds involved had been fully described.

Aflatoxins, produced by *Aspergillus flavus* and *Asp. parasiticus*, were shown to be carcinogenic to rats, but uncertainty still exists about the extent of their carcinogenicity to man although their toxicity is not in dispute. Some countries such as the USA assume that they are carcinogenic and ban their presence in food and feedstuffs, at least to the limit of analytical methods, while in the UK an effort is made to control their formation in susceptible foods and to keep levels low in animal feeds.

Many more mycotoxins such as patulin, citrinin and vomitoxin have been isolated since 1959 and they produce a wide range of symptoms in man and his domestic animals. Recognition of the acute toxic effects caused by mycotoxins allowed for the historical diagnosis of mycotoxicoses in hitherto unexplained epidemics.

Mycotoxins can be ingested either by direct consumption of moldy or previously molded, but apparently unspoiled food, known as 'Primary Mycotoxicosis'; or by the consumption of animal products such as meat or milk from animals fed on contaminated feed, known as 'Secondary Mycotoxicosis'.

Foods which are solid or semi-solid and have a low water content or low a_w are susceptible to mold spoilage since their physiology allows them to compete successfully with other microorganisms under these conditions. Their hyphae can penetrate foods in order to obtain nutrients while at least a part of the organism remains in contact with a plentiful supply of oxygen at the surface.

Examples of food from which mycotoxins have been isolated are wheat, oats, barley, rice, maize, nuts including peanut butter, vegetable oils and coffee.

Traditional IM foods that may be spoiled by molds include cheeses and fermented sausages. The semi-soft cheese Roquefort contains a large amount of fat, up to 5% salt and little moisture which together prevent growth of most bacteria. *Penicillium roquefort* is introduced as a spore inoculation during manufacture of the cheese and its growth provides the sharp flavor and its conidia the characteristic blue-green veins. Some strains of this mold produce a neurotoxin but the conditions in cheese appear to inhibit toxin formation. However, contaminants at the surface of any cheese may form toxins which diffuse through the product. Merely scraping the surface of moldy cheese before eating it is not a sound practice.

Extensive studies concerning the relationship of a_w and temperature on mycotoxin production have been carried out by Northolt and his colleagues working in the Netherlands. Table 12.6 shows the minimal a_w for growth and toxin production for eleven molds in laboratory media and in every example toxin production ceased at higher a_w's than growth. Results for the same strain under different cultural conditions and between strains of one species were very variable. For example, the minimal a_w for aflatoxin production by *Asp flavus* ranged between 0.83 to 0.87, with a minimum growth a_w of 0.80. As yet, there are few data concerning the formation of mycotoxins in IM foods, but a potential for their formation exists.

Pitt[18] reported seven molds as being capable of growth at $0.70 a_w$ or below and the most tolerant species yet studied was *Xeromyces bisporus*, which grows at a minimum a_w of 0.61. This is the lowest a_w at which growth has been reported.

Table 12.6 Minimal a_w for growth and toxin production by molds

Mycotoxin	Mold	Minimal a_w for Growth	Toxin
Aflatoxin	*Aspergillus parasiticus*	0.82	0.87
	Aspergillus flavus	0.80	0.83–0.87
Patulin	*Penicillium patulum*	0.83–0.85	0.95
	Penicillium expansum	0.83–0.85	0.99
	Aspergillus clavatus	0.85	0.99
Ochratoxin	*Aspergillus ochraceus*	0.81	0.88
	Penicillium cyclopium	0.87	0.97
	Penicillium viridicatum	0.83	0.83–0.86
Penicillic acid	*Aspergillus ochraceus*	0.81	0.88
	Penicillium cyclopium	0.87	0.97
	Penicillium martensii	0.83	0.99

Data from Northolt *et al.* [19–23, 24].

IM foods are susceptible to spoilage by molds unless additional preservation factors such as chemical preservatives are used in combination with a reduced a_w. However, unlike crops such as cereals and nuts, where mold growth and mycotoxin formation may occur but visible signs are lost before consumption, IM foods which are spoiled by molds are more likely to be rejected by the consumer due to visible spoilage. This, of course, does *not* constitute a positive assurance of safety.

● Halophilic bacteria

These organisms have an absolute requirement for sodium chloride although some other salts such as potassium chloride can partly replace it. They occur widely but are found in the greatest numbers in saline environments. Because of their specific growth requirements, isolation methods include media supplemented with salt and generally long incubation periods.

The main source of halophiles that spoil salted foods is, as expected, the salt used in the preparation of the product. Crude solar salt, which is still most often used in areas where traditionally dried and salted foods are prepared, is heavily infected with halophiles. In contrast, mined and refined salt is virtually free from these organisms.

The most studied moderate halophiles require 0.2 to 0.5 M NaCl for growth, and at 1M NaCl (approximately $0.97\,a_w$) grow at an optimum rate. Examples include *Vibrio* and *Micrococcus* spp. discussed in a previous section.

The extreme halophiles will not grow at less than about 2.9 M NaCl ($0.90\,a_w$) and are associated with a very distinct ecology. They occur in salt lakes where solar evaporation concentrates the salt and in salt pans used to collect crude solar salt. Extremely halophilic bacteria are usually highly pigmented and can give rise to a red coloration in brines, an effect which has been noted since ancient times. These bacteria are heterotrophic, aerobic and generally red or pink. Gram-staining so disrupts the cell wall that its result can be misleading. However, two broad divisions of extreme halophiles are recognized.

Rod-shaped bacteria which lack a conventional cell wall and which appear to be Gram-negative are placed in the genus *Halobacterium* and include species such as *H. halobium*, *H. salinarium* and *H. cutirubrum*. They grow at an optimum rate in the range 0.86 to $0.81\,a_w$[25] and tolerate saturated salt solutions ($0.75\,a_w$).

Cocci with thick cell walls and which appear to be Gram-positive are given the generic name *Halococcus* although Brown[26] noted that other names such as *Sarcina* and *Micrococcus* have been used. Their a_w tolerance is similar to the *Halobacterium*.

The physiological requirement of halophiles for relatively high concentrations of NaCl in the growth medium restricts their role as spoilage organisms in IM foods to those that contain predominantly NaCl as the a_w controlling solute. Modern IM foods tend to use a balance between salt and sweet flavors to help gain wide acceptability of the product and so traditional IM foods are the more susceptible product types.

Varga *et al.*[27] isolated a red halophile resembling *H. salinarium* from spoiled 'heavy salted' fish produced in Eastern North America. The organism grew at $0.72\,a_w$ when inoculated back into a modern IM food version of the 'heavy salted' product, and survived for over 59 days at $0.71\,a_w$.

● *Staphylococcus aureus*

Staph. aureus is the only human bacterial pathogen able to grow at relatively low water activities, and its water relations and toxins have been extensively studied. It is found on the skin and mucous membranes of warm-blooded animals. Large numbers can be isolated from the nose and from pus in wounds and hence they are easily transmitted into foods during their preparation. Between 20 and 70% of the population are carriers of *Staph. aureus* and approximately 15 to 20% carry enterotoxingenic strains. No other bacterial pathogen is carried over such a wide body area by such a large proportion of normal individuals[28].

At present five serologically different enterotoxins are recognized; enterotoxins A, B, C, D and E, and about 50% of all strains are toxigenic. Approximately 1µg, and occasionally less, enterotoxin is sufficient to cause illness in an otherwise healthy person[29] and symptoms occur 2 to 6 hours after ingesting the contaminated food.

In the UK, *Staph. aureus* is implicated in food poisoning and the number of viable cells isolated from infected food ranges between 10^3 per g and 10^{10} per g but in most cases at least 10^8 per g are present. Data from Gilbert and Wieneke[30] show that in England and Wales nearly 60% of staphylococcal food poisoning outbreaks are the result of eating infected meats, 14% due to infected poultry and nearly 10% due to infected fish and shellfish.

Note that in the UK, epidemiological data are based on numbers of reported outbreaks of food poisoning and not on those of individuals reported as affected by the food poisoning.

Many investigations have been completed on growth and toxin formation by *Staph. aureus* in a variety of growth conditions. It is generally accepted that wherever growth can be detected, toxin may also be formed. *Staph. aureus* remains the single most important food-poisoning organism of concern in low-a_w foods and because of the conflicting results on enterotoxin production, preservation systems should be designed to entirely prevent growth of this organism.

Interactive effect of inhibitors

A better appreciation of the fundamentals of a_w-controlled foods has developed over the last twenty years or so, initially arising from research supported by the US Department of Defense and NASA space programs. Together with proven successes in the pet food industry, this progress has renewed interest in a_w-controlled foods for human consumption.

As an understanding of the nature of these foods has grown the more it has been realized that reduction in a_w may be only one of a number of essential stability factors. The interaction of, for example, pH, redox potential, preservatives, water activity and heat processing may be included in the overall preservation system. This leads to products which are physico-chemically preserved to achieve specific goals, such as reduction in water activity, inhibition of microbes, control of oxidation and prevention of non-enzymic browning.

Water controls the deteriorative chemical reactions which take place in foods through its effect on the concentration, mobility and dissolution of reactants. Non-enzymic browning and lipid oxidation are important spoilage reactions in a_w-controlled foods. A careful choice of product formulations is essential to minimize these reactions in foods which otherwise have a long shelf life as a major benefit.

In practice, low-a_w foods are not made safe and stable with respect to microorganisms by relying on a_w alone because microbial growth has been shown to occur throughout the entire range of a_w's that generally define these foods (0.60 to 0.90 a_w).

Preparation and storage of ingredients

Ingredients should conform to specifications developed according to recognized principles of Good Manufacturing Practice and fitness for their intended use. The nature and relative importance of preservation factors in the overall stability system will dictate whether or not specific types of microorganisms should be included in the specification. Generally, microbiological problems arise because poor quality materials are used together with poor process control and line hygiene.

The chemical and physical nature of ingredients may be critical to achieving microbiologically safe and stable compositionally preserved foods. For example, the level of humectants (water activity lowering agents) in ingredients such as brined vegetables or fruits in syrup may have a major influence on the water activity of the finished product.

Identity checks should be completed on all deliveries of ingredients destined for compositionally preserved foods to positively establish that the material is what it purports to be. Many of the ingredients are likely to be similar in appearance but possess widely differing functional properties, quite apart from the basic need for their safe handling and use. Chemical preservatives may be dry, white powders closely resembling sweetening agents. Confusion in their use would have serious results. Full application of Good Manufacturing Practice for the storage of ingredients is essential.

Processing methods

There are two basic processing methods:

- Adsorption, in which food is dried and then water is added back until the desired level of water activity is achieved.
- Desorption, in which normally hydrated food is mixed with other food components that reduce the water activity to some low equilibrium level, e.g. by the use of humectants such as sugar and salt.

Typically, a combination of these two methods may be employed, such as when meats are diced and then mixed with curing salts and other ingredients to produce certain types of sausages. It is important that the detail of the microbiology together with the physicochemistry of the food is understood in order to optimize product safety. This is because the relationship between water activity and water content differs for the same food prepared by the two different methods just described.

Figure 12.5 shows an idealized sorption isotherm for a hypothetical food. It simply illustrates the water activity versus water content. If curves were developed for a food prepared first by desorption and then by absorption, two curves would result as illustrated in Figure 12.6. The importance of this phenomenon, known as hysteresis, is that microorganisms are inhibited at higher water activities when food is prepared by adsorption.

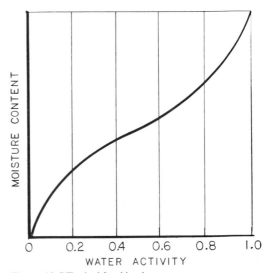

Figure 12.5 Typical food isotherm

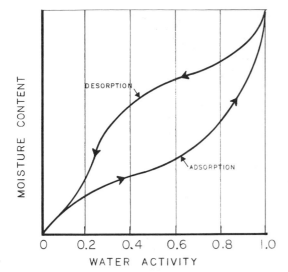

Figure 12.6 Typical moisture sorption isotherm showing hysteresis effect

Where products result from a combination of the two process, the same order of ingredient addition must always be followed during manufacture to insure that the resulting product possesses the microbiological stability it was designed to achieve.

Another reason for understanding the sorption isotherm of the food is to optimize its water activity buffering capacity. This relates to the food's capacity to adsorb limited quantities of water without causing a significant shift in its water activity. For products with little or no primary packaging and those which are multi-serve, there is a real risk that they will pick up water (moisture) from the environment. If these foods are formulated so that their water activity lies on the steep section of the sorption isotherm then relatively large changes in water content will result in minor ones to the water activity. Hence the safety of the product will be protected.

● Multifactor preservation systems

As previously noted, low water activity foods are not made safe and stable by using a_w alone. In practice, these foods are preserved by a combination of factors although the main effect may be from a lowered a_w. In general, other preservation factors have an increasingly inhibitory effect on microorganisms as the minimum a_w for growth is approached. Knowledge of these interactions allows for product development using two or more preservation factors each poised below the level required to inhibit if it were used alone.

Leistner and Rodel[31] described the combined effect (*Hurdeneffekt* or 'Hurdle effect') of a_w,

temperature, pH, *Eh*, nitrite and competitive microflora in extending the shelf life of meat products. The latter wrote about the idea of 'hurdle effects' to inhibit microorganisms using the summation of several preservation factors each used below its own microbiocidal level. It should be remembered that these 'hurdles' apply to specific product/processing/packaging situations.

– Initial numbers of microorganisms

The initial number of microorganisms present in many low water activity foods directly influences the efficiency of the preservation system. The likelihood that a product will remain stable increases as the number of initial organisms present decreases. Unlike heat preserved foods where all viable organisms present are destroyed, low-a_w foods contain organisms which remain viable long after manufacture. Good Manufacturing Practice dictates that ingredients, manufacturing, packing and the production environment should be designed to minimize the microbial contamination of finished product.

– Heat treatments

Heat treatments to reduce initial numbers of cells in the product have been extensively studied for low-a_w foods. Much of the work has involved *Salmonellae*, which do not grow in low-a_w foods but may remain viable for very long periods. As their infective dose can be small it is important to prevent them from being present in the product.

One of the technological problems in producing safe and stable low a_w foods is that organisms become increasingly resistant to the lethal effects of heat processing as a_w is reduced. Resistance is related to removal of water from the cell and is further influenced by the type of solute present. Heat processing should be applied to appropriate ingredients or to intermediates during manufacture when the a_w is close to 1.0 and the product then made quickly and hygienically. Recovery of heat injured organisms is inhibited at the reduced a_w of finished product.

– Solute effects

A microorganism in a low-a_w environment must adapt or die. Part of the adaptation is to adjust thermodynamically by lowering its internal a_w to regain hypertonicity in relation to the surrounding medium. This means that those solutes retained within the cell increase in concentration. Solutes already within the cell may be joined by solutes diffusing in, being transported in or by solutes produced as a response to the environment. If the cell survives beyond this initial adaptation its enzymes must continue to function. To be able to do this, solutes present at such increased concentrations cannot be excessively inhibitory. These accumulated solutes have been termed 'compatible solutes' (although more properly thought of as 'relatively uninhibitory').

Different humectants may inhibit particular microorganisms to a varying extent based on solute effects at any given a_w. Knowledge of the specific effects should be developed to insure optimum design of the process.

– pH

Movement of pH away from the optimum for growth of any microorganism causes decreased growth rate and increased sensitivity to other inhibitory factors.

– Preservatives

Preservatives and other antimicrobials have been used to stabilize foods for centuries and there is a wealth of knowledge and literature on the subject. National legislation in most countries define types, applications and levels of use. Preservatives that are suitable for use in low-a_w foods include the polyhydric alcohols such as glycerol and sorbitol; acid type inhibitors such as benzoate and sorbate, esters of 4-hydroxybenzoate (parabens); sulfites and antibiotics such as pimaricin and nisin. In principle, legislation very properly limits the use of preservatives to prevent them being used to disguise poor sanitary conditions. Where they are used, they are one factor in a multi-factor preservation system.

Experience has shown that effective preservation systems cannot be easily extrapolated from one food formulation to another. The efficiency of a preservation system can be reduced in a complex food system which means that the stability of a product must be investigated by experiment to optimize the formulation. Campden RA's *Guidelines for microbiological challenge testing* [32] give a good example of this in their 'Worked example – ambient stable product'.

Mixtures of preservatives that act synergistically are preferred to single-component systems because a wider spectrum of microorganisms is inhibited and the risk of developing preservative resistant strains is reduced. It is worth noting that undesirable preservative flavors are generally not additive.

Apart from the influence pH has by itself, it is of critical importance when acid inhibitors are involved. Preservatives, such as sorbic acid and benzoic acid, dissociate according to the pH of the system as defined by their dissociation constant pK_a. As it is the undissociated form that has the significant antimicrobial property, lower pH values result in more of the preservative in the system being active. It is a problem with some published papers that insufficient attention is paid to dissoci-

ation and hence the true amount of active preservative present.

Microorganisms may only grow in the presence of free water molecules, and therefore a preservative must be present in the water phase of the food to be effective. As the undissociated molecule of acid type inhibitors dissolve in lipids, high fat formulations can be difficult to preserve. The distribution of a preservative between fat and water is defined by its partition coefficient, which changes little with varying temperature. Partition coeeficients for preservatives remain approximately the same in a variety of vegetable oils; they are somewhat lower in crystallized fat. In the presence of sodium chloride the solubility of sorbic acid in water decreases considerably, and therefore more of the acid enters the lipid phase of products. Sugar, however, does not have this effect. Non-ionic emulsifiers have a complexing tendency with preservatives which reduce their concentration in the aqueous phase.

It is theoretically possible to calculate the level of preservatives required using physical and chemical data obtained from products. This is complicated in practice by possible loss of preservative onto the solid phase of products, e.g. by capillary effects, hydrogen bonding and ionic interactions. Therefore, it is essential that the efficiency of a preservation system is developed and validated using a well-developed challenge test.

Process control and validation

● Control of critical factors

Successful process control requires thorough planning and investigation into the factors critical to product safety and stability. The first investigation is into the preservation system of the product. For compositionally preserved foods it is becoming more feasible to predict the safety of a product based on compositional analysis. This is made possible through the use of data bases and mathematical modeling of the behavior of pathogens, which can be integrated to assist development of the preservation system. However, once the proposed product formulation has been arrived at, thorough challenge testing with microorganisms of Public Health significance and with food spoilage types should be completed.

From this, the HACCP study (see Chapter 3) should be extended to the whole of the product/process/packaging system.

All critical factors should properly be controlled by fail-safe systems. If such means are not available, then a system of two or more controls which are functionally independent from others in the series should be used. This aims to insure that the chances of a critical factor failing and going undetected are minimal.

Examples of critical factors found for low water activity foods include the obvious ones of time and temperature for heat processes. Less obvious factors include the order of ingredient addition, confirming that production equipment is dry, i.e. cleaning operations have not left residues of rinse water; and the time of mixing for critical ingredients such as preservatives.

- Measurement of water activity

As it is generally not possible to measure the a_w of the final batch of a product on-line, the desired a_w is achieved through control of ingredient type, addition, etc. Finished product samples from every batch should be taken for audit analysis and the results kept on file with other pertinent records to demonstrate that the product is safe.

Measurement of a_w is a specialized procedure, although not unduly difficult if care is taken. Procedures can involve simple, utilitarian equipment or complex physical and electrical devices. Naturally, some methods are more accurate and precise than others but an understanding of their strengths and weaknesses will help an operator attain up to as good as ±1.0% in the measurement of a_w.

The following methods are in regular use in food research laboratories:

1. Vapor Pressure Manometer (VPM)
2. Simple gravimetric and hygrometric methods
3. Dew point
4. Freezing point
5. Fiber hygrometers
6. Electric hygrometers

Electric hygrometers are the most commonly used instruments for routine process confirmation. The electrical resistance of a stabilized hygroscopic material held in a small hermetically sealed chamber containing a sample of the food is recorded. The resistance of the material, which is usually lithium chloride, is related to its degree of hydration which,

in an equilibrated food sample chamber, is equivalent to the ERH and hence the a_w of the food. The sensor is connected to a potentionmeter which measures the resistance; the output can go to an analogue or digital display and a recorder.

Sensors should be calibrated routinely using saturated salt solutions as standards. Absolute standards for salt solutions have yet to be accepted universally. However, Table 12.7 shows suggested international standard water activity values for saturated salt solutions as proposed by Resnick et al.[33]. These have been found most satisfactory in day-to-day use for process confirmation. The following six critical statements should be made when measured a_w values are quoted:

1. A description of the method.
2. Temperature of measurement.
3. Standard values of salts used in calibration.
4. Reference to the source of standard values.
5. Repeatability of the method.
6. Precision of the quoted value, e.g. ±0.01 a_w.

Packaging materials

There are no special comments to make about the packaging for low-a_w foods. The range of types is vast and indeed some products have no primary packaging at all other than the physical nature of the outer layer of the product, e.g. some types of sausage.

Coding and labeling

There are no special comments to make about coding and labeling.

Storage and distribution

Obviously, low-a_w foods will deteriorate if they are allowed to come into contact with water (including moisture and high humidity). Storage temperature has a significant effect on the spoilage of these foods and, in general, as the temperature moves away from the optimum for growth of an organism the probability that it will develop in the food decreases. A number of studies on commercial products have shown that shelf life can be significantly extended by using low storage temperatures for these products. Temperatures higher than the optimum for growth can result in salts 'protecting' organisms which is a reminder not to assume general rules of preservation but to practically evaluate any selected preservation system.

Table 12.7 Suggested international standard water activity values for saturated salt solutions at 25°C (from Resnick[33]

Saturated salt solution	Most accepted a_w[a]	Mean a_w[a]
NaBr	0.577	0.576
NaCl	0.753	0.753
$(NH_4)_2SO_4$	0.801	0.803
KCl	0.843	0.844
$BaCl_2 2H_2O$	0.902	0.902
KNO_3	0.925	0.929
K_2SO_4	0.972	0.971

[a] Statistically computed; refer to paper for details.

Special considerations

These have been discussed in previous sections, but whatever the formulation and processing systems chosen, it is essential to be able to demonstrate a positive assurance of safety. This, in practice, means challenge testing – see Chapter 9.B for further information and Campden Technical Manual 20[32]. Remember also that low-a_w confers increased heat resistance to *Salmonella* spp. and other organisms.

Apart from exceptionally a_w tolerant members of the *Enterobacteriaceae,* most food-borne bacterial pathogens are unable to grow in the range of 0.90 to 0.95 a_w.

References

1. Christian, J. H. B. (1980) Reduced water activity. In *Microbial Ecology of Foods, Vol. 1. Factors affecting life and death of microorganisms*, pp. 70–91. International Commission on Microbiological Specifications for Foods, New York: Academic Press
2. Scott, W. J. (1953) Water relations of *Staphylococcus aureus* at 30°C. *Australian Journal of Biological Sciences*, **6**, 549–564
3. Leistner, L. and Rodel, W. (1976) Inhibition of microorganisms in food by water activity. In *Inhibition and Inactivation of Vegetative Microbes*, edited by F. A. Skinner and W. B. Hugo, pp. 219–237. Society for Applied Bacteriology Symposium Series no. 5, London: Academic Press
4. Christian, J. H. B. (1954) The influence of nutrition on the water relations of *Salmonella oranienberg*. *Australian Journal of Biological Sciences*, **8**, 75–82
5. Christian, J. H. B. (1955) The water relations of growth and respiration of *Salmonella oranienberg* at 30°C. *Australian Journal of Biological Sciences*, **8**, 490–497
6. Horner, K. J. and Anagnostopoulus, G. D. (1975) Effect of water activity on heat survival of *Staphylococcus aureus, Salmonella typhimurium* and *Salm. senftenburg*. *Journal of Applied Bacteriology*, **38**, 9–17
7. Corry, J. E. L. (1974) The effect of sugars and polyols on the heat resistance of *Salmonellae*. *Journal of Applied Bacteriology*, **37**, 31–43
8. Corry, J. E. L. (1976) The safety of intermediate moisture foods with respect to *Salmonella*. In *Intermediate Moisture Foods*, edited by R. Davies, G. G. Birch, and K. J. Parker, pp. 215–238. London: Applied Science
9. Marshal, B. J., Ohye, D. F. and Christian, J. H. B. (1971) Tolerance of bacteria to high concentrations of NaCl and glycerol in the growth medium. *Applied Microbiology*, **21**, 363–364
10. Stringer, W. C., Bibkie, M. E. and Naumann, H. D. (1969) Microbial profiles of fresh beef. *Food Technology*, **23**, 97–102
11. Twedt, R. M., Spaulding, P. L. and Hall, H. E. (1969) Morphological, cultural, biochemical and serological comparisons of Japanese strains of *Vibrio parahaemolyticus* with related cultures isolated in the United States. *Journal of Bacteriology*, **98**, 511–518
12. Nickelson, R. and Vanderzant, C. (1971) *Vibrio parahaemolyticus* – a review. *Journal of Milk and Food Technology*, **34**, 447–452
13. Lanigan, G. W. (1963) Silage bacteriology. 1. Water activity and temperature relationships of silage strains of *Lactobacillus plantarum, Lactobacillus brevis* and *Pediococcus cereviseae*. *Australian Journal of Biological Sciences*, **16**, 606–615
14. Troller, J. A. and Christian, J. H. B. (1978) *Water Activity and Food*. London: Academic Press
15. Jakobsen, M. and Trolle, G. (1979) The effect of water activity on growth of *Clostridia*. Nordisk Veterinarmedicin, **31**, 206–213
16. Ohye, D. F. and Christian, J. H. B. (1966) Combined effect of temperature, pH and water activity on growth and toxin production by *Cl. botulinum* types A, B and E. In *Botulism*, edited by M. Ingram and T. A. Roberts, pp. 217–223. London: Chapman and Hall
17. Baird-Parker, A. C. and Freame, B. (1967) Combined effect of water activity, pH and temperature on the growth of *Clostridium botulinum* from spore and vegetative cell inocula. *Journal of Applied Bacteriology*, **30**, 420–429
18. Pitt, J. I. (1975) Xerophilic fungi and the spoilage of foods of plant origin. In *Water Relations of Foods*, edited by R. B. Duckworth, pp. 273–307. London: Academic Press
19. Northolt, M. D., Verhulsdonk, C. A. H., Soentoro, P. S. S. and Paulsch, W. E. (1976) Effect of water activity and temperature on aflatoxin production by *Aspergillus parasiticus*. *Journal of Milk and Food Technology*, **39**, 170–174
20. Northolt, M. D., Van Egmond, H. P. & Paulsch, W. E. (1977) Differences between *Aspergillus flavus* strains in growth and aflatoxin B1 production in relation to water activity and temperature. *Journal of Food Protection*, **40**, 778–781
21. Northolt, M. D., Van Egmond, H. P. and Paulsch, W. E. (1978) Patulin production by some fungal species in relation to water activity and temperature. *Journal of Food Protection*, **41**, 885–890
22. Northolt, M. D., Van Egmond, H. P. and Paulsch, W. E. (1979) Penicillic acid production by some fungal species in relation to water activity and temperature. *Journal of Food Protection*, **41**, 476–484
23. Northolt, M. D., Van Egmond, H. P. and Paulsch, W. E. (1979) Ochratoxin A production by some fungal species in relation to water activity and temperature. *Journal of Food Protection*, **42**, 485–490
24. Northolt, M. D. and Bullerman, L. B. (1982) Prevention of mould growth and toxin formation through control of environmental conditions, *Journal of Food Protection*, **45**, 519–562
25. Lanyi, J. K. (1976) Membrane structure and salt dependence in extremely halophilic bacteria. In *Extreme Environments* edited by M. R. Heinrich, pp. 295–303. London: Academic Press
26. Brown, A. D. (1976) Microbial water stress. *Bacteriological Reviews*, **40**, 803–846
27. Varga, S., Sims, G. G., Mickalik, P. and Regier, L. W. (1979) Growth and control of halophilic microorganisms in salt minced fish. *Journal of Food Science*, **44**, 47–50
28. Williams, R. E. O. (1963) Healthy carriage of *Staphylococcus aureus*: its prevalence and importance. *Bacteriological Reviews*, **27**, 56–71
29. Bergdoll, M. S. (1970) Enterotoxins. In *Microbial Toxins*, Vol. III edited by J. O. Cohen, pp. 301–331. Chichester: John Wiley
30. Gilbert, R. J. and Wieneke, A. A. (1973) Staphylococcal food poisoning with special reference to the

detection of enterotoxins in foods. In *The Microbiological Safety of Foods*, edited by B. C. Hobbs and J. H. B. Christian, pp. 273–284. London: Academic Press

31. Leistner, L. and Rodel, W. (1975) The significance of water activity for microorganisms in meats. In *Water Relations of Foods*, edited by R. B. Duckworth, pp. 309–323. London: Academic Press

32. Campden RA (1987) Technical Manual No. 20. *Guidelines for Microbiological Challenge Testing*, Campden Food Preservation Research Association, Chipping Campden, Gloucestershire GL55 6LD

33. Resnik, S. L., Favetto, G., Chirife, J. and Ferro Fontan, C. (1984) A world survey of water activity of selected saturated salt solutions used as standards at 25°C, *Journal of Food Science*, **49**, 510–513

Further reading

Beuchat, L. R. (ed) (1987). *Food and Beverage Mycology*, 2nd edn. New York: an AVI Book published by Van Nostrand Reinhold

Erickson, L. E. (1982) Recent developments in intermediate moisture foods. *Journal of Food Protection*, **45**, 484–491

Hocking, A. D. (1988) Moulds and yeasts associated with foods of reduced water activity: ecological interactions. In *Food Preservation by Moisture Control*, edited by C. C. Seow, T. T. Teng and C. H. Quah, pp. 57–72. London: Elsevier Applied Science

Labuza, T. P. (1980) The effect of water activity on reaction kinetics of food deterioration. *Food Technology*, **34**, 36

Rockland, L. B. and Nishi, S. K. (1980) Influence of water activity on food product quality and stability. *Food Technology*, **34**, 42

Rockland, L. B. and Beuchat, L. R., (1987). *Water Activity: Theory and Applications to Food*. New York: Marcel Dekker

Schwimmer, S. (1980) Influence of water activity on enzyme reactivity and stability. *Food Technology*, **34**, 64

Troller, J. A. (1980) Influence of water activity on microorganisms in foods. *Food Technology*, **34**, 76

Troller, J.A., Bernard, T.A. and Scott, V.N. (1984) Measurement of water activity. In *Compendium of Methods for the Microbiological Examination of Foods*, 2nd edn, edited by M. L. Speck, pp. 124–134. Washington DC: American Public Health Association

Troller, J. A. (1986) Water relations of food-borne bacterial pathogens – An updated review. *Journal of Food Protection*, **49**, 656

Chapter 13
Criteria for ingredients and finished products

13.A Food safety strategy and the use of criteria

Background

To achieve the aim of producing safe wholesome food with a satisfactory shelf life, it is important for the processor to understand precisely what criteria are and what their proper use can achieve. Mossel[1] sets forth the position with clarity, scholarship and logic in a paper which distills experiences over many years spent at the forefront of Public Health food microbiology. It is therefore required reading for both management and technologists. He explains why the examination of final food samples alone does nothing to make the food safe with the analogy ' – as little as taking (i.e. measuring) blood pressure will in itself relieve hypertension'.

The approach which does work is described by Mossel[1] as the strategy advocated as long ago as the 1930s by S. C. Prescott and K. F. Meyer in the USA and Sir Graham Wilson in the UK. This is the strategy of taking active preventive measures to achieve product safety under defined conditions of production and distribution. This requires positive management action which Mossel calls 'Longitudinally Integrated Safety Assurance' or LISA.

Mossel[1] describes criteria as 'microbiological target values' and sets forth a six-point approach for their development together with a seven-point (or 'seven-P') program for the management strategy of applying the LISA principle.

It is therefore axiomatic that the LISA concept integrates the Hazard Analysis Critical Control Point (HACCP) system with the establishment and use of criteria for ingredients and finished products along with other management tools such as education, training, audit of facilities and operations together with Quality Assurance and applicable regulatory controls.

This chapter aims to help the processors' microbiologist, particularly in the early stages of developing sound criteria or target values. The final section (13.D) is included, for the convenience of the user, as the separate but 'Quality'-related issue of ingredient defects and foreign materials. The concern with defects and foreign materials relates to the hazards which their presence cause either because of potential harm to the consumer or because their presence is objectionable in a food for aesthetic or other reasons.

Microbiological criteria for safety and 'quality'

For the purpose of this chapter a criterion (plural, criteria) is a statement, number, rule or yardstick by which a judgment or decision can be made on the safety and/or stability or quality of a food or process. It is therefore part of an answer to questions such as 'Is this ingredient adequate for our purpose to make a specific product of consistent, reliable quality?' or 'What microbiological status should this product have in order to deliver safety, wholesomeness and the required stability in the specified market?'

Microbiological safety (see ICMSF[2]) is related to the presence or absence of pathogenic microorganisms or their toxins, the number of pathogens present and the ease with which they or their toxins can be controlled or destroyed.

Microbiological quality (see ICMSF[2]) is related to the presence or absence of microorganisms that are capable of causing spoilage or are indicative of the application or otherwise of Good Manufacturing Practice (GMP) and to the effectiveness of measures to destroy or limit the numbers of such organisms.

When applied, criteria should distinguish between an acceptable and unacceptable ingredient or product or between acceptable and unacceptable food processing and handling practices, and should be written to both allow and encourage good decisions to be made.

References

1. Mossel, D. A. A. (1989) Adequate protection of the public against food-transmitted diseases of microbial aetiology. *International Journal of Food Microbiology*, **9**, 271–294
2. ICMSF (1986) *Microorganisms in Foods, Volume 2 – Sampling for Microbiological Analysis: Principles and Specific Applications,* 2nd edn. Toronto: University of Toronto Press

13.B Microbiological criteria

Introduction – the HACCP foundation

The key question is 'What is needed to define microbiological requirements for a specific ingredient, process or product?' What follows is an outline of some good ways in which to seek the answer.

Microbiological criteria or 'yardsticks' are used because of management concern with Public Health hazards, as well as 'Quality' and regulatory concerns. The following terms are used in the sense in which they are defined and used in the Campden Technical Manual 19[1], which is entitled *Guidelines to the Establishment of HACCP*. These terms and their meanings are:

- Critical Control Point (CCP): a location, stage, operation or ingredient which, if not controlled, provides a threat to consumer safety or product acceptability.
- Concern: an expression of the seriousness of a failure to control a critical point, derived from knowledge of a hazard and the risk of it occurring.

Four levels of concern are used in TM 19. These are *all* based on an expert judgment which requires knowledge and experience.

(a) High Concern – without control there is a life-threatening risk.
(b) Medium Concern – there is a threat to the consumer or product which *must* be controlled.
(c) Low Concern – there is little threat to consumer or product although it may be advantageous to control it.
(d) No concern – there is no threat to the consumer.

It is important to realize that concerns occur at all stages of the production of the food, i.e. from the effect of agricultural operations on ingredients, through manufacture and distribution to the preparation and use (or misuse) of the food by the consumer.

- Good Manufacturing Practice (GMP): a specified and recorded method of operation designed to minimize hazards and capable of control and being monitored.
- Hazard: a potential to cause harm to the consumer (safety) or the product (spoilage). This term includes both specific threats, e.g. the presence of 'staphylococcal enterotoxin', and operational malpractice, e.g. allowing cross contamination of raw and cooked meats. Note that Campden Technical Manual 19 gives useful lists of microbiological and non-microbiological hazards including foreign material (foreign bodies).
- Risk: is the chance (probability) that a hazard will be realized (occur).

Warning The usefulness (value) of any Hazard Analysis depends both on how thoroughly it is done and on the rigor of its application. Both aspects require management commitment as well as technical expertise.

It is essential to be aware of both the uses and the limitations of microbiological analysis in hazard analysis and control of critical points (food control). An understanding of the biology and in particular the ecology of the microorganism(s) is required together with an appreciation of the statistical procedures (see Jarvis[2], Chapter 10).

With regard to criteria it is important to appreciate the different perspective of the processor and the regulatory authority. The processor will use a microbiological criterion as one of a number of ways of exercising control and validation (audit) of processes and products. The microbiological criterion will be seen in the context of historical data and is a measure that a soundly established system is under control, e.g. in the assessment of trends both for safety and 'quality'. By contrast, the regulatory authority is concerned only with safety, and may well be in a 'port-of-entry' situation with a decision dependent on test data from only a limited number of samples.

Remember Because of sampling difficulties and the inherent variations of biological systems, numerical criteria cannot be as precise as, for example, determinations of salt (NaCl) content. Understanding their real (true) meaning requires experienced, expert judgment.

Non-technical problems

In addition to the technical problems there are always other factors, such as management pressure to 'produce some limits right now' or a similar desire for action by politicians. Under these pressures, the technologist must maintain his integrity and honesty in dealing with the facts as they are known. Management must realize that there is no substitute for detailed knowledge of products and processes over a period of time and that in the absence of such knowledge only provisional, often very provisional, answers can be given.

It is worth remembering that there is still truth in the remark of Dr Sedgwick who was known as the 'Father of American Public Health Bacteriology'. He stated: 'Standards are often the guess of one worker, easily seized upon, quoted and re-quoted until they assume a semblance of authority.'

The need for the establishment of criteria

Criteria are usually established as part of a program to control a perceived hazard but it is essential, in the words of the National Research Council book[3], 'a criterion can be shown to be effective and practical'. This is obvious good sense and fits perfectly into the LISA strategy (see Section 13.A). They add:

> The criterion must accomplish its objective, i.e., adequately measure the contaminants of concern, be technically attainable under commercial conditions by Good Manufacturing Practices, and be administratively feasible. There are additional factors that should be considered before the need for a specific microbiological criterion can be (regarded as) established. These should include:

1. evidence of a hazard to health based on epidemiological data or a hazard analysis;
2. the nature of the natural and commonly acquired microflora of the ingredient or food and the ability of the food to support microbial growth;
3. the effect of processing on the microflora of the food;
4. the potential for microbial contamination (or re-contamination) and/or growth during processing, handling, storage, and distribution;
5. the category of consumers at risk;
6. the state in which food is distributed, e.g. frozen, refrigerated, heat processed, etc.;
7. potential for abuse at the consumer level;
8. spoilage potential, utility (suitability depending on how the food will be used), and (the existence of) GMPs;
9. the manner in which the food is prepared for ultimate consumption, i.e. heated or not;
10. reliability of methods available to detect and/or quantify the microorganisms(s) and toxins of concern; and
11. the costs/benefits associated with the application of the criterion.

Note: the words in parentheses have been added by way of explanation for users outside the USA.

A discussion of the development of criteria is given later in this section. Remember that responsible microbiologists are reluctant to be dogmatic about a new product or process until a HACCP analysis has been made in sufficient detail to deliver a response with adequate confidence. When time is a severe constraint, most microbiologists will make use of the network of contacts which they have built up over time to get some idea of what is 'usual' or regarded as 'conventional wisdom'. This obviously is 'provisional' or 'very provisional' for a specific application.

Uses of criteria

Criteria are used when there is a need to make an assessment, e.g. of:

- The safety of a food, ingredient or process

- Compliance with Good Manufacturing Practice (GMP), i.e. as part of a verification program. Be aware that compliance with GMP is primarily assessed by compliance with properly established procedures, examination of records and calibration of instruments (related to a HACCP analysis) of a product, processing, packaging or distribution system.
- As part of a development program
- The utility (suitability) of a food or ingredient for a particular purpose
- The keeping quality or shelf life of some perishable foods

Components of a microbiological criterion

When developing a criterion it will be found helpful to make reference to Chapter 3 of the National Research Council book[3]. In a good criterion, at least the following information should be included:

- The identity of the food or ingredient
- The application of the criterion, e.g. 'for use in pickles and sauces' *or* prohibition(s) on use
- The microorganism(s), toxin(s) or contaminant(s) of concern
- The analytical method(s) to be used for detection, enumeration or quantification
- The sampling plan(s)
- The microbiological or other, e.g. temperature, limits

In practice, there will almost certainly be more, often much more, information than this in the working document.

Types of microbiological criteria

Three types or categories of criteria are internationally recognized and there appears to be no need for additional terms. These are 'Standards', 'Guidelines' and 'Specifications'. These can be defined as follows.

Standards

These are legislative and mandatory. They are embodied in a law, ordinance or administrative regulation. Failure to comply is a violation subject to the enforcement policy of the regulatory agency having jurisdiction and must therefore be expected to result in prosecution. Marshall[4] states that 'the attitude to the standard and the required compliance of the product by the enforcement authorities may vary considerably between countries'. Processors, however, should properly and wisely assume that the law will be fully enforced.

Guidelines

These are *not* legal standards although regulatory agencies may use them. The National Research Council[3] states:

A microbiological guideline is a criterion that often is used by the food industry or a regulatory agency to monitor a manufacturing process. Guidelines function as alert mechanisms to signal whether microbiological conditions prevailing at critical control points or in the finished product are within the normal range. Hence, they are used to assess processing efficiency at critical control points and conformity with Good Manufacturing Practices. A microbiological guideline is an advisory criterion in that a given lot of food exceeding the limit for a nonpathogenic organism would not be taken off the market or even downgraded. Guidelines may be mandatory, however, in the sense that food company management and/or regulatory agencies may demand that the conditions responsible for persistent microbiological deficiencies be corrected without delay.

The need for appropriate management response explains why 'In-house' guidelines may vary from company to company even for the same product description.

Specifications

These are technical requirements which form the basis for a commercial transaction. The National Research Council states[3]:

A microbiological specification is a microbiological criterion that is used as a purchase requirement whereby conformance with it becomes a condition of purchase between buyer and vendor of a food or ingredient. A microbiological specification can be either mandatory or advisory.

They may become mandatory in the sense that materials which do not comply are not bought.

It is worth remembering that initially some or all parts of a specification will be regarded as 'provisional'. After experience in use and any necessary modification, it will be regarded as 'established' or alternatively as 'transitional'. The description 'transitional' is an apt one because no specification should be expected to remain unchanged over the decades.

Notes on specifications

Contents

In addition to microbiological data there will be other sections in the specification. These usually include:

- Identity. This should be a full and sufficient description of the material and definition of pertinent properties. Be aware that when specifications prove inadequate, i.e. they are met but fail to deliver what is required, it is usually because the definition of properties is incomplete or imprecise. Where variations in origin, processing or composition are intended these must be explicitly stated.

- Purity and Composition. The usual headings in this section are:

(a) Chemical – criteria which give compositional requirements and assay and/or analytical data
(b) Physical – which gives properties such as size, color, weight, density, viscosity. It may include sensory criteria, e.g. flavor.
(c) Microbiological criteria
(d) Defects and foreign materials. No one wants these but some are inherent in the materials, e.g. discolored seeds or stones in legume seeds (pulses). Sampling plans, analytical methods (for visual defects this is usually careful sorting) and numerical limits are needed for the same reasons as in microbiological criteria.
(e) Sanitary status. This is usually measured by filth tests, e.g. mold count, rot and insect fragment count. Microbiological assessment of sanitary status is conveniently included under 'microbiological criteria'.

- Legal requirements
- Packaging. This describes the net weight or volume and container required. Additional to the cost, consider the risk of the package contaminating the product when it is used, e.g. fiber or paper sacks have to be opened and some fibers or paper must be expected to pass into the product stream. The question must therefore be asked, 'What is the efficiency of the removal equipment?' Consider also the possibility and cost/benefit of having material in quantities which can be used directly, e.g. 2 bags per batch.
- Transport. Shapton[5] writes that this section specifies the conditions under which the material is to be transported and whether, for instance, insulated or refrigerated vehicles are required. In some cases it may be necessary to specify any substances which *must not* have been or be transported in the same vehicle or container.
- Storage. This section is often distributed only for internal use but all or part of it may be issued when materials are kept in store outside the factory premises. It sets forth the conditions of storage. These include, for example, temperature limits for cold stores and product, the length of time the material may be held in stock, and the frequency and type of checks made to insure that the material remains in good condition.

Where fumigation or spraying is required, it is important to specify the materials to be used and the method of application. In some cases it may also be desirable to specify the constructional materials of the warehouse and the precise way in which the ingredient material is to be stored.

- Uses. This section is for internal use. If grading is done then the usage for each grade should be given. This grading has no Public Health implication and is purely a matter of economic good sense. For example, a batch of sugar with a high thermophilic spore load is unsuitable for hot can vended soups but, if thermophiles were the only problem, may be used in jelly, pickles or sauces.

If grading is practiced, then it is essential that everyone ordering, storing and handling the material treats each grade as a separate and distinct ingredient. If this is not done there is a serious risk that the wrong grade will be used in the final product.

Changes to specifications

As well as permanent changes resulting from alterations to product, processes or packaging, changes may have to be made when an 'emergency' occurs. These must not affect the safety or stability of the product. As an example, when making canned soups, size gradings may have to be altered or alternative varieties used during a poor vegetable season. If changes have to be made, it is important to inform all suppliers *immediately*. Do not just accept out-of-specification materials as this will rapidly discredit the whole specification system. *Caution* Size-grading alterations or changes in particulate geometry or consistency (viscosity) may adversely affect the thermal process of a 'Low Acid' canned food.

Usefulness of supporting data for guidelines and specifications

Valuable supporting data are available for microbiological criteria from non-microbiological sources. These include:

- Sensory tests. First signs of incipient spoilage are often best detected by this method. Sometimes heating then cooling the product before the sample is tested will enhance the 'off odor' or 'off flavor'.
- Physical/Chemical changes, e.g. pH shift, increasing acidity or rancidity, can be useful because the tests can be done quickly.
- Records and logs of the production process, especially when designed to show trends, can provide very high quality information which supplements microbiological testing. Records are also important for Quality Audit purposes.

- Prompt information from the vendor of changes in processes or product composition which *might* affect product safety or stability enables the specification to be promptly updated.
- Vendor audits should assess management attitudes as well as the quality systems which are in place (see Chapter 2). Audit data must be of high quality to be really useful; if not they are likely to be positively misleading.

Development of criteria

It should be understood that criteria are an attempt to predict what ought to happen in a complex situation. It demands an adequate knowledge of microbial ecology under *real life* conditions. The complexity of the process is illustrated by the National Research Council[3] chapter headings, 'Selection of Foods for Criteria Related to Safety; Selection of Pathogens as Components of Microbiological Criteria; and Selection of Indicator Organisms and Agents as Components of Microbiological Criteria'. The International Commission on Microbiological Specification for foods [ICMSF] has produced a two-volume text[6,7] on Microbial Ecology of Foods which is invaluable for the technologist developing criteria.

It is not surprising, because of the complexity of the situations and the scale of human and fiscal resources required, that development of sound, widely agreed criteria needs collaborative action by industry, governments and academia on a national and international scale.

The ICMSF in their book *Microorganisms in Foods* – 2[8] discusses the application of criteria to international trade, and also the problems of 'port-of-entry' examination. They conclude:

> Ideally the control of microorganisms in foods is at the point of production, processing, or preparation for consumption (see (their) Chapter 7, Control at Source). However, for much food in international trade there is no knowledge of control at source or of the conditions used during processing and handling. Therefore, there remains a need for criteria to assess the acceptability of foods at port of entry.

Before recommending a criterion for a product, the Commission decided that each product must meet the following conditions:

1. The product must be in international trade.
2. There must be good epidemiological evidence that the product, or product group, has been implicated in food-borne disease and/or may have an inadequate shelf-life if Good Commercial Practice (GCP) has not been followed.
3. There must be good evidence, or good reason to believe, that the application of a criterion

will reduce the health risk in food and/or effectively assess adherence to GCP.

Note that the ICMSF term Good Commercial Practice has essentially the same meaning as Good Manufacturing Practice (GMP).

They also point out the considerable practical difficulties of 'port-of-entry' examination. However, the problems of safety or quality assessment which face the company importing an ingredient produced in a number of locations and assembled in a single shipment are not made fully explicit. In such instances the best assurance is likely to be a well-deserved reputation of the agent for his knowledge of the sources and, above all, his personal integrity.

Development of specifications within a company

Management should be aware that the development of sound specifications requires a considerable technical input. Within a company, the technologist or microbiologist is often asked for specification criteria for a new ingredient or product to be purchased, made, or co-packed. This essentially requires answers to the three questions:

- Is it safe?
- Is it stable – with the desired 'Quality' attributes?
- Are appropriate Good Manufacturing Practices (GMPs) in place?

To answer these questions a good strategy is required, which is given below:

- Put the proposed specification criteria into the context of the LISA (Longitudinally Integrated Safety Assurance) strategy.
- Use HACCP in sufficient detail to determine what is necessary to deliver the required positive assurance of safety or assessment of 'quality'. Be sure that you have a clear understanding of the intended uses of the food and of the possibilities of misuse.
- Check out details of pertinent associated quality systems including audit systems and record keeping.

When this has been done, there should be a full understanding both of what criteria should achieve, or deliver and of the wider context in which specifications are set. The aims and intentions of the criteria will therefore be clear. To reach this stage, firm management decisions and commitment are required in marketing, production and technical areas.

Given good data over a sufficient period of time, the numbers should fall into place easily, whether the specification is for an ingredient or a finished product. However, these good data may not be available, and the microbiologist has to produce 'provisional' criteria from material available at the time.

It is important for management to appreciate and accept that there are no instant mathematical (or other) 'fixes' in the development and use of specification criteria. Development of these criteria requires an understanding of microbial ecology and their use means making a judgment about a particular lot or process and full use of all other pertinent information that is available. This is the reason experienced microbiologists will need to do much more testing on a lot or process if microbiological test data are the *only* information available than would be the case if microbiological data are supported by evidence of sound GMPs being consistently applied as shown by process records and audit reports.

With regard to 'safety' criteria, the microbiologist must take a 'conservative' position and require a positive and demonstrable assurance of safety.

With regard to 'quality' criteria, it is a sound principle to attempt to relate the true total costs of making the wrong decision – which are usually hard to estimate but which include product recall and loss of reputation – to the costs of testing – which are usually accurately known. In developing 'provisional' criteria the microbiologist will use information from any useful and trustworthy (reliable) source, e.g. literature, research associations, government and international sources including Codex Alimentarius, trade and individual companies, university and polytechnic departments as well as personal contacts built up over a period of time.

These sources of information will be used to:

- Find out what are typical testing requirements and results for the ingredient or product. Be aware that this can be difficult because published data are scattered and not always as unambiguous, detailed and specific as is desirable .
- Select an appropriate sampling plan, taking care to be sure that the meaning and implications of the plan are thoroughly understood.
- Determine pertinent numbers and kinds of organisms, taking care to understand the advantages and limitations of the chosen microbiological technique(s).

Notes:

1. Remember that the Data Table is simply a starting point in the complex process of developing a reliable microbiological criterion.
2. The information in the Data Table is not necessarily used or recommended by Heinz companies nor necessarily applicable to any specific product/process/packaging combination.
3. As it is only one of a number of sources of initial information for the development of an appropriate criterion, it is essential to take professional

microbiological advice before use. Remember that the data given is necessarily incomplete, possibly out-of-date and may not be suitable for *your* intended application.

Warning Always check the current legislative position both in the country of manufacture and of sale – they may be different.

Collection and handling of microbiological data

This may be conveniently considered in the following stages:

- Establish legislative requirements (if any) in both the country of manufacture and of sale (if these are different).
- Check, preferably directly with the appropriate technical department, that you have up-to-date supplier or customer requirements (if applicable).
- Gather published information. Be sure that it is fully applicable for your ingredient(s) product(s) and/or process(es). Codex Alimentarius data are limited but useful and publication of new material is continuing.

It is often found that available data are insufficient, incomplete or not wholly applicable but that an answer is required quickly. At this stage, remember to keep clear the important difference between 'safety' and 'quality' aspects. Do not forget that a finished product criterion which is suitable for one group of customers may not be suitable for all. Thus it is usual for groups of special concern, e.g. small babies, the elderly or immuno-compromised, to have appropriate criteria specified for their use.

- Use other sources of information, e.g. research institutes or associations, university departments or polytechnics. Listen to advice, but assess it carefully as the accountability (responsibility) is yours alone.
- Draft the outline (initial) criterion for the specification.
- Run preliminary tests, preferably on several lots to see whether there are likely to be any major problems.
- Develop the 'Provisional' criterion. This should be discussed with the supplier or customer as appropriate.
- Document the 'Provisional' criterion and plan to monitor its use. It may be useful to have 'Provisional' and 'Established' (or 'Transitional') criteria on different colored paper.
- When the professional judgment is that sufficient experience has been obtained to amend or finalize the criterion it may be issued as 'Established' or 'Transitional'.
- Review the criterion when there is a change in formulation, processing or use.

Brief notes on organisms commonly used in criteria

The following groups of organisms are most usually considered when developing criteria.

APC

APC or Aerobic Plate Count is also known (usually incorrectly) as the total viable count (TVC), although the latter should include anaerobes. With experience of its application it is a simple, useful assessment method and guide to sanitary status. Remember, that if there is a 'kill step' in the process the APC at later stages represents subsequent recontamination and growth. The Direct Microscopic Count (DMC) can be useful in these situations. It is also important to remember that the APC gives the number of colony-forming units (cfu) and it is preferable to report the results as cfu per g or ml rather than simply as a number.

Coliforms

This group has no taxonomic validity but is a working concept used by food microbiologists. Its extent and use have been controversial and its inclusion needs careful consideration. It is usually regarded as an indicator for extrinsic pathogens, as an indication of poor sanitation or of a missed pasteurization step.

It is useful to remember that its original use was that of indicator organisms in water supplies. The group was easily recognized in neutral-red bile salt lactose (MacConkey) agar or broth. Other media, e.g. VRBA, are now usually employed with foods. As coliforms exist in high numbers in faeces and can survive in stored water longer than most bacterial pathogens its use as an indicator was sound. When the test is applied to foods, in which the organisms may grow rather than merely survive, the significance is less clear, especially when coliforms of non-fecal origin are involved. As Mehlman (in Speck[9]) remarks, 'It should be emphasized that the term "fecal coliform" like the term "coliform" has no taxonomic validity'.

The basis for the identification of the 'fecal coliform' group is usually considered to be growth at 44.5°C in 24 hours in lauryl sulfate tryptose (LST) or *Escherichia coli* (EC) broths, accompanied by the production of indole in tryptone broth, also at 44.5°C.

In Europe, the trend has been to use '*Enterobacteria*' counts rather than 'coliform', using either violet-red bile agar with 1% glucose (VRBGA) or brilliant green (BG) bile salt broth with glucose replacing lactose.

Escherichia coli

Escherichia coli are traditionally differentiated on the basis of the IMVIC tests (Type I, $++--$ and Type II, $-+--$). The adequacy of this differentiation is dependent on methodology and additional tests are often used. (See Mehlman – Chapter 25 in Speck[9].) Historically, *E. coli* has been considered more important in developing countries but is now recognized as of increased importance in industrialized countries. A number of types of pathogenic *E. coli* are recognized i.e. 'classical enteropathogenic *E. coli*' (or EEC/EPEC), enterotoxigenic *E. coli* (ETEC), enteroinvasive *E. coli* (EIEC) and enterohaemorrhagic *E. coli* (EHEC) or verotoxigenic *E. coli* (VTEC). A battery of tests is required for identification, including serotyping which is beyond the expertise of the normal food microbiological laboratory. It is probably best to regard the isolation of appreciable or unusual numbers of *E. coli* with concern.

Salmonella spp.

The usual test is for presence in 25 g sample units, with increased stringency being given by increasing the number of samples taken. Bulking of samples may be done but there is a practical limit of about 100 g. When monitoring a continuous process, there is sometimes a preference for taking frequent small samples and bulking. Most processors are primarily concerned with presumptive presence, and take 'root-cause corrective (remedial) action' on finding a positive test result.

Typing, e.g. by sera and phages, becomes important to insure that a contaminant has been found and eliminated. *Note:* Some tables are printed in a way which suggests that absence of *Salmonella* spp. in 1 g (0 per g) is the limit but this seems an unlikely criterion. Remember that 'absent' or 'negative' are meaningless terms without quoting the unit quantity from which it is supposed to be absent.

Staphylococcus aureus

It is usually considered that the presence of *Staph. aureus* in conditions which allow growth for 4 hours is cause for serious concern. This is because some strains of *Staph. aureus* produce enterotoxins during growth, in which instance large numbers (10^5 or 10^6 per ml or g) and toxin may be present. State-of-the-art test kits for toxin are available and may well be used by 'Home Office' (head office) or research laboratories. For Quality Assurance purposes, counting on a selective medium, e.g. Baird-Parker agar, is likely to be adequate.

It must be remembered when assessing a count that large numbers should be regarded as potentially serious, whether they are viable or dead, e.g. from a Direct Microscopic Count, since they may belong to a toxigenic strain.

Absence of large numbers of viable organisms is not an assurance that the food is necessarily safe. A 'kill step' after growth and toxin production has occurred still leaves the heat resistant toxin in the material.

For routine Quality Assurance purposes, it would be prudent to regard large or unusual numbers of coagulase and thermonuclease positive *Staphylococci* as potentially serious and needing urgent action.

Clostridia spp.

Both *Cl. botulinum* and *Cl. perfringens* produce toxins, and it is usually considered that toxin production is linked with growth and sporulating cells respectively. Because tests for botulinum toxin require the use of mice and specific anti-toxins they are not normally done by food microbiology laboratories. In some countries, e.g. the UK, there are very strict rules on the use of animals and licenses are required for such work. Counting of *Cl. perfringens* is usually done as a presumptive count for sulfite-reducing anaerobes on a selective medium followed where necessary by other tests. Harmon and Duncan in Speck[10] state that 'hundreds of thousands (10^5) or more per g in a suspect food supports a diagnosis of perfringens poisoning when substantiated by clinical and epidemiological evidence'. They warn of a loss of viability of *Cl. perfringens* cells when foods are frozen or held under prolonged refrigeration before laboratory examination.

Fecal streptococci

Fecal streptococci are also known as enterococci (USA) and 'Group D' or 'Lancefield Group D' streptococci. *Streptococcus faecalis* and *S. faecium* are the species usually found when using sodium azide as the selective agent, e.g. in KF streptococcal agar. Note that the presence of *Pediococci* from starter cultures, e.g. in fermented cured sausage, may interfere with this test. The group has been considered as an indicator of fecal contamination which is more persistent than coliforms, being tolerant of 6.5% NaCl and resistant to freezing. In a dairy plant this group is regarded as an indicator of poor sanitation (hygiene).

Bacillus cereus

Bacillus cereus is a widespread organism which in large numbers, e.g. over 10^6 per g, may cause illness either of a diarrheal or vomiting type. In the UK it is commonly associated with boiled or fried rice which has been held at growth temperatures. However, it

is a widespread organism and of concern, e.g where an instant dried soup might be made and kept hot in a vacuum flask for some hours before consumption. Note that some other *Bacillus* species have occasionally been associated with food poisoning but always in large numbers (around 10^6 per g).

Vibrio parahaemolyticus

Vibrio parahaemolyticus was first reported in Japan as being transmitted on raw fish. In other countries, e.g. the USA and the UK, cooked crustaceans which have been cross contaminated and temperature abused have been implicated. Methods of analysis are given by Twedt *et al.* in Chapter 30 of Speck[11].

Listeria monocytogenes

Currently (1990) methodology is developing but it is clear that:

* *Listeria monocytogenes* is widely distributed
* Only a small proportion of the *Listeria* population is actively pathogenic but tests for pathogenicity are, as yet, only preliminary. Large numbers seem needed to cause disease, although immuno-compromised persons may well be affected by lower numbers. (See profile in Chapter 10.)
* Incubation at 30°C for 48 hours is an effective enrichment procedure, although cold enrichment at 4°C for at least a week is also possible. However, this long incubation was first done for clinical purposes, and is not useful in the food industry.

The usual requirement is absence in 25 g (similar to *Salmonella* spp.) although absence from 10 g is sometimes cited.

Other pathogens

Other pathogens are discussed in Chapter 4 of the NRC book[4]. However, it seems inevitable that organisms other than those currently considered as pathogens will be incriminated. A study of microbial ecology leads to the conclusion that new product/processing/packaging/distribution developments provide new opportunities for microbial growth. These must be expected to result in new forms of spoilage and sometimes pathogenicity.

As an example, Fricker and Tompsett[12] examined fish, raw and cooked meats and prepared salads and concluded that *Aeromonas* spp. were extremely common contaminants of foodstuff and that some of the strains appeared to be virulent.

Yeasts and molds

Yeasts and molds are primarily regarded as spoilage organisms and indicators of poor handling although some molds produce mycotoxins. (See Chapter 41 on toxigenic fungi by Stoloff in Speck[13].)

The Howard Mold count (AOAC method) is described by Pusch *et al.* in Chapter 5 of Speck[14]. The performance and interpretation of the count is not easy and requires experience. Because disintegration of the hyphae affects the count but not the mass of fungal material, attempts have been made to measure fungal mass chemically (see Robertson *et al*[15]). Mold count standards exist for a number of fruits and fruit products, e.g. tomato ketchup (see Marshall[4]), but are omitted from the Data Tables as reference must be made to current standards for a specific product.

Other spoilage organisms

Depending on the substrate and the environment a great range of organisms may be involved in spoilage and there is sometimes a recognizable succession as spoilage progresses. It is worth noting that some spoilage organisms cause changes, e.g. a shift in pH, enabling growth of other microorganisms (possibly pathogenic, e.g. *Cl. botulinum*) on substrates which would not originally allow growth. There are useful chapters in Speck[16] on methodology for Psychrotrophs, Thermodurics, Lipolytics, Proteolytics, Halophiles, Osmophiles, Pectinolytics, Acid producers and Spore formers.

Some ways in which limits (targets) are expressed

In the literature there is a variety of ways in which limits or targets are expressed. Often they are expressed per g or ml and this meaning is clear. They can also be expressed as absent in 0.1 g or ml which is approximately equivalent to less than 10 per g or ml. Similarly, absent in 0.01 g or ml is roughly equivalent to absent in 100 g or ml.

A widely accepted attribute sampling scheme is that of ICMSF[8], where:

- n = the number of sample units required for testing. These should be chosen independently 'at random'.
- c = the maximum allowable number of adverse results, or defective sample units (2-class plan), or marginally acceptable sample units (3-class plan). When more than this number are found in the sample the lot is 'rejected'. (Note that rejection does not necessarily mean destruction.)
- m = a microbiological limit which in a 2-class plan separates 'good' quality from 'defective' quality or in a 3-class plan separates 'good' quality from 'marginally acceptable' quality. Values equal to or less than m represent an acceptable product.
- M = a microbiological limit which in a 3-class plan separates 'marginally acceptable' quality

from 'defective' quality. Values above M are therefore unacceptable.

Alternatively, some commercial specifications or guidelines may be given using the terms Red, Amber and Green. These are analogous to traffic color light signals but have the meaning:

Green: satisfactory. Values are typical of those obtained when the process or system is under control.

Green – Amber: acceptable but not satisfactory. Values are typical of those obtained when the process or system is moving out of control. Corrective action is required immediately.

Amber – Red: not acceptable. A red condition means that a non-standard, unacceptable or unsuitable product is likely to result. In other words, *Do not use*.

These categories may be used with either an attribute or a variable sampling scheme.

Notes on a variable sampling scheme

As Jarvis[2] states, 'Two types of sampling schemes can be used in Acceptance Sampling: one is for variables and the other for attributes'. An attribute scheme is one which is generally used in a 'port-of-entry' situation. The assumption is that each unit in a lot is 'acceptable' or 'defective' (2-class) or 'acceptable', 'marginally defective' or 'defective' (3-class). This is widely used as no previous knowledge of material from a particular source is needed.

A variable scheme can be used provided that counts are reasonably high, say 10^3 per g or ml or more, and follow a normal distribution with 95% of all test results being within $\pm 2 \times$ standard deviation of the mean. Alternatively, provided that the log of the count is approximately normally distributed, a variable scheme may be used.

It should be remembered that to establish a normal or log normal distribution requires a substantial amount of work. If results from this work do not show normal or log normal distribution then a variable scheme cannot be used. Where it can be used it is arguable that a better decision is obtained than with an attribute scheme for the same amount of work. The 'Red, Amber, Green' nomenclature can also be used. Variable schemes may be appropriate for Good Manufacturing Practice (GMP) work but not for Public Health testing. Use of the variable scheme is described by Kilsby *et al.* [17] who use a symbol C_m which is stated to be in practice very similar to the ICMSF limit 'm'. Their symbol C is similar in meaning to the ICMSF limit 'M'. When using this scheme, k_1 and k_2 values may be taken from Malcolm[18]. For further information on sampling schemes see references[8, 19, 20, 21].

When you have developed an established criterion, remember the wise words of NAS-NRC Publication 1195, written in 1964[22]: 'It is recognized that what is good commercial practice today may not be good practice tomorrow and that knowledge of what constitutes health hazard may change. Criteria must therefore be subjected to re-evaluation of usefulness and adequacy as experience and technological knowledge dictate.' Events of the past 27 years have shown how perceptive and right these authors were.

Using criteria when limits are exceeded

From time to time limits will be exceeded: as the National Research Council[3] states at the start of Chapter 7, 'In general, the decision taken relates to the purpose for which the criterion was established'. It also relates to how carefully the criterion was established. An important consideration is whether the latest results, which exceed the limits, are part of a steady trend. If this is so, then why was not effective corrective action taken earlier?

Clearly, if the criterion is a standard (i.e. has legislative authority) then the action to be taken is also likely to be defined by the regulatory authority.

The National Research Council[3] identifies and discusses in Chapter 7 the following decision categories:

- Evidence of existence of a direct health hazard.
- Evidence that a direct health hazard could develop.
- Indications that a product was not produced under conditions assuring safety.
- Indications that a raw material (ingredient) may adversely affect shelf life.
- Evidence that a critical control point is not under control.

When limits are exceeded in a way which suggests that safety may be compromised then the proper and ethical disposal is clear – the material *cannot* be used.

If the 'Quality' limits are significantly exceeded, i.e. if they are over '*M*' or the 'Red' limit, the material *will not* be used because 'Quality' criteria (standards) cannot be met. This does not necessarily mean that the material is 'unsafe' but it is definitely unsuitable for its intended purpose.

The difficult or 'gray area' is when 'Quality' limits are slightly exceeded, i.e. they are between '*m*' and '*M*' or in the 'Amber' range. The decision here depends on the number and kinds of organisms involved and their significance for the process and product. The original HACCP analysis report should give valuable guidance. Good record keeping should also be able to furnish data on any similar occurrences in the past; what disposal was given and whether any adverse consequences were known or suspected.

The Total Quality Management (TQM) philosophy requires that positive steps should be taken on an on-going basis to seek to improve the quality of such decisions. This involves both the microbiologist who is accountable for the quality of advice and the signatory of the disposal document (Quarantine Form) who is accountable for the disposal decision.

It may be added that if the criterion is part of an 'incentive' sampling scheme and above the limits the lot should properly be returned to the supplier. It is important to enforce this, otherwise the supplier will not believe that the purchasing specification is a serious document.

Under the special circumstances where grading is practiced this is usually an internal matter for the company, and the supplier is unlikely to be informed. As, by definition, there is no Public Health nor increased spoilage risk, it is purely a matter of economics whether a material which exceeds a specific criterion is used for a different purpose where it complies as 'satisfactory'.

References

1. Campden RA (1987) Technical Manual No. 19. *Guidelines to the Establishment of Hazard Analysis Critical Control Point (HACCP)*, Campden Food Preservation Research Association, Chipping Campden, Gloucestershire GL55 6LD
2. Jarvis, B. (1989) *Statistical Aspects of the Microbiological Analysis of Foods*, Progress in Industrial Microbiology, Vol. 21. Amsterdam: Elsevier Science.
3. NAS (1985) *An Evaluation of the Role of Microbiological Criteria for Foods and Food Ingredients*. National Academy of Sciences, Washington, DC: National Academy Press
4. Marshall, J.P. (1986) *Food Legislation Surveys No.9. Microbiological Standards for Foodstuffs*, 2nd edn. Leatherhead: Leatherhead Food Research Association
5. Shapton, D.A. (1986) Canned and bottled food products. In *Quality Control in the Food Industry*, Vol. 3, edited by S. M. Herschdoerfer, 2nd edn, pp. 261–322. London: Academic Press
6. ICMSF (1980) *Microbial Ecology of Foods Vol. 1 – Factors Affecting Life and Death of Microorganisms*. International Commission on Microbiological Specifications for Foods, New York: Academic Press
7. ICMSF (1980) *Microbial Ecology of Foods, Vol 2 – Food Commodities*. International Commission on Microbiological Specifications for Foods, New York: Academic Press
8. ICMSF (1986) *Microorganisms in Foods Vol. 2 – Sampling for Microbiological Analysis: Principles and Specific Applications*. International Commission on Microbiological Specifications for Foods, 2nd edn. Toronto: University of Toronto Press
9. Mehlman, I. J. (1984) Coliforms, fecal coliforms,

Escherichia coli and enteropathogenic *E. coli*. In *Compendium of Methods for the Micro-biological Examination of Foods,* 2nd edn, edited by M. L. Speck, pp. 265–285. Washington, DC: American Public Health Association

10. Harmon, S. M. and Duncan, C. L. (1984) *Clostridium perfringens*. In *Compendium of Methods for the Microbiological Examination of Foods,* 2nd edn, edited by M. L. Speck, pp. 483–495. Washington, DC: American Public Health Association

11. Twedt, R. M., Madden, J. M. and Colwell, R. R. (1984) Vibrio. In *Compendium of Methods for the Microbiological Examination of Foods,* 2nd edn, edited by M. L. Speck, pp. 368–385. Washington, DC: American Public Health Association

12. Fricker, C. R. and Tompsett, S. (1989) Aeromonas in foods: a significant cause of food poisoning? *International Journal of Food Microbiology, 9,* 17–23

13. Stoloff, L. (1984) Toxigenic fungi. In *Compendium of Methods for the Microbiological Examination of Foods,* 2nd edn, edited by M. L. Speck, pp. 557–572. Washington, DC: American Public Health Association

14. Pusch, D. J., Busta, F. F., Noats, W. A., Bandler, R. and Cichowicz, S. M. (1984) Direct Microscopic count. In *Compendium of Methods for the Microbiological Examination of Foods,* 2nd edn, edited by M. L. Speck, pp. 84–98. Washington, DC: American Public Health Association

15. Robertson, A., Patel, N. and Allinson, S. (1987) Technical Memorandum No. 456. *Immuno Chemical Determination of Mold Contamination in Tomato Products – a Feasibility Study,* Campden Food Preservation Research Association, Chipping Campden, Gloucestershire GL55 6LD

16. Speck, M. L. (ed) (1984) *Compendium of Methods for the Microbiological Examination of Foods,* 2nd edn. Washington, DC: American Public Health Association

17. Kilsby, D. C., Aspinall, L. J. and Baird-Parker, A. C. (1979) A system for setting numerical microbiological specifications for foods. *Journal of Applied Bacteriology, 46,* 591–599

18. Malcolm, S. (1984) A note on the use of the non-central-distribution in setting numerical microbiological specifications for foods. *Journal of Applied Bacteriology, 57,* 175–177

19. Kilsby, D. C. (1982) Sampling schemes and limits. In *Meat Microbiology,* edited by M. H. Brown, pp. 387–421. London: Applied Science

20. Kilsby, D. C. and Baird-Parker, A. C. (1983) Sampling programs for the microbiological analysis of foods. In *Food Microbiology Advances and Prospects,* edited by T. A. Roberts and F. A. Skinner, pp. 307–315. Society for Applied Bacteriology Symposium Series No. 11, London: Academic Press

21. Smelt, J. P. P. M. and Quadt, J. F. A. (1990) A proposal for using previous experience in designing microbiological sampling plans based on variables. *Journal of Applied Bacteriology, 69,* 504–511

22. NAS – NRC (1964) *An Evaluation of Public Health Hazards from Microbiological Contamination of Foods,* Publication No. 1195. Washington, DC: National Academy of Sciences – National Research Council

13.C Microbiological criteria data tables

These tables have been prepared as a help to the microbiologist who has the task of producing a 'provisional' specification for an unfamiliar ingredient or product. What the microbiologist wants is up-to-date data based on proven application of clear and detailed GMPs to the specific ingredient or product/processing/packaging/distribution system. The data should, at least, enable 95 percentile ranges to be given for appropriate organisms. This is the ideal, but in the absence of such published data, resources are needed within individual companies to develop microbiological criteria which would successfully meet an expertly informed state-of-the-art audit.

The basis for including data was that published standards or guidelines could be regarded as being based on accepted GMPs at a particular time in a specified country. Occassionally an unattributed 'trade' specification has also been included. Tables are therefore not purchasing criteria which would necessarily be used by individual Heinz companies, although due recognition and attention would be given to them during the development of Heinz criteria.

Individual foods are listed as 'Ingredients' or 'Products' although any one food may, in fact, be both.

References for published information are given as numbers within square brackets, and notes are given as letters in round brackets (parentheses). Where no reference is given the source is 'trade' information and as editorial policy is non-attributable.

Where (?) follows a figure, e.g. 1×10^4, it means that the original document was described in the Leatherhead RA reference as 'unclear' or illegible.

Warning to users

Readers are reminded that the tables:

- Are one of several starting points in the development of specific criteria. They *must not be used as criteria without professional advice as they need to be assessed and adapted for individual needs.*
- Are inevitably incomplete. Remember that products with a similar name may be quite different, e.g. Italian salami and German salami.
- Are inevitably not up-to-date. Be aware particularly of emerging pathogens.
- Offer no sampling inspection schemes or methods of examination.

Be aware when using the tables:

- It is the users' responsibility to be clear about the usefulness and meaning of the numbers given.

Always take appropriate professional microbiological and statistical advice before use.

- The tables are *not* all that are needed for the implementation of the Longitudinally Integrated Safety Assurance (LISA) strategy. They are intended to provide background information for microbiologists as they develop LISA for a given process or product.
- That in addition to the microorganisms given in the tables, there will be occasions when viruses, histamine (scombroid poisoning), toxic dinoflagellates (ciguatera and paralytic shellfish poisoning or PSP), *Vibrios* (including *V. cholerae*), as well as *Listeria, Yersinia, Campylobacter* and the emerging pathogens, together with parasites, e.g. *Giardia, Cryptosporidia,* and *Entamoeba* will need to be considered.
- That the user must *always* confirm the up-to-date legislative position both in the country of production and, if different, in the country of use or sale. It is important to consult a legal expert who has local knowledge of the country where foods are to be sold, as important changes in the law and of the views of the agency or agencies having jurisdiction can occur quickly.

Index to the data tables

– Whole, table grade
– Whole, yolk, white, salted or sugared
Entrees – Frozen, containing rice or cornflour as a main ingredient
Fats – Edible
Foods – Miscellaneous, dried
Foods – Miscellaneous, pasteurized in closed containers
Foods – Miscellaneous, ready-to-eat
Foods – Non-packaged, general
Froglegs
Fruit – Dried
– Fresh or frozen
Fruit juices and nectars
Fruit juice – Concentrated
Fruit pulp – Non-heat treated
Gelatin
Gums – Botanic
Ham and other cured smoked meats
Ham – Canned, refrigerated, 'pasteurized hams' and pork shoulder
Herbs, including spices and condiments
Honey
Ice – Prepackaged (frozen water)
Ice cream
Ice cream mix ('preparations')
Ice cream – Regular, vanilla, chocolate or other flavors and novelties – US military criteria
– Ultrapasteurized
Ices – General
– Sherbert, snowball, slush ice products
Maple syrup
Margarine
Mayonnaise
Meals – Meat and poultry entrees
– Not ready-to-eat (to be cooked and prepared) and dried
– Ready cooked, including fillings for pastas and meals with cheese
– Cook-chilled
Meals (finished products) – chilled, entirely heat treated, also ready-to-serve foods served cold
– Chilled, with one or more non-heat treated components
– Frozen, entirely heat treated, also ready-to-serve foods served cold
– Frozen, with one or more heat treated components
– Heated
Meals (or snacks) – Ready cooked, not requiring heating or cooking
– Ready cooked, requiring heating or cooking before consumption
Meat – Barbecue, beef or pork
– Cecinas (corned, dried meat)
– Charcuterie products
– Comminuted, ground or hamburger
– Cooked, other than in USA
– Cooked, US state and city criteria
– Cooked, US states criteria 1977

– Cured or fermented
– Cured, smoked
– Fresh
– Fresh or frozen, Netherlands data
– Fresh or frozen, US states criteria 1977
– Joints and single portions
– Minced, fresh or frozen
– Pate
– Pot pies
– Raw, *not* for use at port-of-entry: in-plant data
– Raw mince or raw sausage at best before date
– Sausages
– Sausage, cooked, standard Dutch quality
– Sausages, fermented Canadian
– Smoked products
Meats – Vacuum packed, sliced, i.e. semi-preserved but perishable products
Milk (cows) – Concentrated or 'evaporated' (UK)
– Concentrated and sweetened or 'condensed' (UK)
– Dried full fat or skim
– Evaporated (Netherlands)
– Flavored
– Gelled and flavored milks with rennet
– Liquid pasteurized
– Liquid raw (not pasteurized)
– Powder for babies
– Sterilized
– Ultra-High Temperature (UHT) treated
Milk (goats) – Liquid raw (not pasteurized)
Mustard
Nuts – Including nut butters and nut meats
Onion rings – Frozen, breaded
Pasta – Frozen and dried
Pasta dumplings – Fresh, manufactured not home-made
Pastry
Patisserie and patisserie cream
Poultry – Cooked entrees, pot pies and poultry rolls
Poultry meat – Cooked, frozen; including chicken and prepared dishes
– Cooked, including chicken
– Cooked, mechanically separated
– Cured and/or smoked
Poultry meat and products – Dehydrated
Poultry meat – Frozen diced for cannery use
– Raw, mechanically separated
Poultry meat, or chicken – Raw
Pot pies – Frozen
Potato – Stuffed, baked, frozen
Potato patties and hash browns – Frozen
Pudding powders
Salads – Stuffed tomatoes etc., to be eaten cold
– various types, US states data
Sandwiches – pre-packed
Seafood – Bivalves (shellfish) cleaned
– Bivalve molluscs, fresh and frozen
– Crabmeat, cooked, chilled and frozen
– Crustacea, France, criteria
– Crustaceans, cooked, frozen

- Crustaceans, raw, frozen
- Fish conserves
- Fish, lightly salted and smoked, e.g. salmon, haddock
- Fish pulp
- *Fried fishcakes, frozen*
- Fresh, frozen or cold smoked *not* crustaceans or molluscs
- Fresh fish and also crustacea, molluscs, marine or fresh water mammals
- Fresh fish fillets or sliced fish
- *Molluscs, fried and frozen*
- *Mussels (imported, Netherlands)*
- *Mussels, cooked and frozen*
- *Mussels and scallops, pre-cooked*
- *Octopus, boiled*
- *Prawns, cooked, frozen (imported, UK)*
- *Pre-cooked breaded fish*
- Preparations based on raw minced fish
- *Raw, breaded, frozen fish*
- Raw or partly cooked marine foods
- Rock lobster
- *Ready-to-eat*
- *Semi-preserved*
- Shellfish, fresh or frozen
- Shellfish live, bivalves and sea urchins

Soft drinks – Miscellaneous

Soya flours, concentrates, isolates and proteins
Soya milk – Dried
Soya protein – Texturized
Soups – Dried
– *Dried, broths or bouillon*
– *Dried, instant*
– *Dried, to be cooked before eating*
Spices, herbs or condiments
Sugar and Molasses
Tea and derivatives
Tomato ketchup and sauces
Topping – Dessert and bakery products, frozen or dehydrated
Vegetables – Dried
– Fresh
– Frozen
– Prepared
Vegetable protein – hydrolyzed
Vinegar – Pasteurized
Water – Drinking (potable)
– *Mineral*
– *Natural mineral*
– Treated and untreated
Whey – Dried
Yeast
Yogurt

Data tables

Product name: Baby food – Biscuits (shelf stable)

Organism	Numbers	Notes
Coliform or Enterobacteria	$m = 10, M = 10^2$ [1](a)	(a) ICMSF
Salmonella spp.	$m = 0$ in 25 g [1](a)	

Product name: Baby-food – Dried, requiring reconstitution, cooking not specified

Organism	Numbers	Notes
APC	1×10^4 to 5×10^5 in 2/5 samples [3](a–d)	(a) Netherlands
Coliform or Enterobacteria	0 to 50 in 2/5 samples [3](a, b, d)	(b) Chile
E. coli or Fecal coliform	0 in 0.1 g [3](c)	(c) Thailand
Salmonella spp.	0 in 25 g to 0 in 50 g [3](a, d)	(d) Finland
Staphylococcus aureus	0 in 1 g [3](a)	
B.cereus (spores)	1×10^2 [3](a)	
Yeasts	1×10^2 [3](b)	
Yeasts and Molds	1×10^2 [3](a)	

Product name: Baby-food – Dried or instant, requiring reconstitution and cooking

Organism	Numbers	Notes
APC	5×10^4 [3](a) to $m = 10^5, M = 10^6$ [1](b)	(a) Yugoslavia
Coliform or Enterobacteria	$m = 10, M = 10^2$ [1](b), [3](a)	(b) ICMSF
E. coli or Fecal coliform	0 in 0.01 g [3](a)	
Salmonella spp.	0 in 50 g [3](a); 0 in 25 g [1] (b)	
Staphylococcus aureus	0 in 0.01 g [3](a)	
Clostridia (*perfringens*)	0 in 0.01 g [3](a)	
Fecal Streptococci	As β-haemolytic, 0 in 0.01 g [3](a)	
Proteus spp.	0 in 0.01 g [3](a)	
Molds	2×10^2 [3](a)	

Product name: Baby food – Dried or instant, requiring reconstitution but not cooking

Organism	Numbers	Notes
APC	$m = 10^3, M = 10^4$ [5] to $m = 10^4, M = 10^5$ [1](a)	(a) Not for fermented products, ICMSF
Coliform or Enterobacteria	$m = 1.8, M = 20$ [5] to $m = 10, M = 10^2$ (d), [3](b)	(b) Yugoslavia
Salmonella spp.	0 in 50 g [3](b); 0.in 25 g [1], [5](c)	(c) Reference [5] applies also to
Staphylococcus aureus	0 in 0.01 g [3](b)	powdered infant Enterobacteria
Clostridia (*perfringens*)	0 in 0.01 g [3](b)	formula
Fecal Streptococci	As β-haemolytic, 0 in 0.01 g [3](b)	(d) ICMSF
Proteus spp.	0 in 0.01 g [3](b)	
Molds	10^2 [3](b)	

Product name: Baby food – Vegetable based for infants and children

Organism	Numbers	Notes
APC	1×10^2 [3](a)	(a) Saudi Arabia, includes requirement
Yeasts	10 [3](a)	that pathogens be absent
Molds	10 [3](a)	

Product name: Bakery products – Frozen, to be cooked; low-acid or high a$_w$ fillings

Organism	Numbers	Notes
Salmonella spp.	0 in 25 g [1](a)	(a) ICMSF – includes meat pies and
Staphylococcus aureus	$m = 10^2$ (estimated), $M = 10^4$ [1](a)	pizzas

Product name: Bakery products – Frozen, ready-to-eat; low-acid or high a$_w$ fillings

Organism	Numbers	Notes
Salmonella spp.	0 in 25 g [1](a)	(a) ICMSF
Staphylococcus aureus	$m = 10^2$ (estimated), $M = 10^4$ [1](a)	

Ingredient name: Beef – Cheek meat

Organism	Numbers	Notes
APC	At 35°C 1×10^7 [2](a)	(a) See reference for details (page 172)
Coliform or Enterobacteria	1×10^4 [2](a)	
E. coli or Fecal coliform	1×10^3 [2](a)	
Staphylococcus aureus	1×10^3 [2](a)	

Ingredient name: Beef – Frozen, ground (raw)

Organism	Numbers	Notes
APC	$m = 5 \times 10^6$, $M = 5 \times 10^7$ [6](a)	(a) Frozen raw ground poultry would
Salmonella spp.	0 in 20 g [6](a)	meet this standard except for
Fecal Streptococci	$m = 1 \times 10^2$, $M = 5 \times 10^3$ [6](a)	*Salmonella* spp. [6]

Ingredient name: Beef – Hearts

Organism	Numbers	Notes
APC	At 35°C, 5×10^6 [2](a)	(a) See reference for details (page 172)
Coliform or Enterobacteria	1×10^4 [2](a)	
E. coli or Fecal coliform	1×10^3 [2](a)	
Staphylococcus aureus	1×10^3 [2](a)	

Product name: Beef – 'Roast'

Organism	Numbers	Notes
Salmonella spp.	0 in 25 g [1](a)	(a) ICMSF – includes beef cooked in water baths

Product name: Biscuits – Filled or coated

Organism	Numbers	Notes
APC	1×10^4[3](a)	(a) Spain
Coliform or Enterobacteria	10[3](a); 50[3](b)	(b) Brazil
E. coli or Fecal coliform	0 in 1 g[3](a, b)	
Salmonella spp.	0 in 25 g[3](a, b)	
Staphylococcus aureus	0 in 0.1 g[3](a); 0 in 1 g[3](b)	
Clostridia (*perfringens*)	20[3](b)	
B. cereus	0 in 1 g[3](a)	
Yeasts and Molds	1×10^2[3](a); 2×10^2[3](b)	

Product name: Biscuits – Simple or plain

Organism	Numbers	Notes
APC	1×10^3[3](a)	(a) Spain
Coliform or Enterobacteria	0 in 1 g[3](a);50[3](b)	(b) Brazil
E. coli or Fecal coliform	0 in 1 g[3](a)	
Salmonella spp.	0 in 25 g[3](a)	
Staphylococcus aureus	0 in 1 g[3](a)	
B. cereus	0 in 1 g[3](a)	
Yeasts and Molds	1×10^2[3](b); 2×10^2[3](a)	

Ingredient name: Bone meal – Edible bone meal or bone flour

Organism	Numbers	Notes
APC	$m = 1 \times 10^3, M = 1 \times 10^5$[2](a)	(a) Canadian standard
E. coli or Fecal coliform	$m = 0, M = 10$[2](a)	

Product name: Bouillon – Dried

Organism	Numbers	Notes
APC	1×10^5[3](a)	(a) Argentina standard
Coliform or Enterobacteria	1×10^3[3](a)	(b) Coagulase positive
E. coli or Fecal coliforms	10[3](a)	
Salmonella spp.	0 in 25 g[3](a)	
Staphylococcus aureus	1×10^2[3](a, b)	
Clostridia (*perfringens*)	10 (spores)[3](a)	

Product name: Breads – Speciality, sweet with egg or milk

Organism	Numbers	Notes
Coliform or Enterobacteria	50[3](a)	(a) Brazil
E. coli or Fecal coliforms	0 in 1 g [3](a)	
Salmonella spp.	0 in 25 g[3](a)	
Staphylococcus aureus	0 in 1 g [3](a)	
Yeasts and Molds	2×10^2[3](a)	

Product name: Breadcrumbs

Organism	Numbers	Notes
APC	1×10^5[3](a)	(a) Netherlands, decree
Coliform or Enterobacteria	1×10^2[3](a)	
Yeasts and Molds	1×10^3[3](a)	

Product name: Butter and whipped butter

Organism	Numbers	Notes
APC	1×10^2[2](a); 2×10^5[3](b)	(a) US military criterion
Coliform or Enterobacteria	Usually 10 to 25[3](b)	(b) Chile
E. coli or Fecal coliforms	10[3](c)	(c) France
Fecal Streptococci	10[2](d)	(d) USDA (Fecal streptococci as
Proteolytic or		Enterococci if test requested)
Proteolytic/Lipolytic	1×10^2[2](d); 5×10^2[3](e)	(e) Argentina
Yeasts and Molds	20[2](a, d); 1×10^2[3](b, e)	

Product name: Butter – 'Farm' or made from unpasteurized cream

Organism	Numbers	Notes
APC	5×10^5[3](a); 1×10^6[3](b)	(a) Australia/Papua New Guinea
E. coli or Fecal coliforms	10[3](b)	(b) Switzerland
Staphylococcus aureus	1×10^3[3](b)	
Pseudomonas spp.	10[3](b)	
Molds	1×10^2[3](b)	

Product name: Butter – Salted

Organism	Numbers	Notes
APC	2.5×10^4[3](a)	(a) New Zealand guidelines data
Coliform or Enterobacteria	50[3](a)	(b) Coagulase positive
E. coli or Fecal coliforms	50[3](a)	
Salmonella spp.	0 in 25 g[3](a)	
Staphylococcus aureus	0 in 1 g[3](a, b)	
Yeasts and Molds	50[3](a)	

Ingredient name: Butterfat or Butter oil or Milk fat

Organism	Numbers	Notes
APC	5×10^3[2](a)	(a) USA military criteria (milk fat)
Coliform or Enterobacteria	0 in 1 g[3](b); 10[2](a)	(b) France
Staphylococcus aureus	0 in 1 g[3](b)	
Anaerobes	5×10^2[3](b)	
Yeasts and Molds	20[2](a)	

Ingredient name: Buttermilk – Dried

Organism	Numbers	Notes
APC	5×10^4[2](a); 2×10^5[2](b)	(a) US extra grade (b) US standard grade

Product name: Buttermilk – Liquid

Organism	Numbers	Notes
APC	1×10^4[3](a)	(a) Argentina
Coliform or Enterobacteria	10[2](b)	(b) USA military criterion
Yeasts and Molds	1×10^2[3](a)	

Ingredient name: Canned foods ingredients

Organism	Numbers	Notes
Thermophilic spore count	[2](a) Aerobic $m = 125$ per 10 g, $M = 150$ per 10 g Aerobic 'flat sour' $m = 50$ per 10 g, $M = 75$ per 10 g Anaerobes H $_2$S negative 2/5 samples with no more than 4/6 +ve tubes in each positive sample unit. Anaerobes H $_2$S positive 3/5 samples with no more than 5 spores in 10 g in each positive sample unit	(a) US military criteria – applies to starches, flours, cereals, alimentary pasta, sugars and dried milks

Ingredient name: Casein and caseinates

Organism	Numbers	Notes
APC	3×10^4[2](a, d) to 1×10^5[2](b)	(a) USDA extra grade
Coliform or Enterobacteria	0 in 0.1 g[2](a, d) to 2 in 0.1 g[2](b)	(b) USDA standard grade
Salmonella spp.	0 in 100 g[2](b, c)	(c) Given as 'optional'
Staphylococcus aureus	0 per g[2](b, c)	(d) EC standard
Thermophiles	5×10^3[2](b, d)	
Yeasts and Molds	5 in 0.1 g[2](b, c)	

Ingredient name: Caseinates to make edible ices

Organism	Numbers	Notes
APC	3×10^4[3](a)	(a) Belgium standard
Coliform or Enterobacteria	10[3](a)	(b) Coagulase positive
Salmonella spp.	0 in 25 g[3](a)	
Staphylococcus aureus	0 in 1 g[3](a, b)	
Yeasts and Molds	50[3]	

Ingredient name: Cereals – General and for canned low-acid packs

Organism	Numbers	Notes
APC	Usually about 10^6[2](a)	(a) Normally found at this level on
Thermophilic spore count	(b)	grains
Aerobes	30 per 2 g	(b) For cake flour and other flours used
Anaerobes	10	for canned low-acid packs
Molds	$m = 1 \times 10^2$ to 1×10^4, $M = 1 \times 10^5$[1](c)	(c) ICMSF Figures for m are estimated
Yeasts and Molds	Up to 10^4[2](a)	

Product name: Cereal flours

Organism	Numbers	Notes
APC	5×10^5(?) [3](a) to 1×10^6[3](b)	(a) Brazil
E. coli or Fecal coliforms	0 in 1 g[3](a) to 1×10^2[3](b)	(b) Spain
Salmonella spp.	0 in 25 g[3](a, b)	(c) Sweden
Staphylococcus aureus	0.1 g[3](a)	
Clostridia (perfringens)	20[3](a)	
Yeasts and Molds	1×10^3[3](a, c) to 1×10^4[3](b)	

Product name: Cereals – Puffed or flaked

Organism	Numbers	Notes
APC	5×10^4[3](a)	(a) Brazil
E. coli or Fecal coliforms	0 in 1 g[3](a)	
Salmonella spp.	0 in 25 g[3](a)	
Staphylococcus aureus	0 in 0.1 g[3](a)	
Clostridia (perfringens)	As sulfite-reducing clostridia at 44°C, 20[3](a)	

Product name: Cheese

Organism	Numbers	Notes
E. coli or Fecal coliforms	1×10^2[3](a)	(a) New Zealand guidelines
Salmonella spp.	0 in 25 g[3](a)	(b) Coagulase positive
Staphylococcus aureus	1×10^3[3](a, b)	

Product name: Cheese – Cheddar also Gouda

Organism	Numbers	Notes
E. coli or Fecal coliforms	0 in 0.01 g[3](a)	(a) Australian states
Staphylococcus aureus	0 in 0.01 g to 0 in 0.1 g[3](a)	

Product name: Cheese – Cottage cheese

Organism	Numbers	Notes
Coliform or Enterobacteria	10[2](a) ; $m = 10$, $M = 1 \times 10^3$[2](b)	(a) USA military criteria
Psychrophiles	1×10^2[2](a)	(b) Canada
Yeasts and Molds	10[2](a)	

Product name: Cheese – Curd cheese

Organism	Numbers	Notes
APC	1×10^5 [3](a)	(a) Spain
Coliform or Enterobacteria	1×10^2 [3](a)	
E. coli or Fecal coliforms	10[3](a)	
Salmonella spp.	0 in 25 g [3](a)	
Staphylococcus aureus	1×10^2 [3](a)	

Product name: Cheese – Fresh pasteurized

Organism	Numbers	Notes
Coliform or Enterobacteria	10[3](a)	(a) France
E. coli or Fecal coliforms	1 per g [3](a)	
Salmonella spp.	0 in 25 g [3](a)	
Staphylococcus aureus	10[3](a)	

Product name: Cheese – Hard or semi-soft; made from pasteurized milk

Organism	Numbers	Notes
E. coli or Fecal coliforms	$m = 10^2$, $M = 2 \times 10^3$ [2](a)	(a) Canada
Staphylococcus aureus	$m = 10^2$, $M = 1 \times 10^4$ [2](a)	

Product name: Cheese – Hard or semi-soft: made from unpasteurized milk

Organism	Numbers	Notes
E. coli or Fecal coliforms	$m = 5 \times 10^2$, $M = 2 \times 10^3$ [2](a)	(a) Canada, cheese must be stored under prescribed conditions to allow for die-off of pathogens
Staphylococcus aureus	$m = 1 \times 10^3$, $M = 1 \times 10^4$ [2](a)	

Product name: Cheese – Hard and semi-hard, whether grated or not grated

Organism	Numbers	Notes
APC	1×10^2(?) [3](a)	(a) Brazil standard
E. coli or Fecal coliforms	10[3](a)	
Salmonella spp.	0 in 25 g [3](a)	
Staphylococcus aureus	0 in 0.01 g [3](a)	
Yeasts and Molds	5×10^2 [3](a)	

Product name: Cheese – Processed cheese

Organism	Numbers	Notes
APC	1×10^4 [3](a) to 1×10^5 [3](b)	(a) Denmark
Coliform or Enterobacteria	10[3](a, b)	(b) Chile
E. coli or Fecal coliforms	10[3](c)	(c) Brazil
Salmonella spp.	0 in 25 g [3](c)	
Staphylococcus aureus	0 in 0.01 g [3](c)	

Product name: Cheese – Processed, American, dehydrated

Organism	Numbers	Notes
APC	5×10^4 [2](a)	(a) USA military criteria
Coliform or Enterobacteria	90 [2](a)	

Product name: Cheese – 'Quesillo' (fresh skimmed-milk cheese)

Organism	Numbers	Notes
APC	1.5×10^5 [3](a)	(a) Chile

Product name: Cheese – 'Ricotta' (soft cheese)

Organism	Numbers	Notes
E. coli or Fecal coliforms	1×10^2 [3](a)	(a) Brazil standard
Salmonella spp.	0 in 25 g [3](a)	
Staphylococcus aureus	0 in 0.01 g [3](a)	

Product name: Cheese – Smoked cheese (not processed)

Organism	Numbers	Notes
APC	3×10^4 [3](a, b)	(a) 'Foreign' bacteria
Coliform or Enterobacteria	10 [3](b)	(b) Denmark

Product name: Cheese – Unripened fresh cheese

Organism	Numbers	Notes
Coliform or Enterobacteria	$m = 1 \times 10^2$, $M = 1 \times 10^3$	
E. coli or Fecal coliforms	$m = 10$, $M = 1 \times 10^2$	
Salmonella spp.	0 in 25 g	
Staphylococcus aureus	$m = 1 \times 10^2$, $M = 1 \times 10^3$	

Product name: Chewing gum

Organism	Numbers	Notes
APC	5×10^2 [3](a)	(a) Spain
Coliform or Enterobacteria	0 in 1 g [3](a)	
Yeasts and Molds	50 [3](a)	

Product name: Chicken – Cooked, diced and frozen for cannery use

Organism	Numbers	Notes
APC	Green-amber 10^5; amber-red 10^6	The manufacturing process distributes colonies into a number of colony-forming units (cfus). This results in comparatively high numerical limits under GMPs.
Salmonella spp.	Amber-red 0 in 25 g	
Staphylococcus aureus	Green-amber 10^3; amber-red 10^4	
Clostridia (*perfringens*)	Amber-red 10	

Thermophilic spore formers: Aerobic green-amber 30 per 2 g; amber-red 100 per 2 g
Anaerobic amber-red 10

Ingredient name: Chicken – Raw

Organism	Numbers	Notes
APC	5×10^5 [4](a)	(a) Nebraska Department of Agriculture guideline. Also states '*E. coli*, *Salmonella* spp. and *Staph. aureus* negative'.
Coliform or Enterobacteria	50 [4](a)	
Yeasts	5×10^5 [4](a)	
Molds	5×10^3 [4](a)	

Ingredient name: Chicken – Raw, fresh or frozen during processing

Organism	Numbers	Notes
APC	$m = 5 \times 10^5$, $M = 1 \times 10^7$ [1](a)	(a) ICMSF *not for port-of-entry sampling*

Product name: Cocoa

Organism	Numbers	Notes
APC	5×10^4(?) [3](a)	(a) Brazil
E. coli or Fecal coliforms	0 in 1 g [3](a)	(b) ICMSF
		(c) Canada
Salmonella spp.	0 in 25 g [1](b), [2](c), [3](a)	Canada standard for cocoa. Also used for chocolate and confectionery. USA military criterion for candy and chocolate
Clostridia (*perfringens*)	As sulfite-reducing clostridia at 44°C, 20 [3](a)	
Yeasts and Molds	5×10^2 [3](a)	

Product name: Cocoa – Soluble or cocoa-based drink mixes

Organism	Numbers	Notes
APC	5×10^4(?) [3](a)	(a) Brazil standard
Coliform or Enterobacteria	1×10^2 [3](a)	
E. coli or Fecal coliforms	0 in 1 g [3](a)	
Salmonella spp.	0 in 25 g [3](a)	
Staphylococcus aureus	0 in 0.01 g [3](a)	
Clostridia (*perfringens*)	As sulfite-reducing clostridia at 44°C, 20 [3](a)	

Ingredient name: Coconut – Dessicated (dried and shredded)

Organism	Numbers	Notes
Salmonella spp.	0 in 25 g[1](a), [3](b)	(a) ICMSF[1] – use $n = 10$ where growth is *not* anticipated and $n = 20$ when growth *is* anticipated. (b) New Zealand

Ingredient name: Coconut – Grated

Organism	Numbers	Notes
Coliform or Enterobacteria	1×10^2[3](a)	(a) Brazil
E. coli or Fecal coliforms	0 in 0.1 g[3](a)	
Salmonella spp.	0 in 25 g[3](a)	
Staphylococcus aureus	0 in 1 g[3](a)	
Clostridia (*perfringens*)	As sulfite-reducing clostridia at 44°C, 20[3](a)	
Yeasts and Molds	5×10^2(?) [3](a)	

Product name: Coffee – Instant or roast

Organism	Numbers	Notes
Coliform or Enterobacteria	10[3](a)	(a) Brazil
E. coli or Fecal coliforms	0 in 1 g[3](a)	
Yeasts and Molds	1×10^2(?) [3](a)	

Ingredient name: Corn syrup

Organism	Numbers	Notes
APC	5×10^4[4](a)	(a) Ohio guidelines
Yeasts	1×10^4[4](a)	
Molds	1×10^4[4](a)	

Ingredient name: Cream or liquid cream

Organism	Numbers	Notes
APC	25×10^3[3](a) to 1×10^5[3](b,c) to 2×10^5[3](d)	(a) Switzerland: on leaving producer
Coliform or Enterobacteria	1×10^2[3](e,f)	(b) Switzerland: at time of sale
E. coli or Fecal coliforms	0 in 1 g[3](e)	(c) Argentina: May-September
Staphylococcus aureus	10[3](f)	(d) Argentina: October-April
Yeasts and Molds	20[3](e)	(e) Argentina
		(f) Switzerland

Ingredient name: Cream – Pasteurized

Organism	Numbers	Notes
APC	5×10^4[3](a) to 1×10^5[3](b)	(a) Australian states, Chile, Papua New
Coliform or Enterobacteria	10[3](c) to 50[3](d)	Guinea, South Africa
E. coli or Fecal coliform	0 in 1 g[3](e)	(b) Malaysia, Spain
Salmonella spp.	0 in 25 g[3](f)	(c) Australia states, Chile, Fiji, Papua
Staphylococcus aureus	10[3](g)	New Guinea, South Africa, Spain
		(d) Malaysia
		(e) Argentina, South Africa, Spain
		(f) Spain: or *Shigella* spp.
		(g) Spain: enterotoxigenic

Ingredient name: Cream – Plastic and frozen

Organism	Numbers	Notes
APC	3×10^4[2](a)	(a) USD.A., 1975
Coliform or Enterobacteria	10[2](a)	
Yeasts and Molds	20[2](a)	

Ingredient name: Cream – Raw or unpasteurized

Organism	Numbers	Notes
APC	1.5×10^5[3](a) to 5×10^5[3](b)	(a) New Zealand guidelines
Coliform or Enterobacteria	20[3](c)	(b) South Africa/Papua New Guinea/
E. coli or Fecal coliform	0 in 1 g[3](c) to 1×10^2[3](d) (as fecal coliform)	Australian states
Salmonella spp.	0 in 25 g[3](d)	(c) South Africa
Staphylococcus aureus	1×10^3[3](d)	(d) France

Ingredient name: 'Cream skim'

Organism	Numbers	Notes
APC	2×10^5[3](a)	(a) Finland
Coliform or Enterobacteria	1×10^2[3](a)	

Product name: Cream – Soured, cultured or acidified

Organism	Numbers	Notes
Coliform or Enterobacteria	1×10^3[3](a)	(a) Sweden, soured milk products
E. coli or Fecal coliform	10[2](b), [3](c)	(b) USA military criteria
Salmonella spp.	0 in 10 g[3](a)	(c) Australian states, Papua New
B. cereus	$m = 1 \times 10^3$, $M = 1 \times 10^4$[3](a)	Guinea
Yeasts	1×10^5[3](a, d)	(d) Unless yeast culture is a standard
Yeasts and Molds	10[2](b)	component

Product name: Cream substitute, dry or liquid, non-dairy

Organism	Numbers	Notes
APC	2×10^4[2](a)	(a) USA military criteria
Coliform or Enterobacteria	10[2](a)	

Ingredient name: Cream – Whipped

Organism	Numbers	Notes
APC	1×10^5[3](a); 1×10^6[3](b); 1×10^8[3](c)	(a) Kenya and Zambia for 'dairy whip'
Coliform or Enterobacteria	10[3](a); 1×10^3[3](c)	and cream
E. coli or Fecal coliform	1×10^2[3](b)	(b) Switzerland
Salmonella spp.	Present in 10 g[3](c)	(c) Sweden; refrigerated, at 'best
Staphylococcus aureus	10[3](b)	before' date
B. cereus	$m = 1 \times 10^3, M = 1 \times 10^4$[3](c)	

Product name: Cream-type pies – Frozen

Organism	Numbers	Notes
APC	$m = 5 \times 10^4, M = 1 \times 10^6$[7]	Canadian data. A later survey in the
Coliform or Enterobacteria	$m = 50, M = 1 \times 10^4$[7]	USA found that 98 to 99% of the
E. coli or Fecal coliform	$m = 10, M = 1 \times 10^3$[7]	samples examined met m, and all
Salmonella spp.	0 in 25 g[7]	samples met m for *E. coli* and *Staph.*
Staphylococcus aureus	$m = 1 \times 10^2, M = 5 \times 10^4$[7]	*aureus*. For coliform, 90% of the
Yeasts and Molds	$m = 5 \times 10^2, M = 1 \times 10^4$[7]	samples met M, and 99% met M.
		Results reported by Schwab *et al.*
		(1985) in Journal of Food Protection,
		48, 70-75.

Product name: Delicatessen

Organism	Numbers	Notes
APC	1×10^4[4](a); 1×10^5[4](b, c); 5×10^5[4](d)	(a) West Virginia
Coliform or Enterobacteria	1×10^2[4](a–c)	(b) Iowa
E. coli or Fecal coliform	3[4](a); 10[4](c)	(c) Nebraska
Salmonella spp.	Less than 1 per g [4](a); negative[4](b,c)	(d) LaCrosse city, Wisconsin
Staphylococcus aureus	Negative[4](b, c); 10[4](a)	
Clostridia (*perfringens*)	Negative[4](c); 10[4](a)	
Fecal Streptococci	Not more than 10 (Enterococci) [4](a)	
Yeasts	1×10^5 [4](c)	
Molds	1×10^4 [4](c)	

Product name: Desserts (puddings) – Usually eaten cold

Organism	Numbers	Notes
APC	1×10^6 [3](a); 1×10^5[3](c)	(a) Netherlands
Coliform or Enterobacteria	1×10^3(a); 1×10^2[3](b); 20[3](c)	(b) Brazil – fruit jelly
E. coli or Fecal coliform	0 in 1 g[3](b)	(c) New Zealand guidelines, instant
Salmonella spp.	0 in 25 g [3](b)	puddings
Staphylococcus aureus	5×10^2[3](a); 1×10^2[3](c)	
Clostridia (perfringens)	1×10^2[3](c)	
B. cereus	1×10^2[3](c)	
Yeasts and Molds	1×10^3 [3] (a, b)	

Product name: Dietetic foods – Biscuits

Organism	Numbers	Notes
Coliform or Enterobacteria	$m = 10$, $M = 1 \times 10^2$[1](a)	(a) ICMSF
E. coli or Fecal coliform	2 [3](b)	(b) New Zealand
Salmonella spp.	0 in 25 g [1](a), [3]b	

Product name: Dietetic foods – Canned

Organism	Numbers	Notes
'Sterility test'	Must pass this test and have received a 'Botulinum cook'[3](a)	(a) Spain

Product name: Dietetic foods – Cooked hospital meals at time of serving

Organism	Numbers	Notes
APC	$m = 1 \times 10^3$, $M = 1 \times 10^4$[8]	(a) Direct Microscopic Count
DMC	$m = 1 \times 10^5$, $M = 1 \times 10^6$[8]	
Coliform or Enterobacteria	$m = 1$, $M = 10$[8]	
Staphylococcus aureus	$m = 10$, $M = 10^2$[8]	
Fecal Streptococci	$m = 10^2$, $M = 10^3$[8] (Lancefield group D)	

Product name: Dietetic foods – Dried and instant requiring reconstitution

Organism	Numbers	Notes
APC	1×10^4[3](a); 5×10^4[3](b,g); 5×10^4[3](f); $m = 10^4$, $M = 10^5$ [1](c), [3](d)	(a) New Zealand, applies to dried product
Coliform or Enterobacteria	0 in 0.1 g[3](g); $m = 10$, $M = 1 \times 10^2$[1](c), [3](d); 1×10^2[3](b);0 in 0.01 g[3](f)	(b) France (c) ICMSF – for Salmonella, $n = 60$
E. coli or Fecal coliform	0 in 1 g[3](f); 1 per g[3](b,d); 2[3](a)	(d) Brazil; APC not applicable to
Salmonella spp.	0 in 25 g [3](a–c); 0 in 30 g[3](f); 0 in 250 g [3](d)	fermented products
Staphylococcus aureus	0 per 1 g [3](a), (b); 0 in 0.1 g as DNase and coagulase +ve [3](f)	(e) Brazil; B. cereus if for children under 12 months
Clostridia (perfringens)	As anaerobic sulfite reducers (spores and vegetative forms), 1×10^2[3](b)	(f) Spain APC not applicable to fermented products
B. cereus	$m = 10$, $M = 1 \times 10^2$[3](e) $m = 1 \times 10^2$, $M = 1 \times 10^3$(d)	(g) Thailand
Molds	3×10^2[3](b)	
Yeasts and Molds	3×10^2[3](f); 1×10^3[3](b); $m = 1 \times 10^2$, $M = 1 \times 10^3$[3](d)	

Product name: Dietetic foods – For infants and children; dried biscuit (FAO/WHO)

Organism	Numbers	Notes
Coliform or Enterobacteria	$m = 3, M = 20$ [2]	(a) Quantity not stated in reference
Salmonella spp.	0 (in 25 g) [2](a)	

Product name: Dietetic foods – For infants and children; dried instant products (FAO/WHO)

Organism	Numbers	Notes
APC	$m = 10^3, M = 10^4$ [2]	(a) Quantity not stated in reference
Coliform or Enterobacteria	$m = 3, M = 20$ [2]	
Salmonella spp.	0 (in 25 g) [2](a)	

Product name: Dietetic foods for infants and children; Dried products requiring heating before consumption (FAO/WHO)

Organism	Numbers	Notes
APC	$m = 10^4, M = 10^5$ [2]	(a) Quantity not stated in reference
Coliform or Enterobacteria	$m = 10, M = 1 \times 10^2$ [2]	
Salmonella spp.	0 (in 25 g) [2](a)	

Product name: Dietetic foods – Ready-to-eat (other than canned or aseptically packed)

Organism	Numbers	Notes
APC	$m = 1 \times 10^4, M = 1 \times 10^5$ [3](a,b)	(a) Brazil
Coliform or Enterobacteria	$m = 1 \times 10^2, M = 1 \times 10^3$ [3](a)	(b) Not for fermented products
E. coli or Fecal coliform	0 in 1 g [3](a)	(c) For children under 12 months
Salmonella spp.	0 in 250 g [3](a)	
Staphylococcus aureus	0 in 1 g [3](a)	
Clostridia (*perfringens*)	$m = 1$ per g, $M = 10$ [3](a)	
B. cereus	$m = 10, M = 1 \times 10^2$ [3](a, c)	
	$m = 1 \times 10^2, M = 1 \times 10^3$ [3](a)	

Product name: Dietetic milk products – Not heat treated, may need reconstitution

Organism	Numbers	Notes
APC	5×10^4 [3](a)	(a) France, levels based on 1 g dried
Coliform or Enterobacteria	5 [3](a)	food, 10 g reconstituted food or 10 g
E. coli or Fecal coliform	1 [3](a)	food as liquid
Salmonella spp. or *Shigella* spp.	0 in 25 g [3](a)	
Staphylococcus aureus	0 per g [3](a)	
Clostridia (*perfringens*)	As sulfite reducers (spores and vegetative forms), 10 [3](a) (perfringens) *Cl. perfringens* 0 in 1 g [3](a)	
Yeasts and Molds	1×10^2 [3](a)	

Ingredient name: Egg – Dried whole, for acetic acid sauces

Organism	Numbers	Notes
APC	1×10^5	
Salmonella spp.	0 in 25 g	
Yeasts and Molds	1×10^2	

Ingredient name: Egg – Dried whole, yolk or albumen

Organism	Numbers	Notes
APC	2.5×10^4 [2](b); $m = 5 \times 10^4$, $M = 10^6$ [1](a); 1×10^5 [3](c–e), [4](f)	(a) ICMSF criteria for pasteurized liquid or frozen egg. 'n' for *Salmonella* varies with level of concern
Coliform or Enterobacteria	$m = 10$, $M = 10^3$ [1](a); absent in 1/3 samples with max 1×10^2 [3](c); 10 [2](b), [3](d); 50 [3](g)	(b) US military criterion
Salmonella spp.	Absent [3](d); 0 in 25 g [1](a), [2](b) [3](e); 0 in 50 g [3](c)	(c) Austria
Yeasts	10 [3](d)	(d) Spain
Molds	10 [3](d)	(e) West Germany
		(f) Ohio Department of Agriculture
		(g) Malaysia

Ingredient name: Egg – Liquid or frozen, pasteurized

Organism	Numbers	Notes
APC	$m = 5 \times 10^4$, $M = 10^6$ [1](a). Other criteria usually 1×10^5 [3] with range of 1×10^4 [3](b) to 5×10^6 [4](c)	(a) ICMSF 'n' for *Salmonella* varies with level of concern
Coliform or Enterobacteria	$m = 10$, $M = 10^3$ [1](a). Other criteria usually 10 [3] with range up to 5×10^3 [4](c)	(b) New Zealand guidelines
Salmonella spp.	0 in 25 g [1](a) which is usual [3] with range of 0 in 20 g [3](d) to 0 in 50 g [3](e)	(c) Ohio Department of Agriculture guidelines
Yeasts and Molds	10 [3](b), 1.5×10^3 [3](d)	(d) Luxembourg; Spain, which in addition has yeast and mold count
		(e) Austria

Ingredient name: Egg – Liquid, unpasteurized

Organism	Numbers	Notes
APC	Usually 1×10^6 [3]	For microbiological safety use *pasteurized* egg
Coliform or Enterobacteria	Usually 0 in 0.01 g [3]	(a) Belgium, Luxembourg, Netherlands
Salmonella spp.	0 in 20 g [3](a); 0 in 25 g [3](b); 0 in 50 g [3](c)	(b) New Zealand guidelines
		(c) Austria

Ingredient name: Egg – Whole, table grade

Organism	Numbers	Notes
APC	1.5×10^4 [2](a)	(a) USA military criteria
Salmonella spp.	0 in 25 g [2](a)	
Yeasts and Molds	50 [2](a)	

Ingredient name: Eggs – Whole, yolk, white, salted or sugared

Organism	Numbers	Notes
APC	$5 \times 10^4[2](a)$	(a) USA military criteria
Salmonella spp.	0 in 25 g[2](a)	
Yeasts and Molds	50[2](a)	

Product name: Entrees – Frozen containing rice or cornflour as a main ingredient

Organism	Numbers	Notes
B. cereus	$m = 10^3$ (estimated), $M = 10^4[1](a)$	(a) ICMSF

Ingredient name: Fats – Edible

Organism	Numbers	Notes
Salmonella spp.	1 in 50 g[3](a) includes *Salm. arizona*	(a) Spain
Yeasts (Lipolytic)	$1 \times 10^2[3](a)$	
Molds	$1 \times 10^2[3](a)$	

Product name: Foods – Miscellaneous, dried

Organism	Numbers	Notes
APC	(?)[3] (a)	(a) Brazil
Coliform or Enterobacteria	(?)[3](a)	
E. coli or Fecal coliform	10[3](a)	
Salmonella spp.	0 in 25 g[3](a)	
Staphylococcus aureus	0 in 0.01 g[3](a)	
Clostridia (*perfringens*)	As sulfite-reducing Clostridia at 44°C, 20[3](a)	
Yeasts and Molds	$5 \times 10^2[3](a)$	

Product name: Foods – Miscellaneous, pasteurized in closed containers

Organism	Numbers	Notes
APC	$1 \times 10^5[3](a)$	(a) Switzerland
E. coli or Fecal coliform	10[3](a)	
Staphylococcus aureus	10[3](a)	
Molds	$1 \times 10^2[3](a)$	

Product name: Foods – Miscellaneous, ready-to-eat

Organism	Numbers	Notes
APC	1×10^5[3](a) to $m = 5 \times 10^5$, $M = 5 \times 10^6$[9](b)	(a) Singapore
Coliform or Enterobacteria	50[3](a)	(b) National Health Institute
E. coli or Fecal coliform	$m = 20$, $M = 2 \times 10^2$[9](b)	microbiology criteria
Salmonella spp.	0 in 25 g[9](b) to 0 in 20 g[3](d)	Guidelines for food 1985, quoted in
Staphylococcus aureus	For toxin see note(c). $m = 1 \times 10^2$, $M = 1 \times$	article
	10^3[9] to 1×10^4[3](d, e)	(c) Switzerland data quoted in [3] page
Clostridia (perfringens)	1×10^4[3](d)	54 gives Staph. enterotoxin A limit of 1
B. cereus	$m = 1 \times 10^2$, $M = 1 \times 10^3$[9] to 1×10^4[3](d)	µg/kg and Staph. enterotoxin B limit of
Ps. aeruginosa	1×10^4[3](d)	10µg/kg, for 'products in general'
Brucella spp.	0 in 10 g[3] (d)	(d) Switzerland, includes foodstuffs in
S. typhi, S. paratyphi,	0 in 50 g[3](d)	general
Shigella, spp.		(e) Not dried food
Vibrio cholerae	0 in 1 g[3](d)	

Product name: Foods – Non-packaged, general

Organism	Numbers	Notes
APC	1×10^5[4](a)	(a) New Hampshire guideline
Coliform or Enterobacteria	1×10^2[4](a)	
E. coli or Fecal coliform	10[4](a)	
Staphylococcus aureus	1×10^2[4](a)	
Clostridia (perfringens)	1×10^3[4](a)	
Fecal Streptococci	As enterococci, 5×10^3[4](a)	
B. cereus	5×10^3[4](a)	
Yersinia spp.	5×10^3[4](a)	

Ingredient name: Frog legs

Organism	Numbers	Notes
APC	5×10^5[3](b)	(a) Canada
E. coli or Fecal coliform	10[3](b)	(b) New Zealand, guidelines for
Salmonella spp.	0 in 1 g[3](c); 0 in 25 g[2](a),[3](b)	uncooked frozen legs
Staphylococcus aureus	1×10^3[3](b)	(c) France, also for shelled snails,
Clostridia (perfringens)	1×10^3[3](c)	chilled or frozen

Ingredient name: Fruit – Dried

Organism	Numbers	Notes
APC	1×10^5[4](a)	(a) Rhode Island Deparment of Health
Coliform or Entobacteria	1×10^2[3](b),[4](a)	guidelines
E. coli or Fecal coliform	0 in 1 g[3](b)	(b) Brazil
Salmonella spp.	0 in 25 g[3](b)	
Yeasts and Molds	1×10^2 (?)[3](b)	

Ingredient name: Fruit – Fresh or frozen

Organism	Numbers	Notes
APC	5×10^5[3](a)	(a) Israel
Coliform or Enterobacteria	5×10^2[3](a)	(b) Enterococci
E. coli or Fecal coliform	2×10^2[3] (c); $m = 1 \times 10^2$, $M = 1 \times 10^3$[1](d)	(c) Brazil
Salmonella spp.	0 in 50 g[3](a) to 0 in 25 g[3](c)	(d) ICMSF, appropriate if pH is over
Staphylococcus aureus	1×10^2[3](a)	4.5, m is estimated
Fecal Streptococci	1×10^3[3](a, b)	

Product name: Fruit juices and nectars

Organism	Numbers	Notes
APC	1×10^2[3](a, b)	Howard mold count figures for USA
Coliform or Enterobacteria	0 in 10 ml[3] (a, b); 1×10^2[3](e)	quoted [3] pages 76,77
E. coli or Fecal coliform	0 in 1 g or ml[3](c–e)	(a) Tomato juice, Portugal
Yeasts and Molds	1×10^2[3](b, d, e); 20[3](c)	(b) Pineapple juice, Portugal
		(c) Fruit nectar, chemically preserved, Brazil
		(d) Fruit nectar, other, Brazil
		(e) Fruit juice, whole, Brazil

Ingredient name: Fruit juice – Concentrated

Organism	Numbers	Notes
APC	(?)[3](a)	(a) Brazil
Coliform or Enterobacteria	20[3](a)	
E. coli or Fecal coliform	0 in 1 g[3](a)	
Yeasts and Molds	1×10^2[3](a)	

Ingredient name: Fruit pulp – Non-heat treated

Organism	Numbers	Notes
Coliform or Enterobacteria	1×10^2[3](a)	(a) Brazil
E. coli or Fecal coliform	0 in 1 g[3](a)	
Salmonella spp.	0 in 25 g[3](a)	
Yeasts and Molds	1×10^5[3](a)	

Ingredient name: Gelatin

Organism	Numbers	Notes
APC	$m = 5 \times 10^3$, $M = 10^5$[2](a); other criteria range from 3×10^3[4](b) through usual 5×10^3 to 1×10^4[3] to 1×10^5[3](c)	(a) Canada
		(b) Georgia Department of Agriculture guideline
Coliform or Enterobacteria	$m = 10$, $M = 1 \times 10^3$[2](a); other criteria range from 1 to 50[3]	(c) Singapore
		(d) Kenya, Zambia, Namibia
E. coli or Fecal coliform	0 in 1 g[3](d)	(e) France
Salmonella spp.	0 in 25 g[1] is usual [3]	
Staphylococcus aureus	$m = 10^2$, $M = 10^4$[1]	
Clostridia (perfringens)	$m = 10^2$, $M = 10^4$[1]. Other criteria [3](e), anaerobic sulfite reducers (spores) 1 per g	

Ingredient name: Gums – Botanic

Organism	Numbers	Notes
APC	1×10^5 (a)	(a) For carob powder used in acetic acid sauces (b) Untreated gums may have counts up to 10^8, pretreated gums have lower counts. Counts up to 10^5 of *Staph. aureus* have been found [2] page 297

Ingredient name: Ham and other cured smoked meats

Organism	Numbers	Notes
APC	1×10^4 [3](c, d); 1×10^6 [3](a)	(a) New Zealand guidelines for bacon
Coliform or Enterobacteria	10 [3](d)	(b) New Zealand guidelines for cured
E. coli or Fecal coliform	0 in 1 g [3](d); 1×10^2 [3](a–c)	or fermented meats
Salmonella spp.	0 in 25 g [3](a–d)	(c) Brazil for raw ham
Staphylococcus aureus	0 in 1 g [3](d); 1×10^2 [3](a, b); 0 in 0.01 g [3](c)	(d) France, whole cooked ham
Clostridia (*perfringens*)	As anaerobic sulfite reducers, 0 in 1 g [3](d); as sulfite-reducing clostridia at 44°C, 20 [3](c), 1×10^2 [3](a, b)	
B. cereus	1×10^3 [3](a, b)	
Yeasts and Molds	1×10^2 (?) [3](c)	

Product name: Ham – Canned, refrigerated, 'pasteurized hams' and pork shoulder

Organism	Numbers	Notes
Step 1	Visual check for swells and defective cans. If any found, go to step 2.	(a) ICMSF scheme see [1] pages 145–146
Step 2	Check air temperature between cans. If 10°C or more, go to step 3.	
Step 3	APC $m = 10^3$, $M = 10^4$	

Ingredient name: Herbs, including spices and condiments

Organism	Numbers	Notes
APC	5×10^5 [3](a); 1×10^5 [4](b)	(a) New Zealand guidelines for natural
Coliform or Enterobacteria	1×10^2 [4](b)	spices
E. coli or Fecal coliform	0 per g [4](c); 1×10^2 [3](a); 2×10^2 (d)	(b) Massachusetts Department of
Salmonella spp.	0 in 25 g [3](c),(d); 0 in 50 g [3] (a)	Health guidelines
Thermophilic spore count		(c) Brazil for 'spices and seasonings'
for cannery use	Aerobic 30 per 2 g, anaerobic 10	(d) Brazil for 'pot herbs'

Ingredient name: Honey

Organism	Numbers	Notes
APC	1×10^4[3](a) to 5×10^4[4](b)	(a) Spain
Coliform or Enterobacteria	0 in 1 g[3](a)	(b) Ohio Department of Agriculture
E. coli or Fecal coliform	0 in 1 g [3](a)	Guidelines
Salmonella spp. or		
Shigella spp.	0 in 25 g [3](a)	
Yeasts	1×10^4[4](b)	
Molds	1×10^2[3](a); 1×10^4[4](b)	

Product name: Ice-prepacked (frozen water)

Organism	Numbers	Notes
Coliform or Enterobacteria	10 per 100 ml [2](a)	(a) Canadian standard for water from which the ice is made

Product name: Ice cream

Organism	Numbers	Notes
APC	$m = 5 \times 10^4$, $M = 2.5 \times 10^5$[2](g) Range from 1×10^4[3](a) to $m = 10^5$, $M = 10^6$[2](b)	(a) Japan
Coliform or Enterobacteria	$m = 10$, $M = 1 \times 10^3$[2](b); $m = 1 \times 10^2$, $M = 1 \times 10^3$[2](g)	(b) Canada, also applies to ice milk
E. coli or Fecal coliform	0 in 1 g[3](c)	(c) Australian states, Papua New Guinea
Salmonella spp.	0 in 25 g[2](g), [3](h)	(d) Sweden
Staphylococcus aureus	$m = 10^2$, $M = 10^3$[3](d)	(e) Austria
Fecal Streptococci	As enterococci, 1×10^2[3](e); $m = 10^3$, $M = 10^4$[3](d)	(f) Netherlands for ice made from fermented milk. Tests made on liquified and air free sample
Yeasts	1×10^2[3](e)	(g) FAO/WHO specification quoted [2]
Molds	1×10^2[3](e)	(h) Usual *Salmonella* spp. limit:
Yeasts and Molds	1×10^2[3](f)	Sweden [3] unsuitable if present in 10 g

Ingredient name: Ice cream mix ('preparations')

Organism	Numbers	Notes
APC	$m = 1 \times 10^3$, $M = 1 \times 10^4$[3](c); $m = 2.5 \times 10^4$, $M = 1 \times 10^5$[2](a); 3×10^4[2](b)	(a) FAO/WHO specification quoted [2]
Coliform or Enterobacteria	$m = 12$, $M = 110$[3](c); $m = 10$, $M = 1 \times 10^2$[2](a); 10[2](b)	(b) Military specification for dehydrated mix [2]
Salmonella spp.	0 in 25 g [3](c), [2](b)	(c) Italy
Staphylococcus aureus	$m = 0.3$, $M = 12$[3](c)	

Product name: Ice cream – Regular, vanilla, chocolate or other flavors and novelties-USA military criteria

Organism	Numbers	Notes
APC	5×10^4[2](a–c)	(a) Regular vanilla
Coliform or Enterobacteria	10[2](a); 20[2](b, c)	(b) Regular chocolate or other flavors
		(c) Novelties

Product: Ice cream – Ultra-pasteurized

Organism	Numbers	Notes
APC	$m = 10, M = 1 \times 10^2$ [3](a)	(a) Finland
Salmonella spp.	0 in 25 g [3](a)	

Product: Ices – General

Organism	Numbers	Notes
APC	1×10^5 [3](a)	(a) Netherlands standard based on
Coliform or Enterobacteria	0 in 0.1 ml [3](a)	liquified and air-free samples
Salmonella spp.	0 in 25 ml [3](a)	
Staphylococcus aureus	As coagulase positive, 0 in 0.1 ml [3](a)	

Product: Ices – Sherbert, snowball, slush ice products

Organism	Numbers	Notes
APC	1×10^5 [3](a)	(a) West Indies
Coliform or Enterobacteria	10 [3](a)	
Yeasts and Molds	50 [3](a)	

Product name: Maple syrup

Organism	Numbers	Notes
APC	5×10^4 [4](a)	(a) Ohio, guidelines
Yeasts	1×10^4 [4](a)	
Molds	1×10^4 [4](a)	

Product name: Margarine

Organism	Numbers	Notes
APC	3.6×10^2 [4](a); 2.5×10^4 [3](b); 5×10^4 [3](c)	(a) Georgia guidelines
Coliform or Enterobacteria	0 in 1 g [3](d); 10 [3](e,f); 25 [3](c); 50 [3](b)	(b) New Zealand guidelines
E. coli or Fecal coliform	0 in 1 g [3](e); 50 [3](b)	(c) Chile
Salmonella spp.	0 in 25 g [3](b)	(d) Portugal
Staphylococcus aureus	As coagulase positive, 0 in 1 g [3](b)	(e) Argentina
Yeasts	1×10^2 [3](c)	(f) Israel
Yeasts and Molds	50 [3](b,e); 1×10^2 [3](f)	

Product name: Mayonnaise

Organism	Numbers	Notes
APC	1×10^3[3](a); 5×10^4[3](b, c)	(a) Argentina
Coliform or Enterobacteria	0 in 0.1 g or 10[3](a, b, d); 25[3](c)	(b) Poland: flavored mayonnaise APC
E. coli or Fecal coliform	0 in 1 g[3](a)	(c) Chile
Salmonella spp.	0 in 20 g[3](b)	(d) Netherlands
Staphylococcus aureus	0 in 0.1 g[3](b)	
Yeasts	10[3](b); 1×10^2 [3](c)	
Molds	10[3](b)	
Yeasts and Molds	20[3](a)	

Product name: Meals – Meat and poultry entrees

Organism	Numbers	Notes
APC	35°C Not more than 1 of 5 over 5×10^4[2](a)	(a) USDA advisory criteria
Coliform or Enterobacteria	Not more than 1 of 5 over 10[2](a)	
E. coli or Fecal coliform	Not more than 1 of 5 positive[2](a)	
Salmonella spp.	None of 5 positive[2](a)	
Staphylococcus aureus	Not more than 1 of 5 positive[2](a)	

Product name: Meals – Not ready-to-eat (to be cooked and prepared) and dried

Organism	Numbers	Notes
APC	$m = 7.5 \times 10^4$, $M = 1.5 \times 10^5$[2](b)	(a) Switzerland
E. coli or Fecal coliform	$m = 3$, $M = 20$[2](b)	(b) Military criteria for dehydrated
Staphylococcus aureus	1×10^5[3](a)	cooked foods
Clostridia (*perfringens*)	1×10^5[3](a)	
B. cereus	1×10^5[3](a)	

Product name: Meals – Ready cooked, including fillings for pastas and meals with cheese

Organism	Numbers	Notes
APC	3×10^5[3](a); 5×10^6[4](b)	(a) France
Coliform or Enterobacteria	1×10^3[3](a); 5×10^3[4](b)	(b) Pre-cooked, ready-to-eat, Ohio
E. coli or Fecal coliform	10[3](a)	Department of Agriculture
Salmonella spp.	0 in 25 g[3](a)	
Staphylococcus aureus	1×10^2[3](a); 1×10^3[4](b)	
Clostridia (*perfringens*)	30[3](a)	

Product name: Meals – Cook-chilled

Organism	Numbers	Notes
APC	48h at 37°C 1×10^5[3](a)	(a) UK guidelines
E. coli or Fecal coliform	10[3](a)	
Salmonella spp.	0 in 25 g[3](a)	
Staphylococcus	As coagulase +ve, 1×10^2[3](a)	
Clostridia (*perfringens*)	1×10^2[3](a)	

Product name: Meals (finished products) – Chilled, entirely heat treated, also ready-to-serve foods served cold

Organism	Numbers	Notes
APC	$m = 1 \times 10^8$ [3](a)	(a) Sweden, 'suitable with remarks' has
Coliform or Enterobacteria	$m = 1 \times 10^3$ (at 37°C) [3](a)	been regarded as m; 'unsuitable' as M
Salmonella spp.	M = present in 10 g [3](a)	
Staphylococcus aureus	$m = 1 \times 10^2$, $M = 1 \times 10^3$ [3](a)	
Clostridia (perfringens)	$m = 1 \times 10^2$, $M = 1 \times 10^3$ [3](a)	
Fecal Streptococci	$m = 1 \times 10^4$, (enterococci) [3](a)	
B. cereus	$m = 1 \times 10^3$, $M = 1 \times 10^4$ [3](a)	
Molds	$m = 1 \times 10^3$ [3](a)	

Product name: Meals (finished products) – Chilled, with one or more non-heat treated components

Organism	Numbers	Notes
APC	$m = 1 \times 10^8$ [3](a)	(a) Sweden, 'suitable with remarks' has
Coliform or Enterobacteria	$m = 1 \times 10^4$ (at 37°C) [3](a)	been regarded as m; 'unsuitable' as M
Salmonella spp.	M = present in 10 g [3](a)	
Staphylococcus aureus	$m = 1 \times 10^2$, $M = 1 \times 10^3$ [3](a)	
Clostridia (perfringens)	$m = 1 \times 10^2$, $M = 1 \times 10^3$ [3](a)	
B. cereus	$m = 1 \times 10^3$, $M = 1 \times 10^4$ [3](a)	
Molds	$m = 1 \times 10^3$ [3](a)	

Product name: Meals (finished products) – Frozen, entirely heat treated, also ready-to-serve foods served cold

Organism	Numbers	Notes
APC	$m = 5 \times 10^5$, $M = 1 \times 10^7$; $M = 1 \times 10^5$ in 3/5 samples, $M = 1 \times 10^6$ in 2/5 samples [3] (b) to $m = 1 \times 10^6$ [3](a)	(a) Sweden, 'suitable with remarks' has been regarded as m; unsuitable as M (b) Some Australian states
Coliform or Enterobacteria	$m = 1 \times 10^2$ (at 37°C) [3](a); $m = 10^3$; $M = 10^4$	(c) Expressed in Australia as none in
E. coli or Fecal coliform	$m = 9$, $M = 70$ [3](b) to $m = 10$, $M = 10^2$	0.01 g and none in 0.001 g
Salmonella spp.	M = present in 10 g [3](a); 0 in 25 g [3](b)	
Staphylococcus aureus	$m = 20$, $M = 10^2$; $m = 1 \times 10^2$, $M = 1 \times 10^3$ [3](a–c)	
Clostridia (perfringens)	$m = 1 \times 10^2$, $M = 1 \times 10^3$ [3](a)	
Fecal Streptococci	$m = 1 \times 10^4$ (enterococci) [3](a)	
B. cereus	$m = 1 \times 10^3$, $M = 1 \times 10^4$ [3](a)	
Molds	$m = 1 \times 10^3$ [3](a)	

Product name: Meals (finished products) – Frozen, with one or more heat treated components

Organism	Numbers	Notes
APC	$m = 1 \times 10^8$ [3](a)	(a) Sweden, 'suitable with remarks' has
Coliform or Enterobacteria	$m = 1 \times 10^3$ (at 37°C) [3](a)	been regarded as m; unsuitable as M
Salmonella spp.	M = present in 10 g [3](a)	
Staphylococcus aureus	$m = 1 \times 10^2$; $M = 1 \times 10^3$ [3](a)	
Clostridia (perfringens)	$m = 1 \times 10^2$; $M = 1 \times 10^3$ [3](a))	
B. cereus	$m = 1 \times 10^3$; $M = 1 \times 10^4$ [3](a)	
Molds	$m = 1 \times 10^3$ [3](a)	

Product name: Meals (finished products) – Heated

Organism	Numbers	Notes
APC	$m = 1 \times 10^5$ [3](a)	(a) Sweden, 'suitable with remarks' has
Coliform or Enterobacteria	$m = 1 \times 10^2$ (at 37°C) [3](a)	been regarded as m; unsuitable as M
Salmonella spp.	$M =$ present in 10 g [3](a)	
Staphylococcus aureus	$m = 1 \times 10^2$; $M = 1 \times 10^3$ [3](a)	
Clostridia (*perfringens*)	$m = 1 \times 10^2$; $M = 1 \times 10^3$ [3](a)	
B. cereus	$m = 1 \times 10^3$; $M = 1 \times 10^4$ [3](a)	
Molds	$m = 1 \times 10^3$ [3](a)	

Product name: Meals (or snacks) – Ready cooked, not requiring heating or cooking

Organism	Numbers	Notes
APC	1×10^4 [3] (a–c)	(a) Spain
Coliform or Enterobacteria	10[3](a) or 0 in 0.1 g [3](b)	(b) Netherlands, includes snacks, pasta
E. coli or Fecal coliform	0 in 0.1 g [3](a); 1×10^4 [3](e) *	and rice dishes
Salmonella spp.		(c) APC for Spain 72 h at 30 to 32°C
or *Shigella* spp.	0 in 50 g [3](a)	(d) DNA-ase positive, Spain and
Staphylococcus aureus	0 in 0.1 g [3](a, b, d); 1×10^4 [3](e, f)	coagulase positive Netherlands
Clostridia (*perfringens*)	As sulfite-reducing anaerobes, 50[3](a); 1×10^4 [3](e)	(e) Switzerland
		(f) Not dried foods
Fecal Streptococci	1×10^2 [3](a) (Lancefield group D)	* This is the figure given
Brucella spp.	Absent in 10 g (especially in milk and products made from unpasteurized milk) [3](e)	
Ps. aeruginosa	1×10^4 [3](e)	
B. cereus	1×10^4 [3](e)	

Product name: Meals (or snacks) – Ready-cooked, requiring heating or cooking before consumption

Organism	Numbers	Notes
APC	1×10^5 [3](c, e); 5×10^5 [3](a); 1×10^6 [3](b, d)	(a) New Zealand, context implies re-heating or cooking
Coliform or Enterobacteria	1×10^3 [3](b)	(b) Spain
E. coli or Fecal coliform	20[3](a); 1×10^2 [3](b)	(c) Netherlands, includes snacks, pasta
Salmonella spp.	0 in 50 g [3](b, f); 0 in 25 g [3](a)	and rice dishes
Staphylococcus aureus	1×10^2 [3](a, b, g); 5×10^2 [3](c, g)	(d) APC for Spain, 72h at 30 to 32°C
Clostridia (*perfringens*)	1×10^2 [3](a); 1×10^3 [3](b)	(e) If to be heated by consumer after
Fecal Streptococci	1×10^3 [3](b) (Lancefield group D)	purchase except if it contains raw beans
B. cereus	1×10^2 [3](a)	or sprouts
		(f) Or *Shigella* spp
		(g) DNAse positive Spain, coagulase positive Netherlands

Product name: Meat – Barbecue, beef or pork

Organism	Numbers	Notes
APC	At 35°C not more than 1 of 5 over 5×10^4 [2](a)	(a) USDA advisory criteria
Coliform or Enterobacteria	Not more than 1 of 5 over 10 [2](a)	
E. coli or Fecal coliform	Not more than 1 of 5 positive [2](a)	
Salmonella spp.	None of 5 positive [2](a)	
Staphylococcus aureus	Not more than 1 of 5 positive [2](a)	
Clostridia (*perfringens*)	Not more than 1 of 5 at 1×10^3 or more [2](a)	

Product name: Meat – Cecinas (corned, dried meat)

Organism	Numbers	Notes
APC	$1 \times 10^5 [3](a)$; $8 \times 10^5 [3](b)$	(a) Chile, other than raw
Coliform or Enterobacteria	$2 \times 10^2 (a)$	(b) Chile, raw
		Both also state no *E. coli*, *Salmonella* spp. *Arizona*, *Staph. aureus*

Product name: Meat – Charcuterie products

Organism	Numbers	Notes
APC	$3 \times 10^5 (d)$	[3]France standards
Coliform or Enterobacteria	$1 \times 10^3 (d)$	(a) Dried and eaten as such
E. coli or Fecal coliform	$10(d)$; $1 \times 10^2 (a)$; $1 \times 10^3 (b, c)$,	(b) Eaten after cooking
Salmonella spp.	0 in 25 g (a, d)	(c) Cured products, raw, salted dried,
Staphylococcus aureus	$1 \times 10^2 (d)$; $5 \times 10^2 (a, c)$; $1 \times 10^3 (b)$	sliced or unsliced
Clostridia (*perfringens*)	As anaerobic sulfite reducers, 30(d); 50(a, c); $1 \times 10^2 (b)$	(d) Cooked charcuterie, sliced or unsliced, force meat

Ingredient: Meat – Comminuted, ground or hamburger

Organism	Numbers	Notes
APC	1×10^6 to usual around 5×10^6 to 1×10^7	There are a number of figures in
Coliform or Enterobacteria	1×10^2 to 1 or 2×10^3 to 2×10^4	references [3] and [4] but no ranges are
E. coli or Fecal coliform	Usual 50	given
Salmonella spp.	0 in 25 g	
Staphylococcus aureus	1×10^2 to 1×10^3	
Clostridia (*perfringens*)	1×10^2	

Product name: Meats – Cooked, other than in USA

Organism	Numbers	Notes
APC	$3 \times 10^5 (a)$; $1 \times 10^6 (b)$; $m = 1 \times 10^7 (c, d)$, $m = 1 \times 10[8](e)$	Data taken from reference [3]
Coliform or Enterobacteria	$50(b)$; $m = 1 \times 10^3 (c, e)$	(a) France, roast or stuffed meat slices
E. coli or Fecal coliform	$10(a)$; $m = 1 \times 10^3 (e)$	(b) Malaysia, Singapore
Salmonella spp.	0 in 25 g (a); M = present in 10 g (c–e)	(c) Sweden, meat in pieces; *not* vacuum
Staphylococcus aureus	$1 \times 10^2 (a)$; $M = 1 \times 10^3$ (c–e)	packed or gas packed m = suitable with
Clostridia (*perfringens*)	$10(a)$; $M = 1 \times 10^3$ (c–e)	remarks, M = unsuitable
Fecal Streptococci	$m = 1 \times 10^4$ (c), $m = 1 \times 10^5$ (d, e), (enterococci)	(d) Sweden, meat in pieces, vacuum or gas packed
		(e) Sweden, meat sliced, whether or not vacuum packed or gas packed

Product name: Meats cooked – US state and city criteria

Organism	Numbers	Notes
APC	5×10^4(a); 1×10^5(e); 1×10^6(b),(d); 5×10^6(c)	Data taken from reference [4]
Coliform or Enterobacteria	10(a, b, d); 1×10^2(d, e); 1.5×10^3(c)	(a) Massachusetts, meat cooked
E. coli or Fecal coliform	Negative(e)	(b) North Dakota, meat food products
Salmonella spp.	Negative (b, d, e)	heat treated
Staphylococcus aureus	Negative (e); 10(b); 1×10^3(d)	(c) Ohio, roast
Clostridia (*perfringens*)	Negative (e)	(d) Tennessee, processed meats
Fecal Streptococci	1×10^3(d)	(e) New York City, ready-to-eat meat
Hemolytic staphylococcus	1×10^2(e)	containing gravies, sauces
Yeasts and Molds	1×10^3(d)	

Product name: Meats – Cooked: US states' criteria, 1977

Organism	Numbers	Notes
APC	Range of 7; 1×10^4 to usual 1×10^6 to highest 5×10^6	See reference [10]
Coliform or Enterobacteria	Range of 2; 10 to 50	
E. coli or Fecal coliform	Range of 5; 10 usual to 50	

Product name: Meat – Cured or fermented

Organism	Numbers	Notes
E. coli or Fecal coliform	1×10^2[3](a)	(a) New Zealand guidelines
Salmonella spp.	0 in 25 g[3](a)	(b) Australian states
Staphylococcus aureus	1×10^2[3](a), [3](b) or 0 in 0.01 g[3](b)	
Clostridia (*perfringens*)	1×10^2[3](a)	
B. cereus	1×10^3[3](a)	

Product name: Meat – Cured

Organism	Numbers	Notes
APC	1×10^6[4](b)	(a) France, charcuterie, raw, salted/
Coliform or Enterobacteria	50[4](b)	dried, sliced or not
E. coli or Fecal coliform	10[4](b) to 1×10^3[3](a)	(b) Nebraska, meat products smoked
Salmonella spp.	0 in 25 g[3](a); Negative [4](b)	and/or heat treated
Staphylococcus aureus	10[4](b) to 5×10^2[3](a)	
Clostridia (*perfringens*)	Negative [4](b); 50[3](a)	
Yeasts	1×10^6[4](b)	
Molds	1×10^5[4](b)	

Ingredient name: Meat – Fresh

Organism	Numbers	Notes
APC	3×10^4[3](a); 2×10^5[3](b); 1×10^6[3](c); 5×10^6[4](d),[2](e)	(a) Brazil, raw prepared or not
Coliform or Enterobacteria	1×10^3[4](d)	(b) Greece, cut on day of production
E. coli or Fecal coliform	20[3](b); 50[4](d),[2](e)	(c) Australian states, chopped or
Salmonella spp.	0 in 50 g[3](c); negative[4](d)	minced meat, uncooked
Staphylococcus aureus	10[3](b); 1×10^3[4](d)	(d) Tennessee, fresh red meats
Clostridia (*perfringens*)	10[3](b) to 20[3](a)	(e) Abandoned Oregon standard for
Fecal Streptococci	1×10^3[4](d)	red meat, fresh or frozen
Yeasts and Molds	1×10^3[4](d)	

Ingredient name: Meats – Fresh or frozen, Netherlands data

Organism	Numbers	Notes
APC	$5/100 \text{ cm}^2$(a); 6.5(b); $6/100 \text{ cm}^2$(c); 7(d)	Counts obtained under GMP, numerical data as *log* 95th percentile of cfu
Coliform or Enterobacteria	$3/100 \text{ cm}^2$(a), 4(b); $4/100 \text{ cm}^2$(c); 5 to 6(d)	ND means 'no survey' (i.e. not done)
E. coli or Fecal coliform	70% +ve/10 cm^2, 20% +ve/1 cm^2(a); ND(b); 50% +ve/1 cm^2(c); ND(d)	For details see reference [8] (a) Wholesale meats, quarters and carcasses, counts per cm^2 swab
Salmonella spp.	Presence or absence test $0/100 \text{ cm}^2$(a) $1/100 \text{ cm}^2$(c); 0 (d)	(b) Frozen boneless, counts per ml of hand-pressed drip
Staphylococcus aureus	ND(a); ND(b); ND(c); 2 to 3(d)	(c) Consumer size cuts, counts per cm^2 swab
		(d) Minced meat, counts per g. Except for *Salmonella*, count are very dependent on type and origin of meat, particularly whether beef or pork, and also whether prepared in a factory or butcher's shop

Ingredient name: Meat – Fresh or frozen: US states' criteria 1977

Organism	Numbers	Notes
APC	Range of 16; 1×10^5 to usual 5×10^6 to highest 1×10^7	See reference [10]
Coliform or Enterobacteria	Range of 8; 1×10^2 to 1×10^4	
E. coli or Fecal coliform	Range of 11; 0 in 1 g to usual 50 to highest 1×10^3	

Ingredient name: Meat – Joints and single portions

Organism	Numbers	Notes
APC	5×10^4[3](a)	(a) France, vacuum packed or aerobic; chilled or frozen
E. coli or Fecal coliform	1×10^2[3](a); 3×10^2[3](b)	(b) France, single portions; chilled or frozen
Salmonella spp.	0 in 25 g[3](a, b)	
Staphylococcus aureus	1×10^2[3](b)	
Clostridia (*perfringens*)	2[3](a); 10[3](b)	

Ingredient name: Meat – Minced, fresh or frozen

Organism	Numbers	Notes
APC	5×10^5 to 1×10^6	There are a number of figures given in references [3] and [4], so ranges are given
Coliform or Enterobacteria	1×10^2	
E. coli or Fecal coliform	50 to 1×10^2	
Salmonella spp.	0 in 25 g is usual	
Staphylococcus aureus	1×10^2 to usual 1×10^3	
Clostridia (*perfringens*)	10 to 30	

Product name: Meat – Pate

Organism	Numbers	Notes
APC	1×10^6 [3](a)	(a) Australian states
Salmonella spp.	0 in 25 g [3](a), [1](b)	(b) ICMSF, $n = 20$

Product name: Meat – Pot pies

Organism	Numbers	Notes
APC	At 35°C, not more than 1 of 5 over 5×10^4 [2](a)	(a) USDA advisory criteria
Coliform or Enterobacteria	Not more than 1 of 5 over 10 [2](a)	
E. coli or Fecal coliform	Not more than 1 of 5 positive [2](a)	
Salmonella spp.	None of 5 positive [2](a)	
Staphylococcus aureus	Not more than 1 of 5 positive [2](a)	
Clostridia (*perfringens*)	Not more than 1 of 5 at 1×10^3 [2](a)	

Ingredient name: Meat – Raw, *not* for use at port of entry: in plant data

Organism	Numbers	Notes
APC	10^2 to 10^4 per cm^2 [2](a) $m = 10^5$, $M = 10^6$ [1](b) $m = 10^6$, $M = 10^7$ [1](c) $m = 5 \times 10^5$, $M = 10^7$ [1](d) 10^6 to 10^7 per cm^2 [2](e)	(a) Usual range of fresh dressed beef, pork and lamb in USA. Most organisms are *not* psychrotrophic (b) Carcass meat before chilling (c) Chilled carcass and chilled edible offal (d) Carcass meat frozen; boneless beef, veal, pork or mutton frozen; and edible offal frozen (e) Usually acceptable sensory quality for *vacuum packed* beef. Similar cuts stored in air permeable films have 'off odors'

Ingredient name: Meat – Raw mince or raw sausage at 'best before' date

Organism	Numbers	Notes
APC	$m = 1 \times 10^8$(a)	See reference [3] Sweden, m = suitable
Coliform or Enterobacteria	$m = 1 \times 10^5$ at 37°C	with remarks, M = unsuitable
	$m = 1 \times 10^4$ at 44°C	(a) A somewhat increased level is
Salmonella spp.	M = present in 10 g	acceptable in vacuum packed meat
Staphylococcus aureus	$M = 1 \times 10^3$	
Clostridia (*perfringens*)	$M = 1 \times 10^4$	

Product name: Meat – Sausages

Organism	Numbers	Notes
APC	1×10^5 [3](a, e); 1×10^6[4](b);	(a) Portugal, frankfurters, loose or
	1.5×10^7[4](c); 1×10^6(?) [3](d)	packed
Coliform or Enterobacteria	5×10^3[4](b); 1×10^4[4](c)	(b) Ohio guidelines for Bologna,
E. coli or Fecal coliform	0 in 1 g[3](a); 50[3](e); 1×10^2(?) [3](d)	Frankfurters, Braunschweiger and
Salmonella spp.	0 in 25 g[3](a),(d),(e)	smoked sausage
Staphylococcus aureus	0 in 1 g[3](a); 0 in 0.01 g[3](d, e)	(c) Ohio guidelines, beef, pork sausage
Clostridia (*perfringens*)	0 in 0.1 g[3](a); 20[3](d, e)	(d) Brazil, fresh sausages
	Clostridial sulfite reducers (spores),	(e) Brazil, other sausages
	0 in 0.01 g[3](a)	
Yeasts and Molds	1×10^2(?) [3](d); 5×10^2[3](e)	

Product name: Meat – Sausage, cooked, standard Dutch quality

Organism	Numbers	Notes
APC	10^5 aerobic; 10^5 anaerobic	Reference values, sampled when
Coliform or Enterobacteria	10	leaving the factory, with a core
Staphylococcus aureus	1×10^2	temperature not exceeding 7°C. See
Clostridia (*perfringens*)	1×10^3	reference [8] page 94
Fecal Streptococci	1×10^3 (Lancefield group D streptococci)	

Product name: Meat – Sausages, fermented Canadian

Organism	Numbers	Notes
E. coli or Fecal coliform	$m = 1 \times 10^2$, $M = 2 \times 10^3$(a)	Produced under good hygenic
	$m = 10$, $M = 1 \times 10^2$(b)	practices.
Salmonella spp.	0 in 25 g(a, b)	See reference [11]
Staphylococcus aureus	$m = 2.5 \times 10^2$, $M = 1 \times 10^4$(a)	(a) Non-heat treated (raw)
	$m = 50$, $M = 5 \times 10^2$(b)	(b) Heat treated
Campylobacter and *Yersinia*	0 in 25 g (a, b)	

Product name: Meat – Smoked products

Organism	Numbers	Notes
APC	1×10^5[4](a)	(a) Rhode Island
E. coli or Fecal coliform	Negative [4](a)	

Product name: Meats – Vacuum packed, sliced, i.e. semi-preserved but perishable products

Organism	Numbers	Notes
APC	10^6; anaerobic colony count 10^5; Gram-negative rods 10^3	After 3 days' incubation at 17°C products should be acceptable sensorially. These reference values apply (see reference [8])
Coliform or Enterobacteria	10^2	
Staphylococcus aureus	10^2	
Clostridia (*perfringens*)	10^2	
Fecal Streptococci	10^3, (Lancefield group D)	
B. cereus	10^3	

Product name: Milk (cows) – Concentrated or 'evaporated' (UK)

Organism	Numbers	Notes
APC	1×10^5 [3](a, b) 5 spores [3](c)	(a) Japan
Coliform or Enterobacteria	10 [3](b)	(b) Spain 'concentrated milk'
		(c) France

Product name: Milk (cows) – Concentrated and sweetened or 'condensed' (UK)

Organism	Numbers	Notes
APC	1×10^3 [2](a); 1×10^4 [3](b, c); 3×10^4 [3](d); 5×10^4 [3](e)	(a) USDA grade
Coliform or Enterobacteria	0 in 1 g [3](b, d); 10 [2](a), [3](f) or 0 in 0.1 g [3](c)	(b) Spain
Yeasts	5 [2](a); 3×10^2 [3](f)	(c) Thailand
Molds	5 [2](a)	(d) France
Yeasts and Molds	10 [3](c)	(e) Japan
		(f) Chile

Product name: Milk (cows) – Dried, full fat or skim

Organism	Numbers	Notes
APC	$m = 3 \times 10^4$, $M = 3 \times 10^5$ [1] $m = 5 \times 10^4$, $M = 2 \times 10^5$ [2](a)	There are many national criteria – examples indicate range
DMC	4×10^7 [2](b) to 7.5×10^7 [2](c)	(a) Proposed IDF specification 1982
Coliform or Enterobacteria	$m = 10$, $M = 1 \times 10^2$ [1], [2](a)	(b) US military criteria for dried whole milk Premium grade
E. coli or Fecal coliform	0 in 1 g [3](d); 0 in 5 g [3](e) to 90 [3](f)	(c) US military criteria for dried whole milk Extra grade
Salmonella spp.	0 in 25 g [1](g), [2](a) to 0 in 100 g [2](h)	(d) South Africa
Staphylococcus aureus	10 [3](i) (j)	(e) Argentina
Clostridia (*perfringens*)	1×10^2	(f) India
Thermophilic spore count	Aerobes 30 per 2 g, anaerobes 10(k)	(g) ICMSF alter the stringency of the test by changing *n* according to level of concern and whether population is 'high-risk'
		(h) US military criteria for milk, non-fat, dry
		(i) Spain
		(j) Switzerland
		(k) Cannery ingredient specification
		DMC means Direct Microscopic Count

Product name: Milk (cows) – Evaporated (Netherlands)

Organism	Numbers	Notes
APC	1×10^2 after 5 days at 30°C in original pack [3](a)	(a) Netherlands – applies to any fat level

Product name: Milk (cows) – Flavored

Organism	Numbers	Notes
APC	$m = 5 \times 10^4$, $M = 10^6$ [2](a); 2×10^4 [2](b)	(a) Canada for milk and partly skimmed milk and skimmed milk with added milk solids
Coliform or Enterobacteria	10 [2](b)	(b) US military criteria for milk (plain or chocolate), cream, half and half, filled

Product name: Milks (cows) – Jelled and flavored milks with rennet

Organism	Numbers	Notes
APC	1×10^3 [3](a)	(a) France
Coliform or Enterobacteria	10 [3](a)	
E. coli or Fecal coliform	1 [3](a)	
Salmonella spp.	0 in 25 g [3](a)	

Ingredient name: Milk (cows) – Liquid, pasteurized

Organism	Numbers	Notes
APC	5×10^3 [3](a) to more usual range of 5×10^4 and 1×10^5	Many countries have standards, see reference [2] and [3]
Coliform or Enterobacteria	5 [3](b) to more usual 10 to 50 [3](c)	Examples given show a range of values
E. coli or Fecal coliform	0 in 1 ml [3](d), (f) to 0 in 0.1 ml [3](e)	(a) Argentina, 'certified pasteurized milk'
Salmonella spp.	0 in 250 ml [3](f)	(b) Papua New Guinea
Staphylococcus aureus	10 [3](f) to 1×10^2 [3](g)	(c) Malaysia
		(d) Kenya, South Africa, Turkey, Zambia
		(e) Thailand
		(f) France – on last date for consumption of packaged ('conditionne pasteurized milk')
		(g) Switzerland 'special milk' at time of sale to customer (*Note* pasteurization status not certain from context)

Ingredient name: Milk (cows) – Liquid raw (not pasteurized)

Organism	Numbers	Notes
APC	1×10^4[3](a) to 5×10^4[3](b) to usual range of 1 or 1.5×10^5 to 5×10^5, to 1 or 2×10^6[3](c) to 4×10^6[3](d)	Many countries have standards, see references [2] and [3] Examples given show a range of values
Coliform or Enterobacteria	50 in 100 ml [3](e) to usual range of 10 to 1×10^2 to 1×10^3	(a) Argentina (b) Northern Ireland
E. coli or Fecal coliform	0 in 0.01 ml [3](g); 10 in 100 ml [3](e)	(c) Canada, for manufacturing
Salmonella spp.	0 in 1 liter [3](f)	(d) Japan, for manufacturing (also for
Streptococci (beta haemolytic)	0 in 0.1 ml [3](f)	goat's milk) (e) Finland
Streptococci agalactiae (serological group B)	0 in 1 ml [3](h)	(f) France, on last date for consumption (g) South Africa – on milk for further processing (h) West Germany

Product name: Milk (cows) – Powder for babies

Organism	Numbers	Notes
APC	1×10^4[3](a)	(a) Italy
Coliform or Enterobacteria	0 in 1 g [3](a)	
Salmonella spp.	0 in 25 g [3](a)	
Staphylococcus aureus	0 in 1 g [3](a)	

Product name: Milk (cows) – Sterilized

Organism	Numbers	Notes
APC	12 [3](a); 1×10^2[3](b–d)	(a) Italy
Thermophiles	12 [3](a)	(b) Holland, after 5 days at 30°C (c) Spain, spores (d) France, 10 in 0.1 ml

Product name: Milk (cows) – Ultra high temperature (UHT) treated

Organism	Numbers	Notes
APC	1×10^2[3](a, b); 1.1×10^2[3](c); 1×10^3[3](d)	(a) France and UK, 10 in 0.1 ml
Thermophiles	1.1×10^2[3](c)	(b) Spain, spores
Yeasts and Molds	1×10^2[3](d)	(c) Italy (d) Netherlands, yeast and mold count after 10 days at 24°C in original pack

Ingredient name: Milk (goats) – Liquid raw (not pasteurized)

Organism	Numbers	Notes
APC	5×10^4[3](a); 1.5×10^5[3](b); 4×10^6[3](c)	(a) Scotland (also for sheep's milk)
Coliform or Enterobacteria	0 in 0.1 ml [3](b); 0 in 0.01 ml [3](a)	(b) Australian states (c) Japan, for manufacturing (also for cow's milk)

Product name: Mustard

Organism	Numbers	Notes
APC	1×10^5 [3](a)	(a) Spain
Coliform or Enterobacteria	10[3](a)	
Salmonella spp. or		
Shigella spp.	0 in 25 g[3] (a)	

Ingredient name: Nuts – Including nut butters and nut meats

Organism	Numbers	Notes
E. coli or Fecal coliform	Less than 0.36 per g in 2 of 10 samples or in less than 20% of cases if more than 10 samples taken[3](a)	(a) USA, for nut meats (b) For peanut and other nut butters; ICMSF use $n = 10$ if product is to be used without treatment to destroy microbes. Use $n = 20$ if used in a high-moisture food (c) USA for nuts, includes shelled and unshelled (d) US military criteria for peanut butter
Salmonella spp.	0 in 25 g[1](b)	
% 'Moldy' nuts	5 to 15 according to type[3](c)	
Aflatoxin (ppb)	20[2](d)	

Product name: Onion rings – Frozen, breaded

Organism	Numbers	Notes
APC	1×10^6 [4](a)	(a) Georgia guidelines
Coliform or Enterobacteria	1.1×10^3 [4](a)	
E. coli or Fecal coliform	Negative [4](a)	
Salmonella spp.	Negative [4](a)	
Staphylococcus aureus	Negative [4](a)	

Product name: Pasta – Frozen and dried

Organism	Numbers	Notes
APC	3×10^4 [3](b); 1×10^5 (?) [3](d)	(a) ICMSF state 'best estimate' for m
Coliform or Enterobacteria	50[3](d)	(b) Thailand, for flavored noodles and vermicelli
E. coli or Fecal coliform	0 in 1 g[3](d); 3[3](b)	(c) Dried spaghetti, for cannery use
Salmonella spp.	0 in 25 g[3](d)	(d) Brazil, pasta or macaroni dry, with or without filling
Staphylococcus aureus	$m = 10^2$ [1](a), $M = 10^4$ [1](a)	
Clostridia (*perfringens*)	20[3](d)	
Thermophilic spore count	Aerobic, average of 5 samples 30 per 2 g; anaerobic, 10(c)	
Molds	1×10^2 [3](b)	
Yeasts and Molds	Average of 5 samples 1×10^2(c); 5×10^2 [3](d)	

Product name: Pasta dumplings – Fresh, manufactured not home made

Organism	Numbers	Notes
APC	$m = 1 \times 10^5$, $M = 1 \times 10^6$ [12](a)	(a) Italian Ministry of Health limits quoted in paper
Salmonella spp.	0 in 25 g [12](a)	
Staphylococcus aureus	$m = 1 \times 10^2$, $M = 1 \times 10^3$ [12](a)	
Clostridia (perfringens)	$m = 1 \times 10^2$, $M = 1 \times 10^3$ [12](a)	

Product name: Pastry

Organism	Numbers	Notes
APC	24 h at 37°C, 1×10^5 [3](a)	(a) Singapore
Coliform or Enterobacteria	50 [3](a)	

Product name: Patisserie and patisserie cream

Organism	Numbers	Notes
APC	3×10^5 [3](a)	(a) France
Coliform or Enterobacteria	1×10^3 [3](a)	(b) Spain, patisserie products
E. coli or Fecal coliform	0 in 1 g [3](b); 1 [3](a)	
Salmonella spp.	0 in 25 g [3](a); Salmonella or Shigella, 0 in 30 g [3](b)	
Staphylococcus aureus	0 in 0.1 g [3](b); 1×10^2 [3](a)	
Clostridia (perfringens)	10 [3](a); 1×10^3 sulfite reducers in products with meat [3](b)	
Yeasts and Molds	5×10^2 in cereal-based products [3](b)	

Product name: Poultry – Cooked poultry entrees, pot pies and poultry rolls

Organism	Numbers	Notes
APC	At 35°C, not more than 1 of 5 over 5×10^4	USDA advisory criteria, see reference [2]
Coliform or Enterobacteria	Not more than 1 of 5 over 10	
E. coli or Fecal coliform	Not more than 1 of 5 positive	
Salmonella spp.	None of 5 positive	
Staphylococcus aureus	Not more than 1 of 5 positive	
Clostridia (perfringens)	Not more than 1 of 5 at 1×10^3 or more	

Ingredient name: Poultry meat – Cooked, frozen; including chicken and prepared dishes

Organism	Numbers	Notes
APC	5×10^5 [3](c)	(a) ICMSF, to be reheated – Salmonella $n = 5$; for ready to eat, Salmonella $n = 10$
E. coli or Fecal coliform	20 [3](c)	(b) ICMSF, use where products are packaged or repackaged after processing. $n = 5$ if temperature abuse is unlikely, use $n = 10$ if likely
Salmonella spp.	0 in 25 g [1](a)	(c) New Zealand, cooked chicken pieces (take away/fast food)
Staphylococcus aureus	1×10^2 [3](c); $m = 10^3$, $M = 10^4$ [1](b)	

Ingredient name: Poultry meat – Cooked including chicken

Organism	Numbers	Notes
APC	1×10^4 [3](a)	(a) South Africa – includes all poultry
Salmonella spp.	No salmonellae, shigellae or escherichiae [3](a)	intended for human consumption
Staphylococcus aureus	0 in 10 g [3](a)	
Clostridia (*perfringens*)	10 [3](a)	

Ingredient name: Poultry meat – Cooked, mechanically separated

Organism	Numbers	Notes
APC	3×10^5 [3](a)	(a) France
E. coli or Fecal coliform	10 [3](a)	
Salmonella spp.	0 in 25 g [3](a)	
Clostridia (*perfringens*)	As anaerobic sulfite reducers, 30 [3](a)	

Ingredient name: Poultry meat – Cured and/or smoked

Organism	Numbers	Notes
Salmonella spp.	0 in 25 g [1](b)	(a) ICMSF, If either packaged or
Staphylococcus aureus	$m = 10^3$, $M = 10^4$ [1](a)	repackaged after processing. $n = 10$
		(b) ICMSF, n = 10 aureus

Product name: Poultry meat and products – Dehydrated

Organism	Numbers	Notes
APC	$m = 7.5 \times 10^4$, $M = 1.5 \times 10^5$ [2](b)	(a) ICMSF, $n = 10$
E. coli or Fecal coliform	$m = 3$, $M = 20$ [2](b)	(b) Military criteria for chicken and
Salmonella spp.	0 in 25 g [1](a)	chicken products; chicken diced,
		compressed; chicken with rice

Ingredient name: Poultry meat – Frozen diced, for cannery use

Organism	Numbers	Notes
APC	Green-amber 10^5, amber-red 10^6(a)	Green-amber is similar to m, amber-red
Salmonella spp.	0 in 25 g	to M
Staphylococcus aureus	Green-amber 10^3, amber-red 10^4(a)	(a) The process of manufacture
Clostridia (*perfringens*)	Amber-red 10	distributes the original colonies into a
Thermophilic spore	Aerobic, green-amber 30 per 2 g, amber-red $1 \times$	larger number of colony-forming units
formers:	10^2 per 2 g	(cfus). This results in comparatively
	Anaerobic, amber-red 10	high numerical limits

Ingredient name: Poultry meat – Raw, mechanically separated

Organism	Numbers	Notes
APC	1×10^6[3](a)	(a) France
E. coli or Fecal coliform	5×10^3[3](a)	
Salmonella spp.	0 in 1 g[3](a)	
Staphylococcus aureus	1×10^3[3](a)	
Clostridia (*perfringens*)	Anaerobic sulfite reducers, 1×10^2[3](a)	

Ingredient name: Poultry meat, or chicken – Raw

Organism	Numbers	Notes
APC	10^3 to 10^4 per cm^2 [2] (a); $m = 5 \times 10^5$, $M = 10^7$[1](b); 1×10^6[4](e); 5×10^6[4](f)	(a) Usual range in USA for fresh processed carcasses
Coliform or Enterobacteria	50[4](f); 1×10^3[4](e)	(b) ICMSF, raw chicken during processing *not for port-of-entry use*
E. coli or Fecal coliform	Negative [4](f)	(c) France, whole poultry chilled or frozen
Salmonella spp.	Negative [4](f); 0 in 25 g[3](c),(d)	(d) New Zealand, fresh poultry
Staphylococcus aureus	Negative[4](f)	(e) Iowa
Yeasts	5×10^5[4](f)	(f) Nebraska
Molds	5×10^3[4](f)	

Product name: Pot pies – Frozen

Organism	Numbers	Notes
APC	1×10^5[4](a)	(a) Nebraska, guideline applies to frozen pot pies and frozen vegetables
Coliform or Enterobacteria	1×10^2[4](a)	
E. coli or Fecal coliform	10[4](a)	
Salmonella spp.	Negative[4](a)	
Staphylococcus aureus	1×10^2[4](a)	
Clostridia (*perfringens*)	Negative[4](a)	
Yeasts	1×10^5[4](a)	
Molds	1×10^4[4](a)	

Product name: Potato – Stuffed, baked, frozen

Organism	Numbers	Notes
APC	5×10^4[4](a)	(a) Georgia, guidelines
Coliform or Enterobacteria	3[4](a)	
E. coli or Fecal coliform	Negative [4](a)	
Staphylococcus aureus	Negative[4](a)	

Product name: Potato patties and hash browns – Frozen

Organism	Numbers	Notes
APC	5×10^4[4](a)	(a) Georgia, guidelines
Coliform or Enterobacteria	1×10^2[4](a)	
E. coli Fecal coliform	Negative[4](a)	
Salmonella spp.	Negative[4](a)	
Staphylococcus aureus	Negative[4](a)	

Product name: Pudding powders

Organism	Numbers	Notes
APC	5×10^4[3](a)	(a) Luxembourg, 'no pathogens'
Coliform or Enterobacteria	0 in 0.1 g[3](a)	
Salmonella spp.	0 in 25 g[3](a)	
Staphylococcus aureus	Coagulase positive, 0 in 0.1 g[3](a)	
Yeasts and Molds	1×10^2[3](a)	

Product name: Salads – Stuffed tomatoes etc., to be eaten cold

Organism	Numbers	Notes
APC	1×10^4[3](a, b); 4×10^4[3](h); 1×10^5[3](g); m $= 1 \times 10^7$[3](c–f)	(a) Netherlands
Coliform or Enterobacteria	1×10^2[2],[3](g, h); 1×10^3[3](a); $m = 1 \times 10^4$[3](c–e)	(b) Netherlands *unless inherent to the product* See also reference [2] page 293 where an APC of 1×10^5 would be less
Salmonella spp.	M = present in 10 g[3](c–e)	than load on raw vegetables (e.g. 10^6)
Staphylococcus aureus	1×10^2[3](g, h); $M = 1 \times 10^3$[3](c–e)	but high for gelatin salads or products
Yeasts	$m = 1 \times 10^4$[3](c, d)	made entirely of cooked ingredients
Yeasts and molds	20[3](g)	(c) Sweden, 'suitable with remarks' has been regarded as *m*; unsuitable as *M*.
		(d) Sweden, mayonnaise based, e.g. potato salad, shellfish salad
		(e) Sweden, raw vegetables without mayonnaise
		(f) Sweden, higher if sour milk is used
		(g) New Zealand guidelines, cole slaw, bean salad (take away/fast food)
		(h) New Zealand guidelines, potato salad

Product name: Salads – Various types, US states data

Organism	Numbers	Notes
APC	1×10^5[4](a, b); 5×10^6[4](c)	(a) Georgia, including meat
Coliform or Enterobacteria	1×10^2[4](b); 1×10^3[4](a); 5×10^3[4](c)	(b) New Jersey, eight types including poultry, egg, fish, pasta and
Salmonella spp.	Negative in 1 g[4](a, b)	vegetable(s)
Staphylococcus aureus	1×10^2[4](a, b); 1×10^3[4](c)	(c) Ohio, four types – meat, egg, potato and pasta
Shigella spp.,		
Enteropathogenic		
E. coli (EEC)	Negative in 1 g[4](b)	

Product name: Sandwiches – Prepacked

Organism	Numbers	Notes
APC	1×10^6[4](a)	(a) North Carolina: Cumberland County Board of Health
Coliform or Enterobacteria	1×10^3[4](a)	
E. coli or Fecal coliform	1×10^2[4](a)	
Salmonella spp.	Negative[4](a)	
Staphylococcus aureus	1×10^3[4](a)	
Yeasts and molds	5×10^4[4](a)	

Ingredient name: Seafood – Bivalves (shell fish), cleaned

Organism	Numbers	Notes
APC	5×10^5 per ml [3](a); 1×10^5[3](b)	(a) Spain
E. coli or Fecal coliform	500 per liter [3] (a); 0.5 per g within 24h of depuration [3](b)	(b) Australian states
Salmonella spp.	0 in 25 ml [3](a)	
Fecal Streptococci	As Lancefield group D 1×10^2[3](a)	
V. parahaemolyticus	1×10^2[3](a)	

Ingredient name: Seafood – Bivalve molluscs, fresh and frozen

Organism	Numbers	Notes
APC	$m = 5 \times 10^5$[1](a); 5×10^4[4](b); 1×10^5[4](c), [3](f); 5×10^5[2](d)	Includes mussels, clams, oysters and scallops
Coliform or Enterobacteria	230 per 100 g [2](d), [3](f); 2.4[4](b); 50[4](c)	(a) ICMSF recommend use of 25°C, see reference [1] page 186
E. coli or Fecal coliform	2.3 per 100 g [4](b, c); 4 per ml of flesh and body fluid [3](e); $m = 16$[1]	(b) Georgia guideline, clams, mussels and oysters fresh or frozen
Salmonella spp.	Negative [4](b, c)	(c) Georgia guidelines, scallops, breaded–frozen
Staphylococcus aureus	Negative [4](b, c)	(d) US military criteria for oysters, fresh (culled) and frozen (shucked)
		(e) Italy, edible molluscs
		(f) Australian states, oysters fresh, frozen and packaged fresh

Product name: Seafood – Crabmeat, cooked, chilled and frozen

Organism	Numbers	Notes
APC	2.5×10^4[4](d); 1×10^5[4](c, e) $m = 10^5$, $M = 10^6$[1](a)	(a) ICMSF recommend the use of 25°C, see reference [1] page 186
Coliform or Enterobacteria	1×10^2[4](c)	(b) If appropriate, ICMSF
E. coli or Fecal coliform	50 per 100 g [4](e); $m = 11$, $M = 5 \times 10^2$[1]; negative [4](c, d)	(c) Georgia guidelines, crabmeat – fresh cooked
Salmonella spp.	Negative [4](c)	(d) Maryland, crabmeat – pasteurized
Staphylococcus aureus	$m = 10^3$[1]; negative [4](c)	(e) Virginia, crabmeat
V. parahaemolyticus	$m = 10^2$, $M = 10^3$[1](b)	

Ingredient name: Seafood – Crustacea, France, criteria

Organism	Numbers	Notes
APC	1×10^3(a); 1×10^5(b, c)	See reference [3], order of 21 December 1979
E. coli or Fecal coliform	1 per g(a, b); 10(c)	(a) All crustacea, including shrimps or prawns, whole cooked or raw, chilled or frozen
Salmonella spp.	0 in 25 g (a–c)	(b) Whole crustacea, cooked, refrigerated, other than shrimps or prawns
Staphylococcus aureus	1×10^2(c)	(c) Prawns or shrimps, cooked, shelled and refrigerated, and shelled and chilled or frozen
Clostridia (perfringens)	As anaerobic sulfite reducers 2(a, b); 10(c)	

Product name: Seafood – Crustaceans, cooked, frozen

Organism	Numbers	Notes
APC	1×10^5[4](d); $m = 1 \times 10^5$, $M = 1 \times 10^6$[2](a); $m = 5 \times 10^5$, $M = 10^7$[1](b)	(a) FAO/WHO, precooked frozen shrimp and prawn.
Coliform or Enterobacteria	20[4](d)	Sample size for *Salmonella* not given in
E. coli or Fecal coliform	$m = 11$, $M = 5 \times 10^2$[1]; negative[4](d)	reference.
Salmonella spp.	0 (in 25 g) [2](a); 0 in 25 g[1](c); negative[4](d)	(b) ICMSF recommend the use of 25°C,
Staphylococcus aureus	$m = 5 \times 10^2$, $M = 5 \times 10^3$[2](a); $m = 10^3$[1]; negative[4](d)	see reference [1] page 186 (c) If appropriate, ICMSF
V. parahaemolyticus	$m = 10^2$, $M = 10^3$[1](c)	(d) Georgia guideline, for shrimp, peeled, cooked

Ingredient name: Seafood – Crustaceans, raw, frozen

Organism	Numbers	Notes
APC	1×10^4[4](c); 1×10^5[4](e); 5×10^5[3](g); 1×10^6[4](d, f); $m = 10^6$, $M = 10^7$[1](a)	(a) ICMSF recommend use of 25°C, see ref. [1] page 186
Coliform or Enterobacteria	Negative[4](c); 50[4](d); 1.1×10^3[4](f)	(b) If appropriate, ICMSF
E. coli or Fecal coliform	36 per 100 g[4](e); $m = 11$, $M = 5 \times 10^2$[1]; 20[3](g); negative[4](c),(d),(f)	(c) Georgia guideline, deviled crabs – frozen or cooked
Salmonella spp.	0 in 25 g[1](b); negative[4](c, d, f); no *Salmonellae*[3](g)	(d) Georgia guideline, deviled crabs fresh, uncooked
Staphylococcus aureus	1×10^2[3](g); $m = 10^3$, $M = 10^4$[1](b); negative[4](c, d, f)	(e) Maryland, crabmeat, fresh (f) Georgia guideline, shrimp, breaded, frozen, raw
V. parahaemolyticus	$m = 10^2$, $M = 10^3$[1](b)	(g) Papua New Guinea, prawns and shrimps

Product name: Seafood – Fish conserves

Organism	Numbers	Notes
APC	1×10^4(?) [3](a); 5×10^5(?) [3](c)	(a) Brazil, raw
E. coli or Fecal coliform	1×10^2[3](a); 20[3](c)	(b) Brazil, heated treated and packed –
Salmonella spp.	0 in 25 g[3](a, c)	after incubation for 10 days at 35°C
Staphylococcus aureus	0 in 0.01 g [3](a); 0 in 0.1 g [3](c)	should show no microbiological
Clostridia (*perfringens*)	As sulfite-reducing clostridia at 44°C, 20[3](c)	changes to package or any changes in the characteristics of the product
Yeasts and molds	1×10^3(?) [3](a); 5×10^2[3](c)	(c) Brazil, other fish conserves

Ingredient name: Seafood – Fish, lightly salted and smoked, e.g. salmon, haddock

Organism	Numbers	Notes
APC	1×10^6[3](a)	(a) France
Coliform or Enterobacteria	0 in 1 g[3](a)	
Salmonella spp.	0 in 25 g[3](a)	
Staphylococcus aureus	1 per g[3](a)	
Clostridia (*perfringens*)	As anaerobic sulfite reducers, 0 in 1 g[3](a)	

Ingredient name: Seafood – Fish pulp

Organism	Numbers	Notes
Coliform or Enterobacteria	1×10^3 [3](a)	(a) Chile. No *E. coli*, *Staph. aureus*, *Salmonella* or *Arizona*

Ingredient name: Seafood – Fresh, frozen or cold smoked *not* crustaceans or molluscs

Organism	Numbers	Notes
APC	$m = 5 \times 10^5$, $M = 10^7$ [1](a); 3×10^6(?) [3](c); 1×10^6 [4](d), [3](e)	(a) ICMSF recommend APC at 25°C as 35°C can be about one-tenth that of 25°C. See reference [1] page 186
E. coli or Fecal coliform	10 [3](e); m = 11, $M = 500$ [1]; 1×10^2 [3](c, e); negative [4](d)	(b) If appropriate, ICMSF
Salmonella spp.	0 in 25 g [1](b), [3](c)	(c) Brazil, fish
Staphylococcus aureus	10 [3](e); $m = 10^3$, $M = 10^4$ [1](b); 0 in 0.01 g [3](c)	(d) Rhode Island guidelines for fresh seafood
Clostridia (*perfringens*)	20 [3](c)	(e) Chile, fish and shellfish. No *Salmonella* or *Arizona*
V. parahaemolyticus	$m = 10^2$, $M = 10^3$ [1](b)	

Ingredient name: Seafood – Fresh fish, and also crustacea, molluses, marine or fresh water mammals

Organism	Numbers	Notes
APC	1×10^6 [3](a)	(a) Spain
Coliform or Enterobacteria	2.5×10^2 [3](a)	
E. coli or Fecal coliform	As *E. coli* type 1, 0 in 1 g [3](a)	
Salmonella spp.	0 in 25 g [3](a)	
Staphylococcus aureus	As coagulase and DNAse positive, 0 in 5 g [3](a)	
Fecal Streptococci	As enterococci, 1×10^3 [3](a)	

Ingredient name: Seafood – Fresh fish fillets or sliced fish

Organism	Numbers	Notes
APC	1×10^4 [3](a); 1×10^5 [3](b); 10^6 [8](c)	(a) France, sliced fish, breaded (crumbed) or not or fish fillets, frozen or chilled
Coliform or Enterobacteria	10^3 [8](c)	(b) France, sliced fish, breaded or not, or fish fillets, refrigerated
E. coli or Fecal coliform	1 per g [3](a); 10 [3](b), [8](c)	(c) For fresh fish fillets; APC ref. value may be exceeded by a maximum of 2/10 samples but none should exceed 10%. See reference [8] page 93
Salmonella spp.	0 in 25 g [3](a, b), [8](c)	
Staphylococcus aureus	1×10^2 [3](a, b); 10^2 [8](c)	
Clostridia (*perfringens*)	As anaerobic sulfite reducers, 2 [3](a); 10 [3](b)	
V. parahaemolyticus	0 in 25 g [8](c)	

Product name: Seafood – Fried fishcakes, frozen

Organism	Numbers	Notes
APC	1×10^4 [4](a)	(a) Georgia, guideline
Coliform or Enterobacteria	Negative [4](a)	
E. coli or Fecal coliform	Negative [4](a)	
Salmonella spp.	Negative [4](a)	
Staphylococcus aureus	Negative [4](a)	

Product name: Seafood – Molluscs, fried and frozen

Organism	Numbers	Notes
APC	1×10^4[4](a)	(a) Georgia guidelines, clams and
Coliform or Enterobacteria	Negative [4](a)	scallops
E. coli or Fecal coliform	Negative [4](a)	
Salmonella spp.	Negative [4](a)	
Staphylococcus aureus	Negative [4](a)	

Ingredient name: Seafood – Mussels (imported, Netherlands)

Organism	Numbers	Notes
E. coli or Fecal coliform	3 per ml flesh [3](a)	(a) For import into the Netherlands
Salmonella spp.	0 in 25 ml flesh [3](a)	

Product name: Seafood – Mussels, cooked and frozen

Organism	Numbers	Notes
APC	1×10^4[3](a)	(a) Spain
Coliform or Enterobacteria	10[3](a)	
E. coli or Fecal coliform	0 in 1 g [3](a)	
Salmonella spp.		
or Shigella spp.	0 in 25 g[3](a)	
Staphylococcus aureus	As enterotoxigenic staphylococci, 1×10^2[3](a)	

Product name: Seafood – Mussels and scallops, precooked

Organism	Numbers	Notes
APC	1×10^6[3](a)	(a) France
E. coli or Fecal coliform	10[3](a)	
Salmonella spp.	0 in 25 g[3](a)	
Staphylococcus aureus	1×10^2[3](a)	
Clostridia (*perfringens*)	As anaerobic sulfite reducers, 30[3](a)	

Product name: Seafood – Octopus, boiled

Organism	Numbers	Notes
APC	1×10^5[3](a)	(a) Japan
Coliform or Enterobacteria	No coliforms[3](a)	

Product name: Seafood – Prawns, cooked, frozen (imported, UK)

Organism	Numbers	Notes
APC	1×10^5[3](a), 1×10^6[3](b)	British Frozen Foods Federation
E. coli or Fecal coliform	10[3]	guidelines
Salmonella spp.	0 in 25 g [3]	(a) Release unconditionally
Staphylococcus aureus	1×10^3[3]	(b) Release with a warning to use
		immediately on thawing

Product name: Seafood – Pre-cooked breaded fish

Organism	Numbers	Notes
APC	2.5×10^4[4](a); $m = 5 \times 10^5$, $M = 10^7$[1](c)	(a) Georgia guideline for fried,
Coliform or Enterobacteria	10[4](a)	breaded, frozen fish
E. coli or Fecal coliform	$m = 11$, $M = 5 \times 10^2$[1]; negative[4](a)	(b) ICMSF, if appropriate
Salmonella spp.	Negative[4](a)	(c) ICMSF recommend the use of 25°C,
Staphylococcus aureus	$m = 10^3$, $M = 10^4$[1](b); negative [4](a)	see reference [1] page 186

Ingredient name: Seafood – Preparations based on raw minced fish

Organism	Numbers	Notes
APC	5×10^5[3](a)	(a) France
E. coli or Fecal coliform	1×10^2[3](a)	
Salmonella spp.	0 in 25 g[3](a)	
Staphylococcus aureus	1×10^2[3](a)	
Clostridia (perfringens)	As anaerobic sulfite reducers 10[3](a)	

Product name: Seafood – Raw, breaded, frozen fish

Organism	Numbers	Notes
APC	1×10^5[4](a); $m = 8.5 \times 10^4$, $M = 5.4 \times 10^5$[3](b)	(a) Georgia, guidelines
Coliform or Enterobacteria	1×10^2[4](a); $m = 23$, $M = 2.3 \times 10^2$[3](b)	(b) US recommendations for frozen fish
E. coli or Fecal coliform	Negative [4](a)	sticks
Salmonella spp.	Negative [4](a)	
Staphylococcus aureus	Negative[4](a)	

Ingredient name: Seafood – Raw or partly cooked marine foods

Organism	Numbers	Notes
APC	48h at 35°C, 1×10^6[3](a, b)	(a) South Africa, also no *Salmonella*,
E. coli or Fecal coliform	As *E. coli* type 1, 10 per 100 g[3](a) or 5×10^2	*Shigella*, or *Vibrio cholerae*
	per 100 g[3](c)	(b) South Africa, not applicable to
Staphylococcus aureus	Coagulase +ve, 10 per g[3](a)	oysters, mussels or clams
		(c) South Africa, for oysters, mussels or
		clams

Product name: Seafood – Ready-to-eat

Organism	Numbers	Notes
APC	5×10^4 (24 h at 37°C) [3](a)	(a) Singapore – cooked crab meat,
	1×10^5 (24 h at 37°C) [3](b)	prawns, shrimps, molluscs
Coliform or Enterobacteria	50 (48 h at 37°C) [3](a, b)	(b) Singapore – fish

Ingredient name: Seafood – Rock lobster

Organism	Numbers	Notes
APC	5×10^5 [3] (a)	(a) New Zealand guidelines
E. coli or Fecal coliform	20[3](a)	
Salmonella spp.	0 in 25 g[3](a)	
Staphylococcus aureus	1×10^2[3](a)	

Product name: Seafoods – Semi-preserved

Organism	Numbers	Notes
APC	1×10^4[3](a), 1×10^5[3](b)	(a) France, pasteurized
Coliform or Enterobacteria	0 in 1 g[3](a, b)	(b) France, non-pasteurized, e.g.
Salmonella spp.	0 in 25 g[3](a, b)	rollmop herrings, red herrings,
Staphylococcus aureus	0 in 1 g[3](a, b)	anchovies in salt or oil
Clostridia (*perfringens*)	As anaerobic sulfite reducers, 0 in 1 g[3] (a, b) to 10 for anchovies in brine	

Ingredient name: Seafood – Shellfish, fresh or frozen

Organism	Numbers	Notes
APC	At 35°C 5×10^5[2](a),[3](h) At 20°C; 1×10^5[1](c)	(a) US National Shellfish Sanitation Program (NSSP) for product received at wholesale market.
Coliform or Enterobacteria	2.3[3](h); 2.4×10^3 per 100 g [4](b); usual range 2 to 10 per g or ml[1](d)	Used also by individual states. See reference [2] page 249-255. Note fecal
E. coli or Fecal coliform	230 per 100 g[2](a); 39 per ml flesh and body liquid[3](g)	coliform by MPN, also New Zealand guidelines[3]
Salmonella spp.	0 in 25 g [1](e) ,[3](h)	(b) Massachusetts guidelines per 100 g
Paralytic shellfish poisoning (PSP) toxin	80µg per 100 g[2](f)	(c) Denmark – Molluscs, see reference [1] page 186 (d) European countries, molluscs, see reference [1] page 186 (e) France, molluscs, see reference [1] page 186 (f) US state action level, see reference [2] page 252 (g) Italy, *E. coli* in fish other than lamellibranch molluscs (h) New Zealand guidelines

Ingredient name: Seafood – Shellfish, live, bivalves and sea-urchins

Organism	Numbers	Notes
E. coli or Fecal coliform	3×10^2 per 100 ml[3](a)	(a) France, fecal coliform specified
Salmonella spp.	0 in 25 g [3](a)	ICMSF[1] page 188, gives APC at 25°C
Fecal Streptococci	2.5×10^3 per 100 ml[3](a)	at harvest as 10^2 to 10^5

Product name: Soft drinks – Miscellaneous

Organism	Numbers	Notes
APC	$2 \times 10^2[4](c)$; $5 \times 10^3[3](a, d)$; $1 \times 10^4[3](b)$	(a) Portugal
Coliform or Enterobacteria	0 in 1 ml[3](b)	(b) Spain, sugar syrup, fruit juice/pulp
E. coli or Fecal coliforms	0 in 10 ml[3](a)	syrups, flavored syrups, fantasy syrups
Salmonella spp.	0 in 25 g[3](b)	(c) Ohio, carbonated beverages
Yeasts	50[4](c); $5 \times 10^2[4](d)$	(d) Ohio, pasteurized cider and juices
Molds	1 in 10 ml[3](b); 50[4](c); $1 \times 10^3[4](d)$	

Ingredient name: Soya flours, concentrates, isolates and proteins

Organism	Numbers	Notes
APC	$5 \times 10^4[3](a, b)$; $m = 1 \times 10^5$, $M = 10^7[3](c)$; $2 \times 10^5[3](d)$	(a) Netherlands, soya products for bread
Coliform or Enterobacteria	0 in 1 g[3](b); 0 in 0.1 g[3](a); $1 \times 10^2[3](d)$	(b) Brazil, protein concentrate, soya isolate
E. coli or Fecal coliforms	10[3](d); $m = 10$, $M = 1 \times 10^2[3](c)$	(c) Italy, soya flours and proteins
Salmonella spp.	0 in 50 g [3](b, d); 0 in 25 g[1](e); 0 in 1 g[3](c)	(d) Brazil, soya flour – defatted
Molds	$m = 10^2$ to 10^4, $M = 10^5[1](e)$	(e) ICMSF, soya flours, concentrates
Yeasts and Molds	$1 \times 10^2[3](a)$; $1 \times 10^3[3](b, d)$	and isolates; m for mold count is estimate based on limited data

Product name: Soya milk – Dried

Organism	Numbers	Notes
APC	$1 \times 10^5[3](a)$	(a) Thailand

Ingredient name: Soya protein – texturized

Organism	Numbers	Notes
APC	$2 \times 10^4[3](a)$	(a) Brazil
Coliform or Enterobacteria	0 in 1 g[3](a)	
Salmonella spp.	0 in 50 g[3](a)	
Thermophilic spores	$1.5 \times 10^2[3](a)$	
Yeasts and Molds	$1 \times 10^2[3](a)$	

Product name: Soups – Dried

Organism	Numbers	Notes
APC	$3 \times 10^5[3](a)$	(a) France
Coliform or Enterobacteria	$1 \times 10^3[3](a)$	
E. coli or Fecal coliforms	10[3](a)	
Salmonella spp.	0 in 25 g[3](a)	
Staphylococcus aureus	$1 \times 10^2[3](a)$	
Clostridia (*perfringens*)	As anaerobic sulfite reducers, 30[3](a)	

Product name: Soups – Dried, broths or bouillon

Organism	Numbers	Notes
APC	1×10^5[3](a)	(a) Argentina
Coliform or Enterobacteria	1×10^3[3](a)	(b) Thailand, also 'no pathogens or
E. coli or Fecal coliforms	3[3](b); 10[3](a)	toxins'
Salmonella spp.	0 in 25 g [3](a)	
Staphylococcus aureus	1×10^2[3](a)	
Clostridia (*perfringens*)	10[3](a)	
Molds	1×10^2[3](b)	

Product name: Soups – Dried, instant

Organism	Numbers	Notes
APC	1×10^5[3](a, b, d)	(a) New Zealand
Coliform or Enterobacteria	1×10^3[3](b)	(b) Argentina 'cooked products'
E. coli or Fecal coliforms	10[3](b–d); 20[3](a)	(c) Spain
Salmonella spp.	0 in 25 g [3](b, c)	(d) Switzerland (ready-to-eat)
Staphylococcus aureus	1×10^2[3](c, d); 1×10^3[3](a, b)	
Clostridia (*perfringens*)	10[3](b, c); 1×10^2[3](a)	
B. cereus	1×10^2[3](a)	

Product name: Soups – Dried, to be cooked before eating

Organism	Numbers	Notes
APC	1×10^6[3](a)	(a) New Zealand
E. coli or Fecal coliform	1×10^2[3](a, d)	(b) Spain
Salmonella spp.	0 in 25 g[3](a–c)	(c) Argentina
Staphylococcus aureus	1×10^2[3](a–c); 1×10^3[3](d)	(d) Switzerland
Clostridia (*perfringens*)	10[3](b, c); 1×10^2[3](a)	
B. cereus	1×10^3[3](a)	

Ingredient name: Spices, herbs or condiments

Organism	Numbers	Notes
APC	5×10^5[3](c); 1×10^6[4](d)	(a) Spain, spices and condiments;
Coliform orEnterobacteria	5×10^2[4](d)	clostridia given as anaerobic sulfite-
E. coli or Fecal coliforms	Negative[4](d); 10[3](a); 1×10^2[3](c);	reducing spore formers
	2×10^2[3](b)	(b) Brazil, pot herbs
Salmonella spp.	0 in 25 g [3](a, b); 0 in 50 g[3](c)	(c) New Zealand, spices, natural;
Staphylococcus aureus	1×10^2[3](c)	clostridia given as *Cl. perfringens*
Clostridia (*perfringens*)	1×10^2[3](c); 1×10^3[3](a)	(d) Rhode Island guideline

Ingredient name: Sugar and molasses

Organism	Numbers	Notes
APC	1×10^2 per 10 g [4](b); 2×10^2 per 10 g [4](c); 5×10^4(d)	(a) US specification (1971) for sugar to be used in soft drinks
E. coli or Fecal coliforms	0 in 1 g [3](e)	(b) Ohio guidelines, bottlers' sugar, liquid
Salmonella spp.	0 in 25 g [3](e)	(c) Ohio guidelines, bottlers' sugar, granulated
Yeasts	10 per 10 g [2](a); 10 [4](b, c); 1×10^4 [4](d)	(d) Ohio guidelines, dry or liquid sugar
Molds	10 per 10 g [2](a); 10 [4](b, c); 1×10^4 [4](d)	(e) Brazil, molasses, hard brown sugar
Yeasts and Molds	5×10^2 [3](e)	

Product name: Tea and derivatives

Organism	Numbers	Notes
APC	At $31 \pm 1°C$, 1×10^6 [3](a)	(a) Spain
E. coli or Fecal coliform	10 [3](a, b)	(b) Brazil
Salmonella spp. or *Shigella* spp.	0 in 25 g [3](a)	
B. cereus	1×10^3 [3](a)	
Molds	1×10^4 [3](a)	

Product name: Tomato ketchup and sauces

Organism	Numbers	Notes
APC	1×10^4 [3](a)	(a) Spain, tomato ketchup
Coliform or Enterobacteria	10 [3](a)	(b) Spain, 'tomato frito' sauce
Salmonella spp. or *Shigella* spp.	0 in 25 g [3](a, b)	(c) Howard mold count 40 to 50% of fields is specified by several countries – see reference [3]
B. cereus	10 [3](b)	

Product name: Topping – Dessert and bakery products, frozen or dehydrated

Organism	Numbers	Notes
APC	1×10^4 [2](a)	(a) US military criteria
Coliform or Enterobacteria	10 [2](a)	
Salmonella spp.	0 in 25 g [2](a)	

Ingredient name: Vegetables – Dried

Organism	Numbers	Notes
APC	1×10^5 [1](a) 3×10^5 [3](b)	(a) Rhode Island (b) New Zealand guidelines (c) ICMSF state that *m* is an estimate
Coliform or Enterobacteria	1×10^2 [4](a)	
E. coli or Fecal coliform	1×10^2 [3](b); $m = 10^2$, $M = 10^3$ [1](c); negative [4](a)	
Salmonella spp.	0 in 25 g [3](b)	
Staphylococcus aureus	1×10^2 [3](b)	
Clostridia (*perfringens*)	Negative [4](a)	
B. cereus	1×10^3(b)	

Ingredient name: Vegetables – Fresh

Organism	Numbers	Notes
APC	10^6[2](c)	Consider the effect of irrigation water
E. coli or Fecal coliform	10[3](b); 2×10^2[3](a)	and/or washing
Salmonella spp.	0 in 25 g[3](a, b)	(a) Brazil, applies to vegetables, fruit
		('verduras', 'legumes'), roots,
		rhizomes, tubers, mushrooms
		(b) Brazil, unprocessed 'general' food
		of vegetable origin
		(c) See note on page 293, which states
		that the natural flora of raw vegetables
		can easily exceed this figure

Ingredient name: Vegetables – Frozen

Organism	Numbers	Notes
APC	1×10^5[4](a); 5×10^5[3](c)	Remember that 'clumping' of
Coliform or Enterobacteria	1×10^2[4](a); 5×10^2[3](c)	microorganisms considerably affects
E. coli or Fecal coliform	0 in 1 g[3](f); 10[4](a), [3](d); 50[3](e);	the count in frozen vegetables
	$m = 10^2$, $M = 10^3$[1](b)	(a) Nebraska guideline, applies to
Salmonella spp.	Negative[4](a); 0 in 50 g[3](c); 0 in 25 g[3](d–f)	frozen pot pies and frozen vegetables
Staphylococcus aureus	1×10^2[4](a), [3](c, e)	(b) ICMSF state *m* is an estimate
Clostridia (*perfringens*)	Negative[4](a); 20[3](e, f)	(c) Israel, frozen fruits and vegetables
Fecal Streptococci	As Enterococci, 1×10^3[3](c)	(d) Brazil, unprocessed foods of
Yeasts	1×10^5[4](a)	vegetable origin
Molds	1×10^4[4](a)	(e) Brazil, partly prepared foods of
		vegetable origin
		(f) Brazil, prepared foods of vegetable
		origin

Product name: Vegetables – Prepared

Organism	Numbers	Notes
APC (Psychrotropic)	10^5[8](a)	(a) The value of 10^5 may be exceeded
E. coli orFecal coliform	10[8]	by a maximum of 2 out of 10 samples,
Fecal Streptococci	As Lancefield group D, 10[8]	but neither should be over 10^6

Ingredient name: Vegetable protein – Hydrolyzed

Organism	Numbers	Notes
APC	5×10^4[3](a)	(a) Brazil
Coliform or Enterobacteria	0 in 1 g[3](a)	
Salmonella spp.	0 in 50 g[3](a)	
Yeasts and molds	1×10^3[3](a)	

Ingredient name: Vinegar – Pasteurized

Organism	Numbers	Notes
APC	30[3](a)	(a) Netherlands, 'sterile' vinegar should have no microorganisms

Ingredient name: Water – Drinking (potable)

Organism	Numbers	Notes
APC	At 37°C, $m = 10$[2](a); $m = 5$, $M = 20$[2](b) At 22°C, $m = 1 \times 10^2$[2] (a); $m = 20$, $M = 1 \times 10^2$[2](b)	Reference [2] gives these EEC standards. Note all pathogenic bacteria, viruses, algae or parasites to be absent
Coliform or Enterobacteria	10 per 100 ml[3](d)	(a) Tap water
E. coli or Fecal coliform	$M = 1$[2](c); 0 in 100 ml[3](d)	(b) Bottled water
Salmonella spp.		(c) By MPN
or Shigella spp.	0 in 5 liters[3](e)	(d) Malaysia Fecal coliform
Clostridia (*perfringens*)	As sulfite-reducing clostridia, $M = 1$[2](c)	(e) Switzerland

Product name: Water – Mineral

Organism	Numbers	Notes
APC	37°C for 24h, 50 per ml[3] (a); 5×10^2 per ml; 1×10^2 per ml (source water only)[3](b)	(a) Argentina, also no pathogens or parasites
Coliform or Enterobacteria	0 in 250 ml[3](a); 0 per 100 ml[3](b); 2.2 per 100 ml[3](c)	(b) Switzerland (c) Thailand
E. coli or Fecal coliform	0 in 250 ml[3](a); none [3](c)	
Fecal Streptococci	As enterococci, 0 in 250 ml[3](a); 0 in 100 ml[3](b)	
Pseudomonas aeruginosa	0 in 250 ml[3](a); 0 in 100 ml[3](b)	

Product name: Water – Natural mineral

Organism	Numbers	Notes
APC	At 37°C 20 per ml [2] (a); 5 per ml [2](a, b) At 22°C 1×10^2 per ml[2] (a); 20 per ml[2] (a, b)	(a) Reference [2] gives these EEC standards
Coliform or Enterobacteria	Including *E. coli*; absent in 250 ml[2] (a, b)	Parasites and pathogenic
Clostridia	As sulfite-reducing anaerobes, absent in 50 ml[2](a, b)	microorganisms should be absent (b) Codex standards also given in
Fecal Streptococci	Absent in 250 ml[2](a, b)	reference [2] are essentially the same
Pseudomonas aeruginosa	Absent in 250 ml[2] (a, b)	but add APC at source 21°C for 72h and
pH	3.5 max[1](c)	37°C for 24h (c) ICMSF gives this pH for *carbonated* waters

Ingredient name: Water – Treated and untreated

Organism	Numbers	Notes
APC Coliform or Enterobacteria Fecal Streptococci	$20[3](a); 1 \times 10^2[3](b); 3 \times 10^2[3](c, d)$ 0 in 100 ml [3](d) As enterococci, 0 in 100 ml [3](d)	Switzerland data (a) After treatment (b) At source (before treatment) (c) In the distribution network (d) Applies to treated and untreated water

Ingredient name: Whey – Dried

Organism	Numbers	Notes
APC Coliform or Enterobacteria	$5 \times 10^4[2](a); 1 \times 10^5[3](b);$ $2 \times 10^5[3](c)$ $5[3](b); 10[2](a); 25[3](c)$	(a) US extra grade (b) France, Hatmaker 'A' (c) France, Hatmaker 'B'

Ingredient name: Yeast

Organism	Numbers	Notes
Coliform or Enterobacteria *E. coli* or Fecal coliform *Salmonella* spp. Rope spores	0 in 0.1 g [3](a) 0 in 1 g [3](a) 0 in 25 g [1](b); 0 in 50 g [3](a) Type I, $1 \times 10^2[2](c)$; Type II, $2 \times 10^2[2](c)$	(a) Brazil, 'biological yeast cultures' (b) ICMSF, $n = 20$ (c) US military criteria for baker's yeast, compressed Type I and baker's yeast, active, dry, Type II

Product name: Yogurt

Organism	Numbers	Notes
Coliform or Enterobacteria *E. coli* or Fecal coliform *Staphylococcus aureus* Viable lactic acid bacteria Yeasts Molds Yeasts and molds	$10[3](b, c, e)$ 0 in 10 ml [3](d); 2 out of 5 replicates of 0.1 ml may contain *E. coli* [3](a); 1 per g of fecal coliforms [3](b) 0 in 25 g [3](b) $1 \times 10^6 \, minimum [3](a)$ $1 \times 10^2[3](e)$ $1 \times 10^2[3](e)$ $1 \times 10^3[3](c)$	(a) New Zealand guidelines (b) France, for fermented milks (yogurt, Kefir) (c) Switzerland, for fermented milks except Kefir (d) New South Wales, yogurt (e) Australian states

References

1. ICMSF (1986) *Microorganisms in Foods 2,* 2nd edn. Toronto: University of Toronto Press
2. NAS (1985) *An Evaluation of the Role of Microbiological Criteria for Foods and Food Ingredients.* National Academy of Sciences, Washington, DC: National Academy Press
3. Marshall, J. P., (1986) *Food Legislation Surveys Number 9. Microbiological Standards for Foodstuffs,* 2nd edn. Leatherhead: The British Food Manufacturing Industries Research Association
4. Wehr, H. M., (1982) Attitudes and policies of governmental agencies on microbial criteria for foods – an update. *Food Technology, 36,* (9), 45–54 + 92
5. Collins-Thompson, D. L., Weiss, K. F., Riedel, G. W. and Charbonneau, S. (1980) Microbiological guidelines and sampling plans for dried infant cereals and powdered infant formula from a Canadian National Microbiological Survey. *Journal of Food Protection,* **43**, 613–616
6. Mates, A. (1983) Microbiological survey of frozen ground meat and a proposed standard. *Journal of Food Protection,* **46**, 87–89
7. Todd, E. C. D., Jarvis, G. A. , Weiss, K. F., Reidel, G. W. and Chardonneau, S. (1983) Microbiological quality of frozen cream-type pies sold in Canada. *Journal of Food Protection,* **46**, 34–40
8. Mossel, D. A. A. (1982) *Microbiology of Foods,* 3rd edn, The University of Utrecht, Faculty of Veterinary Medicine, Utrecht, The Netherlands, 96
9. Hosell, S. and Randall, B. (1987) Bacteriological quality of takeaway lunches in Christchurch. *Food Technology in New Zealand,* **22**, 32–33
10. Wehr, H. M. (1978) Attitudes and policies of state governments. *Food Technology, 32,* (1), 63–67
11. Warburton, D. W., Weiss, K. F., Purvis, U. and Hill, R. W. (1987) The microbiological quality of fermented sausage produced under good hygienic practices in Canada. *Food Microbiology,* **4**, 187–197
12. Trovatelli, L. D., Scheisser, A., Massa, S., Cesaroni, D. and Poda, G. (1988) Microbiological quality of fresh pasta dumplings sold in Bologna and the surrounding district. *International Journal of Food Microbiology,* **7**, 19–24

13.D Ingredients – defects and foreign materials

Background

While perfection is an ideal, experience shows that all biological materials are variable and ingredient materials are likely to have some proportion of defects and foreign materials. The term 'defect' needs to be clearly understood. It means an undesirable property for a particular purpose. For example, a large size of navy bean may be wanted for one product but not another. If it is not wanted it is a 'defect'. More usually, defects are thought of as, for example, bruised, damaged, blemished or discolored fruit. Here, it is understood that some defective material is inevitable in a consignment and a criterion is needed which separates the acceptable from the unacceptable for a particular process and product.

Information on specific defect criteria is not readily available compared with that for microbiological criteria. Usually the source of the material comes from associations of producers, local commodity trade associations or from government sources which describe grades applicable to agricultural produce from an area, province, state or country. One way of looking at this is to regard them as statements of Good Agricultural or Commercial Practice. Note that the term 'Standard' does not mean what it does microbiologically and may be nearer to a 'grade'. As examples of this material, consider navy beans. Northarvest standards for Dry Edible Beans (Minnesota and North Dakota) provided by North Central Bean Dealers Association and Northarvest Growers Association have been published (November 26, 1985). Other standards have been issued by Michigan Department of Agriculture – Food Division (January 22, 1987), the US Department of Agriculture (August 1, 1989) and by the Canadian Grain Commission (August 1, 1980) among others. Criteria developed by users with participation of producers are much less common. However, UK users in association with the Campden Food Preservation Research Association produced the Campden Standard for Dry Pea Beans for Canning dated July 1982. The drafting group included producers' representatives from Michigan and Ontario.

While such a collaborative enterprise is very welcome and deserves every encouragement, in most circumstances the user must develop his own criteria for his products and processes. Because of the variety of ingredients and circumstances only a few examples are given in the defects and foreign material data table to indicate the kinds of criteria in use. It is usual in a specification to consider defects and foreign material together.

Development of criteria – defects and foreign materials

With the present state-of-the-art it should be appreciated that first-hand practical experience of an ingredient or commodity is invaluable in developing criteria. Quite simply, there is no substitute for experience. This is valuable not only in determining the significance of a defect or foreign material but also in warning of the occasional but potentially serious event which needs to be documented.

The HACCP approach is the obvious one to use in assessing what are the likely defects and foreign materials and their significance for the product/process/packaging system in which the ingredient will be used. It is important to distinguish between those which affect safety and those which influence sensory properties (or elegance factors). Sensory properties, of course, are not trivial if they affect sales but safety is of prime importance.

When making the HACCP analysis it is necessary to consider both the inherent properties of the material, e.g. stones in navy beans, foreign seeds in lentils, blemished fruit and vegetables, and also the contribution of the ingredient packaging system. This includes a possibility of, for example, splinters from wooden lug boxes or crates; fibers, pieces of paper or laminate from sacks; staples or pieces of string from sack closures; dirt on the outside of sacks; swarf from opening of cans; entrapped plastic wrapper in blocks of frozen meat. HACCP analysis must link into other analyses, e.g. be aware of the possibility of materials picking up taints from previous loads in the truck, railcar or container used for transportation of the ingredient material.

It is critically important to develop the HACCP analysis in sufficient detail to deliver what is expected of it. Failure to do so will lead to a false sense of security and hence inability to take the necessary preventive action.

As a very simple example of the level of inquiry that must be made, consider one small aspect of the use of apples for making apple puree. The size range of apples may seem to be of no consequence. However, black specks – which are unwanted – may be formed from the disintegration of calyces when the apples are pulped. The proportion of calyx material to the whole is clearly less with larger apples, so the size range is, in fact, significant.

Another instance of the need for careful consideration is meat for manufacturing purposes; gristle and bones are the main inherent defects although others, e.g. hide, hair, blood clots and excess fat, should be considered. As an example of foreign material which has been found on meat delivered to the processor, be aware that the tickets used to identify the carcass, which may be of paper (e.g. raffle or cloakroom tickets), or plastic tags or of collagen, may occasionally be left on meat together with, occasionally, a skewer or meat hook.

As an example of the very unexpected and most unusual occurrence, it has been known for green (toxic) woody nightshade berries to be present among a delivery of frozen peas. An instance is also known of glass from a bottle (thrown over a hedge?) being included in a root vegetable after harvesting.

Having decided on the defect or foreign material, determined its significance and assessed the control option(s), a criterion requires a sampling plan to be devised. It is really important to appreciate the limitations on what the chosen plan will deliver in relation to the perceived need in order to establish the cost/benefits.

Next, the method of examination or analysis must

be considered and its accuracy determined under practical conditions. Remember that large-scale visual inspection may give different results on a day-to-day basis to those achieved on a single experimental trial. Be sure that any numerical criteria are realistically related to the amount of material which is examined.

Although a discussion on methods is outside the scope of this book there is a good summary of the uses of Microscopic Mold Counts by the NRC[1].

As a result of these studies, the criterion or criteria can be developed. These should not be regarded as permanent since circumstances change. For example, three decades ago in the UK, sand was sometimes a problem in imported tomato paste or 'puree'. It is not so now.

Notes on data table

Note that the material in the data table is not necessarily used or recommended by Heinz. It may not be suitable for a particular process or product but it shows some ways in which criteria may be presented.

Index to defects data table

Apples, dessert – fresh (Worcester variety)
Apple pulp, sulfited
Apples, solid pack, canned
Cabbage, white – fresh
Carrots – fresh
Red lentils, unpolished, split
Sweet corn – frozen
Sultanas – pre-cleaned

Defects and foreign material data table

Data	Notes or comment
Apples, dessert – fresh (Worcester variety)	
Size – number of apples averaged on a single bulk box not greater than 160 per 13.6 kg (30 lb.)	Size can also be defined as a minimum diameter with a suitable, say 5%, tolerance
Defects – 5% by weight maximum for bruising, disease, insect damage, etc. requiring removal by trimming. Hail damage is acceptable providing the skin is intact and has not formed a scab	

Defects and foreign material data table (continued)

Data	Notes or comment
Apple pulp sulfited	
Dark specks, number per sample of 20 g: <41 – delivery acceptable 41 to 80 – delivery acceptable but supplier advised of fault > 80 – delivery rejectable	Grade on green/amber/red system Green Amber Red
Apples, solid pack, canned	
Defects on each 2.75 kg (6 lb.) apple shall not exceed: (a) Whole seeds (or equivalent pieces) – 1 (b) Calyces (or equivalent pieces) – 1 (c) Pieces of skin in aggregate – 325 mm^2 (0.5 in^2) (d) Hard core material (not larger than 10 mm across) – 1 piece (e) Foreign matter – nil (f) Rot or visible mold – nil (g) Insect or rodent contamination – nil	For use in canned junior foods
Cabbage, white – fresh	
Size – Not less than 90% by weight shall weigh at least 2 Kg (4.5 lb)	Size is given because the cabbage will be de-cored and chopped so a smaller size could affect the white/light green leaf ratio.
Defects – Not more than 15% by weight shall be affected by 'burst' or blow, flowering shoots or damage caused by decay, freezing, bruising, insect attack disease or mechanical damage	

Defects and foreign material data table (continued)

Data	Notes or comment
Carrots – fresh	These will be cut or diced by the processor. There is a size requirement which is not quoted here as the main interest is defects and foreign material
Defects – Deliveries should be free from frost damage and there shall not be more than 3% by weight suffering from blemishes caused by soft rot, clayburn, cavity and water mark, including not more than 1% in which the soft rot affects more than 15% of the individual	
Deliveries free from adhering clay not removable by standard washing methods	
Deliveries shall have not more than 2% by weight of cull material consisting of broken or crushed carrots, foliage tops over 13 mm (0.5 in) long, weeds, loose soil, string and harmless foreign bodies.	
Red lentils – unpolished split	
Discolored cotyledons – not more than 6% by weight Broken lentils shall not exceed 5%	This specification refers to lentils (*Lens esculenta Moench*) of Turkish, Ethiopian, Syrian or Lebanese origin only. Intended for use in canned goods
A 50 kg (110 lb) bag shall contain not more than: (a) 30 stones (b) 1 g of small black seeds (cockle) (c) 75 g of wheat/barley (d) 4 g of mud balls	
Shall be free from evidence of insect or rodent contamination and from foreign material other than that specified above	

Defects and foreign material data table (continued)

Data	Notes or comment
Sweet corn – frozen	
Defects shall not exceed:	
(a) Husk – $7\,cm^2$ ($1\,in^2$) in 10 kg (22 lb) (total husk)	For use in canned food. Method of examination, large-scale visual sorting
(b) Pieces of cob – 2 pieces in 10 kg (22 lb)	
(c) Silk (cornsilk) – 35 cm (13.8 in) silk in 1 kg (2.2 lb)	
(d) Damaged kernel – 2% max.	
(e) Extruding kernels – 7% max.	
(f) Total defects – 15% max.	
Sultanas – pre-cleaned	
Defects – In each 50 kg (110 lb) sample (i.e. 4 × 12.5 kg. cartons) – shall not exceed:	

Defects and foreign material data table

Data	Notes or comment
(a) damaged or blemished berries (by weight) – 5%	
(b) pieces of major stalk – 5 individuals	(b) major – stiff pieces over 1 mm diameter and 12 mm (0.5 in) length
(c) pieces of minor stalk – 150 mm (6 in) in total (aggregate) length	(c) minor stalk – pliable pieces 1 mm diameter or less + stiff pieces less than 12 mm long
(d) cap stems per 100 g (3.5 oz) portion – 15	
(e) stones, extraneous vegetable matter, foreign material, rodent, bird, insect or mite contamination, clearly visible mold – nil	

Reference

1. NAS (1985) *An Evaluation of the Role of Microbiological Criteria for Foods and Food Ingredients.* National Academy of Sciences, Washington, DC: National Academy Press

Index

445